AF356385

Molecular Mechanisms and Therapies of Pancreatic Cancer

Molecular Mechanisms and Therapies of Pancreatic Cancer

Editors

Donatella Delle Cave
Claudio Luchini

Basel • Beijing • Wuhan • Barcelona • Belgrade • Novi Sad • Cluj • Manchester

Editors
Donatella Delle Cave
Institute of Genetics and
Biophysics Adriano
Buzzati-Traverso (IGB-ABT)
CNR
Naples
Italy

Claudio Luchini
Department of Diagnostics and
Public Health, Section of
Pathology
University of Verona
Verona
Italy

Editorial Office
MDPI AG
Grosspeteranlage 5
4052 Basel, Switzerland

This is a reprint of articles from the Special Issue published online in the open access journal *International Journal of Molecular Sciences* (ISSN 1422-0067) (available at: https://www.mdpi.com/journal/ijms/special_issues/6L2059972N).

For citation purposes, cite each article independently as indicated on the article page online and as indicated below:

Lastname, A.A.; Lastname, B.B. Article Title. *Journal Name* **Year**, *Volume Number*, Page Range.

ISBN 978-3-7258-2289-8 (Hbk)
ISBN 978-3-7258-2290-4 (PDF)
doi.org/10.3390/books978-3-7258-2290-4

© 2024 by the authors. Articles in this book are Open Access and distributed under the Creative Commons Attribution (CC BY) license. The book as a whole is distributed by MDPI under the terms and conditions of the Creative Commons Attribution-NonCommercial-NoDerivs (CC BY-NC-ND) license.

Contents

About the Editors

Donatella Delle Cave

Donatella Delle Cave is a postdoctoral researcher at the "Institute of Genetics and Biophysics A. Buzzati-Traverso" at the CNR in Naples, Italy. In 2017, she obtained her PhD in "Biochemical and Biotechnological Sciences" at the "University Vanvitelli" in Naples, Italy, and during that period her project was focused on the study of the antitumorigenic effects of S-Adenosylmethionine in breast cancer cells. During her postdoctoral studies, she acquired comprehensive and interdisciplinary knowledge of cancer biology, in particular gastrointestinal tumors, with a specific focus on pancreatic ductal adenocarcinoma (PDAC) and colorectal cancer (CRC). By using a multidisciplinary approach and cutting-edge strategies—combining omics analyses with cell biology techniques, three-dimensional models (spheroids and patient-derived organoids), CRISPR/Cas9 technology, drug delivery nanosystems and in vivo mouse models (both xenograft and orthotopic)—Delle Cave has identified specific markers which characterize the subpopulation of cancer stem cells within these tumors. Delle Cave's research has been published in several manuscripts, in which she figures as both the first and corresponding author. As a young postdoctoral researcher, the skills she acquired allowed her to win two fellowships financed by the "Fondazione Italiana per la Ricerca sulle Malattie del Pancreas" (FIMP), two fellowships supported by "Fondazione Umberto Veronesi" (FUV), a research grant financed by FIMP, several travel grants, and a "best poster" award.

Claudio Luchini

Claudio Luchini, MD, PhD, is Associate Professor of Pathology at the University of Verona and the staff pathologist in charge of pancreatic cancer diagnosis. He was a research scholar at the Indiana University School of Medicine in Indianapolis (2014) and a research fellow at the Sol Goldman Pancreatic Cancer Research Center of the Johns Hopkins University in Baltimore (2014–2015). He co-authored more than 280 publications and scientific contributions, mainly about pancreatic tumors. In addition, he has participated in the drafting of national and international guidelines regarding molecular pathology and pancreatic tumors. He is also a member of the directive board of the Italian College of Professors of Pathology, and Chair of the Education Committee of the Pancreato-Biliary Pathology Society (PBPS-USCAP).

International Journal of
Molecular Sciences

Editorial

Emerging Therapeutic Options in Pancreatic Cancer Management

Donatella Delle Cave

Institute of Genetics and Biophysics 'Adriano Buzzati-Traverso', CNR, 80131 Naples, Italy;
donatella.dellecave@igb.cnr.it

Citation: Delle Cave, D. Emerging Therapeutic Options in Pancreatic Cancer Management. *Int. J. Mol. Sci.* **2024**, *25*, 1929. https://doi.org/10.3390/ijms25031929

Received: 30 January 2024
Accepted: 4 February 2024
Published: 5 February 2024

Copyright: © 2024 by the author. Licensee MDPI, Basel, Switzerland. This article is an open access article distributed under the terms and conditions of the Creative Commons Attribution (CC BY) license (https://creativecommons.org/licenses/by/4.0/).

Pancreatic ductal adenocarcinoma (PDAC) is a devastating disease with a 5-year survival rate of <8% [1–3]. PDAC is characterized by dense desmoplastic stroma, which can constitute up to 90% of the tumor bulk, consisting of non-cellular and cellular components [4,5]. Pancreatic stellate cells (PSCs) and cancer-associated fibroblasts (CAFs) are the principal cellular components responsible for extracellular matrix (ECM) deposition and remodeling, and they play critical roles in cancer progression and treatment resistance [6]. PSC cells are present in a normal pancreas in a quiescent state and are principally involved in vitamin A storage within the cytoplasm [7]. After an injury, they activate and convert into a myofibroblast-like phenotype defined by α-SMA (alpha smooth muscle actin) expression, responsible for different ECM components' secretion, in particular, fibronectins, laminins, and collagen, thus contributing to the highly fibrotic state of PDAC tumors [8–10]. CAFs in PDAC are most often divided into three subtypes, with different tumor-supporting capacities: myofibroblast-type CAFs (myCAFs), inflammatory CAFs (iCAFs), and antigen-presenting CAFs (apCAFs) [11–13]. Local approaches such as radiation therapy, hyperthermia, microwave or radiofrequency ablation, irreversible electroporation, and high-intensity focused ultrasound are capable of modifying the tumor microenvironment (TME) and ECM composition and structure, potentially enhancing chemotherapy [14]. Recently, immunotherapy has become a novel and promising alternative approach to target tumors [15–17]. The objective of immunotherapy is to (re)activate the immune system against cancer. Several innovative immunotherapies have been explored in pancreatic ductal adenocarcinoma (PDAC), with a focus on activating T cells either by blocking inhibitory signals or by enhancing their antitumor activity [18,19]. Stouten et al. focused their attention on the interactions between the immunosystem and CAFs, which can significantly influence immunotolerance and tumor growth [20]. The existing literature supports the role of CAFs in contributing to immunotherapy resistance; yet, the underlying mechanisms remain inadequately explored. Pathways associated with activated CAFs (iCAFs) are primarily considered pro-tumorigenic, although not exclusively. CAFs participate in desmoplasia, but the depletion of myCAFs can paradoxically lead to tumor progression. This presents a significant challenge for current and future therapeutic endeavors. At the moment, the combination of therapies targeting CAFs with immunotherapies has shown promising effects in murine models and certain clinical trials of PDAC patients [21,22]. PDAC typically progresses silently and without noticeable symptoms, leading to a challenging prognosis and poor outcomes [23,24]. Consequently, there is a pressing need to enhance early diagnosis and detect reliable biomarkers. In this context, one of the most promising approaches is the detection of exosomes in the bloodstream [25,26]. PDAC-associated microRNAs (miRNAs) packaged within exosomes could be used as diagnostic markers for the early detection of PDAC [27]. In this regard, Makler and colleagues identified four distinct differentially expressed miRNAs between plasma exosomes harvested from PDAC patients and those from control patients: miR-93-5p, miR-339-3p, miR-425-5p, and miR-425-3p, with an area under the curve (AUC) of the receiver operator characteristic curve (ROC) of 0.885, a sensitivity of 80%, and a specificity of 94.7%, which is comparable to the CA19-9 standard

PDAC marker diagnostic [28]. Recently, the composition of the tumor microbiome has emerged as a novel prognostic factor for PDAC, as it differs from one patient to another and in response to chemotherapy [29–32]. Merali et al. demonstrated the existence of a bile microbiome signature in patients with PDAC who experienced obstructive jaundice caused by the disease, and the identification of specific bacteria in the bile has the potential to facilitate the detection and stratification of PDAC [33]. Deregulation of key signaling pathways in cancer, as well as altered genes expression, have critical functions in tumor progression [34–38]. Stukas et al. demonstrated that the inhibition of aryl hydrocarbon receptor (AHR) expression in PDAC sensitizes cells to gemcitabine, the gold standard treatment for pancreatic cancer, through the ELAVL1-DCK pathway [39–43]. Zuccolini et al. evaluated the effects of two different Ca^{2+}-gated K^+ channel KCa3.1 (commonly known as IK) blockers, namely Clotrimazole and Senicapoc, on metastatic melanoma and PDAC [44–47]. Although both inhibitors reduced the viability and migration of the tumor cells, neither of them altered the intracellular Ca^{2+} concentration in PDAC. In conclusion, the studies discussed in this editorial provide valuable insights into novel therapeutic strategies and personalized approaches for the treatment of PDAC. Further research in this field is needed to improve our knowledge regarding pancreatic cancer progression and to develop personalized treatments, with the aim of improving patient outcomes.

Funding: D.D.C. was supported by Fondazione Umberto Veronesi (FUV) and Fondazione Italiana per la ricerca sulle Malattie del Pancreas (FIMP).

Conflicts of Interest: The author declares that the research was conducted in the absence of any commercial or financial relationships that could be construed as a potential conflict of interest.

References

1. Sarantis, P.; Koustas, E.; Papadimitropoulou, A.; Papavassiliou, A.G.; Karamouzis, M.V. Pancreatic Ductal Adenocarcinoma: Treatment Hurdles, Tumor Microenvironment and Immunotherapy. *World J. Gastrointest. Oncol.* **2020**, *12*, 173–181. [CrossRef]
2. Arnold, M.; Abnet, C.C.; Neale, R.E.; Vignat, J.; Giovannucci, E.L.; McGlynn, K.A.; Bray, F. Global Burden of 5 Major Types of Gastrointestinal Cancer. *Gastroenterology* **2020**, *159*, 335–349.e15. [CrossRef]
3. Hosein, A.N.; Dougan, S.K.; Aguirre, A.J.; Maitra, A. Translational Advances in Pancreatic Ductal Adenocarcinoma Therapy. *Nat. Cancer* **2022**, *3*, 272–286. [CrossRef]
4. Hosein, A.N.; Brekken, R.A.; Maitra, A. Pancreatic Cancer Stroma: An Update on Therapeutic Targeting Strategies. *Nat. Rev. Gastroenterol. Hepatol.* **2020**, *17*, 487–505. [CrossRef] [PubMed]
5. Valkenburg, K.C.; de Groot, A.E.; Pienta, K.J. Targeting the Tumour Stroma to Improve Cancer Therapy. *Nat. Rev. Clin. Oncol.* **2018**, *15*, 366–381. [CrossRef] [PubMed]
6. Cave, D.D.; Di Guida, M.; Costa, V.; Sevillano, M.; Ferrante, L.; Heeschen, C.; Corona, M.; Cucciardi, A.; Lonardo, E. TGF-B1 Secreted by Pancreatic Stellate Cells Promotes Stemness and Tumourigenicity in Pancreatic Cancer Cells through L1CAM Downregulation. *Oncogene* **2020**, *39*, 4271–4285. [CrossRef] [PubMed]
7. Wu, Y.; Zhang, C.; Jiang, K.; Werner, J.; Bazhin, A.V.; D'Haese, J.G. The Role of Stellate Cells in Pancreatic Ductal Adenocarcinoma: Targeting Perspectives. *Front. Oncol.* **2021**, *10*, 621937. [CrossRef] [PubMed]
8. Cave, D.D.; Buonaiuto, S.; Sainz, B.; Fantuz, M.; Mangini, M.; Carrer, A.; Di Domenico, A.; Iavazzo, T.T.; Andolfi, G.; Cortina, C.; et al. LAMC2 Marks a Tumor-Initiating Cell Population with an Aggressive Signature in Pancreatic Cancer. *J. Exp. Clin. Cancer Res.* **2022**, *41*, 315. [CrossRef] [PubMed]
9. Öhlund, D.; Franklin, O.; Lundberg, E.; Lundin, C.; Sund, M. Type IV Collagen Stimulates Pancreatic Cancer Cell Proliferation, Migration, and Inhibits Apoptosis through an Autocrine Loop. *BMC Cancer* **2013**, *13*, 154. [CrossRef] [PubMed]
10. Puls, T.J.; Tan, X.; Whittington, C.F.; Voytik-Harbin, S.L. 3D Collagen Fibrillar Microstructure Guides Pancreatic Cancer Cell Phenotype and Serves as a Critical Design Parameter for Phenotypic Models of EMT. *PLoS ONE* **2017**, *12*, e0188870. [CrossRef] [PubMed]
11. Geng, X.; Chen, H.; Zhao, L.; Hu, J.; Yang, W.; Li, G.; Cheng, C.; Zhao, Z.; Zhang, T.; Li, L.; et al. Cancer-Associated Fibroblast (CAF) Heterogeneity and Targeting Therapy of CAFs in Pancreatic Cancer. *Front. Cell Dev. Biol.* **2021**, *9*, 655152. [CrossRef] [PubMed]
12. Zhang, T.; Ren, Y.; Yang, P.; Wang, J.; Zhou, H. Cancer-Associated Fibroblasts in Pancreatic Ductal Adenocarcinoma. *Cell Death Dis.* **2022**, *13*, 897. [CrossRef] [PubMed]
13. Boyd, L.N.C.; Andini, K.D.; Peters, G.J.; Kazemier, G.; Giovannetti, E. Heterogeneity and Plasticity of Cancer-Associated Fibroblasts in the Pancreatic Tumor Microenvironment. *Semin. Cancer Biol.* **2022**, *82*, 184–196. [CrossRef] [PubMed]

14. De Grandis, M.C.; Ascenti, V.; Lanza, C.; Di Paolo, G.; Galassi, B.; Ierardi, A.M.; Carrafiello, G.; Facciorusso, A.; Ghidini, M. Locoregional Therapies and Remodeling of Tumor Microenvironment in Pancreatic Cancer. *Int. J. Mol. Sci.* **2023**, *24*, 12681. [CrossRef] [PubMed]

15. Robert, C. A Decade of Immune-Checkpoint Inhibitors in Cancer Therapy. *Nat. Commun.* **2020**, *11*, 3801. [CrossRef]

16. Panchal, K.; Sahoo, R.K.; Gupta, U.; Chaurasiya, A. Role of Targeted Immunotherapy for Pancreatic Ductal Adenocarcinoma (PDAC) Treatment: An Overview. *Int. Immunopharmacol.* **2021**, *95*, 107508. [CrossRef]

17. Wong, S.K.; Beckermann, K.E.; Johnson, D.B.; Das, S. Combining Anti-Cytotoxic T-Lymphocyte Antigen 4 (CTLA-4) and - Programmed Cell Death Protein 1 (PD-1) Agents for Cancer Immunotherapy. *Expert Opin. Biol. Ther.* **2021**, *21*, 1623–1634. [CrossRef]

18. Chouari, T.; La Costa, F.S.; Merali, N.; Jessel, M.-D.; Sivakumar, S.; Annels, N.; Frampton, A.E. Advances in Immunotherapeutics in Pancreatic Ductal Adenocarcinoma. *Cancers* **2023**, *15*, 4265. [CrossRef]

19. Timmer, F.E.F.; Geboers, B.; Nieuwenhuizen, S.; Dijkstra, M.; Schouten, E.A.C.; Puijk, R.S.; De Vries, J.J.J.; Van Den Tol, M.P.; Bruynzeel, A.M.E.; Streppel, M.M.; et al. Pancreatic Cancer and Immunotherapy: A Clinical Overview. *Cancers* **2021**, *13*, 4138. [CrossRef]

20. Stouten, I.; Van Montfoort, N.; Hawinkels, L.J.A.C. The Tango between Cancer-Associated Fibroblasts (CAFs) and Immune Cells in Affecting Immunotherapy Efficacy in Pancreatic Cancer. *Int. J. Mol. Sci.* **2023**, *24*, 8707. [CrossRef]

21. Chen, Y.; McAndrews, K.M.; Kalluri, R. Clinical and Therapeutic Relevance of Cancer-Associated Fibroblasts. *Nat. Rev. Clin. Oncol.* **2021**, *18*, 792–804. [CrossRef] [PubMed]

22. Bockorny, B.; Semenisty, V.; Macarulla, T.; Borazanci, E.; Wolpin, B.M.; Stemmer, S.M.; Golan, T.; Geva, R.; Borad, M.J.; Pedersen, K.S.; et al. BL-8040, a CXCR4 Antagonist, in Combination with Pembrolizumab and Chemotherapy for Pancreatic Cancer: The COMBAT Trial. *Nat. Med.* **2020**, *26*, 878–885. [CrossRef] [PubMed]

23. Lin, K.; Lin, A.N.; Lin, S.; Lin, T.; Liu, Y.X.; Reddy, M. A Silent Asymptomatic Solid Pancreas Tumor in a Nonsmoking Athletic Female: Pancreatic Ductal Adenocarcinoma. *Case Rep. Gastroenterol.* **2017**, *11*, 624–632. [CrossRef] [PubMed]

24. Principe, D.R.; Underwood, P.W.; Korc, M.; Trevino, J.G.; Munshi, H.G.; Rana, A. The Current Treatment Paradigm for Pancreatic Ductal Adenocarcinoma and Barriers to Therapeutic Efficacy. *Front. Oncol.* **2021**, *11*, 688377. [CrossRef]

25. Nakamura, K.; Zhu, Z.; Roy, S.; Jun, E.; Han, H.; Munoz, R.M.; Nishiwada, S.; Sharma, G.; Cridebring, D.; Zenhausern, F.; et al. An Exosome-Based Transcriptomic Signature for Noninvasive, Early Detection of Patients with Pancreatic Ductal Adenocarcinoma: A Multicenter Cohort Study. *Gastroenterology* **2022**, *163*, 1252–1266.e2. [CrossRef] [PubMed]

26. Xu, B.; Chen, Y.; Peng, M.; Zheng, J.H.; Zuo, C. Exploring the Potential of Exosomes in Diagnosis and Drug Delivery for Pancreatic Ductal Adenocarcinoma. *Int. J. Cancer* **2023**, *152*, 110–122. [CrossRef] [PubMed]

27. Makler, A.; Narayanan, R.; Asghar, W. An Exosomal miRNA Biomarker for the Detection of Pancreatic Ductal Adenocarcinoma. *Biosensors* **2022**, *12*, 831. [CrossRef]

28. Makler, A.; Asghar, W. Exosomal miRNA Biomarker Panel for Pancreatic Ductal Adenocarcinoma Detection in Patient Plasma: A Pilot Study. *Int. J. Mol. Sci.* **2023**, *24*, 5081. [CrossRef]

29. Stower, H. Predicting Pancreatic Cancer Survival via the Tumor Microbiome. *Nat. Med.* **2019**, *25*, 1330. [CrossRef]

30. Bangolo, A.I.; Trivedi, C.; Jani, I.; Pender, S.; Khalid, H.; Alqinai, B.; Intisar, A.; Randhawa, K.; Moore, J.; De Deugd, N.; et al. Impact of Gut Microbiome in the Development and Treatment of Pancreatic Cancer: Newer Insights. *World J. Gastroenterol.* **2023**, *29*, 3984–3998. [CrossRef]

31. Bharti, R.; Grimm, D.G. Current Challenges and Best-Practice Protocols for Microbiome Analysis. *Brief. Bioinform.* **2021**, *22*, 178–193. [CrossRef]

32. Guo, W.; Zhang, Y.; Guo, S.; Mei, Z.; Liao, H.; Dong, H.; Wu, K.; Ye, H.; Zhang, Y.; Zhu, Y.; et al. Tumor Microbiome Contributes to an Aggressive Phenotype in the Basal-like Subtype of Pancreatic Cancer. *Commun. Biol.* **2021**, *4*, 1019. [CrossRef] [PubMed]

33. Merali, N.; Chouari, T.; Terroire, J.; Jessel, M.-D.; Liu, D.S.K.; Smith, J.-H.; Wooldridge, T.; Dhillon, T.; Jiménez, J.I.; Krell, J.; et al. Bile Microbiome Signatures Associated with Pancreatic Ductal Adenocarcinoma Compared to Benign Disease: A UK Pilot Study. *Int. J. Mol. Sci.* **2023**, *24*, 16888. [CrossRef]

34. Cave, D.D.; Rizzo, R.; Sainz, B. The Revolutionary Roads to Study Cell–Cell Interactions in 3D In Vitro Pancreatic Cancer Models. *Cancers* **2021**, *13*, 930. [CrossRef] [PubMed]

35. Yokobori, T.; Nishiyama, M. TGF-β Signaling in Gastrointestinal Cancers: Progress in Basic and Clinical Research. *J. Clin. Med.* **2017**, *6*, 11. [CrossRef] [PubMed]

36. Adamopoulos, C.; Cave, D.D.; Papavassiliou, A.G. Inhibition of the RAF/MEK/ERK Signaling Cascade in Pancreatic Cancer: Recent Advances and Future Perspectives. *Int. J. Mol. Sci.* **2024**, *25*, 1631. [CrossRef]

37. Seton-Rogers, S. Fibroblasts Shape PDAC Architecture. *Nat. Rev. Cancer* **2019**, *19*, 418. [CrossRef]

38. Özdemir, B.C.; Pentcheva-Hoang, T.; Carstens, J.L.; Zheng, X.; Wu, C.-C.; Simpson, T.R.; Laklai, H.; Sugimoto, H.; Kahlert, C.; Novitskiy, S.V.; et al. Depletion of Carcinoma-Associated Fibroblasts and Fibrosis Induces Immunosuppression and Accelerates Pancreas Cancer with Reduced Survival. *Cancer Cell* **2014**, *25*, 719–734. [CrossRef]

39. Stukas, D.; Jasukaitiene, A.; Bartkeviciene, A.; Matthews, J.; Maimets, T.; Teino, I.; Jaudzems, K.; Gulbinas, A.; Dambrauskas, Z. Targeting AHR Increases Pancreatic Cancer Cell Sensitivity to Gemcitabine through the ELAVL1-DCK Pathway. *Int. J. Mol. Sci.* **2023**, *24*, 13155. [CrossRef]

40. Koltai, T.; Reshkin, S.J.; Carvalho, T.M.A.; Di Molfetta, D.; Greco, M.R.; Alfarouk, K.O.; Cardone, R.A. Resistance to Gemcitabine in Pancreatic Ductal Adenocarcinoma: A Physiopathologic and Pharmacologic Review. *Cancers* **2022**, *14*, 2486. [CrossRef]
41. Amrutkar, M.; Gladhaug, I. Pancreatic Cancer Chemoresistance to Gemcitabine. *Cancers* **2017**, *9*, 157. [CrossRef] [PubMed]
42. Thierry, C.; Françoise, D.; Marc, Y.; Olivier, B.; Rosine, G.; Yves, B.; Antoine, A.; Jean-Luc, R.; Sophie, G.-B. FOLFIRINOX versus Gemcitabine for Metastatic Pancreatic Cancer. *N. Engl. J. Med.* **2011**, *9*, 1817–1825.
43. Chen, C.; Zhao, S.; Zhao, X.; Cao, L.; Karnad, A.; Kumar, A.P.; Freeman, J.W. Gemcitabine Resistance of Pancreatic Cancer Cells Is Mediated by IGF1R Dependent Upregulation of CD44 Expression and Isoform Switching. *Cell Death Dis.* **2022**, *13*, 682. [CrossRef] [PubMed]
44. Zuccolini, P.; Barbieri, R.; Sbrana, F.; Picco, C.; Gavazzo, P.; Pusch, M. IK Channel-Independent Effects of Clotrimazole and Senicapoc on Cancer Cells Viability and Migration. *Int. J. Mol. Sci.* **2023**, *24*, 16285. [CrossRef]
45. Schnipper, J.; Dhennin-Duthille, I.; Ahidouch, A.; Ouadid-Ahidouch, H. Ion Channel Signature in Healthy Pancreas and Pancreatic Ductal Adenocarcinoma. *Front. Pharmacol.* **2020**, *11*, 568993. [CrossRef]
46. Motawi, T.M.K.; Sadik, N.A.H.; Fahim, S.A.; Shouman, S.A. Combination of Imatinib and Clotrimazole Enhances Cell Growth Inhibition in T47D Breast Cancer Cells. *Chem. -Biol. Interact.* **2015**, *233*, 147–156. [CrossRef]
47. Hofschröer, V.; Najder, K.; Rugi, M.; Bouazzi, R.; Cozzolino, M.; Arcangeli, A.; Panyi, G.; Schwab, A. Ion Channels Orchestrate Pancreatic Ductal Adenocarcinoma Progression and Therapy. *Front. Pharmacol.* **2021**, *11*, 586599. [CrossRef] [PubMed]

Disclaimer/Publisher's Note: The statements, opinions and data contained in all publications are solely those of the individual author(s) and contributor(s) and not of MDPI and/or the editor(s). MDPI and/or the editor(s) disclaim responsibility for any injury to people or property resulting from any ideas, methods, instructions or products referred to in the content.

International Journal of
Molecular Sciences

Article

Anakinra-Loaded Sphingomyelin Nanosystems Modulate In Vitro IL-1-Dependent Pro-Tumor Inflammation in Pancreatic Cancer

Marcelina Abal-Sanisidro [1,2,3,†], Michele De Luca [4,5,†], Stefania Roma [4,5], Maria Grazia Ceraolo [4,5], Maria de la Fuente [1,2,3,6,*,‡], Lucia De Monte [4,5,‡] and Maria Pia Protti [4,5,*,‡]

[1] Nano-Oncology and Translational Therapeutics Group, Health Research Institute of Santiago de Compostela (IDIS), SERGAS, 15706 Santiago de Compostela, Spain; marce.abal.san@gmail.com
[2] University of Santiago de Compostela (USC), 15782 Santiago de Compostela, Spain
[3] Biomedical Research Networking Center on Oncology (CIBERONC), 28029 Madrid, Spain
[4] Tumor Immunology Unit, Istituto di Ricovero e Cura a Carattere Scientifico (IRCCS) San Raffaele Scientific Institute, 20132 Milan, Italy; michele.deluca@istitutotumori.mi.it (M.D.L.); roma.stefania@hsr.it (S.R.); ceraolo@ingm.org (M.G.C.); demonte.lucia@hsr.it (L.D.M.)
[5] Division of Immunology, Transplantation and Infectious Diseases, IRCCS San Raffaele Scientific Institute, 20132 Milan, Italy
[6] DIVERSA Technologies S.L., Edificio Emprendia, Campus Sur, 15782 Santiago de Compostela, Spain
[*] Correspondence: maria.de.la.fuente.freire@sergas.es (M.d.l.F.); protti.mariapia@hsr.it (M.P.P.)
[†] These authors contributed equally to this work.
[‡] These authors also contributed equally to this work.

Abstract: Pancreatic cancer is a very aggressive disease with a dismal prognosis. The tumor microenvironment exerts immunosuppressive activities through the secretion of several cytokines, including interleukin (IL)-1. The IL-1/IL-1 receptor (IL-1R) axis is a key regulator in tumor-promoting T helper (Th)2- and Th17-type inflammation. Th2 cells are differentiated by dendritic cells endowed with Th2-polarizing capability by the thymic stromal lymphopoietin (TSLP) that is secreted by IL-1-activated cancer-associated fibroblasts (CAFs). Th17 cells are differentiated in the presence of IL-1 and other IL-1-regulated cytokines. In pancreatic cancer, the use of a recombinant IL-1R antagonist (IL1RA, anakinra, ANK) in in vitro and in vivo models has shown efficacy in targeting the IL-1/IL-1R pathway. In this study, we have developed sphingomyelin nanosystems (SNs) loaded with ANK (ANK-SNs) to compare their ability to inhibit Th2- and Th17-type inflammation with that of the free drug in vitro. We found that ANK-SNs inhibited TSLP and other pro-tumor cytokines released by CAFs at levels similar to ANK. Importantly, inhibition of IL-17 secretion by Th17 cells, but not of interferon-γ, was significantly higher, and at lower concentrations, with ANK-SNs compared to ANK. Collectively, the use of ANK-SNs might be beneficial in reducing the effective dose of the drug and its toxic effects.

Keywords: pancreatic ductal adenocarcinoma; sphingomyelin nanosystems; IL-1; IL-1RA; TSLP; IL-17; IFN-γ

Citation: Abal-Sanisidro, M.; De Luca, M.; Roma, S.; Ceraolo, M.G.; de la Fuente, M.; De Monte, L.; Protti, M.P. Anakinra-Loaded Sphingomyelin Nanosystems Modulate In Vitro IL-1-Dependent Pro-Tumor Inflammation in Pancreatic Cancer. *Int. J. Mol. Sci.* **2024**, *25*, 8085. https://doi.org/10.3390/ijms25158085

Academic Editors: Donatella Delle Cave and Claudio Luchini

Received: 20 May 2024
Revised: 9 July 2024
Accepted: 16 July 2024
Published: 24 July 2024

Copyright: © 2024 by the authors. Licensee MDPI, Basel, Switzerland. This article is an open access article distributed under the terms and conditions of the Creative Commons Attribution (CC BY) license (https://creativecommons.org/licenses/by/4.0/).

1. Introduction

Pancreatic ductal adenocarcinoma (PDAC) is a very aggressive disease with a dismal prognosis (expected to become the second leading cause of cancer mortality by 2030) [1,2]. The tumor microenvironment exerts an immunosuppressive activity, which supports tumor cell growth, invasion, and metastases, and it is sustained by a complex cross-talk between tumor, stromal, and immune cells in the tumor microenvironment [3–9].

The cytokines of the interleukin (IL)-1 group are inflammatory mediators frequently upregulated in a variety of cancers, and their production is often associated with poor prognosis, with several mechanisms accounting for their pro-tumorigenic effects [10,11].

In PDAC, the activation of *Kras* promotes the production of inflammatory cytokines, including IL-1, which activates NF-κB to promote tumor cell survival and proliferation, increased invasive and metastatic behavior, and angiogenesis [5,12–14]. In addition, IL-1 cytokines exert important immunomodulatory functions in the tumor microenvironment, where IL-1α and IL-1β derived from tumor cells and/or tumor-associated macrophages act on IL-1 receptor (IL-1R)-expressing cells, including cancer-associated fibroblasts (CAFs), with the secretion of several tumor-promoting cytokines (including IL-6, IL-8, and transforming growth factor (TGF)-β) [15–18] and CD4+ T helper (Th) cell effectors [19].

In the context of PDAC, we previously reported that the ratio of Th2/Th1 cells is a predictor of poor prognosis after surgery in chemo-naïve patients [20]. Specifically, we found that differentiation of Th2 cells was due to activation of myeloid dendritic cells (DCs) with Th2-polarizing capability that was dependent on the thymic stromal lymphopoietin (TSLP) secreted by tumor-derived IL-1-activated CAFs [20–23], pointing to a relevant indirect role for IL-1 in driving Th2-type inflammation in PDAC. IL-1 is also a relevant cytokine for human Th17 cell differentiation [24–27]. Th17 cells were increased in PDAC compared to normal pancreatic tissue [28], and high levels of IL-17 and Th17 cells in the tumor were associated with worse clinical correlates in PDAC patients [28–33]. These data highlight the relevance to PDAC of an indirect and a direct role of IL-1 in the differentiation of pro-tumor Th2 and Th17 cells, respectively.

Considering the key role of IL-1 in tumor-promoting inflammation in PDAC, IL-1 inhibition by a recombinant IL-1R antagonist (IL-1RA) (anakinra, ANK) has been tested in in vitro and in vivo preclinical studies, demonstrating that ANK is an effective anti-tumor agent [23,34–36]. Based on these results, clinical trials in PDAC patients with metastatic disease as well as in the neoadjuvant setting using ANK in combination with chemotherapy are ongoing/have been planned (NCT02021422, NCT02550327, NCT04926467). Preliminary reports showed the safety and feasibility of this therapeutic approach with toxicities expected from the combination regimen [37,38].

As toxicity due to ANK may represent a problem in these combination regimens, the targeting of ANK to the tumor might impact effective intratumor drug concentrations and reduce toxicities. In this context, the use of nanoparticles for drug delivery may improve drug uptake and/or drug tolerance [39].

Here, we prepared sphingomyelin nanosystems (SNs) loaded with ANK (ANK-SNs) and used them in vitro to test their capacity to inhibit IL-1-induced CAF activation and cytokine secretion and Th17 cell differentiation and function in comparison to the free drug.

2. Results

2.1. Preparation and Characterization of ANK-SNs

We prepared ANK-SNs by adapting the ethanol injection method that we previously described [40,41]. Preparation of unloaded SNs (from now on referred to as SNs), composed of vitamin E (VitE), sphingomyelin (SM), and surfactant lipids, was followed by the dropwise addition of the recombinant protein ANK, as represented in Figure 1A. In order to be loaded in SNs, ANK was isolated from its pharmaceutical form Kineret® by using a desalting column (Figure 1B).

Polyethylene glycol (PEG)ylated lipids have been extensively used to link biomolecules (such as proteins, peptides, and antibodies) to the surfaces of nanostructures [39,42,43]. Indeed, PEG presents features in drug delivery, such as prolonged blood circulation time and reduced potential interactions with blood components (opsonization), that give stealth properties to nanosystems, thus improving their therapeutic efficacy [39,42,43].

To display ANK on the surface of SNs, our strategy was to chemically conjugate it to a PEGylated stearic carbon chain containing a chemically active functional group to perform a covalent reaction (DSPE-PEG(2k)-X; X = NHS, DBCO, or Maleimide) [44,45]. In the case of DSPE-PEG(2k)-DBCO, in order to implement SPAAC click chemistry, ANK was modified with a linker to incorporate the highly reactive group azide ($-N_3$) into its structure. For that, we reacted the lysines, which have a terminal group ($-NH_2$), with the -NHS ester present in the linker, and N3-ANK modification was verified by high-performance liquid chromatography with mass spectroscopy (HPLC-MS) (Supplementary Figure S1).

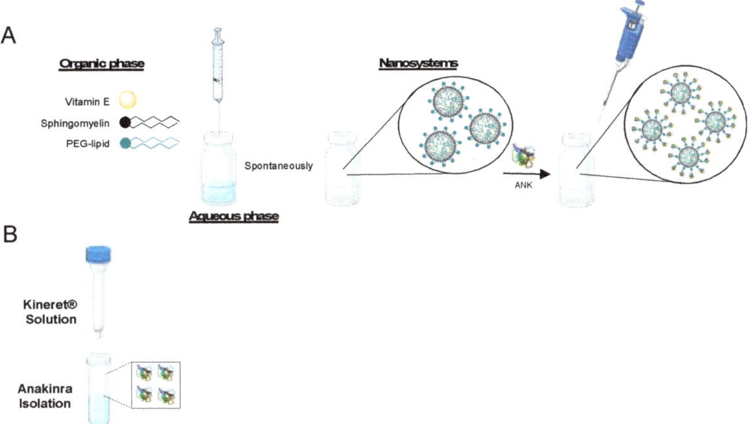

Figure 1. Preparation of the anakinra-loaded sphingomyelin nanosystems (ANK-SNs). (**A**) Schematic representation of the single-step SN preparation by the ethanol injection method, followed by the dropwise addition of ANK solution in PBS at physiological pH. (**B**) ANK isolation from its pharmaceutical form (Kineret®) by using desalting columns. (Created with BioRender.com).

We obtained three SN formulations, differing one from the other in the surfactant lipid linker used in its preparation (i.e., VitE/SM/DBCO, VitE/SM/Maleimide, and VitE/SM/NHS), and we performed their physicochemical characterization in terms of size and surface charge by dynamic light scattering (DLS) and laser Doppler anemometry (LDA). All formulations showed a small particle size, a monodisperse population, and a negative zeta potential (Table 1 and Figure 2A). Interestingly, particles of this size (<200 nm) are more likely to accumulate in tumors thanks to the enhanced permeation retention effect [46]. Furthermore, nanosystems with negatively charged surfaces can reduce undesirable clearance by the reticuloendothelial system and improve blood compatibility, making their delivery to tumor sites more efficient [47]. The negatively charged surface also makes them more biocompatible in comparison to positively charged nanosystems, which interact with cell structures by electrostatic interactions [47,48]. The percentage of ANK loading was highest (92%) for the formulation composed of VitE/SM/DBCO/ANK, whereas it was similar, i.e., 72% and 75%, for the formulations VitE/SM/Maleimide/ANK and VitE/SM/NHS/ANK, respectively (Table 1).

Table 1. Physicochemical characterization of sphingomyelin nanosystems (SNs) (composed of VitE/SM/PEG-lipid-X) and anakinra-loaded sphingomyelin nanosystems (ANK-SNs) (composed of VitE/SM/PEG-lipid-X-ANK) measured by dynamic light scattering (DLS) and laser Doppler anemometry (LDA) *.

Formulation	Size (nm)	PdI	ZP (mV)	% ANK
VitE/SM/DBCO	55 ± 1	<0.2	−26 ± 1	-
VitE/SM/DBCO/ANK	162 ± 2	<0.1	−28 ± 1	92 ± 2
VitE/SM/Maleimide	61 ± 2	<0.2	−28 ± 1	-
VitE/SM/Maleimide/ANK	163 ± 7	0.2	−22 ± 1	72 ± 1
VitE/SM/NHS	60 ± 3	0.2	−21 ± 4	-
VitE/SM/NHS/ANK	136 ± 2	0.2	−23 ± 1	75 ± 2

* Results are expressed as mean ± SD, n = 5. Abbreviations used: nm = nanometer, PdI = polydispersity index, ZP = zeta potential in millivolts (mV). Percentage of ANK associated with SNs was quantified by BCA protein assay.

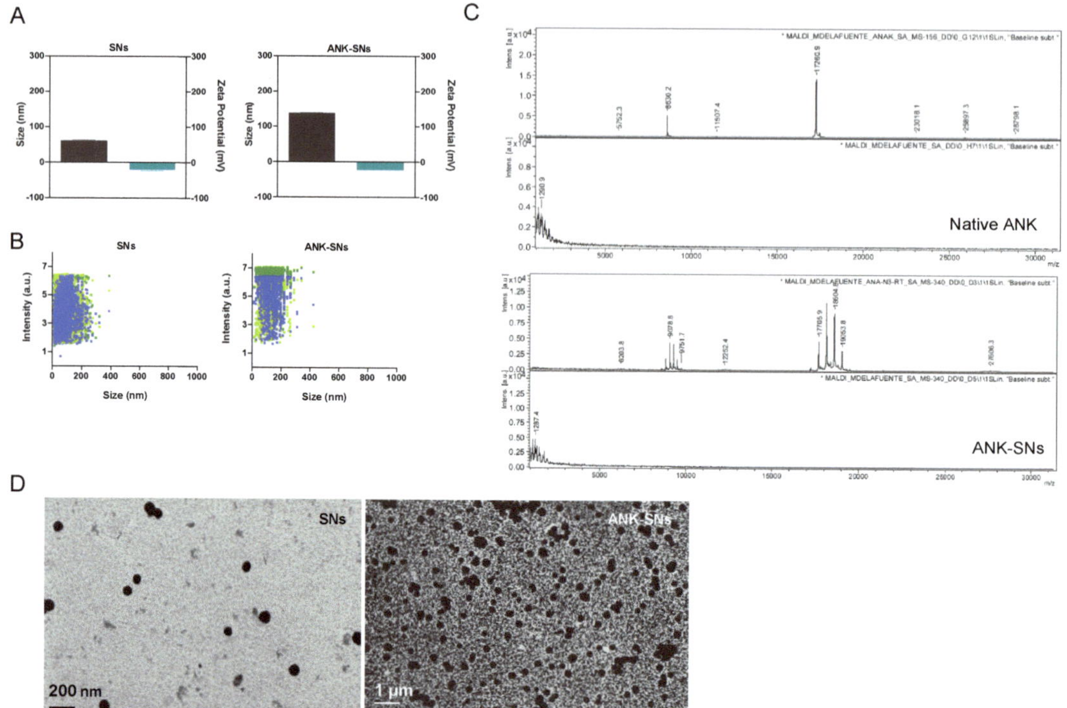

Figure 2. Physicochemical characterization of sphingomyelin nanosystems (SNs) (VitE/SM/NHS) and anakinra-loaded sphingomyelin nanosystems (ANK-SNs) (VitE/SM/NHS/ANK) using several analytical techniques. (**A**) Size (nm) (black columns) and surface charge (mV) (turquoise columns) of SNs and ANK-SNs, measured by dynamic light scattering (DLS) and dynamic light scattering (LDA). Data are expressed as mean ± SD (at least n = 5). (**B**) Complementary physicochemical characterization by nanoparticle tracking analysis (NTA) of SNs and ANK-SNs (n ± 5) is represented as size vs. light scattering intensity (arbitrary unit a.u.). (**C**) Matrix-assisted laser desorption/ionization (MALDI) time-of-flight (TOF) analysis of native ANK (calculated exact mass of 17.260 kDa) and ANK-SNs (calculated mean mass ~20 kDa). (**D**) Representative field emission scanning electron microscopy (FESEM) images of SNs (left) and ANK-SNs (right). SN scale bar: 200 nm. ANK-SN scale bar: 1 μm.

As in the in vitro assays used to address the nanosystem biological functions (see below), after preliminary testing with all three formulations, VitE/SM/NHS/ANK was chosen for subsequent experiments; we performed full characterization of the nanosystems with this formulation. In addition to DLS, nanosystems were characterized by nanoparticle tracking analysis (NTA). NTA is based on the combination of laser light scattering microscopy with a charge-coupled device camera that visualizes and analyzes individual tracks of nanoparticles moving under Brownian motion, allowing the calculation of particle size and concentration [49,50]. Compared to SNs, ANK-SNs presented a higher mean particle size as well as span value due to the loading of the ANK protein and the interaction of the nanoparticles with the phosphate-buffered saline (PBS) (Figure 2B and Table 2). D-values D10, D50, and D90 representative of the particle size diameter at 10, 50, and 90% cumulative distribution are reported in Table 2. Collectively, these results well correlate with DLS data.

We also performed matrix-assisted laser desorption/ionization (MALDI) time-of-flight (TOF) spectrometry to confirm an increase in ANK mass due to the covalent bond to DSPE-PEG(2k)-NHS (estimated ~2.8 kDa) (Supplementary Figure S2). Indeed, for ANK-SNs, we observed an increment in mass (~20 KDa) compared to the one expected for the native form of ANK (17.3 kDa) (Figure 2C). Other nanosystem components, such as VitE and SM, were not detected by this technique, as they are not coupled to the protein.

Table 2. Physicochemical properties of sphingomyelin nanosystems (SNs) (composed of VitE/SM/PEG-lipid-NHS) and anakinra-loaded sphingomyelin nanosystems (ANK-SNs) (composed of VitE/SM/PEG-lipid-NHS-ANK) determined by nanoparticle tracking analysis (NTA) *.

Formulation	Mean Size (nm)	D10	D50	D90	Span	Concentration (Particles/mL)
SNs	87 ± 2	49 ± 5	85 ± 1	129 ± 3	0.7	5.4×10^{11}
ANK-SNs	133 ± 6	96 ± 2	132 ± 2	178 ± 3	0.9	6.2×10^{11}

* Mean particle size (diameter), D-values (D10, D50, D90), calculated SPAN value and sample concentration in particles per milliliter (particle/mL) (mean $\pm$ SD, n = 5).

Lastly, we used field emission scanning electron microscopy (FESEM) to check the particle distribution and the spatial disposition. As shown in Figure 2D, representative images clearly indicated the characteristic morphology and spherical shape of the SNs, as we previously reported [40]. In the case of ANK-SNs, we observed a slight increment in the particle size (Figure 2D, right panel), already reported by DLS and NTA data. The spherical distribution did not seem to be affected by the loading of the ANK protein, pointing out the good stability of the nanosystems.

Collectively, as verified and confirmed by DLS, NTA, MALDI-TOF, and FESEM analyses, ANK protein was successful and efficiently loaded at the SN surface with an expected increment in size of about 60–65 nm.

2.1.1. ANK-SN Stability over Time and in Biorelevant Media

We assessed the colloidal stability of nanosystems under different storage conditions using DLS. We showed that size and polydispersity index (PdI) for both SNs and ANK-SNs were stable for at least one month (Figure 3A). Stability was also tested in different culture conditions. SNs and ANK-SNs were incubated in Dulbecco's modified eagle medium (DMEM) and RPMI culture media supplemented with 10% fetal bovine serum (FBS) at 37 °C, and particle size and PdI were monitored for 48 hours (h). We found that the two parameters remained stable in culture with both media (Figure 3B,C). This stability can be explained by the adsorption of serum proteins to the nanosystem surface due to charge balance. ANK-SNs also seemed to be stabilized thanks to the electrostatic interaction between ANK and serum proteins.

Collectively, we found that the nanosystems maintained their colloidal properties over time and in different culture media, and we proceeded to study their internalization and cytotoxicity in cell systems in vitro.

2.1.2. In Vitro Internalization and Viability Assays of the Nanosystems in Tumor Cells

Before using ANK-SNs in cell systems relevant to the study of immunomodulation of Th2 and Th17 pro-tumor inflammation in PDAC, we challenged our nanosystems with PDAC cells and looked at their internalization and cytotoxic potential.

We performed internalization studies by confocal microscopy using the PDAC cell line L3.6pl. For these experiments, nanosystems were prepared as described above but using fluorescent-labeled TopFluor® SNs. As shown in Figure 4A, we detected high fluorescent signals at the two time points tested (i.e., 2 h and 4 h). Internalization seemed more efficient when we used ANK-SNs compared to SNs, possibly related to the interaction of the ANK protein with components present at the cell membrane. Interestingly, we previously found *IL1R1* mRNA expression in several PDAC cell lines and high *IL1R1* expression was found in the L3.6pl precursor cell line L3.3 (The Human Protein Atlas), suggesting specific interaction of ANK with its receptor, namely IL-1R1.

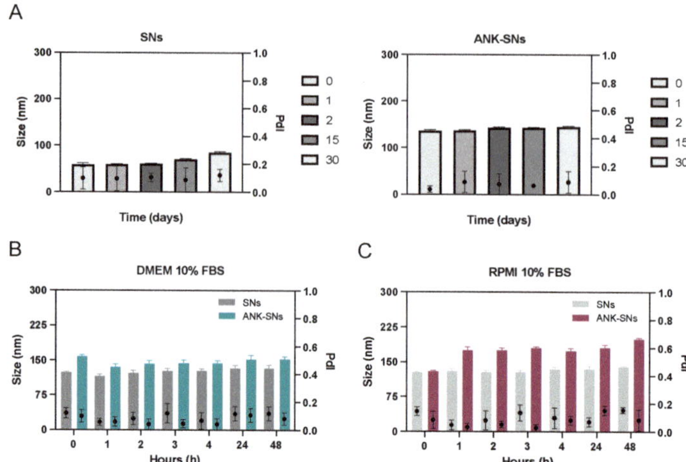

Figure 3. Stability in size and polydispersity index (PdI) of sphingomyelin nanosystems (SNs) and anakinra-loaded sphingomyelin nanosystems (ANK-SNs) over time. (**A**) Stability up to 30 days of SNs and ANK-SNs stored at 4 °C. Stability upon incubation up to 48 h in DMEM (**B**), and RPMI (**C**), supplemented with 10% FBS at 37 °C and under orbital stirring (300 rpm).

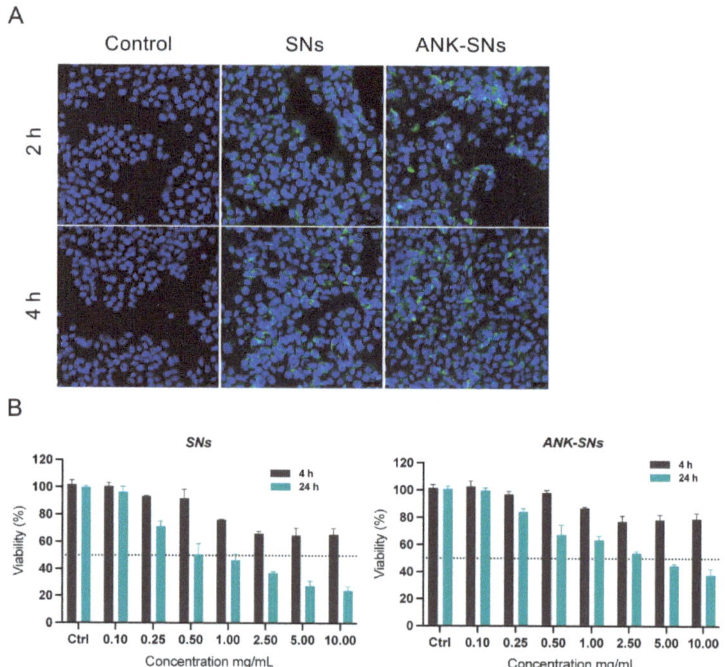

Figure 4. Internalization and cytotoxicity assays of nanosystems using pancreatic ductal adenocarcinoma (PDAC) cells. (**A**) Representative confocal microscopy images after 2 h and 4 h treatment in L3.6p cells. Fluorescent-labeled TopFluor® sphingomyelin nanosystems (SNs) and TopFluor® anakinra-loaded sphingomyelin nanosystems (ANK-SNs) are in green. Cell nuclei stained with Hoechst 33342 are in blue. (**B**) Cell viability determined by Alamar Blue™ assay after treatment with SNs and ANK-SNs (0.1 mg/mL to 10 mg/mL) for 4 h and 24 h. Data are expressed as means ± SD. The dotted line is set at 50% of viability.

For cell viability assays, we treated the L3.6pl cells for 4 h and 24 h at increasing concentrations of nanosystems (0.1 mg/mL to 10 mg/mL corresponding to 2.2 µg/mL to 220 µg/mL of ANK), and cytotoxicity was revealed by Alamar Blue^{TM} fluorescent assay. As shown in Figure 4B, at lower concentrations (0.10–0.50 mg/mL), the nanosystems showed little cytotoxicity. At higher concentrations, decreased viability was seen in both nanosystems, although cytotoxicity was lower for ANK-SNs compared to SNs. These results defined the toxic concentrations (i.e., 0.25–0.5 mg/mL) used to guide the following experiments.

2.2. ANK and ANK-SNs Are Equivalent in Down-Modulating Cytokine Secretion by IL-1-Activated CAFs

We then moved to compare the effect of ANK versus ANK-SNs in inhibiting the secretion of tumor-promoting cytokines in the tumor microenvironment that depends on IL-1/IL-1R signaling. CAFs are a relevant cell population in the PDAC stroma that express IL-1R1 [51], and when stimulated with IL-1α and/or IL-1β, they secrete several pro-tumor cytokines [36,52,53], including TSLP, which is responsible for pro-tumor Th2 inflammation in PDAC [20,23]. Indeed, we previously reported that CAFs obtained from primary PDAC samples and activated in vitro with IL-1α and IL-1β (or IL-1-containing supernatants from PDAC microenvironment cultures) release TSLP, which in turn activates myeloid DCs with Th2-polarizing capability [20]. Importantly, we also found that interfering in vitro and in vivo with the IL-1/IL-1R signaling pathway with ANK inhibited TSLP secretion by activated CAFs [23].

In this study, we tested the inhibitory activity of ANK-SNs compared to free ANK by using the in vitro system depicted in Figure 5A. CAFs were stimulated with IL-1α + IL-1β in the absence and in the presence of ANK, ANK-SNs, or SNs (i.e., to evaluate the basal level of cytokine modulation by the SNs alone), and after 2 days, the supernatant was collected for the detection of the tumor-promoting TSLP, IL-8, IL-6, and TGF-β cytokines [20,23,36,52,53]. To identify the optimal concentrations of ANK to use in these experiments, we created a dose–response curve with different ANK concentrations and used the release of TSLP as a readout. We found that 5 µg/mL gave the maximum inhibitory effect, whereas 0.1 µg/mL did not show inhibitory effects (Figure 5B), and we decided to use these two concentrations in the following experiments. We then tested the three nanosystem formulations (i.e., VitE/SM/DBCO, VitE/SM/Maleimide, and VitE/SM/NHS) to detect the best-performing one(s) using the two selected concentrations. We found that at 5 µg/mL, the VitE/SM/NHS/ANK and VitE/SM/DBCO/ANK formulations inhibited TSLP release at levels comparable to ANK, whereas VitE/SM/Maleimide was much less efficient (Figure 5C, left panel, and Supplementary Figure S3A). In addition, when we considered the effect of the SNs alone and in comparison to the other formulations, VitE/SM/NHS did not show any stimulatory effect (Figure 5C, left panel, and Supplementary Figure S3A), and it was chosen for the following experiments. In the same experimental setting, we also measured the release of pro-tumor IL-8, IL-6, and TGF-β. In the case of IL-8, the results were very similar to TSLP (Figure 5D, left panel), whereas in the case of IL-6 (Figure 5E, left panel) and TGF-β (Figure 5F, left panel), the inhibition induced by ANK-SNs at 5 µg/mL was slightly inferior to the one obtained with ANK. The results of a larger set of experiments verified that ANK and ANK-SNs equally inhibited the release of the different cytokines (Figure 5C–F, right panels).

We confirmed in CAFs that treatment with the nanosystems did not induce relevant cytotoxicity at the concentrations (i.e., 0.1 and 5 µg/mL) used in experiments shown in Figure 5B (Supplementary Figure S4A). In addition, an analysis of internalization of the nanosystems by confocal microscopy showed that down-modulation of the IL-1R1 expression at the CAF surface after treatment was significantly higher with ANK SNs compared to SNs (Supplementary Figure S5A), demonstrating the specific interaction of IL-1R1 with ANK-SNs at the CAF surface.

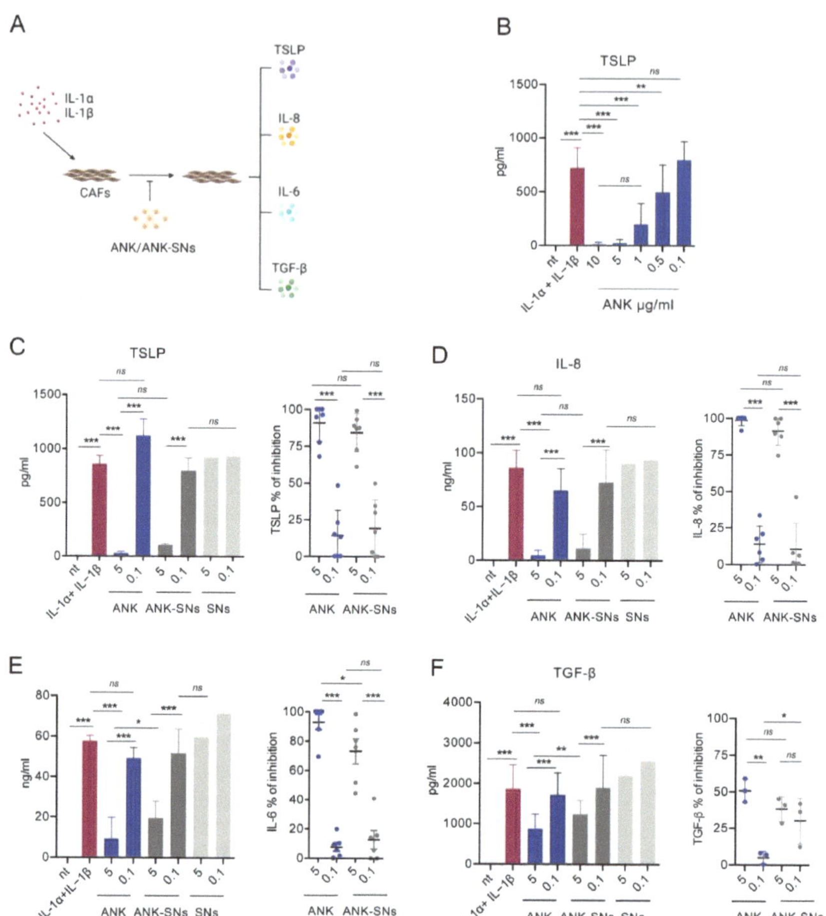

Figure 5. Anakinra (ANK) and anakinra-loaded sphingomyelin nanosystems (ANK-SNs) are equivalent in down-modulating cytokine secretion by IL-1-activated cancer-associated fibroblasts (CAFs). (**A**) In vitro experimental model. CAFs were treated with recombinant IL-1α + IL-1β, mimicking tumor-derived IL-1, in the absence or in the presence of ANK, ANK-SNs, or SNs. Secretion of TSLP, IL-8, IL-6, and TGF-β was measured after culture. (Created with BioRender.com). (**B**) Dose–response curve to detect the best concentrations of ANK for inhibiting TSLP secretion by CAFs (n = 3). ANK was added at the indicated concentrations. Untreated CAFs were used as negative (nt) control; IL-1α + IL-1β-treated CAFs were used as positive (IL-1α + IL-1β) control. (**C–F**) ANK, ANK-SNs, and SNs (VitE/SM/NHS formulation), tested for their capacity to down-modulate cytokine secretion, were added at the indicated concentrations and based on the titration curve shown in B. Negative and positive controls used were as in B. (**C**) TSLP. *Left*, TSLP secretion with controls (n = 2). *Right*, TSLP percentage inhibition of cumulative experiments (n = 7). (**D**) IL-8. *Left*, IL-8 secretion with controls (n = 2). *Right*, IL-8 percentage inhibition of cumulative experiments (n = 6). (**E**) IL-6. *Left*, IL-6 secretion with controls (n = 2). *Right*, IL-6 percentage inhibition of cumulative experiments (n = 6). (**F**) TGF-β. *Left*, TGF-β secretion with controls (n = 2). *Right*, TGF-β percentage inhibition of cumulative experiments (n = 3). Data are mean ± SEM from the indicated number (n) of independent experiments. Significance was calculated by one-way ANOVA test and Newman–Keuls post-test. Values were considered significantly different for * $p < 0.05$, ** $p < 0.01$, *** $p < 0.001$; ns = not significant.

Collectively, we found that ANK-SNs did not have cytotoxic effects, were efficiently internalized through specific interaction at the cell surface with IL-1R1, and down-modulated the secretion by CAFs of TSLP, IL-8, IL-6, and TGF-β, thus dampening their tumor-promoting functions.

2.3. ANK-SNs Are Superior to Free ANK in Down-Modulating the Secretion of IL-17, but Not of Interferon-γ (IFN-γ), by In Vitro Differentiated Th17 Cells

Th17-type inflammation has been reported to favor tumor progression in PDAC [28,29,31–33]. Due to the role of IL-1 in the differentiation of human Th17 cells [24–27], we evaluated the effect of ANK-SNs added in culture during in vitro Th17 differentiation from naïve CD4$^+$ T cells. We followed the experimental scheme described in [27] and depicted in Figure 6A. Naïve CD4$^+$ T cells were activated with anti-CD2, anti-CD3, and anti-CD28 antibody-coated beads and cultured in the presence of the polarizing cytokines (i.e., IL-1β, IL-23, IL-6, and TGF-β) and in the absence or the presence of ANK, ANK-SNs, and SNs. After 5 days of culture, Th17 cells were restimulated with the antibody-coated beads for 24 h, and the supernatant was collected for IL-17 and IFN-γ detection. To identify the best concentration of ANK to be used to inhibit IL-17 secretion by Th17 cells, we created dose–response titration curves (Figure 6B). Based on these experiments, we decided to use 10 µg/mL (with maximum inhibitory effect) and 0.5 µg/mL (without inhibitory effect) for the initial testing with the three nanosystem formulations. Similar to the results obtained with CAFs, the VitE/SM/NHS/ANK formulation showed the best performance (Figure 6C and Supplementary Figure S3B). Indeed, VitE/SM/DBCO/ANK was inferior to VitE/SM/NHS/ANK in inhibiting IL-17 secretion, and in the case of VitE/SM/Maleimide/ANK, an inhibitory function was already present with SNs alone (Supplementary Figure S3B, right). Importantly, VitE/SM/NHS/ANK at both 10 and 0.5 µg/mL induced inhibition of IL-17 secretion that was significantly higher compared with the one obtained with ANK (Figure 6C, left and right). As Th17 cells, in addition to IL-17, may secrete IFN-γ, which is a relevant anti-tumor cytokine [54], we measured the release of this cytokine in the same experimental setting. Both ANK and ANK-SNs inhibited IFN-γ secretion at comparable levels, although inhibition was much lower compared to that of IL-17 (Figure 6D, left and right). Lastly, as we found that 5 µg/mL was the best ANK-SN concentration to use in the CAF experiments, we verified the efficacy of this dose also in the Th17 model. We also confirmed that at 5 µg/mL, ANK-SNs were superior to ANK in inhibiting IL-17 secretion by Th17 cells (Figure 6E), and importantly at levels even higher than the ones obtained with ANK at 10 µg/mL (median level 80% compared to 60%) (Figures 6E and 6C, right panel). As reported above, no significant difference between ANK-SNs and ANK in inhibiting IFN-γ secretion by Th17 was observed (Figure 6F).

We confirmed in Th17 cells that nanosystems did not induce relevant cytotoxicity in any conditions and concentrations used, when tested with both colorimetric and flow cytometry assays (Supplementary Figure S4B–D). Confocal microscopy showed internalization of the nanosystems with down-modulation of the IL-1R1 expression on Th17 cells that was significantly higher after treatment with ANK-SNs compared to SNs (Supplementary Figure S5B), demonstrating a specific interaction of ANK-SNs with IL-1R1.

Collectively, we found that nanosystems did not affect Th17 cell viability, were internalized through ANK/IL-1R1 interaction, and exerted their modulatory effects primarily by inhibiting pro-tumor IL-17, while preserving anti-tumor IFN-γ.

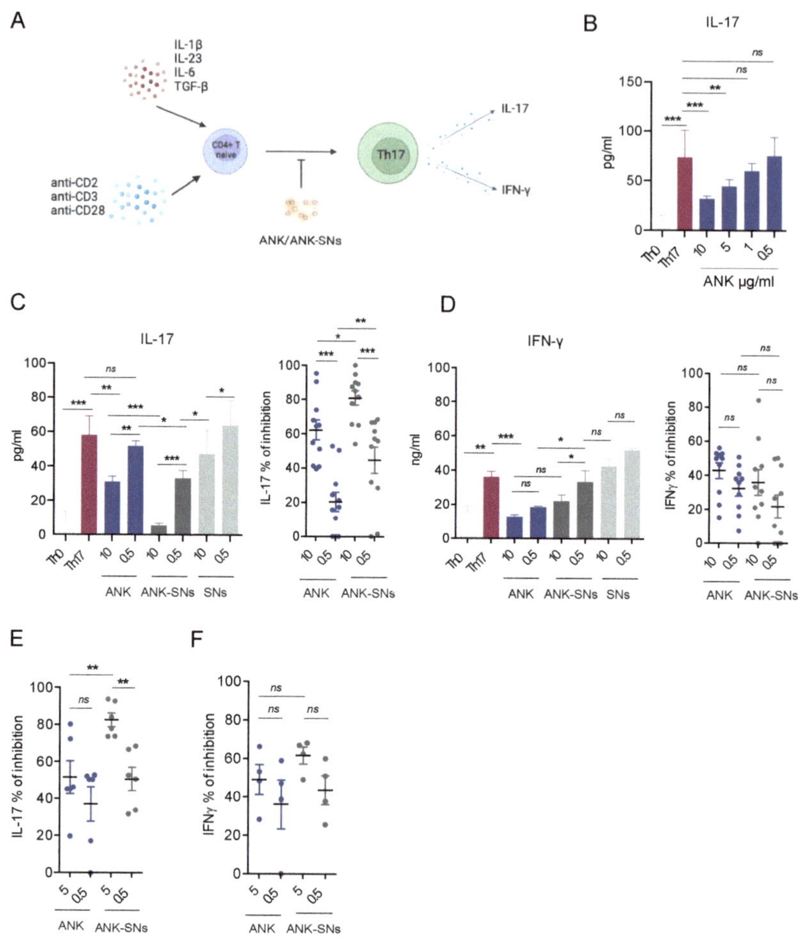

Figure 6. Anakinra-loaded sphingomyelin nanosystems (ANK-SNs) are superior to free anakinra (ANK) in down-modulating the secretion of IL-17, but not of IFN-γ, by in vitro differentiated Th17 cells. (**A**) In vitro experimental model. Naïve CD4$^+$ T cells were differentiated towards Th17 cells by treatment with anti-CD2/CD3/CD28-coated beads and IL-1β, IL-23, IL-6, and TGF-β, and in the absence or in the presence of ANK, ANK-SNs, or SNs. On day 5, Th17 cells were collected and restimulated with anti-CD2/CD3/CD28-coated beads, and IL-17 and IFN-γ secretion levels were measured by ELISA. (Created with BioRender.com). (**B**) Dose–response curve to determine the best concentrations of ANK for inhibiting IL-17 secretion in Th17 cells (n = 3). ANK was added at the indicated concentrations. Th0 (i.e., naïve CD4$^+$ T cells activated with anti-CD2/CD3/CD28-coated beads only) were used as a negative control. Th17 cells were used as a positive control. (**C,D**) ANK, ANK-SNs, and SNs (VitE/SM/NHS formulation), tested for their capacity to down-modulate cytokine secretion by Th17 cells, were added at the indicated concentrations and based on the titration curve shown in (**B**). (**C**) IL-17. *Left*, IL-17 secretion with controls (n = 2). *Right*, IL-17 percentage inhibition of cumulative experiments (n = 11) (**D**) IFN-γ. *Left*, IFN-γ secretion with controls (n = 2). *Right*, IFN-γ percentage inhibition of cumulative experiments (n = 10). (**E**) IL-17 percentage inhibition of cumulative experiments using 5 μg/mL of ANK/ANK-SNs (n = 6). (**F**) IFN-γ percentage inhibition of cumulative experiments using 5 μg/mL of ANK/ANK-SNs (n = 4). Data are mean ± SEM from the indicated number (n) of independent experiments. Significance was calculated by one-way ANOVA test and Newman–Keuls post-test. Values were considered significantly different for * $p < 0.05$, ** $p < 0.01$, *** $p < 0.001$.

3. Discussion

Protein-based therapeutics are an important class of medicines, accounting for more than 130 pharmaceutics approved for clinical use in the European Union and the United States, to treat a wide variety of human conditions, including cancer [55–57]. However, they suffer from enzymatic degradation, short half-life, and physicochemical instability with a loss of bioactivity, hampering their clinical efficacy [39]. To overcome these issues, their delivery can be improved by combining therapeutic proteins with nanotechnology, and lipidic and polymeric nanocarriers have been described for this purpose [39,58,59].

In this work, we used lipidic nanosystems that enable covalent binding of the protein ANK to the modified hydrophobic carbon chains disposed onto the surface. Previous extensive characterization of these nanosystems, composed of the two natural compounds, i.e., VitE and SM, revealed suitable physicochemical properties, very high biocompatibility in vitro and in vivo, and colloidal stability during storage and in biological media, all relevant properties for clinical translation [40]. All these properties were confirmed for the nanosystems loaded with ANK, described here.

ANK has been used to treat autoimmune and autoinflammatory disorders with clinical benefit, especially for autoinflammatory diseases [60]. However, because of its short half-life, ANK needs daily administration to maintain its efficacy, with relevant local side effects at the subcutaneous injection site that may lead to drug discontinuation [61]. The ANK-SN formulations described here might help in targeting the drug to the relevant site of action (i.e., the tumor) with increased efficacy and clinical benefit, but also increased tolerability.

Due to the role of IL-1 in tumor progression [10,11], ANK has been tested in preclinical in vitro and in vivo settings with encouraging results [23,34–36], and clinical trials have planned the incorporation of ANK into treatment protocols [62]. ANK was used in small pilot studies of combination regimens in advanced metastatic colorectal cancer with long-lasting tumor stabilization [63] and in HER2-negative metastatic breast cancer with a reduction in or stabilization of the tumor volume [64]. In PDAC, ANK was tested in combination with the highly toxic FOLFIRINOX regimen in advanced disease [37], and in combination with gemcitabine, nab-paclitaxel, and cisplatin in patients with non-metastatic disease prior to resection, resulting in reduced local cancer spread compared with historical controls receiving chemotherapy alone [38]. Collectively, these clinical trials revealed potential clinical benefit for ANK, and loading of the drug in the nanosystems described here might greatly increase the therapeutic efficacy while reducing side effects, not only in PDAC but also in other tumors where the IL-1/IL-1R axis is relevant.

The aim of the present study was to evaluate the efficacy of ANK as a free drug compared to the drug loaded in the nanosystems in two in vitro models of Th2- and Th17-type inflammation relevant to PDAC progression.

Concerning Th2 inflammation, we showed that ANK-SNs, at both 5 and 0.1 μg/mL, were equal to the free drug in inhibiting TSLP secretion by CAFs, suggesting that the drug loaded in the nanosystems maintained its efficacy. In addition, other tumor-promoting cytokines released by CAFs were also inhibited by ANK-SNs, including IL-8, IL-6, and TGF-β (although at levels slightly inferior to ANK). Interestingly, these two last cytokines are also relevant inducers of Th17 differentiation [27], suggesting a comprehensive effect in the tumor microenvironment.

Concerning Th17 cell differentiation, we found that ANK-SNs were superior to the free drug in inhibiting IL-17 secretion at both 10 and 5 μg/mL. In addition, the inhibition of IL-17 secretion by Th17 cells with ANK-SNs at 5 μg/mL reached a mean of about 80% that was still superior to the inhibition obtained with ANK alone at 10 μg/mL, meaning that a relevant effect was observed at a much lower drug concentration. This is important when considering drug toxicity.

Another important aspect is the performance of ANK-SNs when considering the IFN-γ secreted by Th17 cells. In different cancer contexts, Th17 cells have been described with either pro- or anti-tumor functions [65]. This paradoxical behavior of Th17 cells has been attributed to their plasticity potential that enables their conversion into other types of

CD4$^+$ T cells or their acquisition of polyfunctional activity, such as Th17/T regulatory and Th17/Th1. In these last polyfunctional Th17 cells, the release of IFN-γ might contribute to anti-tumor activity. In our in vitro system, we found that, although the inhibition of IL-1 signaling by both ANK and ANK-SNs reduced the secretion of both IL-17 and IFN-γ, the inhibition of IFN-γ was not significantly different between the two drug compositions (mean 40%) and was similar at all concentrations tested. This means that ANK-SNs compared to ANK were more efficacious in inhibiting IL-17 secretion while maintaining an equal much lower level of inhibition of anti-tumor IFN-γ.

In conclusion, using two in vitro model systems of primary CAFs and differentiated Th17 cells, we have demonstrated that ANK loaded in nanosystems maintains (and importantly, ANK-SNs are more efficacious than the free drug even at lower concentrations) their capacity to downregulate the secretion of cytokines that are indirectly or directly relevant in PDAC for the establishment of Th2- and Th17-type inflammation, respectively.

We showed that these nanosystems do not present significant toxicity in vitro and are stable in culture, thus guaranteeing their biocompatibility and encouraging safe application in medical practice. Altogether, our results encourage further in vivo testing of ANK loaded in these nanosystems, which could offer advantages for patients with PDAC and other neoplastic diseases where treatment with ANK has already shown positive clinical effects.

Mouse models recapitulating the tumor-promoting Th2- and Th17-type inflammation, as reported in the human disease, have been described [29,66,67], and they could offer good models for exploring the new pharmaceutical preparations, comprising any additional/different side effects compared to the ones described for ANK, described here.

As IL-1 family cytokines have been described in mouse models to play important roles in regulating host–microbiome dialogue at barrier sites [68], in future studies, understanding whether and how therapeutic inhibition of these cytokines may impact microbiome dysbiosis might provide further opportunities to improve patients' outcome.

4. Materials and Methods

4.1. Materials

Vitamin E (VitE, DL-α-Tocopherol) was purchased from Sigma (Merck, Darmstadt, Germany, EU). Sphingomyelin (SM, Lipoid E SM) was kindly provided by Lipoid GmbH (Ludwigshafen, Germany, EU). Surfactant lipids, being DSPE-PEG(2k)-X (X = DBCO, Maleimide, NHS), were obtained from Avanti$^®$ Polar Lipids (Alabaster, AL, USA). Fluorescent labeling lipid, C11 TopFluor$^®$ Sphingomyelin, was also acquired from Avanti$^®$ Polar Lipids. Azido-dPEG$^®_8$-NHS ester was provided by Merck (Darmstadt, Germany, EU). HPLC-grade ethanol (EtOH) and dimethyl sulfoxide were purchased from Thermo Fisher Scientific (Rockford, IL, USA). DMEM was obtained from Sigma, RPMI 1640 Medium, FBS and penicillin/streptomycin from Thermo Fisher Scientific, Iscove's modified Dulbecco's medium (IMDM) and X-VIVO^{TM} 20 Medium both from Lonza (Walkersville, MD, USA).

4.2. Preparation of ANK and Azide-Modified ANK (N3-ANK) and Characterization

The recombinant protein ANK was purified from the other components present in the pharmaceutical form Kineret$^®$, namely salts, using a desalting column (CentriPure 10, emp Biotech GmbH, Berlin, Germany, EU), and eluted with PBS (2 mM, pH 7.4), following the manufacturer instructions. The concentration of the obtained ANK protein was determined by the BCA Protein Assay Kit (Thermo Fisher Scientific, Rockford, IL, USA).

The ANK protein (0.058 μmol) was also modified with Azido-dPEG$^®_8$-NHS linker (1.16 μmol) (Sigma, Merck, Darmstadt, Germany, EU) in PBS (10 mM). The reaction was performed at room temperature (RT) for 6 h, with storage at 4 °C. Linker excess was removed by using spin filters (Amicon$^®$ Ultra Centrifugal Filters, 10 k, 0.5 mL, Merck, Darmstadt, Germany, EU) in a refrigerated microcentrifuge (5417R; Eppendorf), as previous reported [40]. N3-ANK modification was characterized by HPLC-MS (TimsTof Pro, Bruker, Billerica, MA, USA; coupled to the Elute UHPLC) by dissolving 5 pmol of sample in ultrapure water. The analysis method used was positive electrospray ionization, acquisition range 500–3000 m/z, chromatographic separation with column C4, and the mobile phases employed were A, water/formic acid 0.1%, and B, acetonitrile/formic acid 0.1%; the chromatographic separation time was 15 minutes (min). MALDI-TOF analysis was carried out in Ultraflex III TOF/TOF (Bruker, Billerica, MA, USA); the analysis method chosen was MALDI linear mode ionization using a sinapinic acid matrix.

4.3. Preparation of SNs, ANK-SNs, and TopFluor®-SM-Labeled ANK-SNs

SNs and ANK-SNs composed of VitE, SM, and surfactant lipids were prepared by adapting the ethanol injection method, as previously described [40]. Briefly, 5 mg of VitE, 0.5 mg of SM, and 0.05 mg of surfactant lipids were dissolved in 100 µL of ethanol. SNs were instantaneously formed by injecting the organic phase into 0.9 mL of ultrapure water, at RT. ANK-SNs were then formed by dropwise adding ANK solution (14 µM) in PBS under magnetic stirring for up to 5 min and incubated at RT for 6 h followed by an overnight incubation at 4 °C.

Fluorescent-labeled SNs were prepared by admixing TopFluor®-SM (4.5 µg) with the other compounds of the organic phase (final volume of 100 µL). The organic phase was then injected in 0.9 mL of ultrapure water and immediately formed at RT.

4.4. Physicochemical Characterization of the Nanosystems

Particle size and PdI were determined by DLS and Z-potential values by LDA, using a Zetasizer NanoZS® (Malvern Instruments, Worcestershire, UK). Samples were diluted to 1:10 with ultrapure water, and the measurements were performed at 25 °C with a detection angle of 173°. SN and ANK-SN formulations were evaluated in biological media to ensure the maintenance of their properties during in vitro assays. Formulations were incubated at 37 °C with 10% FBS-supplemented DMEM and RPMI 1640 under orbital shaking. Additionally, extended characterization was performed by NTA. For NTA, the nanosystems were diluted at 1:1000 (v/v) in ultrapure water using a NanoSight NS3000 system (Malvern Instrument, UK) with a laser operating at λ = 488 nm and 200 mW power. Data collection was settled with 3 repeats/60 s capture time, with both shutter and gain manually determined for each sample. NTA 2.0 Build 127 software was used for measurement and subsequent data analysis.

SNs and ANK-SNs were analyzed by MALDI-TOF. The analysis was conducted using Ultraflex III TOF/TOF (Bruker) equipment; the method used was MALDI linear mode ionization using a sinapinic acid matrix with previous sample ZipTip desalting.

SNs and ANK-SNs were evaluated for particle size and morphology distribution by FESEM, using a ZEISS FESEM ULTRA Plus microscope (Carl Zeiss Micro Imaging, GmbH, Germany, EU) configured with InLens and STEM mode, operating at 20 kV. First, 20 µL (0.55 mg/mL) of a filtered nanosystem was stained with phosphotungstic acid (2% w/v) at a ratio of 1:1. Next, 10 µL was placed on a carbon-coated grid and left for 30 s, twice. The grid was allowed to dry and then repeatedly washed with filtered ultrapure water. Finally, it was let dry until its visualization.

4.5. Establishment of CAF Cell Lines

Primary cultures of CAFs were obtained from PDAC surgical samples, as we previously described [20]. Briefly, tumor pieces were put in culture in IMDM medium (Lonza, Walkersville, MD, USA) plus 10% FBS, and CAFs obtained by outgrowth. CAFs were characterized for expression of the fibroblast-activation protein (FAP) marker and the absence of expression of the EpCAM epithelial marker by flow cytometry, as described below.

CAFs at 3rd-4th culture passage were stably transfected to express the human telomerase reverse transcriptase gene (hTERT) by lentiviral infection using hTERT Cell Immortalization KIT (ALSTEM, Richmond, CA, USA), according to manufacturer's instructions. Seventy-two hours after infection, puromycin (Sigma) selection was applied at 1.5 µg/mL final concentration. Puromycin-resistant colonies were checked for hTERT expression by real-time qPCR. CAF cell lines were periodically tested for Mycoplasma contamination using the MycoBlue Mycoplasma Detector kit (D101, Vazyme, Nanjing, China).

4.6. Effects of Nanosystems on Cytokine Secretion by CAFs

CAFs were seeded at 3×10^4 cells/well in IMDM 10% FBS in flat-bottom 96-well plates and starved overnight without serum. The next day, the medium was replaced with IMDM 2% FBS alone or with 20 ng/mL of each recombinant human IL-1α + IL-1β (R&D Systems, Minneapolis, MN, USA). In the inhibition experiments, ANK (Kineret®, Amgen Europe, Breda, The Netherlands), ANK-SNs and SNs were added at the indicated concentrations. After 4 h of incubation, the medium was replaced with IMDM 2% FBS for 72 h. Cell viability was tested as described below and is reported in Supplementary Figure S4A. TSLP, IL-6, IL-8, and TGF-β secretion in CAF supernatants was measured by ELISA (TSLP and TGF-β, R&D Systems; IL-6 and IL-8, Mabtech, Nacka Strand, Sweden, EU).

4.7. Effects of Nanosystems on Th17 Cell Differentiation and Cytokine Secretion

Peripheral blood mononuclear cells (PBMCs) were obtained by Ficoll-Hypaque (Cytiva, Malborough, MA, USA) gradient stratification from healthy donor buffy coats. Naïve CD4$^+$ T cells were magnetically isolated from PBMCs after two rounds of Naïve CD4$^+$ T cell Isolation Kit (Miltenyi Biotec, Bergish Gladbach, Germany, EU) and plated at 1×10^6 cells/well in 48-well plates containing X-VIVO 20 in the presence of anti-CD2-, anti-CD3-, and anti-CD28-coated beads (Miltenyi Biotec, Bergish Gladbach, Germany, EU). The following cytokines were added: IL-1β (10 ng/mL) (R&D Systems), IL-6 (20 ng/mL) (PeproTech, Cranbury, NJ, USA), TGF-β (1 ng/mL) (PeproTech), and IL-23 (50 ng/mL) (PeproTech or R&D Systems). Th0 cells were obtained by stimulation of purified naïve CD4$^+$ T cells with the antibody-coated beads in the absence of the polarizing cytokines. In inhibition experiments, ANK, ANK-SNs, and SNs were added at the indicated concentrations. After 5 days, cells were collected and washed extensively, and their viability was determined as described below and is reported in Supplementary Figure S4B–D. CD4$^+$ T cells (1×10^6/mL) were restimulated for 24 h with anti-CD2-, anti-CD3-, and anti-CD28-coated beads (one bead per cell). The secretion of IL-17 and IFN-γ in culture supernatants was detected by ELISA (Mabtech, Nacka Strand, Sweden, EU).

4.8. Cell Viability

The L3.6pl PDAC cell line, originally described in [69], was kindly supplied by Dr. Bruno Sainz Jr (Madrid, Spain, EU). L3.6pl cells were cultured in RPMI 1640 media (Invitrogen, Thermo Fisher Scientific, Waltham, MA, USA) containing 10% FBS and 50 units/mL penicillin/streptomycin in a humidified incubator at 37 °C, and cultures were tested for Mycoplasma contamination periodically, as above. L3.pl6 cells were seeded at a density of 2.5×10^4 cell/well into black-wall clear-bottom 96-well plates (Sigma) and incubated in the absence or the presence of increasing concentration of SNs or ANK-SNs for 4 h and 24 h. Cell viability was determined using a 1:10 dilution AlamarBlue™ assay. Fluorescence was read using a Victor Nivo™ multimode plate reader (Thermo Fisher Scientific, Waltham, MA, USA) at 530 nm excitation and 590 nm emission. Data were normalized to negative control 100% viability, and the reagent itself was used as a fluorescence background.

CAFs were stimulated with pro-inflammatory cytokines, and Th17 cells were differentiated and cultured both in the absence and in the presence of inhibitory stimuli, as described above. At the end of stimulation/differentiation, CAFs were washed with PBS 100 µL of IMDM 10% FBS culture medium was replaced, and Th17 cells were collected and cultured in 100 µL X-VIVO 20. Cell viability was evaluated by the addition of 20 µL/well of CellTiter 96 Aqueous One Solution Cell Proliferation Assay (PROMEGA, Madison, WI, USA) for 4 h. The absorbance was recorded at 490 nm and 650 nm to subtract the background. The viability of Th17 cells was also tested by flow cytometry; see below.

4.9. Flow Cytometry

Th17 cells were harvested after in vitro restimulation for 24 h with antibody-coated beads, as described above, and stained with anti-human CD4 (BD Pharmingen™ FITC mouse anti-human CD4, cat. 555346) and diluted 1:1000 with Invitrogen™ LIVE/DEAD™ Fixable Aqua Dead Cell Stain Kit (Invitrogen, Thermo Fisher Scientific, Waltham, MA, USA). Cell viability was assessed by flow cytometry; data were acquired with BD FACSCanto. A gating strategy was adapted to avoid autofluorescence derived from antibody-coated beads (Supplementary Figure S4C). CAFs were characterized by staining with anti-FAP antibody (APC-conjugated mouse anti-human FAP antibody, R&D Systems, cat. FAB3715A) and anti-EpCAM antibody (PE-conjugated mouse anti-human EpCAM/CD326 antibody, BD Biosciences, San Jose, CA, USA, cat. 347198), following the manufacturer's instructions, and data were acquired as above.

4.10. Confocal Microscopy

Internalization of nanosystems in L3.6pl cells was evaluated by confocal laser microscopy (Leica Microscope SP8®, Wetzlar, Germany, EU). First, 6×10^4 cells were seeded in an 8-well chamber slide plate and grew overnight at 37 °C. Fluorescent-labeled TopFluor®-SM SNs and fluorescent-labeled TopFluor®-SM ANK-SNs (0.2 mg/mL) were added for 2 h. The treatment was then removed, cells were fixed for 15 min with 4% paraformaldehyde (PFA) (Thermo Fisher Scientific), and cell nuclei were stained with Hoechst 33342 (Thermo Fisher Scientific).

For CAF experiments, 1×10^5 cells/well were plated on a glass coverslip, previously treated with 0.01% Poly-L-lysine solution (Sigma), and starved overnight. The next day, CAFs were stimulated with IL-1α + IL-1β, as described above, and treated with ± ANK or fluorescent-labeled TopFluor®-SM SNs or fluorescent-labeled TopFluor®-SM ANK-SNs for 4 h, fixed with 4% PFA (Thermo Fisher Scientific) in PBS for 10 min at RT and washed with PBS. Treated CAFs were permeabilized with 0,1% saponin in dH_2O for 15 min. The coverslips were then blocked overnight at 4 °C with PBS 2% BSA (Sigma). Successively, a first incubation with anti-IL-1R1 primary antibody (Invitrogen, Thermo Fisher Scientific, Waltham, MA, USA) for 1 h at RT and a secondary incubation with Alexa Fluor 546 (Invitrogen, Thermo Fisher Scientific, Waltham, MA, USA) for 2 h at RT were applied. Hoechst 33342 (Thermo Fisher Scientific, Waltham, MA, USA) was added to mark the nuclei, and the coverslips were mounted in glycerol. Cells were imaged with an Olympus FV3000 confocal laser scanning microscope.

For Th17 cell experiments, CD4$^+$ naïve T cells were magnetically isolated from PBMCs and plated at 1×10^6 cells/well on a glass coverslip, previously treated with 0,01% Poly-L-lysine solution (Sigma), and stimulated, as described above, with IL-1β, IL-6, TGF-β, and IL-23 in the presence of anti-CD2-, anti-CD3-, and anti-CD28-coated beads ± fluorescent-labeled TopFluor®-SM SNs or fluorescent-labeled TopFluor®-SM ANK-SNs for 5 days. Th17 cells were then fixed with 4% PFA (Thermo Fisher Scientific) in PBS for 10 min at RT, washed with PBS, and blocked overnight at 4 °C with PBS 2% BSA. Successively, a first incubation with anti-IL-1R1 primary antibody (Invitrogen, Thermo Fisher Scientific, Waltham, MA, USA) for 1 h at RT and a secondary incubation with Alexa Fluor 546 (Invitrogen) for 2 h at RT were applied. Hoechst 33342 (Thermo Fisher Scientific, Waltham, MA, USA was added to mark the nuclei, and the coverslips were mounted in glycerol. Cells are imaged with an Olympus FV3000 confocal laser scanning microscope.

For IL-1R1 quantification, area coverage on the cellular surface was determined using ImageJ Software 1.53k. Z-plans were initially acquired and successively stacked to obtain a single image that was converted to 8-bit. After the threshold was applied, the area of the image covered by the fluorochrome linked to IL-1R1 was measured.

4.11. Statistics

Differences between groups were statistically determined using a one-way ANOVA test and Neuman–Keuls post-test. The percentage of inhibition was calculated by considering the values of the positive controls in the case of ANK and SNs in the case of ANK-SNs as 100% expression. All statistical analyses were performed using GraphPad Prism (Version 6.0 software) (GraphPadSoftware, San Diego, CA, USA). A p value < 0.05 was considered to be significant.

Supplementary Materials: The following supporting information can be downloaded at https://www.mdpi.com/article/10.3390/ijms25158085/s1.

Author Contributions: M.d.l.F., L.D.M. and M.P.P. conceived and supervised the study; M.A.-S., M.d.l.F., L.D.M. and M.P.P. designed methodology; M.A.-S., M.D.L. and M.G.C. performed experiments, data collection, and data analysis; S.R. performed flow cytometry analyses; M.d.l.F. and M.P.P. acquired funding; M.A.-S., M.D.L., L.D.M. and M.P.P. wrote the original draft. M.A.-S., M.D.L., L.D.M. and M.P.P. accessed and verified the data. All authors reviewed and/or edited the final manuscript and were responsible for the decision to submit the manuscript. All authors have read and agreed to the published version of the manuscript.

Funding: This work was supported by the Italian Ministry of Health (ERA-NET EURONANOMED III project PANIPAC, grant number JTC20018/041 to MPP), the Italian Association for Cancer Research (AIRC Special Program in Metastatic disease: the key unmet need in oncology, 5 per mille, grant number 22737 to MPP), the Instituto de Salud Carlos III (ISCIII) and the European Regional Development Fund (FEDER) (grant number AC18/00107 to M.d.l.F), the Instituto de Salud Carlos III (ISCIII) (ERA-NET EURONANOMED III project PANIPAC, grant number JTC2018/041 to M.d.l.F), and the Health Research Institute of Santiago de Compostela (IDIS) (predoctoral research fellowship/2020 to M.A.-S).

Institutional Review Board Statement: The study was conducted in accordance with the Declaration of Helsinki, and approved by the Institutional Ethics Committee (Comitato Etico Fondazione Centro San Raffaele, Istituto Scientifico Ospedale San Raffaele, Milan, Italy) approved the study protocol (PDACTREAT).

Informed Consent Statement: Informed consent was obtained from all subjects involved in the study.

Data Availability Statement: The data presented in thus study are available upon request to the corresponding authors.

Conflicts of Interest: M.d.l.F. is the co-founder and CEO of DIVERSA Technologies S.L. The other authors declare that they have no conflicts of interest with the contents of this article.

Abbreviations

ANK: anakinra; ANK-SNs, anakinra loaded-SNs; BSA, bovine serum albumin; CAFs, cancer-associated fibroblasts; DCs, dendritic cells; DLS, dynamic light scattering; DMEM, Dulbecco's modified eagle medium; FAP, fibroblast-activation protein; FBS, fetal bovine serum; FESEM, field emission scanning microscopy; h, hour/hours, HPLC-MS, high-performance liquid chromatography with mass spectroscopy; hTERT, human telomerase reverse transcriptase gene; IFN-γ; interferon-γ; IL, interleukin; IL-1R, IL-1 receptor; IL-1RA, IL-1R antagonist; IMDM; Iscove's modified Dulbecco's medium; LDA, laser Doppler anemometry; MALDI-TOF, matrix-assisted laser desorption/ionization time-of-flight; min, minutes; NTA, nanoparticle tracking analysis; PBMCs, peripheral blood mononuclear cells; PBS, phosphate-buffered saline; PDAC, pancreatic ductal adenocarcinoma; PdI, polydispersity index; PEG, polyethylene glycol; PFA, paraformaldehyde; RT, room temperature; SM, sphingomyelin; SNs, unloaded-nanosystems; TGF-β, transforming growth factor-β; Th, T helper; TSLP, thymic stromal lymphopoietin; VitE, vitamin E.

References

1. Hidalgo, M. Pancreatic cancer. *N. Engl. J. Med.* **2010**, *362*, 1605–1617. [CrossRef] [PubMed]
2. Rahib, L.; Smith, B.D.; Aizenberg, R.; Rosenzweig, A.B.; Fleshman, J.M.; Matrisian, L.M. Projecting cancer incidence and deaths to 2030: The unexpected burden of thyroid, liver, and pancreas cancers in the United States. *Cancer Res.* **2014**, *74*, 2913–2921. [CrossRef] [PubMed]
3. Feig, C.; Gopinathan, A.; Neesse, A.; Chan, D.S.; Cook, N.; Tuveson, D.A. The pancreas cancer microenvironment. *Clin. Cancer Res.* **2012**, *18*, 4266–4276. [CrossRef] [PubMed]

4.   Vonderheide, R.H.; Bayne, L.J. Inflammatory networks and immune surveillance of pancreatic carcinoma. *Curr. Opin. Immunol.* **2013**, *25*, 200–205. [CrossRef] [PubMed]

5.   Wormann, S.M.; Diakopoulos, K.N.; Lesina, M.; Algul, H. The immune network in pancreatic cancer development and progression. *Oncogene* **2014**, *33*, 2956–2967. [CrossRef] [PubMed]

6.   Protti, M.P.; De Monte, L. Immune infiltrates as predictive markers of survival in pancreatic cancer patients. *Front. Physiol.* **2013**, *4*, 210. [CrossRef] [PubMed]

7.   Zheng, L.; Xue, J.; Jaffee, E.M.; Habtezion, A. Role of immune cells and immune-based therapies in pancreatitis and pancreatic ductal adenocarcinoma. *Gastroenterology* **2013**, *144*, 1230–1240. [CrossRef] [PubMed]

8.   Leinwand, J.; Miller, G. Regulation and modulation of antitumor immunity in pancreatic cancer. *Nat. Immunol.* **2020**, *21*, 1152–1159. [CrossRef]

9.   Huber, M.; Brehm, C.U.; Gress, T.M.; Buchholz, M.; Alashkar Alhamwe, B.; von Strandmann, E.P.; Slater, E.P.; Bartsch, J.W.; Bauer, C.; Lauth, M. The Immune Microenvironment in Pancreatic Cancer. *Int. J. Mol. Sci.* **2020**, *21*, 7307. [CrossRef]

10. Garlanda, C.; Mantovani, A. Interleukin-1 in tumor progression, therapy, and prevention. *Cancer Cell* **2021**, *39*, 1023–1027. [CrossRef] [PubMed]

11. Dosch, A.R.; Singh, S.; Nagathihalli, N.S.; Datta, J.; Merchant, N.B. Interleukin-1 signaling in solid organ malignancies. *Biochim. Biophys. Acta Rev. Cancer* **2022**, *1877*, 188670. [CrossRef]

12. Gukovsky, I.; Li, N.; Todoric, J.; Gukovskaya, A.; Karin, M. Inflammation, autophagy, and obesity: Common features in the pathogenesis of pancreatitis and pancreatic cancer. *Gastroenterology* **2013**, *144*, 1199–1209.e1194. [CrossRef]

13. Hausmann, S.; Kong, B.; Michalski, C.; Erkan, M.; Friess, H. The role of inflammation in pancreatic cancer. *Adv. Exp. Med.Biol.* **2014**, *816*, 129–151.

14. Zambirinis, C.P.; Pushalkar, S.; Saxena, D.; Miller, G. Pancreatic cancer, inflammation, and microbiome. *Cancer J.* **2014**, *20*, 195–202. [CrossRef]

15. Zhang, T.; Ren, Y.; Yang, P.; Wang, J.; Zhou, H. Cancer-associated fibroblasts in pancreatic ductal adenocarcinoma. *Cell Death Dis.* **2022**, *13*, 897. [CrossRef]

16. Sun, Q.; Zhang, B.; Hu, Q.; Qin, Y.; Xu, W.; Liu, W.; Yu, X.; Xu, J. The impact of cancer-associated fibroblasts on major hallmarks of pancreatic cancer. *Theranostics* **2018**, *8*, 5072–5087. [CrossRef]

17. Huang, H.; Brekken, R.A. Recent advances in understanding cancer-associated fibroblasts in pancreatic cancer. *Am. J. Physiol. Cell Physiol.* **2020**, *319*, C233–C243. [CrossRef]

18. Geng, X.; Chen, H.; Zhao, L.; Hu, J.; Yang, W.; Li, G.; Cheng, C.; Zhao, Z.; Zhang, T.; Li, L.; et al. Cancer-Associated Fibroblast (CAF) Heterogeneity and Targeting Therapy of CAFs in Pancreatic Cancer. *Front. Cell Dev. Biol.* **2021**, *9*, 655152. [CrossRef]

19. Jain, A.; Song, R.; Wakeland, E.K.; Pasare, C. T cell-intrinsic IL-1R signaling licenses effector cytokine production by memory CD4 T cells. *Nat. Commun.* **2018**, *9*, 3185. [CrossRef] [PubMed]

20. De Monte, L.; Reni, M.; Tassi, E.; Clavenna, D.; Papa, I.; Recalde, H.; Braga, M.; Di Carlo, V.; Doglioni, C.; Protti, M.P. Intratumor T helper type 2 cell infiltrate correlates with cancer-associated fibroblast thymic stromal lymphopoietin production and reduced survival in pancreatic cancer. *J. Exp. Med.* **2011**, *208*, 469–478. [CrossRef]

21. Protti, M.P.; De Monte, L. Cross-talk within the tumor microenvironment mediates Th2-type inflammation in pancreatic cancer. *Oncoimmunology* **2012**, *1*, 89–91. [CrossRef] [PubMed]

22. De Monte, L.; Woermann, S.; Brunetto, E.; Heltai, S.; Magliacane, G.; Reni, M.; Paganoni, A.M.; Recalde, H.; Mondino, A.; Falconi, M.; et al. Basophil recruitment into tumor draining lymph nodes correlates with Th2 inflammation and reduced survival in pancreatic cancer patients. *Cancer Res.* **2016**, *76*, 1792–1803. [CrossRef] [PubMed]

23. Brunetto, E.; De Monte, L.; Balzano, G.; Camisa, B.; Laino, V.; Riba, M.; Heltai, S.; Bianchi, M.; Bordignon, C.; Falconi, M.; et al. The IL-1/IL-1 receptor axis and tumor cell released inflammasome adaptor ASC are key regulators of TSLP secretion by cancer associated fibroblasts in pancreatic cancer. *J. Immunother. Cancer* **2019**, *7*, 45. [CrossRef]

24. Chung, Y.; Chang, S.H.; Martinez, G.J.; Yang, X.O.; Nurieva, R.; Kang, H.S.; Ma, L.; Watowich, S.S.; Jetten, A.M.; Tian, Q.; et al. Critical regulation of early Th17 cell differentiation by interleukin-1 signaling. *Immunity* **2009**, *30*, 576–587. [CrossRef]

25. Revu, S.; Wu, J.; Henkel, M.; Rittenhouse, N.; Menk, A.; Delgoffe, G.M.; Poholek, A.C.; McGeachy, M.J. IL-23 and IL-1beta Drive Human Th17 Cell Differentiation and Metabolic Reprogramming in Absence of CD28 Costimulation. *Cell Rep.* **2018**, *22*, 2642–2653. [CrossRef]

26. Lee, W.W.; Kang, S.W.; Choi, J.; Lee, S.H.; Shah, K.; Eynon, E.E.; Flavell, R.A.; Kang, I. Regulating human Th17 cells via differential expression of IL-1 receptor. *Blood* **2010**, *115*, 530–540. [CrossRef]

27. Volpe, E.; Servant, N.; Zollinger, R.; Bogiatzi, S.I.; Hupe, P.; Barillot, E.; Soumelis, V. A critical function for transforming growth factor-beta, interleukin 23 and proinflammatory cytokines in driving and modulating human T(H)-17 responses. *Nat. Immunol.* **2008**, *9*, 650–657. [CrossRef]

28. He, S.; Fei, M.; Wu, Y.; Zheng, D.; Wan, D.; Wang, L.; Li, D. Distribution and clinical significance of Th17 cells in the tumor microenvironment and peripheral blood of pancreatic cancer patients. *Int. J. Mol. Sci.* **2011**, *12*, 7424–7437. [CrossRef]

29. McAllister, F.; Bailey, J.M.; Alsina, J.; Nirschl, C.J.; Sharma, R.; Fan, H.; Rattigan, Y.; Roeser, J.C.; Lankapalli, R.H.; Zhang, H.; et al. Oncogenic Kras activates a hematopoietic-to-epithelial IL-17 signaling axis in preinvasive pancreatic neoplasia. *Cancer Cell* **2014**, *25*, 621–637. [CrossRef]

30. Chellappa, S.; Hugenschmidt, H.; Hagness, M.; Line, P.D.; Labori, K.J.; Wiedswang, G.; Tasken, K.; Aandahl, E.M. Regulatory T cells that co-express RORgammat and FOXP3 are pro-inflammatory and immunosuppressive and expand in human pancreatic cancer. *Oncoimmunology* **2016**, *5*, e1102828. [CrossRef] [PubMed]

31. Lang, C.; Wang, J.; Chen, L. CD25-expressing Th17 cells mediate CD8(+) T cell suppression in CTLA-4 dependent mechanisms in pancreatic ductal adenocarcinoma. *Exp. Cell Res.* **2017**, *360*, 384–389. [CrossRef] [PubMed]

32. Barilla, R.M.; Diskin, B.; Caso, R.C.; Lee, K.B.; Mohan, N.; Buttar, C.; Adam, S.; Sekendiz, Z.; Wang, J.; Salas, R.D.; et al. Specialized dendritic cells induce tumor-promoting IL-10(+)IL-17(+) FoxP3(neg) regulatory CD4(+) T cells in pancreatic carcinoma. *Nat. Commun.* **2019**, *10*, 1424. [CrossRef] [PubMed]

33. Khan, I.A.; Singh, N.; Gunjan, D.; Gopi, S.; Dash, N.; Gupta, S.; Saraya, A. Increased circulating Th17 cell populations in patients with pancreatic ductal adenocarcinoma. *Immunogenetics* **2023**, *75*, 433–443. [CrossRef]

34. Zhuang, Z.; Ju, H.Q.; Aguilar, M.; Gocho, T.; Li, H.; Iida, T.; Lee, H.; Fan, X.; Zhou, H.; Ling, J.; et al. IL1 Receptor Antagonist Inhibits Pancreatic Cancer Growth by Abrogating NF-kappaB Activation. *Clin. Cancer Res.* **2016**, *22*, 1432–1444. [CrossRef]

35. Takahashi, R.; Macchini, M.; Sunagawa, M.; Jiang, Z.; Tanaka, T.; Valenti, G.; Renz, B.W.; White, R.A.; Hayakawa, Y.; Westphalen, C.B.; et al. Interleukin-1beta-induced pancreatitis promotes pancreatic ductal adenocarcinoma via B lymphocyte-mediated immune suppression. *Gut* **2021**, *70*, 330–341. [PubMed]

36. Dosch, A.R.; Singh, M.; Dai, X.; Mehra, S.; Silva, I.C.; Bianchi, A.; Srinivasan, S.; Gao, Z.; Ban, Y.; Chen, X.; et al. Targeting Tumor-Stromal IL6/STAT3 Signaling through IL1 Receptor Inhibition in Pancreatic Cancer. *Mol. Cancer Ther.* **2021**, *20*, 2280–2290. [CrossRef] [PubMed]

37. Whiteley, A.; Becerra, C.; McCollum, D.; Paulson, A.S.; Goel, A. A pilot, non-randomized evaluation of the safety of anakinra plus FOLFIRINOX in metastatic pancreatic ductal adenocarcinoma patients. *J. Clin. Oncol.* **2016**, *34*, e165750. [CrossRef]

38. Becerra, C.; Paulson, A.S.; Cavaness, K.M.; Celinski, S. Gemcitabine, nab-paclitaxel, cisplatin, and anakinra (AGAP) treatment in patients with non-metastatic pancreatic ductal adenocarcinoma (PDAC). *J. Clin.Oncol.* **2018**, *36*, 449. [CrossRef]

39. Viegas, C.S.; Seck, F.; Fonte, P. An insight on lipid nanoparticles for therapeutic proteins delivery. *J. Drug. Deliver. Technol.* **2022**, *77*, 103839. [CrossRef]

40. Bouzo, B.L.; Calvelo, M.; Martin-Pastor, M.; Garcia-Fandino, R.; de la Fuente, M. In Vitro-In Silico Modeling Approach to Rationally Designed Simple and Versatile Drug Delivery Systems. *J. Phys. Chem. B* **2020**, *124*, 5788–5800. [CrossRef]

41. Bouzo, B.L.; Lores, S.; Jatal, R.; Alijas, S.; Alonso, M.J.; Conejos-Sanchez, I.; de la Fuente, M. Sphingomyelin nanosystems loaded with uroguanylin and etoposide for treating metastatic colorectal cancer. *Sci. Rep.* **2021**, *11*, 17213. [CrossRef] [PubMed]

42. Harris, J.M.; Chess, R.B. Effect of pegylation on pharmaceuticals. *Nat. Rev. Drug Discov.* **2003**, *2*, 214–221. [CrossRef] [PubMed]

43. Suk, J.S.; Xu, Q.; Kim, N.; Hanes, J.; Ensign, L.M. PEGylation as a strategy for improving nanoparticle-based drug and gene delivery. *Adv. Drug Deliv. Rev.* **2016**, *99*, 28–51. [CrossRef] [PubMed]

44. Wang, L.; Jiang, R.; Wang, L.; Liu, Y.; Sun, X.L. Preparation of chain-end clickable recombinant protein and its bio-orthogonal modification. *Bioorg. Chem.* **2016**, *65*, 159–166. [CrossRef] [PubMed]

45. Yoon, H.Y.; Lee, D.; Lim, D.K.; Koo, H.; Kim, K. Copper-Free Click Chemistry: Applications in Drug Delivery, Cell Tracking, and Tissue Engineering. *Adv. Mater.* **2022**, *34*, e2107192. [CrossRef]

46. Fang, J.; Nakamura, H.; Maeda, H. The EPR effect: Unique features of tumor blood vessels for drug delivery, factors involved, and limitations and augmentation of the effect. *Adv. Drug Deliv. Rev.* **2011**, *63*, 136–151. [CrossRef] [PubMed]

47. Duan, X.; Li, Y. Physicochemical characteristics of nanoparticles affect circulation, biodistribution, cellular internalization, and trafficking. *Small* **2013**, *9*, 1521–1532. [CrossRef] [PubMed]

48. Cheng, Q.; Wei, T.; Farbiak, L.; Johnson, L.T.; Dilliard, S.A.; Siegwart, D.J. Selective organ targeting (SORT) nanoparticles for tissue-specific mRNA delivery and CRISPR-Cas gene editing. *Nat. Nanotechnol.* **2020**, *15*, 313–320. [CrossRef]

49. Filipe, V.; Hawe, A.; Jiskoot, W. Critical evaluation of Nanoparticle Tracking Analysis (NTA) by NanoSight for the measurement of nanoparticles and protein aggregates. *Pharm. Res.* **2010**, *27*, 796–810. [CrossRef]

50. Kim, A.; Ng, W.B.; Bernt, W.; Cho, N.J. Validation of Size Estimation of Nanoparticle Tracking Analysis on Polydisperse Macromolecule Assembly. *Sci. Rep.* **2019**, *9*, 2639. [CrossRef]

51. Elyada, E.; Bolisetty, M.; Laise, P.; Flynn, W.F.; Courtois, E.T.; Burkhart, R.A.; Teinor, J.A.; Belleau, P.; Biffi, G.; Lucito, M.S.; et al. Cross-Species Single-Cell Analysis of Pancreatic Ductal Adenocarcinoma Reveals Antigen-Presenting Cancer-Associated Fibroblasts. *Cancer Discov.* **2019**, *9*, 1102–1123. [CrossRef] [PubMed]

52. Tjomsland, V.; Spangeus, A.; Valila, J.; Sandstrom, P.; Borch, K.; Druid, H.; Falkmer, S.; Falkmer, U.; Messmer, D.; Larsson, M. Interleukin 1alpha sustains the expression of inflammatory factors in human pancreatic cancer microenvironment by targeting cancer-associated fibroblasts. *Neoplasia* **2011**, *13*, 664–675. [CrossRef] [PubMed]

53. Kennel, K.; Bozlar, M.; De Valk, A.F.; Greten, F.R. Cancer-Associated Fibroblasts in Inflammation and Antitumor Immunity. *Clin. Cancer Res.* **2023**, *29*, 1009–1016. [CrossRef] [PubMed]

54. Burke, J.D.; Young, H.A. IFN-gamma: A cytokine at the right time, is in the right place. *Semin. Immunol.* **2019**, *43*, 101280. [CrossRef] [PubMed]

55. Leader, B.; Baca, Q.J.; Golan, D.E. Protein therapeutics: A summary and pharmacological classification. *Nat. Rev. Drug Discov.* **2008**, *7*, 21–39. [CrossRef]

56. Dimitrov, D.S. Therapeutic proteins. *Methods Mol. Biol.* **2012**, *899*, 1–26. [PubMed]

57. Lagasse, H.A.; Alexaki, A.; Simhadri, V.L.; Katagiri, N.H.; Jankowski, W.; Sauna, Z.E.; Kimchi-Sarfaty, C. Recent advances in (therapeutic protein) drug development. *F1000Research* **2017**, *6*, 113. [CrossRef]
58. Sivadasan, D.; Sultan, M.H.; Madkhali, O.; Almoshari, Y.; Thangavel, N. Polymeric Lipid Hybrid Nanoparticles (PLNs) as Emerging Drug Delivery Platform-A Comprehensive Review of Their Properties, Preparation Methods, and Therapeutic Applications. *Pharmaceutics* **2021**, *13*, 1291. [CrossRef]
59. Dilliard, S.A.; Siegwart, D.J. Passive, active and endogenous organ-targeted lipid and polymer nanoparticles for delivery of genetic drugs. *Nat. Rev. Mater.* **2023**, *8*, 282–300. [CrossRef]
60. Cavalli, G.; Colafrancesco, S.; Emmi, G.; Imazio, M.; Lopalco, G.; Maggio, M.C.; Sota, J.; Dinarello, C.A. Interleukin 1alpha: A comprehensive review on the role of IL-1alpha in the pathogenesis and treatment of autoimmune and inflammatory diseases. *Autoimmun. Rev.* **2021**, *20*, 102763. [CrossRef]
61. Bettiol, A.; Lopalco, G.; Emmi, G.; Cantarini, L.; Urban, M.L.; Vitale, A.; Denora, N.; Lopalco, A.; Cutrignelli, A.; Lopedota, A.; et al. Unveiling the Efficacy, Safety, and Tolerability of Anti-Interleukin-1 Treatment in Monogenic and Multifactorial Autoinflammatory Diseases. *Int. J. Mol. Sci.* **2019**, *20*, 1989. [CrossRef] [PubMed]
62. Fang, Z.; Jiang, J.; Zheng, X. Interleukin-1 receptor antagonist: An alternative therapy for cancer treatment. *Life Sci.* **2023**, *335*, 122276. [CrossRef]
63. Isambert, N.; Hervieu, A.; Rebe, C.; Hennequin, A.; Borg, C.; Zanetta, S.; Chevriaux, A.; Richard, C.; Derangere, V.; Limagne, E.; et al. Fluorouracil and bevacizumab plus anakinra for patients with metastatic colorectal cancer refractory to standard therapies (IRAFU): A single-arm phase 2 study. *Oncoimmunology* **2018**, *7*, e1474319. [CrossRef]
64. Wu, T.C.; Xu, K.; Martinek, J.; Young, R.R.; Banchereau, R.; George, J.; Turner, J.; Kim, K.I.; Zurawski, S.; Wang, X.; et al. IL1 Receptor Antagonist Controls Transcriptional Signature of Inflammation in Patients with Metastatic Breast Cancer. *Cancer Res.* **2018**, *78*, 5243–5258. [CrossRef]
65. Marques, H.S.; de Brito, B.B.; da Silva, F.A.F.; Santos, M.L.C.; de Souza, J.C.B.; Correia, T.M.L.; Lopes, L.W.; Neres, N.S.M.; Dorea, R.; Dantas, A.C.S.; et al. Relationship between Th17 immune response and cancer. *World J. Clin. Oncol.* **2021**, *12*, 845–867. [CrossRef]
66. Ochi, A.; Nguyen, A.H.; Bedrosian, A.S.; Mushlin, H.M.; Zarbakhsh, S.; Barilla, R.; Zambirinis, C.P.; Fallon, N.C.; Rehman, A.; Pylayeva-Gupta, Y.; et al. MyD88 inhibition amplifies dendritic cell capacity to promote pancreatic carcinogenesis via Th2 cells. *J. Exp. Med.* **2012**, *209*, 1671–1687. [CrossRef] [PubMed]
67. Dey, P.; Li, J.; Zhang, J.; Chaurasiya, S.; Strom, A.; Wang, H.; Liao, W.T.; Cavallaro, F.; Denz, P.; Bernard, V.; et al. Oncogenic KRAS-Driven Metabolic Reprogramming in Pancreatic Cancer Cells Utilizes Cytokines from the Tumor Microenvironment. *Cancer Discov.* **2020**, *10*, 608–625. [CrossRef] [PubMed]
68. Narros-Fernandez, P.; Chomanahalli Basavarajappa, S.; Walsh, P.T. Interleukin-1 family cytokines at the crossroads of microbiome regulation in barrier health and disease. *FEBS J.* **2024**, *291*, 1849–1869. [CrossRef] [PubMed]
69. Bruns, C.J.; Harbison, M.T.; Kuniyasu, H.; Eue, I.; Fidler, I.J. In vivo selection and characterization of metastatic variants from human pancreatic adenocarcinoma by using orthotopic implantation in nude mice. *Neoplasia* **1999**, *1*, 50–62. [CrossRef]

Disclaimer/Publisher's Note: The statements, opinions and data contained in all publications are solely those of the individual author(s) and contributor(s) and not of MDPI and/or the editor(s). MDPI and/or the editor(s) disclaim responsibility for any injury to people or property resulting from any ideas, methods, instructions or products referred to in the content.

International Journal of
Molecular Sciences

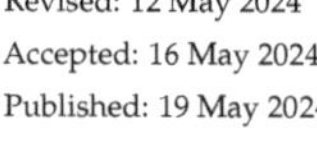

Article

Nanotechnology-Based Strategy for Enhancing Therapeutic Efficacy in Pancreatic Cancer: Receptor-Targeted Drug Delivery by Somatostatin Analog

Xin Gu [1], Joydeb Majumder [1], Olena Taratula [2], Andriy Kuzmov [1], Olga Garbuzenko [1], Natalia Pogrebnyak [1] and Tamara Minko [1,3,*]

[1] Department of Pharmaceutics, Ernest Mario School of Pharmacy, Rutgers, The State University of New Jersey, Piscataway, NJ 08554, USA

[2] Department of Pharmaceutical Sciences, College of Pharmacy, Oregon State University, Portland, OR 97201, USA

[3] Rutgers Cancer Institute of New Jersey, Rutgers, The State University of New Jersey, New Brunswick, NJ 08901, USA

* Correspondence: minko@pharmacy.rutgers.edu; Tel.: +1-848-445-6348; Fax: +1-732-445-3133

Abstract: A novel nanotechnology-based drug delivery system (DDS) targeted at pancreatic cancer cells was developed, characterized, and tested. The system consisted of liposomes as carriers, an anticancer drug (paclitaxel) as a chemotherapeutic agent, and a modified synthetic somatostatin analog, 5-pentacarbonyl-octreotide, a ligand for somatostatin receptor 2 (SSTR2), as a targeting moiety for pancreatic cancer. The cellular internalization, cytotoxicity, and antitumor activity of the DDS were tested in vitro using human pancreatic ductal adenocarcinoma (PDAC) cells with different expressions of the targeted SSTR2 receptors, and in vivo on immunodeficient mice bearing human PDAC xenografts. The targeted drug delivery system containing paclitaxel exhibited significantly enhanced cytotoxicity compared to non-targeted DDS, and this efficacy was directly related to the levels of SSTR2 expression. It was found that octreotide-targeted DDS proved exceptionally effective in suppressing the growth of PDAC tumors. This study underscores the potential of octreotide-targeted liposomal delivery systems to enhance the therapeutic outcomes for PDAC compared with non-targeted liposomal DDS and Paclitaxel-Cremophor® EL, suggesting a promising avenue for future cancer therapy innovations.

Keywords: somatostatin receptor 2 (SSTR2); nanoparticle-based drugs; liposome; paclitaxel; targeted delivery system

Citation: Gu, X.; Majumder, J.; Taratula, O.; Kuzmov, A.; Garbuzenko, O.; Pogrebnyak, N.; Minko, T. Nanotechnology-Based Strategy for Enhancing Therapeutic Efficacy in Pancreatic Cancer: Receptor-Targeted Drug Delivery by Somatostatin Analog. *Int. J. Mol. Sci.* **2024**, *25*, 5545. https://doi.org/10.3390/ijms25105545

Academic Editors: Claudio Luchini and Donatella Delle Cave

Received: 19 April 2024
Revised: 12 May 2024
Accepted: 16 May 2024
Published: 19 May 2024

Copyright: © 2024 by the authors. Licensee MDPI, Basel, Switzerland. This article is an open access article distributed under the terms and conditions of the Creative Commons Attribution (CC BY) license (https://creativecommons.org/licenses/by/4.0/).

1. Introduction

Accounting for 90% of all pancreatic malignancies, pancreatic ductal adenocarcinoma (PDAC) is the fourth leading cause of cancer-related death in the United States, with a five-year survival rate of 8% [1]. Although surgery can be an effective way to treat PDAC, the statistical report suggests that 80% of pancreatic cancer cases are unresectable, and, in many cases, local recurrence can occur after tumor resection [2]. PDAC's complexity is underscored by its genetic diversity, elevated interstitial fluid pressure, and thick tumor stroma, which contribute to a challenging microenvironment. These factors notably amplify its tendency to metastasize and resist drugs, making it difficult for conventional chemotherapy treatments to deliver therapeutic agents effectively [3–5]. Nanotechnology has risen as a critical field in tackling the complexities of diverse cancer types through innovative therapeutic and diagnostic approaches [6–9]. The versatility and potential of this domain have led to its growing prominence, especially in improving treatments for conditions such as PDAC, making substantial progress in recent years [10]. The FDA's approval of nanoformulations like Onivyde® (irinotecan liposome) and Abraxane® (nab-paclitaxel) for PDAC treatment emphasizes the impactful role of nanotechnology. These advanced treatments have outperformed conventional gemcitabine-based therapies in clinical effectiveness [11,12]. However, it is essential to note that these formulations do not specifically target unique cell markers, which suggests an area for further innovation and development in nanoparticle-based therapies.

Liposomes, a type of phospholipid-based vesicles that offer the ability to encapsulate hydrophilic drugs in their internal core and lipophilic drugs within their lipid bilayers, leading to the enhanced solubility and encapsulation efficiency of lipophilic medications [13]. Since their discovery, they have proven to be a productive delivery system for hydrophobic drugs with superior biological compatibility. Moreover, liposomes can evade the reticuloendothelial system (RES) through surface modifications, such as PEG conjugation [14–16]. Additionally, active targeting can be achieved by coupling ligands to cell surface receptors. With their precision engineering, nanoparticles can target tumor-supporting cancer-associated fibroblasts (CAFs) and distinct markers on PDAC cells, such as epidermal growth factor receptor (EGFR), vascular endothelial growth factor receptor (VEGFR), integrins, hyaluronan, transferrin, etc. This targeted approach significantly improves the specificity and effectiveness of treatments [10,17–20]. This targeted approach improves therapeutic outcomes and minimizes damage to healthy cells, reducing the severe side effects typically caused by conventional chemotherapy and radiation. Apart from the markers mentioned above, it has been found that somatostatin receptor 2 (SSTR2) is abundantly expressed in PDAC cells [21–27]. Hence, it may be an attractive target for the related ligand—somatostatin—and its analogs, such as octreotide and lanreotide (Figure 1). Derived primarily from the bark of the *Taxus brevifolia* tree, Paclitaxel (Figure 1) is a potent anti-cancer medication that the FDA has approved for the treatment of ovarian, breast, and non-small cell lung cancer (NSCLC), as well as Kaposi's sarcoma. Its effectiveness in treating these types of cancer is well established. Moreover, it has also been used off-label to address solid tumors, including advanced cervical cancer and metastatic bladder cancer [28,29]. However, commercially available paclitaxel is mainly in the form of Taxol® (Paclitaxel-Cremophor® EL) and Abraxane® (nab-paclitaxel), which cannot be explicitly delivered to the tumor site; therefore, they restrict treatment efficacy and increase the risk of damage to healthy organs. In this study, a modified synthetic somatostatin (SST) analog, 5-pentacarbonyl-octreotide (OCT), was used as an SSTR2-targeting agent, and agent-conjugated liposomal paclitaxel, OCT-DSPE-PEG-Liposome -Paclitaxel (OCT-Lip-PTX), is proposed as a potential model drug for the formulated targeting therapy.

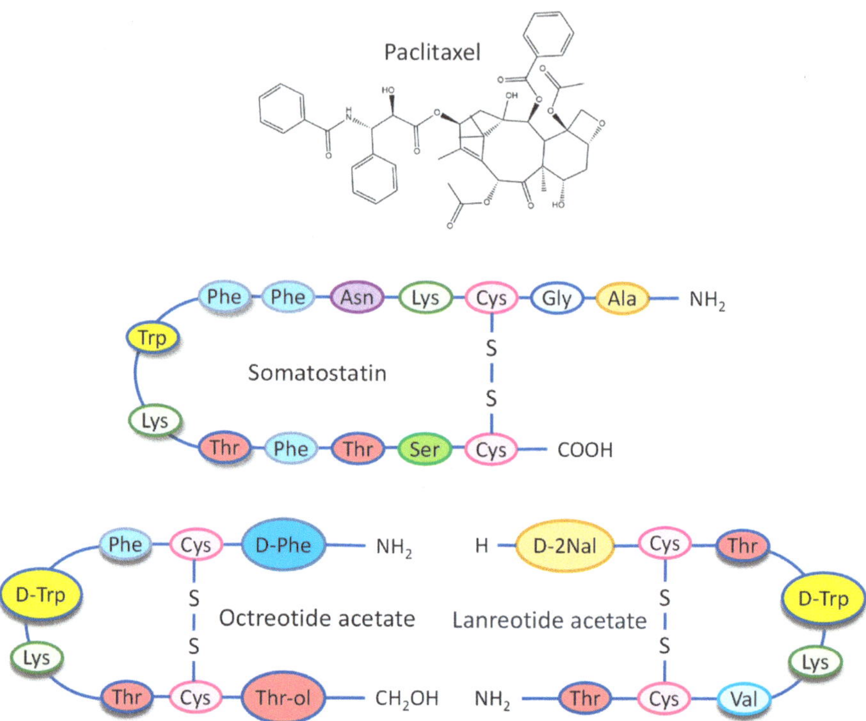

Figure 1. Paclitaxel's structural formula and the amino acid compositions of human somatostatin-14 and its synthetic analogs octreotide and lanreotide. Somatostatin (SST) is presented in two forms of cyclic peptides, 14 and 28 amino acids. The somatostatin synthetic analogs, octreotide, and lanreotide, have 8 amino acids. The four amino acids (Phe, (D)Trp, Lys, Thr) necessary for binding to the SSTR2 receptor are shadowed. The terminal Thr-COOH group in SST analogs is reduced to an alcoholic group, while L-Tryptophan has been replaced by the non-natural enantiomer D-Tryptophan, in order to limit enzymatic degradation and increase the peptide's biological activity.

2. Results

2.1. Synthesis and Characterization of Liposomal Formulations

The size, type and composition of liposomes were selected based on our previous investigations in order to provide the most effective delivery to cancer cells and of the maximal anticancer drug efficacy in vivo [6,9,30,31]. DSPE-PEG$_{2000}$-OCT was synthesized using the methods described below; matrix-assisted laser desorption/ionization time-of-flight mass spectrometry (MALDI-TOF-MS) confirmed the final yield. Data were acquired by ABI-MDS SCIEX 4800 system (SCIEX, Framingham, MA, USA) with linear mid-mass mode from 1.5 kDa to 8 kDa (Figure 2). Figure 3 illustrates the structure of OCT-Lip-PTX, with paclitaxel entrapped between the hydrophobic tails and octreotide linked to liposomes through DSPE-PEG$_{2000}$. Data from the dynamic light scattering (DLS) procedure show that all liposome formulations had sizes of 80 nm to 140 nm (Figure 3), with a slight negative charge (Table 1). It was found that OCT-Lip-PTX showed a comparable releasing profile at various temperatures to its non-targeting counterpart (Figure 4).

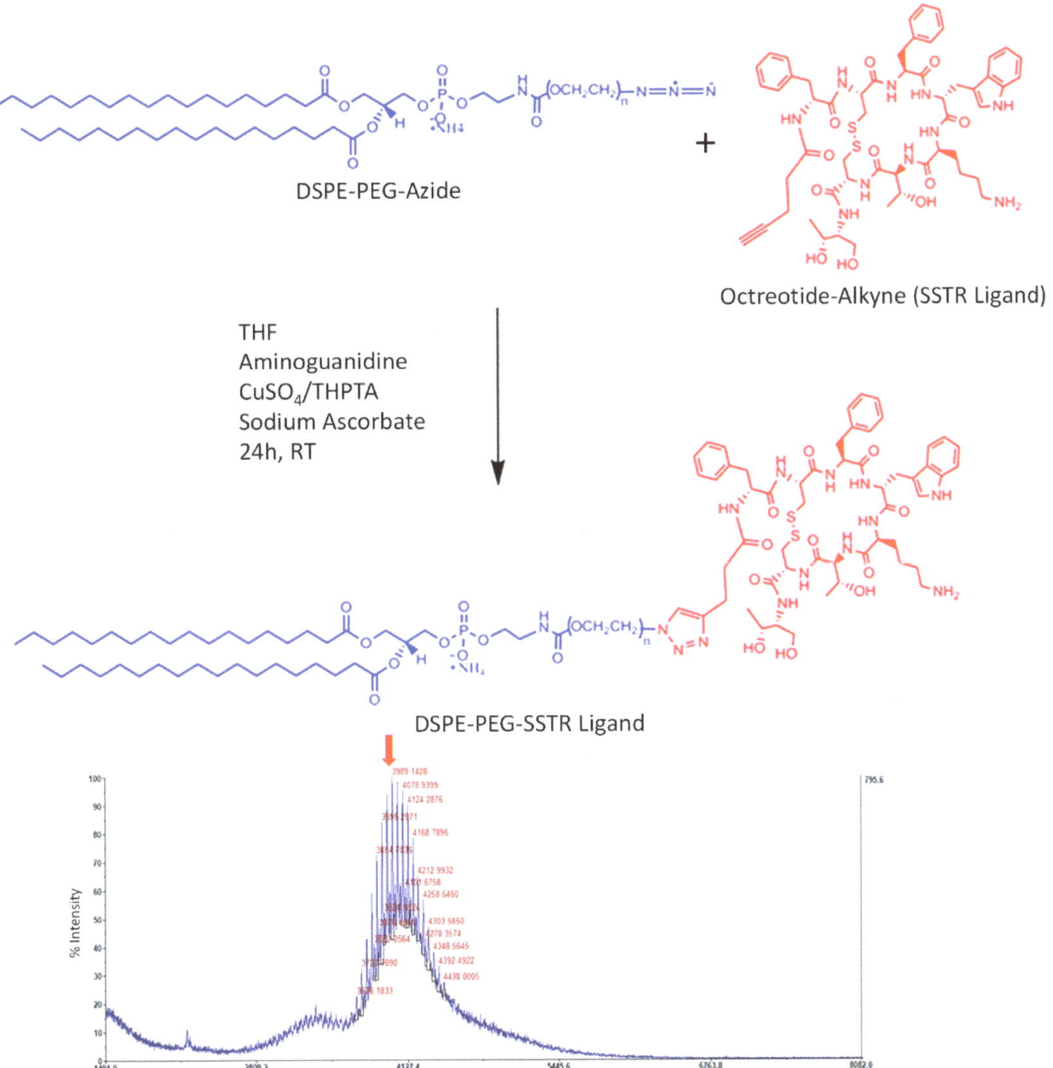

Figure 2. Schematic representation of DSPE-PEG-SSTR ligand synthesis (**top panel**) and representative image of the final product's MALDI-TOF-MS (**bottom panel**). A copper-catalyzed azole-alkyne click chemistry (CuAAC) reaction was performed to conjugate DSPE-PEG-azide with 5-pentynecarbonyl-octreotide (OCT). Both substances were dissolved in tetrahydrofuran (THF) and mixed. After that, the mixture of CuSO4 and tris (benzyltriazolylme-thyl) amine (THPTA) was added dropwise to the reaction mixture and stirred. Next, aminoguanidine and sodium ascorbate were added to the reaction mixture and stirred for 24 h at room temperature (RT). The final product was lyophilized to get the DSPE-PEG2000-OCT ligand in solid powder form. Matrix-assisted laser desorption/ionization time-of-flight mass spectrometry (MALDI-TOF-MS) confirmed the final yield of the ligand. The predicted mass of the resulting compound calculated based on the information provided by manufacturers (with accounting on the polydispersity of PEG), is shown at the MALDI-TOF-MS image as red arrow.

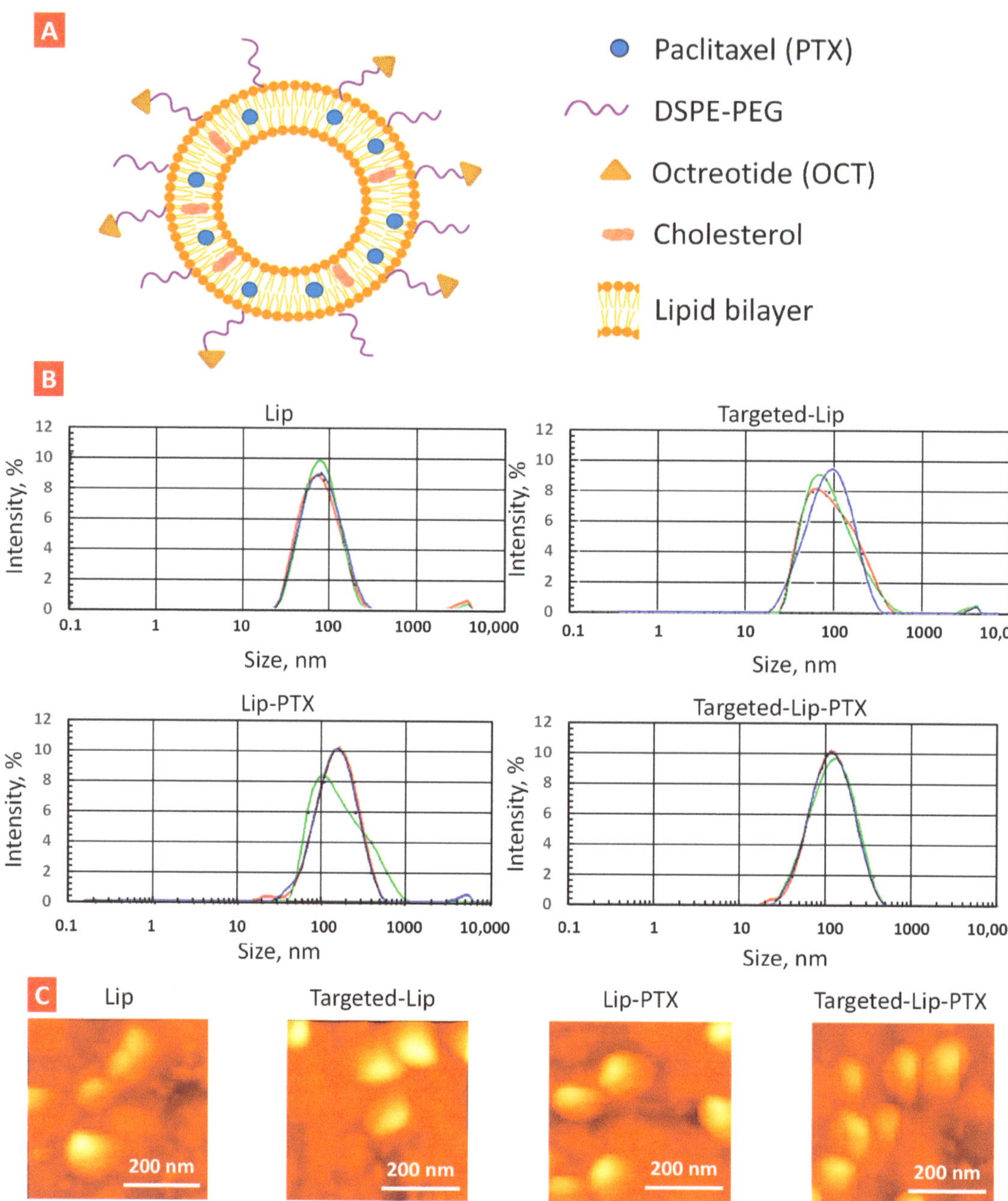

Figure 3. The structure, size distribution and atomic force microscopy (AFM) images of the formulation. (**A**) The assumed structure of the targeted liposomal delivery system containing paclitaxel. (**B**) The size distribution of empty liposomes (Lip), targeted liposomes without drugs (Targeted-Lip), non-targeted liposomes containing paclitaxel (Lip-PTX), and targeted liposomes containing paclitaxel (Targeted-Lip-PTX). The particle size of liposomes was measured by dynamic light scattering using Malvern Zetasizer Nano (Malvern Instruments, UK) Different colors show distribution curves for three individual measurements of the same formulation. (**C**) Representative AFM images of liposomal formulations. Panoramic images were captured in phase contrast mode.

Table 1. Characterization of empty liposomes (Lip), targeted liposomes without the drug (OCT-Lip), non-targeted liposomes containing paclitaxel (OCT-Lip-PTX), and tumor-targeted liposomes containing paclitaxel (OCT-Lip-PTX). Means ± SD are shown.

Formulation	Particle Size, nm	Polydispersity Index	Zeta Potential, mV	Encapsulation Efficiency, %
Lip	88.76 ± 0.09	0.210 ± 0.008	−2.22 ± 1.36	88.76
OCT-Lip	96.10 ± 0.73	0.261 ± 0.008	−1.78 ± 0.07	96.10
Lip-PTX	134.20 ± 0.93	0.248 ± 0.007	−3.20 ± 0.72	94.4
OCT-Lip-PTX	111.30 ± 0.31	0.230 ± 0.002	−3.27 ± 0.56	88.46

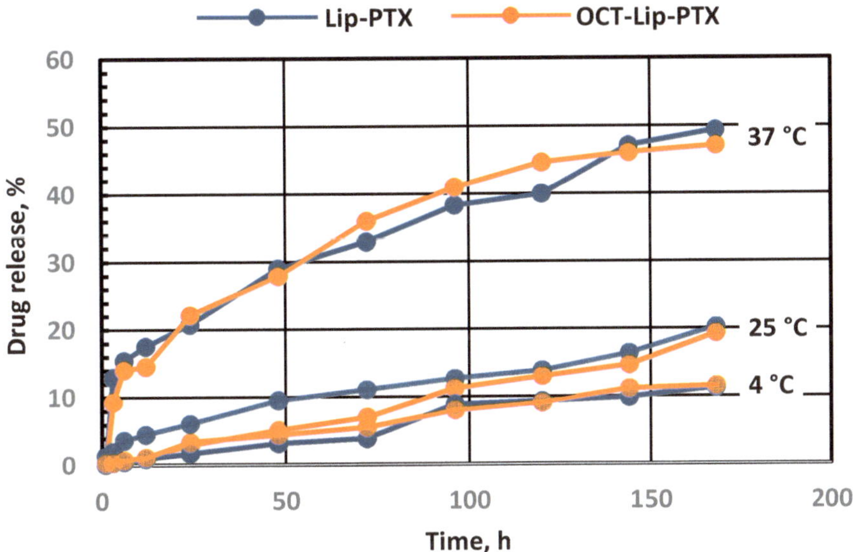

Figure 4. In vitro drug release profiles of liposomal formulations. Non-targeted (Lip-PTX) and cancer-targeted (OCT-Lip-PTX) nanoparticles were introduced into release media that were maintained at three constant temperatures, 4 °C, 25 °C, and 37 °C, over 7 days. Samples were collected at specific time points of 0 h, 0.5 h, 1 h, 2 h, 3 h, 6 h, 8 h, 12 h, 24 h, 48 h, 72 h, 96 h, 120 h, 144 h, and 168 h. After sample collection, an equal volume of 1% Tween 20/PBS solution was replaced at each time point. Free paclitaxel released from the formulations was separated by dialysis, and the bound drug concentration in the liposomes was analyzed using HPLC methods.

2.2. Expression of SSTR2 in Pancreatic Cancer Cells

The results of quantitative reverse transcription polymerase chain reaction (QRT-PCR) assays demonstrated that SSTR2 mRNA is expressed in all three types of investigated pancreatic cancer cell lines, with varying degrees of abundance. It is noteworthy that the expression of SSTR2 was found to be considerably different among the three cells. Specifically, CFPAC-1 cells exhibited almost 4.2-fold higher mRNA expression than MiaPaca-2 cells and 2.5-fold higher mRNA expression than PANC-1 cells (Figure 5). All RNA samples tested positive for internal controls GAPDH and β-Actin.

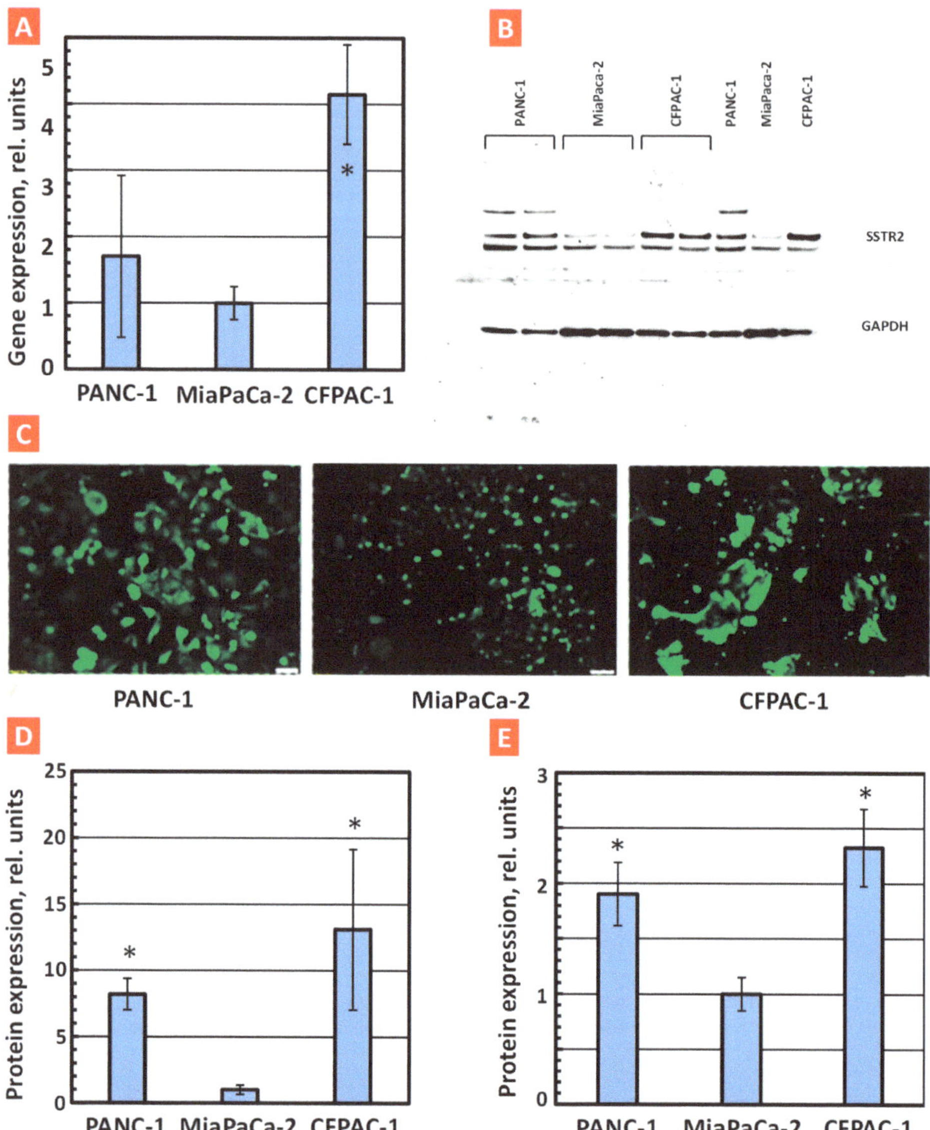

Figure 5. Expression of somatostatin receptors 2 (SSTR2) in PANC-1, MiaPaca-2, and CFPAC-1 human pancreatic cancer cells. (**A**) Gene expression measured by QRT-PCR. (**B**) Representative image of protein expression measured by Western blotting. GAPDH was used as an internal standard. (**C**) Representative image of protein expression measured by immunohistochemistry (scale bars—100 µm). Pancreatic cancer cells were incubated for 24 h with Anti-SSR2antibody (ab9550) followed by Goat Anti-Rabbit IgG H&L (Alexa Fluor® 488, ab150077, green fluorescence). (**D**) Quantitative analysis of Western blotting image. (**E**) Quantitative analysis of immunohistochemistry images. The expression of SSTR2 mRNA/protein in MiaPaca-2 cells was set to 1 unit. Means ± SD are shown. * $p < 0.05$ when compared with MiaPaca-2 cells.

Immunocytochemistry assays were performed to confirm the SSTR2 protein expression qualitatively. All three types of paraformaldehyde-fixed cells were immunostained, and the expression of SSTR2 was observed in CFPAC-1, MiaPaca-2, and PANC-1. CFPAC-1 showed the strongest fluorescence expression among the three cell lines (Figure 5). Then, SSTR2 protein expression was assessed quantitatively by Western blot in cell lysate samples. A significant amount of the expression of endogenous SSTR2 was observed in CFPAC-1 cells in the range of 75 kDa, and a moderate amount of SSTR2 protein was observed in PANC-1 cells compared to MiaPaca-2 cells, which was at a barely visible level (Figure 5). The immunocytochemistry and Western blot results were consistent with the RT-qPCR results, indicating that SSTR2 is much more pronounced in CFPAC-2 cells.

2.3. Cellular Internalization of Liposomes and Intracellular Release of Paclitaxel

The results of the confocal microscope clearly indicated that fluorescence liposomes (red fluorescence) were internalized into the pancreatic cancer cells and localized in the cytoplasm; paclitaxel (green fluorescence) was successfully released from liposomes and localized in the cytoplasm and nuclei (blue fluorescence). Both non-targeted (Lip-PTX) and cancer-targeted (OCT-Lip-PTX) liposomal formulations were internalized by the pancreatic cancer cells and released their PTX payload in the cytoplasm (Figure 6). The fluorescence of labeled PTX tightly packed in liposomes was quenched [32]. Therefore, the appearance of the green fluorescence signal inside the cells testified the release of the drug from liposomes inside the cells. The accumulation of paclitaxel in the cytoplasm facilitated its pharmacological activities, such as microtube targeting.

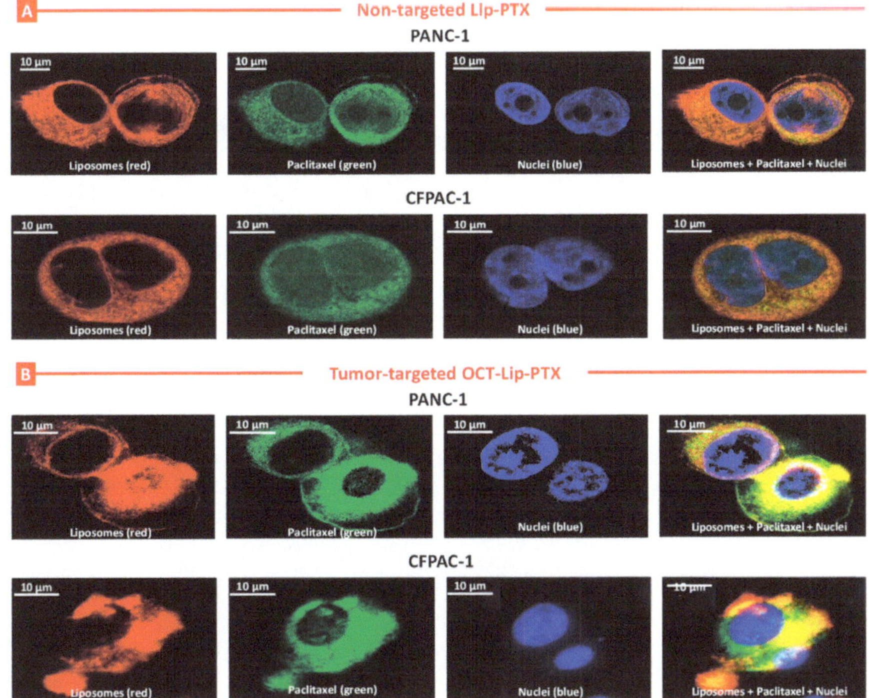

Figure 6. Cellular internalization of non-targeted (Lip-PTX (**A**)) and tumor-targeted (OCT-Lip-PTX (**B**)) liposomes containing paclitaxel by a confocal microscope in Human PANC-1 and CFPAC-1 pancreatic cancer cells. The cells were incubated for 24 h with liposomes (red fluorescence) containing paclitaxel (green fluorescence). Nuclei were labeled with DAPI (blue fluorescence). The superimposition of red and green colors gives yellow or orange colors.

2.4. Cytotoxicity of Different Formulations of Paclitaxel

The IC_{50} values of free non-bound paclitaxel, PEGylated liposomal paclitaxel, and octreotide-targeted PEGylated liposomal paclitaxel against PANC-1, MiaPaca-2, and CFPAC-

1 cells showed the different cytotoxicity of different treatments for all cell lines (Figure 7). After 72 h incubation, the octreotide-conjugated liposomal paclitaxel concentration for 50% inhibition (IC_{50}) of CFPAC-1 cells was approximately 0.04 nM, which was significantly lower than that of free paclitaxel and PEGylated liposomal paclitaxel, with the mean IC_{50} equaling 1.24 nM and 0.38 nM, respectively. Similar results were obtained for both MiaPaca-2 and PANC-1. The average IC_{50} of octreotide-conjugated PEGylated liposomal paclitaxel and paclitaxel against PANC-1 was 98.95 nM and 188.40 nM, respectively. For MiaPaca-2, the IC_{50} for octreotide-conjugated PEGylated liposomal paclitaxel and paclitaxel was 0.59 nM and 2.17 nM, respectively.

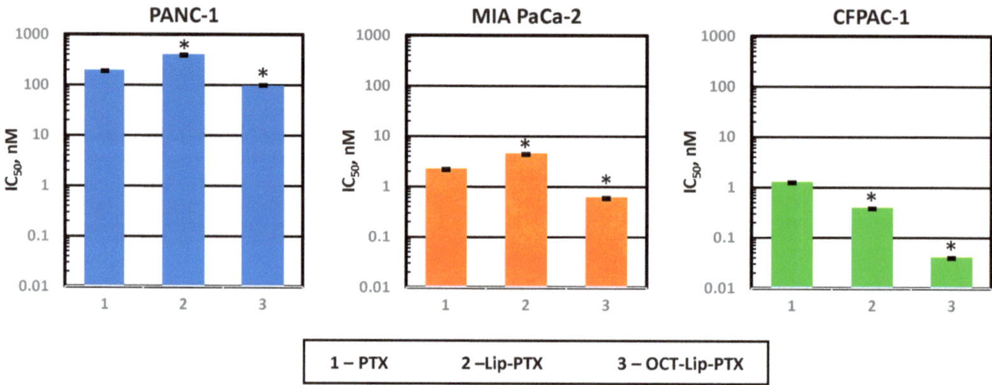

Figure 7. Cytotoxicity of free paclitaxel (PTX), non-targeted (Lip-PTX) and targeted by octreotide (OCT-Lip-PTX) liposomal PTX formulations in different human pancreatic cancer cells. Means ± SD are shown. * $p < 0.05$ when compared with free unbound paclitaxel.

2.5. Comparison of the Antitumor Effect of Free Paclitaxel, Non-Targeted and Pancreatic Cells Targeted Liposomal Formulations of Paclitaxel

The successful uptake of liposomal paclitaxel by PDAC cells and promising in vitro results do not necessarily guarantee exceptional tumor growth suppression in vivo. Therefore, the next step of the present study was focused on investigating the advantages of DDS targeted to somatostatin receptors 2 overexpressed in pancreatic cancer cells when compared with a non-targeted delivery system and clinically used Taxol® (Figure 8). The mice inoculated with MiaPaca-2 and CFPAC-1 cells were treated with the commercially available paclitaxel (PTX), non-targeted liposomal paclitaxel—Lip-PTX—and targeted to SSTR2 OCT-Lip-PTX. All the investigated drug formulations were used in the previously detected maximum-tolerated dose of paclitaxel (MTD, 2.5 mg/kg). The data obtained in the series of mice inoculated with cancer cells demonstrated that the administration of free paclitaxel and untargeted Lip-PTX limited tumor growth. In contrast, the mice with the same type of tumor that were treated with a targeted drug delivery system (OCT-Lip-PTX) demonstrated significantly better outcomes, resulting in the prevention or almost complete elimination of tumors. Similar results were obtained for both MiaPaca-2 and CFPAC-1 types of human pancreatic tumor. OCT-Lip-PTX demonstrated superior efficacy in eliminating PDAC cells, suppressing pancreatic tumor growth, and even reducing tumor volume, compared to the non-targeted liposome formulations currently used in clinic Taxol®. It also should be stressed that CFPAC-1 cells were more resistant to paclitaxel than CFPAC-1 and MiaPaca-2 cells. However, the positive effects of paclitaxel delivery by non-targeted and especially OCT-targeted liposomes were more pronounced in these cells.

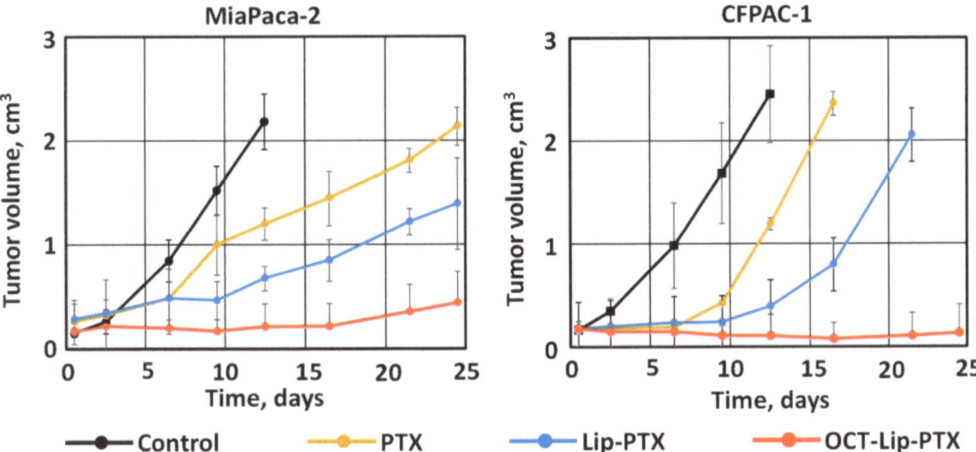

Figure 8. Antitumor activity in mice bearing xenografts of MiaPaca-2 and CFPAC-1 human pancreatic cancer cells. After tumors reached a size of about 0.2 cm^3 (approximately eight days after the subcutaneous injection of cancer cells), mice were treated twice per week within four weeks with the following formulations: Control (mice treated with saline), PTX (Paclitaxel-Cremophor® EL), Lip-PTX (DSPE-PEG-Liposome-Paclitaxel), and OCT-Lip-PTX (5-pentacarbonyl-octreotide)-DSPE-PEG-Liposome-Paclitaxel). Mice were euthanized when the tumor size exceeded 2 cm^3 or at the end of the treatment. Means ± SD are shown.

3. Discussion

The ligand–target SSTR2, chosen for this study, has the potential to enhance potency and minimize undesirable side effects. The studies have shown that the SSTR2 is highly expressed in PDAC cells and tissues, although there may be variations in findings across different studies [21–27,33–35]. This variability could be due to differences in measurement techniques, experimental conditions, or treatment methods with corresponding ligands. SSTRs have been identified in a range of neuroendocrine tumors, such as pituitary adenomas, pancreatic endocrine tumors, small cell lung cancers, medullary thyroid carcinomas, paragangliomas, and carcinoids. Most studies concluded that the level of SSTR2 expression in the tissues affected by pancreatic cancer is considerably higher than that of the adjacent healthy tissues. In particular, it was found that more than 80% of studied patients (88 out of 108) were SSTR2-positive [23]. This expression in SSTR2-positive cancer tissues was 8 to 10 times higher than in non-cancerous pancreatic tissue. Although the clinical relevance of such overexpression is unclear, the difference in the expression in cancerous and normal tissue creates the preconditions for the use of SSTR2 ligands as an effective targeting moiety to provide a cancer-specific delivery of therapeutics and the possible limitation of adverse side effects upon healthy tissues.

Notably, tumors that exhibit neuroendocrine traits and significant SSTR2 expression could be potential candidates for peptide receptor radionuclide therapy (PRRT), either as a standalone treatment or in combination with other therapies like targeted chemotherapy [36–38]. This finding presents exciting treatment possibilities for a specific subset of pancreatic tumors, potentially transforming the management of this condition. Octreotide, a frequently used octapeptide analog of SST, demonstrates a higher binding affinity towards SSTR2 and SSTR5 and an extended half-life of 2 h, in contrast to the short half-life of the native SST, which ranges from 1–3 min. SST activates SHP-1, triggers intracellular pro-apoptotic signals, and promotes apoptosis [39]. This makes octreotide—the first somatostatin analog approved for treating hormone-producing pituitary, pancreatic, and intestinal neuroendocrine tumors—a potential candidate for the tumor-targeted delivery of drugs, highlighting the practical implications of the research [40].

However, previous clinical evaluations demonstrated that the impact of SST analogs on symptom management and decelerating tumor growth in neuroendocrine tumors (NETs) is still limited [41,42]. Despite this, SST analogs have shown substantial potential to be combined with other therapeutic or imaging agents via bioconjugation for the targeted delivery of its payloads to SSTR-positive cancer cells. Sun et al. proposed a targeted delivery system by conjugated cytotoxic camptothecin (CPT) with an SST analog (JF-07-69) via an activated carbamate linker, which demonstrated a 92% inhibition rate in CFPAC-1 tumor models after 4-week treatment and a 56% inhibition rate at low treatment doses (1 mg/pellet) [25]. Further studies involved PTX-OCT conjugates with additional modifications, such as PTX-Lys-OCT, PTX-Phe-OCT, and PTX-Phe-OCT-Lys-PTX, incorporating a succinyl linker [42]. The paclitaxel–octreotide demonstrated significant tumor growth inhibition and reduced microvessel density in A549 NSCLC xenograft models, indicating potent anti-angiogenic effects alongside direct antitumor activity. Conjugating PTX with OCT via a hydrophilic PEG linker significantly enhanced the solubility and SSTR-binding affinity of the conjugate, thereby increasing cytotoxicity against SSTR-overexpressing NCI-H446 cells and demonstrating superior in vivo efficacy and reduced systemic toxicity in xenograft models compared to commercial PTX formulations [43]. However, the targeted delivery of paclitaxel via nanoparticle integration remains limited, but recent research has demonstrated superior antitumor effects in tumor xenograft models. This study proposed a targeting liposome drug delivery system by connecting octreotide to paclitaxel-loaded liposomes via DSPE-PEG linkers for treating SSTR2-positive PDAC tumors.

Our research centered on a novel drug delivery system designed to target tumor cells, explicitly using the abovementioned targeting system. This system employs a dual-targeting strategy that combines the enhanced permeability and retention (EPR) effect for passive targeting with ligand–receptor interactions for active targeting, specifically using octreotide to target SSTR2. The delivery vehicle we developed is an octreotide-DSPE-PEG-liposomal formulation loaded with paclitaxel, engineered to maximize drug delivery efficiency to tumor cells overexpressing SSTR2. In our study, we investigated the role of receptor expression levels in receptor-mediated endocytosis, focusing on PDAC cell lines PANC-1, MiaPaca-2, and CFPAC-1. Among the PDAC cell lines, CFPAC-1 exhibited the highest SSTR2 expression, while MiaPaca-2 showed the lowest. As predicted, our innovative delivery system, octreotide-DSPE-PEG-liposomal paclitaxel, has demonstrated impressive efficiency in administering loaded drugs to targeted tumor cells. The cytotoxicity studies revealed a significant finding; namely, the targeted delivery system most effectively inhibited tumor growth in CFPAC-1 cells, aligning with the cell line's high SSTR2 expression levels. This finding indicates that the targeted system can deliver the drug more effectively to cells with higher receptor expression, leveraging receptor-mediated endocytosis for enhanced drug uptake. The effectiveness of our targeted drug delivery system was further validated through tumor inhibition studies conducted in animal models. The OCT-Lip-PTX-treated tumors showed remarkable control over tumor growth across both cell line groups. Notably, in the group derived from CFPAC-1 cells, which exhibited the highest SSTR2 expression levels, we observed a halt in tumor growth and actual tumor shrinkage. The results underscore the potential of our targeted delivery system to improve drug accumulation within SSTR2-positive tumor cells, thereby enhancing the therapeutic efficacy against cancer while reducing the likelihood of the adverse effects typically associated with nonspecific drug distribution, and not only inhibiting tumor progression but also inducing tumor regression in cases where receptor expression is maximally aligned with the targeting mechanism of the therapy. This targeted approach promises to direct more of the active drug to the tumor site, minimizing unintended interactions with healthy tissue and organs and offering a more efficient and safer cancer treatment strategy.

It should be stressed that different composition of targeted and non-targeted formulations theoretically could affect their pharmacokinetics. However, the primary influence on the pharmacokinetics and distribution of the delivery system is provided by active targeting specifically to cancer cells. Previously, we revealed that targeting nanocarriers to tumor-specific receptors minimizes the influence of the architecture, composition, size, and molecular mass of nanocarriers on the efficacy of cancer treatment [30]. By comparing three types of carriers containing PTX (linear PEG polymer, polyamidoamine (PAMAM) dendrimers, and PEGylated liposomes similar to the present study), we showed that the pharmacokinetics and distribution of all cancer-targeted delivery systems were similar despite the different chemical compositions and architecture of nanoparticles. Consequently, the relatively small differences in the composition of non-targeted and targeted liposomes should have minimal influence on these parameters compared to the targeting itself.

The crucial question that demands further exploration is how to reconcile the limited effect observed in vitro with the significant differences observed in vivo regarding tumor growth delay. This is a typical situation when non-targeted and cancer-targeted systems are investigated both in vitro and in vivo [44]. The improvement of cellular internalization (observed in vitro) is only one, and not the primary, advantage of active tumor targeting by using ligands for the receptors expressed in cancer cells. The main mechanism of enhancing the anticancer activity of tumor-targeted formulations is a dramatic change in the organ distribution of targeted formulations. In addition to the tumor, non-targeted nanotechnology-based formulations accumulate in the liver, kidney, and spleen [30,44]. Their accumulation in tumors is attributed to the so-called enhanced permeability and retention (EPR) effect (passive targeting), leading to the retention of high-molecular-weight substances in extremely vascularized tumors with limited and impaired lymphatic drainage. In contrast, tumor-targeted formulations predominately accumulate in the tumor, leaving healthy organs intact. Therefore, in most cases, tumor-targeted systems demonstrate much more pronounced augmentation of their antitumor efficacy compared to their non-targeted counterpart in vivo than in vitro.

4. Materials and Methods

4.1. Chemical and Reagents

Paclitaxel (TAX) and Tween-80 were purchased from Sigma (St. Louis, MO, USA). Egg phosphatidylcholine (Egg PC), Cholesterol, and DSPE-PEG2000-azide (1,2-distearoyl-sn-glycero-3-phosphoethanolamine-N-[azido (polyethylene glycol)-2000]) were purchased from Avanti Polar Lipids (Alabaster, AL, USA). To create a somatostatin analog, octreotide (H-D-Phe-Cys*-Phe-D-Trp-Lys-Thr-Cys*-Thr-ol) 5-pentynecarboxylic acid was used to replace the N-terminus of the D-Phenylalanine group with an alkyne group. The resulting 5-pentynecarbonyl-octreotide was synthesized by Peptides International, Inc. (Louisville, KY, USA). Fetal bovine serum (FBS), 3-(4,5-dimethylthiazol-2-yl)-2,5-diphenyl tetrazolium bromide (MTT) reagents and dimethyl sulfoxide (DMSO) were purchased from Sigma (St. Louis, MO, USA). Horse serum was obtained from Thermo Fisher Scientific (Waltham, MA, USA). SSTR2 primers were purchased from Invitrogen (Waltham, MA, USA). Anti-SSTR2 primary antibodies were obtained from Santa Cruz Biotechnology (Dallas, TX, USA). Horseradish peroxidase (HRP)-conjugated secondary antibody was purchased from Cell Signaling Technology (Danvers, MA, USA).

4.2. Cell Lines and Culture Conditions

Human pancreatic cancer cell lines PANC-1, MiaPaca-2, and CFPAC-1 were chosen based on the literature reports about their expression of SSTR2, and all cell lines were purchased from the American Type Culture Collection (ATCC, Manassas, VA, USA). PANC-1 cells were cultured in T25 flasks using Dulbecco's modified Eagle medium (DMEM), supplemented with 10% FBS and 1% penicillin-streptomycin solution (10,000 I.U./mL penicillin, 10,000 (µg/mL) streptomycin). MiaPaca-2 cells were also cultured in T25 flasks using Dulbecco's Modified Eagle Medium (DMEM), supplemented with 10% FBS, 2.5% horse serum, and 1% penicillin–streptomycin solution. CFPAC-1 cells were grown in Iscove's modified Dulbecco's medium (IMDM), supplemented with 10% FBS and 1% penicillin–streptomycin solution. All the cells were cultured at 37 °C with humidity control in a CO_2 incubator.

4.3. Expression of SSTR2 mRNA

To measure the gene expression of SSTR2, all three cell lines were incubated in cell medium in T-75 flasks and harvested until they reached 70% confluence. Then, total RNA was extracted from three cell lines using RNeasy® Mini Kits (Qiagen, Valencia, CA, USA) according to the manufacturer's instructions. The extracted mRNA was reverse-transcribed into cDNA using High-Capacity cDNA Reverse Transcription Kits (Applied Biosystems, Carlsbad, CA, USA) with a Veriti™ 96-well thermal cycler (Applied Biosystems, Carlsbad, CA, USA). Finally, the cDNA levels for SSTR2 in PANC-1, MiaPaca-2, and CFPAC-1 cells were measured using QRT-PCR with the Step One Plus System (Applied Biosystems, Carlsbad, CA, USA).

4.4. Expression of SSTR2 Protein

Immunocytochemistry (ICC) images were used to visualize the SSTR2 protein expression. Cells were seeded in 0.01% poly-L-lysine treated 4-well chamber slides. After 24 h of incubation, the media was removed, and the cells were washed with phosphate-buffered saline (PBS). Then, the cells were fixed with 4% paraformaldehyde and permeabilized with methanol. Next, after washing with phosphate-buffered saline with Tween 20 (PBS-T) buffer, the cells were blocked with 1% bovine serum albumin prepared in PBS and incubated in anti-SSTR2 primary antibodies. Then, the cells were incubated with Alexa Fluor 488-conjugated secondary antibodies. Finally, the cells were stained with 4,6-diamidino-2-phenylindole (DAPI) (Sigma Chemical Co., St. Louis, MO, USA) and mounted with ProLong Gold Antifade Mountant (Thermo Fisher Scientific, Waltham, MA, USA). Images were recorded using a $10\times$ objective of a fluorescence microscope.

The SSTR2 protein expression levels were measured by Western blotting. Briefly, cells were seeded into 6-well plates. Then, all culture media were cleared, and each well was washed twice with PBS. Next, cells were treated with ice-cold radioimmunoprecipitation assay (RIPA) buffer, supplemented with Triton X-100, a protease and phosphatase inhibitor cocktail, for 25 min and sonicated in ice water for 1 min before being transferred into microcentrifuge tubes. Cell lysates were then centrifuged at $10,000\times g$ for 10 min at 4 °C, and the supernatant was transferred into new microcentrifuge tubes. The protein was quantified with a Pierce bicinchoninic acid (BCA) protein assay (Pierce, Rockford, IL, USA). Then, 40 μg of proteins were added into each well of 4 to 12% NuPage Bis-Tris gel (Invitrogen, Waltham, MA, USA) for electrophoresis for 50 min at 200 V. Proteins were transferred from the gel to PVDF membranes with an iBlot 2 dry blotting system (Thermo Fisher Scientific, Waltham, MA, USA). Next, the membranes were submerged in 5% non-fat milk with PBS-T (0.05% Tween 20 in PBS buffer) on a rotating shaker for 1 h at room temperature to block non-specific binding, followed by washing with PBS-T. Then, the membranes were incubated overnight at 4 °C in anti-SSTR2 primary antibodies (1:100 dilution, Santa Cruz Biotechnology, Dallas, TX, USA). GAPDH was used as a loading control. After that, membranes were washed three times with PBS-T and incubated in horseradish peroxidase (HRP)-conjugated secondary antibody for 2 h at room temperature. Finally, to develop protein bands, membranes were soaked in SuperSignal™ West Pico PLUS Chemiluminescent Substrate (Thermo Scientific, Waltham, MA, USA) for 10 min and visualized using the BIO-RAD ChemiDoc Imaging System (Bio-Rad Laboratories, Hercules, CA, USA).

A quantitative analysis of Western blotting and ICC images was performed using ImageJ software Version 1.53m (https://imagej.net/ij/, accessed on 19 May 2024).

4.5. Preparation and Characterization of OCT-DSPE-PEG$_{2000}$

A copper-catalyzed azide–alkyne click chemistry (CuAAC) reaction [45] was performed to conjugate DSPE-PEG-Azide with 5-pentynecarbonyl-octreotide (Figure 2). In total, 50.0 mg of DSPE-PEG-Azide (MW 2816.48, 0.017 mmol) was dissolved in 3 mL tetrahydrofuran (THF) in a 10 mL RB flask. Next, 18.0 mg of 5-pentynecarbonyl-octreotide (MW 1098.47, 0.016 mmol) was dissolved in 2 mL of 1:1 THF and methanol and added to the RB flask while stirring. After that, 100 μL of 20 mM $CuSO_4$ was added to 100 μL of 50 mM tris (benzyltriazolylmethyl) amine (THPTA), and 200 μL of this mixture was added dropwise to the reaction mixture and stirred for one minute. Next, 25 μL of 100 mM aminoguanidine was added to the reaction mixture and stirred for one minute for homogeneous mixing. Finally, 25 μL of 100 mM sodium ascorbate was added to this reaction mixture, and the reaction mixture was stirred at 500 rpm for 24 h at room temperature. After that, the reaction mixture was evaporated to dryness at room temperature to get the crude mixture of the product.

The crude reaction mixture obtained was dissolved in 1 mL of water and transferred to a 3.5 kDa molecular weight cut-off (MWCO) dialysis tube (Repligen, Waltham, MA, USA), which was run against 1.5 L of 0.04 (M) EDTA buffer (pH 8.2). The dialysis was run for 36 h, and the buffer was changed after 3, 6, 12, and 24 h. Then, the purified product was transferred into a 5 mL beaker and left at room temperature for 1 h to evaporate the remaining solvent. The final product was lyophilized to get the 'DSPE-PEG$_{2000}$-OCT Ligand' in solid powder form.

The MALDI-TOF-MS analysis of the DSPE-PEG$_{2000}$-OCT purified ligand was redissolved in MeOH to reach a 10 mg/mL concentration. The measurement was carried out using a matrix of 10 mg/mL sinapinic acid (SA) in 2.5 mg/mL of cyano-4-hydroxycinnamic acid (CHCA), 50% acetonitrile, and 0.1% trifluoroacetic acid (TFA), and 5 pM of insulin was used as the external calibration standard.

4.6. Preparation of PEGylated Liposomal Paclitaxel, OCT-PEGylated Liposomal Paclitaxel, and Fluorescent Liposomal Paclitaxel

Paclitaxel-loaded PEGylated liposomes were prepared using the previously developed procedure [31]. Briefly, in untargeted formulation, the liposomes were prepared with lipids in a molar ratio of 51:44:5 mole% of egg yolk phosphatidylcholine (EPC)/Cholesterol/1,2,-distearoyl-sn-glycero-3-phosphoethanolamine-N-aminopolyethelenglycol- 2000 ammonium salt (DSPE-PEG$_{2000}$) and 0.878 mM paclitaxel. For the targeted formulation OCT-PEGylated Liposome Paclitaxel, the lipids were mixed in a molar ratio of 47:41:2:10 of EPC/Cholesterol/DSPE-PEG$_{2000}$/DSPE-PEG-OCT with 0.878 mM of paclitaxel and dissolved in the same solvent. The fluorescent liposomal paclitaxel was formulated with 100 μg of Oregon Green R 488 Paclitaxel (green fluorescence) and 1 mg of Egg Liss Rhod PE (red MAL fluorescence), along with other lipids using the aforementioned method.

The lipids and paclitaxel mixture were dissolved in 4.0 mL of 3:1 chloroform/methanol. Once fully dissolved, the clear liquid was transferred into a 250 mL round-bottom flask and evaporated under reduced pressure at 37 °C on a rotary evaporator (Rotavapor® R-210/R-215, BUCHI Corp., New Castle, DE, USA). The resulting thin layer was rehydrated with 4 mL of a 0.9% saline solution. To create unilamellar liposomes, the mixture was sonicated continuously for 15 min. The resulting liposome product was then transferred into 8 kD MWCO dialysis bags and dialyzed in 0.9% saline while stirring (100 rpm) for 12 h at 4 °C to remove any unentrapped, free paclitaxel. All purified products were stored at 4 °C for further studies.

4.7. Zeta Potential, Particle Size, and Polydispersity Index (PDI)

The particle size, zeta potential, and polydispersity index (PDI) of liposome samples were measured by Malvern Zetasizer Nano (Malvern Instruments, Malvern, UK) at 25 °C. An aliquot of 25 μL of each formulation sample was diluted with 1.5 mL of 0.9% saline in Malvern disposable cuvettes. All measurements were completed in triplicate, and average values were calculated.

4.8. Atomic Force Microscopy (AFM)

The shape of all types of liposomal formulations were studied by atomic force microscopy imaging using the previously described procedure [46]. Briefly, 50 μL of a liposome suspension in water was deposited on pre-cut (∼25 × 25 mm^2) and pre-cleaned Plain

Premium microscope slides (Fisher Scientific Co., Pittsburgh, PA, USA), kept for 10 min at 100% humidity to achieve particles precipitation. Water was removed by dry nitrogen flow and dried samples were subjected to imaging with an atomic force microscope (Nano-R AFM Pacific Nanotechnology Instrument, PNI, Santa Clara, CA, USA) in close contact (tapping) mode using tapping-mode-etched OMCLAC160TS silicon probes (Olympus Optical Co., Tokyo, Japan). The captioning was performed in height mode.

4.9. High-Performance Liquid Chromatography (HPLC) Analysis of Paclitaxel

A Waters HPLC system was utilized to determine the concentration of paclitaxel in the samples. The system was equipped with dual pumps (Waters 1525), a Waters 2487 Dual Absorbance UV detector, and a Waters 717 autosampler. Each sample was injected with 20 µL and run at a 0.5 mL/min flow rate. The samples were examined using a reverse-phase LiChrospher®100 RP-18 column (250 × 4 mm, 5 µm, Merck, Darmstadt, Germany) at room temperature. The detecting wavelength employed was 227 nm. The mobile phase comprised acetonitrile and water (60:40, v/v).

4.10. Entrapment Efficiency and HPLC Measurement

Dialysis was employed as a method to separate free paclitaxel from entrapped drugs. In total, 1 mL crude liposomal suspension, either Lip-PTX or OCT-Lip-PTX, was introduced into dialysis bags with a MWCO of 15 kDa. These bags were then dialyzed in 0.9% saline solution at a temperature of 4 °C for 12 h, with continuous stirring at 100 rpm to ensure consistent diffusion. An aliquot of 25 µL of the crude, or recovered liposome product after dialysis, was dissolved and diluted in 975 µL of the dissolving solvent composed of water, isopropanol, ether, and ethanol (5:2:1:2, $v/v/v/v$). The concentration of the paclitaxel added initially and the concentration of the paclitaxel entrapped were measured by HPLC, and the actual weight of the total paclitaxel before (W_{total}) and after ($W_{entrapped}$) purification can be calculated by multiplying proper dilution factors [47]. Drug entrapment efficiency was calculated by the equation below:

$$EE\% = \frac{W_{entrapped}}{W_{total}} \times 100\%$$

4.11. Paclitaxel Release Profile

An aliquot of 0.8 mL of liposome paclitaxel samples Lip-PTX or OCT-Lip-PTX was added to 15 kDa MWCO dialysis bags individually. The dialysis bags were then submerged in beakers containing 80 mL of 1% Tween 20/PBS solution, with magnetic stirring set to 100 rpm. The study was carried out at three different temperatures, 4 °C, 25 °C, and 37 °C, over 7 days. In total, 1 mL of samples were collected at specific time points of 0 h, 0.5 h, 1 h, 2 h, 3 h, 6 h, 8 h, 12 h, 24 h, 48 h, 72 h, 96 h, 120 h, 144 h, and 168 h. After sample collection, an equal volume of 1% Tween 20/PBS solution was replaced at each time point. As explained in the previous section, the samples were analyzed using HPLC methods.

4.12. Cellular Internalization Study

In order to visualize liposomes, they were labeled with near-infrared cyanine dye Cy5.5 (Amersham Biosciences, Piscataway, NJ, USA) with red fluorescence, as previously described [48]. Paclitaxel, Oregon Green™ 488 Conjugate (Fluotax) with green fluorescence (Thermo Fisher Scientific, Waltham, MA, USA) was used instead of regular PTX in the part of synthesized formulations to visualize drug release from the system inside cancer cells. The fluorescence of Fluotax tightly packed in liposomes was quenched. Therefore, the appearance of the green fluorescence signal inside the cells testified the release of the drug from liposomes inside the cells.

All three cell lines were individually plated in 4 well-chambered Coverglass (Thermo Scientific Nunc, Naperville, IL, USA) and allowed to incubate for 24 h at 37 °C. The culture medium for each formulation was the same (DMEM and IMDM for PANC-1 and CFPAC-1 cells, respectively). Following this, they were treated with fluorescent liposomal paclitaxel, which had been previously prepared with the same volume and concentration of PTX for non-targeted and targeted systems. After 24 h of treatment, the cells were rinsed with Dulbecco's phosphate-buffered saline (DPBS) containing 0.05% Tween 20 and fixed with a 4% formaldehyde solution. Next, 0.5 mL of DPBS was added to each chamber. The cell nuclei were stained by a blue-fluorescent DNA stain 4′,6-diamidino-2-phenylindole (DAPI). The slides were observed using a Confocal Microscope (Leica TCS SP8, CarlZeiss, Jena, Germany), and pictures were captured and analyzed using LAS AF Lite Leica Version 2.6.3 software.

4.13. Cytotoxicity Study

MTT (3-(4,5-dimethylthiazol-2-yl)-2,5-diphenyl tetrazolium bromide) assays were used to measure the cell viability of PANC-1, MiaPaca-2, and CFPAC-1 cells after treatment. Cells were seeded in 96-well plates with 0.1 mL culture media in each well and incubated at 37 °C for 24 h. Then, culture media were removed, and the cells were, respectively, treated with paclitaxel (free paclitaxel), PEGylated liposomal paclitaxel, OCT-DSPE-PEGylated liposomal paclitaxel, or empty liposomes in 0.2 mL of various concentrations. Each treatment mentioned above was prepared in 100 µM, and 12 working solutions in decreasing concentrations were prepared by serial dilutions (1:10). Cells incubated in fresh culture media were used as a control. After 72 h of incubation, treatment solutions were removed, and cells were incubated in 0.12 mL of a 1 mg/mL MTT solution at 37 °C for 3 h. The Formazan crystals were dissolved with a precise reagent (10.5 g SDS in 25 mL dimethylformamide (DMF) + 25 mL deionized (DI) water). The absorbance was measured using Tecan Infinite M200 PRO (Morrisville, NC, USA) at a wavelength of 570 nm. The IC_{50}, a commonly used parameter in assessing cytotoxicity, denotes the concentration of a given treatment that leads to a 50% inhibition of cell proliferation. Based on the results, CFPAC-1 and MiaPaca-2 cells were selected for the following in vivo studies and analysis.

4.14. In Vivo Animal Studies

Animal studies were performed according to the protocols and animal use procedures approved by the Institutional (Rutgers, the State University of New Jersey) Animal Care and Use Committee (IACUC). Male and Female NCr (nu/nu strain), 6–8-week-old mice, weighing about 20 g, were purchased from Taconic Farms, Inc. (Germantown, NY, USA) and were housed in cages under controlled laboratory conditions (temperature 22–24 °C, relative humidity 50 ± 10% and 12 h/12 h light/dark cycle) and allowed free access to a sterilized rodent pellet diet and acidified drinking water. The animals were acclimatized for at least 72 h before any experiments, and "pre-numbered" ear tags were used to identify each mouse.

4.14.1. Maximum Tolerated Doses (MTDs) of Paclitaxel

For the Paclitaxel MTD study, healthy mice received one injection of the clinically used Taxol ® formulation at 5 different concentrations. For each concentration tested, 5 mice were used. The animals were observed for signs of acute toxicity, such as weight loss and abnormal behavior.

For MTD studies, the concentrations (doses) of PTX in CrEL-D tested were 1, 2, 2.5, 3.0, and 3.5 mg PTX/kg in 0.2 mL injection per animal n = 5 per dose). Healthy mice were weighed and then began receiving IV tail injections (~0.2 mL per mouse) of different concentrations of CrEL-D using a q5dx4 schedule. Mice were monitored and weighed on the day of the treatment, the day after, and every other day during the trial. Toxicity was assessed as a percentage of weight loss. The MTD was defined as the highest dose with <15% body weight loss and not causing significant lethality or any prominent observable changes during the trial. The difference in mean body weight was calculated concerning the beginning of treatment (day 1) as follows:

$$(\text{mean body weight on day x} - \text{mean body weight on day 1})/(\text{mean body weight on day 1}) \times 100$$

The dose of paclitaxel was determined to be the maximum tolerated dose and was used in all studied drug formulations.

4.14.2. Animal Model of Pancreatic Cancer and Antitumor Activity

An animal model of human pancreatic cancer xenografts was created by injecting 5×10^6 CFPAC-1 and Mia PaCa-2 cells subcutaneously into the athymic nu/nu mice flanks. When the tumors reached a size of about 0.3–0.4 cm^3, mice were randomly divided into groups (6 animals per group) and treated intraperitoneally with five different formulations: saline (control), empty liposomes, Raxol®, PEGylated liposomal paclitaxel, OCT-DSPE-PEGylated liposomal paclitaxel. The animals were treated twice per week for 4 weeks. The tumor growth was measured by a caliper on the day of the treatment, the day after, and every other day during the trial. Tumor volume was calculated as $d^2 \times D/2$, where d and D are the smallest and widest diameter of the tumor in mm, respectively. According to the approved institutional animal use protocol, the mice were sacrificed when the tumor reached around 1.2–1.3 cm^3. All other measurements were performed 24 h after the treatment. Changes in tumor size were used as an overall marker for antitumor activity.

5. Conclusions

A chemically modified somatostatin analog-conjugated to paclitaxel-loaded liposomes was formulated for the chemotherapy of pancreatic cancer. Based on our studies, this novel formulation displayed high selectivity to SSTR2-expressing tumor cells, significantly inhibited cell proliferation, and increased drug cytotoxicity, demonstrating superior tumor suppression effects compared to the non-targeting formulations, with the possibility of maintaining limited toxicity on healthy organs. Therefore, the novel 5-pentynecarbonyl-octreotide-conjugated liposomal paclitaxel presents a promising delivery system specific to pancreatic cancer cells and tissues with high expression of SSTR2.

While paclitaxel serves as a model drug in this study, various cytotoxic agents such as doxorubicin, methotrexate, camptothecin, carboplatin, cisplatin can also be incorporated into the proposed delivery system for co-encapsulation, given the amphiphilic structure of the liposomal membrane (for lipophilic drugs) and their liquid inner space (for hydrophilic drugs). The use of hydrophilic drugs can potentially improve the drug-loading capacity and stability of liposomes. However, even for the hydrophobic PTX, we registered close to 90% loading capacity and decent stability of the formulations under short-term storage in the refrigerator, overall. In general, the stability of lyophilized lipid-based formulations is excellent, and changes in nanoparticle characteristics, including the drug-loading capacity, are minimal after one freezing/thaw cycle. In contrast, at body temperature, after entering cancer cells, liposomes should release the encapsulated drug in order to induce the death of cancer cells. Consequently, the further enhancement of the stability of carriers under body temperature would decrease their ability to kill cancer cells.

An additional optimization can benefit liposomal formulations concerning their composition, PEGylated lipids and OCT fractions, and paclitaxel loading, among other factors. The main objective of the present research was to demonstrate that targeting a nanotechnology-based drug formulation to pancreatic cancer cells using a somatostatin analog could increase its effectiveness against tumors while drawing attention to this approach. In our laboratory, we are exploring further optimization of the tumor-targeted delivery of anticancer drugs to pancreatic cancer.

The present experimental data show that the cancer-targeted liposomal formulations of anticancer drugs are more toxic when compared with non-targeted formulations and free-non-bound drugs in the same concentration. Our previous detailed investigations of various delivery systems clearly showed that tumor targeting almost eliminates adverse toxic side effects on healthy organs and tissues in vivo [9,30,44]. Based on these results, we do not expect that the proposed approach in the current investigation to pancreatic tumor-targeting can increase the adverse side effects compared to similar liposomal formulations targeted to tumors by the LHRH peptide. However, further toxicologic evaluation of the proposed system is required and is being conducted in our lab.

Due to the complex and heterogeneous nature of PDAC's tumor microenvironment, employing multifunctional nanoparticles that combine tumor-penetrating peptides such as TAT, antennapedia, and iRGD could possibly improve the precision and efficiency of targeting, leading to enhanced intracellular delivery of therapeutic agents. Combining the targeting nano-based carriers with gene-silencing approaches, such as RNAi tailored to the specific genetic makeup of patients, holds considerable promise for advancing personalized medicine [6]. This approach could markedly improve therapeutic outcomes and benefit patients with various genetic profiles. Nonetheless, translating these promising preclinical findings into clinical practice requires further investigation, including the optimization of conjugate design, comprehensive in vivo efficacy and toxicity evaluations, and, ultimately, clinical trials to ascertain their safety and effectiveness in humans.

Author Contributions: Conceptualization, T.M.; methodology, X.G., J.M., O.T., A.K., O.G. and N.P.; formal analysis, T.M. and X.G.; investigation, T.M. and X.G.; writing—original draft preparation, X.G.; writing—review and editing, T.M.; visualization, T.M.; supervision, T.M.; project administration, T.M.; funding acquisition, T.M. All authors have read and agreed to the published version of the manuscript.

Funding: This research was funded in part by the NIH/NCI, grant number R01 CA209818.

Institutional Review Board Statement: The animal study protocol was approved by the Institutional Review Board of Rutgers, the State University of New Jersey (protocol code PROTO999900225, dates approval: 24 February 2024 and 25 March 2021).

Informed Consent Statement: Not applicable.

Data Availability Statement: The data supporting the conclusions of this article will be made available by the authors on request.

Acknowledgments: We are grateful to V. A. Soldatenkov for help in obtaining atomic force microscopy images. Some figures were created using BioRender software (https://app.biorender.com/, accessed on 18 May 2024).

Conflicts of Interest: The authors declare no conflicts of interest.

References

1. *Cancer Facts & Figures 2024*; American Cancer Society: Atlanta, GA, USA, 2024.
2. Moletta, L.; Serafini, S.; Valmasoni, M.; Pierobon, E.S.; Ponzoni, A.; Sperti, C. Surgery for Recurrent Pancreatic Cancer: Is It Effective? *Cancers* **2019**, *11*, 991. [CrossRef] [PubMed]
3. Ansari, M.J.; Bokov, D.; Markov, A.; Jalil, A.T.; Shalaby, M.N.; Suksatan, W.; Chupradit, S.; Al-Ghamdi, H.S.; Shomali, N.; Zamani, A.; et al. Cancer combination therapies by angiogenesis inhibitors; a comprehensive review. *Cell Commun. Signal.* **2022**, *20*, 49. [CrossRef] [PubMed]
4. Pérez-Díez, I.; Andreu, Z.; Hidalgo, M.R.; Perpiñá-Clérigues, C.; Fantín, L.; Fernandez-Serra, A.; de la Iglesia-Vaya, M.; Lopez-Guerrero, J.A.; García-García, F. A Comprehensive Transcriptional Signature in Pancreatic Ductal Adenocarcinoma Reveals New Insights into the Immune and Desmoplastic Microenvironments. *Cancers* **2023**, *15*, 2887. [CrossRef] [PubMed]
5. Yang, J.; Liu, Y.; Liu, S. The role of epithelial-mesenchymal transition and autophagy in pancreatic ductal adenocarcinoma invasion. *Cell Death Dis.* **2023**, *14*, 506. [CrossRef] [PubMed]
6. Garbuzenko, O.B.; Sapiezynski, J.; Girda, E.; Rodriguez-Rodriguez, L.; Minko, T. Personalized Versus Precision Nanomedicine for Treatment of Ovarian Cancer. *Small* **2024**, *Early View*. [CrossRef]
7. Savla, R.; Garbuzenko, O.B.; Chen, S.; Rodriguez-Rodriguez, L.; Minko, T. Tumor-Targeted Responsive Nanoparticle-Based Systems for Magnetic Resonance Imaging and Therapy. *Pharm. Res.* **2014**, *31*, 3487–3502. [CrossRef]
8. Taratula, O.; Kuzmov, A.; Shah, M.; Garbuzenko, O.B.; Minko, T. Nanostructured lipid carriers as multifunctional nanomedicine platform for pulmonary co-delivery of anticancer drugs and siRNA. *J. Control. Release* **2013**, *171*, 349–357. [CrossRef]
9. Zhang, M.; Garbuzenko, O.B.; Reuhl, K.R.; Rodriguez-Rodriguez, L.; Minko, T. Two-in-one: Combined targeted chemo and gene therapy for tumor suppression and prevention of metastases. *Nanomedicine* **2012**, *7*, 185–197. [CrossRef] [PubMed]
10. Liu, H.; Shi, Y.; Qian, F. Opportunities and delusions regarding drug delivery targeting pancreatic cancer-associated fibroblasts. *Adv. Drug Deliv. Rev.* **2021**, *172*, 37–51. [CrossRef]

11. De Luca, R.; Blasi, L.; Alù, M.; Gristina, V.; Cicero, G. Clinical efficacy of nab-paclitaxel in patients with metastatic pancreatic cancer. *Drug Des. Dev. Ther.* **2018**, *12*, 1769–1775. [CrossRef]

12. Frampton, J.E. Liposomal Irinotecan: A Review in Metastatic Pancreatic Adenocarcinoma. *Drugs* **2020**, *80*, 1007–1018. [CrossRef] [PubMed]

13. Nsairat, H.; Khater, D.; Sayed, U.; Odeh, F.; Al Bawab, A.; Alshaer, W. Liposomes: Structure, composition, types, and clinical applications. *Heliyon* **2022**, *8*, e09394. [CrossRef] [PubMed]

14. Beltrán-Gracia, E.; López-Camacho, A.; Higuera-Ciapara, I.; Velázquez-Fernández, J.B.; Vallejo-Cardona, A.A. Nanomedicine review: Clinical developments in liposomal applications. *Cancer Nanotechnol.* **2019**, *10*, 11. [CrossRef]

15. Filipczak, N.; Pan, J.; Yalamarty, S.S.K.; Torchilin, V.P. Recent advancements in liposome technology. *Adv. Drug Deliv. Rev.* **2020**, *156*, 4–22. [CrossRef] [PubMed]

16. Torchilin, V.P.; Omelyanenko, V.G.; Papisov, M.I.; Bogdanov, A.A.; Trubetskoy, V.S.; Herron, J.N.; Gentry, C.A. Poly(ethylene glycol) on the liposome surface: On the mechanism of polymer-coated liposome longevity. Biochim. et Biophys. *Acta (BBA)-Biomembr.* **1994**, *1195*, 11–20. [CrossRef]

17. Khare, V.; Alam, N.; Saneja, A.; Dubey, R.D.; Gupta, P.N. Targeted Drug Delivery Systems for Pancreatic Cancer. *J. Biomed. Nanotechnol.* **2014**, *10*, 3462–3482. [CrossRef]

18. Murata, M.; Narahara, S.; Kawano, T.; Hamano, N.; Piao, J.S.; Kang, J.-H.; Ohuchida, K.; Murakami, T.; Hashizume, M. Design and Function of Engineered Protein Nanocages as a Drug Delivery System for Targeting Pancreatic Cancer Cells via Neuropilin-1. *Mol. Pharm.* **2015**, *12*, 1422–1430. [CrossRef]

19. Nedaeinia, R.; Avan, A.; Manian, M.; Salehi, R.; Ghayour-Mobarhan, M. EGFR as a Potential Target for the Treatment of Pancreatic Cancer: Dilemma and Controversies. *Curr. Drug Targets* **2014**, *15*, 1293–1301. [CrossRef] [PubMed]

20. Sato, N.; Cheng, X.-B.; Kohi, S.; Koga, A.; Hirata, K. Targeting hyaluronan for the treatment of pancreatic ductal adenocarcinoma. *Acta Pharm. Sin. B* **2016**, *6*, 101–105. [CrossRef]

21. Froidevaux, S.; Hintermann, E.; Török, M.; Mäcke, H.R.; Beglinger, C.; Eberle, A.N. Differential regulation of somatostatin receptor type 2 (sst 2) expression in AR4-2J tumor cells implanted into mice during octreotide treatment. *Cancer Res.* **1999**, *59*, 3652–3657.

22. Jiang, J.; Deng, L.; He, L.; Liu, H.; Wang, C. Expression, purification, refolding, and characterization of octreotide-interleukin-2: A chimeric tumor-targeting protein. *Int. J. Mol. Med.* **2011**, *28*, 549–556. [CrossRef] [PubMed]

23. Shahbaz, M.; Ruliang, F.; Xu, Z.; Benjia, L.; Cong, W.; Zhaobin, H.; Jun, N. mRNA expression of somatostatin receptor subtypes SSTR-2, SSTR-3, and SSTR-5 and its significance in pancreatic cancer. *World J. Surg. Oncol.* **2015**, *13*, 46. [CrossRef] [PubMed]

24. Srkalovic, G.; Cai, R.-Z.; Schally, A.V. Evaluation of Receptors for Somatostatin in Various Tumors Using Different Analogs*. *J. Clin. Endocrinol. Metab.* **1990**, *70*, 661–669. [CrossRef] [PubMed]

25. Sun, L.-C.; Mackey, L.V.; Luo, J.; Fuselier, J.A.; Coy, D.H. Targeted Chemotherapy Using a Cytotoxic Somatostatin Conjugate to Inhibit Tumor Growth and Metastasis in Nude Mice. *Clin. Med. Insights Oncol.* **2008**, *2*, 491–499. [CrossRef] [PubMed]

26. Szepeshazi, K.; Schally, A.V.; Halmos, G.; Sun, B.; Hebert, F.; Csernus, B.; Nagy, A. Targeting of cytotoxic somatostatin analog AN-238 to somatostatin receptor subtypes 5 and/or 3 in experimental pancreatic cancers. *Clin. Cancer Res.* **2001**, *7*, 2854–2861.

27. Virgolini, I.; Yang, Q.; Li, S.; Angelberger, P.; Neuhold, N.; Niederle, B.; Scheithauer, W.; Valent, P. Cross-competition between vasoactive intestinal peptide and somatostatin for binding to tumor cell membrane receptors. *Cancer Res.* **1994**, *54*, 690–700.

28. Awosika, A.O.; Farrar, M.C.; Jacobs, T.F. Paclitaxel. In *StatPearls*; StatPearls Publishing: Treasure Island, FL, USA, 2024.

29. Gallego-Jara, J.; Lozano-Terol, G.; Sola-Martínez, R.A.; Cánovas-Díaz, M.; de Diego Puente, T. A Compressive Review about Taxol®: History and Future Challenges. *Molecules* **2020**, *25*, 5986. [CrossRef] [PubMed]

30. Saad, M.; Garbuzenko, O.B.; Ber, E.; Chandna, P.; Khandare, J.J.; Pozharov, V.P.; Minko, T. Receptor targeted polymers, dendrimers, liposomes: Which nanocarrier is the most efficient for tumor-specific treatment and imaging? *J. Control. Release* **2008**, *130*, 107–114. [CrossRef] [PubMed]

31. Pakunlu, R.I.; Wang, Y.; Saad, M.; Khandare, J.J.; Starovoytov, V.; Minko, T. In vitro and in vivo intracellular liposomal delivery of antisense oligonucleotides and anticancer drug. *J. Control. Release* **2006**, *114*, 153–162. [CrossRef]

32. Heneweer, C.; Medina, T.P.; Tower, R.; Kalthoff, H.; Kolesnick, R.; Larson, S.; Medina, O.P. Acid-Sphingomyelinase Triggered Fluorescently Labeled Sphingomyelin Containing Liposomes in Tumor Diagnosis after Radiation-Induced Stress. *Int. J. Mol. Sci.* **2021**, *22*, 3864. [CrossRef]

33. Buscail, L.; Delesque, N.; Estève, J.P.; Saint-Laurent, N.; Prats, H.; Clerc, P.; Robberecht, P.; IBell, G.; Liebow, C.; Schally, A.V. Stimulation of tyrosine phosphatase and inhibition of cell proliferation by somatostatin analogues: Mediation by human somatostatin receptor subtypes SSTR1 and SSTR2. *Proc. Natl. Acad. Sci. USA* **1994**, *91*, 2315–2319. [CrossRef]

34. Delesque, N.; Buscail, L.; Estève, J.P.; Saint-Laurent, N.; Müller, C.; Weckbecker, G.; Bruns, C.; Vaysse, N.; Susini, C. sst2 somatostatin receptor expression reverses tumorigenicity of human pancreatic cancer cells. *Cancer Res.* **1997**, *57*, 956–962.

35. Fisher, W.E.; Doran, T.A.; Muscarella, P.; Boros, L.G.; Ellison, E.C.; Schirmer, W.J. Expression of Somatostatin Receptor Subtype 1–5 Genes in Human Pancreatic Cancer. *JNCI J. Natl. Cancer Inst.* **1998**, *90*, 322–324. [CrossRef]

36. Gradiz, R.; Silva, H.; Carvalho, L.; Botelho, M.F.; Mota-Pinto, A. MIA PaCa-2 and PANC-1—Pancreas ductal adenocarcinoma cell lines with neuroendocrine differentiation and somatostatin receptors. *Sci. Rep.* **2016**, *6*, 21648. [CrossRef]

37. Günther, T.; Tulipano, G.; Dournaud, P.; Bousquet, C.; Csaba, Z.; Kreienkamp, H.-J.; Lupp, A.; Korbonits, M.; Castaño, J.P.; Wester, H.-J.; et al. International Union of Basic and Clinical Pharmacology. CV. Somatostatin Receptors: Structure, Function, Ligands, and New Nomenclature. *Pharmacol. Rev.* **2018**, *70*, 763–835. [CrossRef] [PubMed]

38. Patel, Y.C. Somatostatin and Its Receptor Family. *Front. Neuroendocr.* **1999**, *20*, 157–198. [CrossRef] [PubMed]
39. Barbieri, F.; Bajetto, A.; Pattarozzi, A.; Gatti, M.; Würth, R.; Thellung, S.; Corsaro, A.; Villa, V.; Nizzari, M.; Florio, T. Peptide Receptor Targeting in Cancer: The Somatostatin Paradigm. *Int. J. Pept.* **2013**, *2013*, 926295. [CrossRef]
40. Chatzisideri, T.; Leonidis, G.; Sarli, V. Cancer-targeted delivery systems based on peptides. *Futur. Med. Chem.* **2018**, *10*, 2201–2226. [CrossRef] [PubMed]
41. Paragliola, R.M.; Corsello, S.M.; Salvatori, R. Somatostatin receptor ligands in acromegaly: Clinical response and factors predicting resistance. *Pituitary* **2016**, *20*, 109–115. [CrossRef]
42. Shen, H.; Hu, D.; Du, J.; Wang, X.; Liu, Y.; Wang, Y.; Wei, J.-M.; Ma, D.; Wang, P.; Li, L. Paclitaxel–octreotide conjugates in tumor growth inhibition of A549 human non-small cell lung cancer xenografted into nude mice. *Eur. J. Pharmacol.* **2008**, *601*, 23–29. [CrossRef]
43. Huo, M.; Zhu, Q.; Wu, Q.; Yin, T.; Wang, L.; Yin, L.; Zhou, J. Somatostatin Receptor–Mediated Specific Delivery of Paclitaxel Prodrugs for Efficient Cancer Therapy. *J. Pharm. Sci.* **2015**, *104*, 2018–2028. [CrossRef] [PubMed]
44. Dharap, S.S.; Wang, Y.; Chandna, P.; Khandare, J.J.; Qiu, B.; Gunaseelan, S.; Sinko, P.J.; Stein, S.; Farmanfarmaian, A.; Minko, T.; et al. Tumor-specific targeting of an anticancer drug delivery system by LHRH peptide. *Proc. Natl. Acad. Sci. USA* **2005**, *102*, 12962–12967. [CrossRef] [PubMed]
45. Presolski, S.I.; Hong, V.P.; Finn, M. Copper-Catalyzed Azide–Alkyne Click Chemistry for Bioconjugation. *Curr. Protoc. Chem. Biol.* **2011**, *3*, 153–162. [CrossRef] [PubMed]
46. Chandna, P.; Saad, M.; Wang, Y.; Ber, E.; Khandare, J.; Vetcher, A.A.; Soldatenkov, V.A.; Minko, T. Targeted Proapoptotic Anticancer Drug Delivery System. *Mol. Pharm.* **2007**, *4*, 668–678. [CrossRef] [PubMed]
47. Yang, T.; Cui, F.-D.; Choi, M.-K.; Lin, H.; Chung, S.-J.; Shim, C.-K.; Kim, D.-D. Liposome Formulation of Paclitaxel with Enhanced Solubility and Stability. *Drug Deliv.* **2007**, *14*, 301–308. [CrossRef]
48. Saad, M.; Garbuzenko, O.B.; Minko, T.; Gener, P.; Rafael, D.F.d.S.; Fernández, Y.; Ortega, J.S.; Arango, D.; Abasolo, I.; Videira, M.; et al. Co-delivery of siRNA and an anticancer drug for treatment of multidrug-resistant cancer. *Nanomedicine* **2008**, *3*, 761–776. [CrossRef]

Disclaimer/Publisher's Note: The statements, opinions and data contained in all publications are solely those of the individual author(s) and contributor(s) and not of MDPI and/or the editor(s). MDPI and/or the editor(s) disclaim responsibility for any injury to people or property resulting from any ideas, methods, instructions or products referred to in the content.

International Journal of
Molecular Sciences

Article

PANC-1 Cell Line as an Experimental Model for Characterizing PIVKA-II Production, Distribution, and Molecular Mechanisms Leading to Protein Release in PDAC

Antonella Farina, Sara Tartaglione, Adele Preziosi, Patrizia Mancini, Antonio Angeloni and Emanuela Anastasi *

Department of Experimental Medicine, Sapienza University of Rome, Policlinico Umberto I, 00181 Rome, Italy; antoeffe22@gmail.com (A.F.); sara.tartaglione@uniroma1.it (S.T.); adelepreziosi@gmail.com (A.P.); patrizia.mancini@uniroma1.it (P.M.); antonio.angeloni@uniroma1.it (A.A.)
* Correspondence: emanuela.anastasi@uniroma1.it

Abstract: Pancreatic ductal adenocarcinoma (PDAC) represents a highly aggressive malignancy with a lack of reliable diagnostic biomarkers. Protein induced by vitamin K absence (PIVKA-II) is a protein increased in various cancers (particularly in hepatocellular carcinoma), and it has recently exhibited superior diagnostic performance in PDAC detection compared to other biomarkers. The aim of our research was to identify an in vitro model to study PIVKA-II production, distribution, and release in PDAC. We examined the presence of PIVKA-II protein in a panel of stabilized pancreatic cancer cell lines by Western blot analysis and indirect immunofluorescence (IFA). After quantitative evaluation of PIVKA-II in PaCa 44, H-Paf II, Capan-1, and PANC-1, we adopted the latter as a reference model. Subsequently, we analyzed the effect of glucose addiction on PIVKA-II production in a PANC-1 cell line in vitro; PIVKA-II production seems to be directly related to an increase in glucose concentration in the culture medium. Finally, we evaluated if PIVKA-II released in the presence of increasing doses of glucose is concomitant with the expression of two well-acknowledged epithelial–mesenchymal transition (EMT) markers (Vimentin and Snail). According to our experimental model, we can speculate that PIVKA-II release by PANC-1 cells is glucose-dependent and occurs jointly with EMT activation.

Keywords: PDAC; PIVKA-II; glucose; epithelial to mesenchymal transition

Citation: Farina, A.; Tartaglione, S.; Preziosi, A.; Mancini, P.; Angeloni, A.; Anastasi, E. PANC-1 Cell Line as an Experimental Model for Characterizing PIVKA-II Production, Distribution, and Molecular Mechanisms Leading to Protein Release in PDAC. *Int. J. Mol. Sci.* **2024**, *25*, 3498. https://doi.org/10.3390/ijms25063498

Academic Editors: Claudio Luchini and Donatella Delle Cave

Received: 21 February 2024
Revised: 10 March 2024
Accepted: 13 March 2024
Published: 20 March 2024

Copyright: © 2024 by the authors. Licensee MDPI, Basel, Switzerland. This article is an open access article distributed under the terms and conditions of the Creative Commons Attribution (CC BY) license (https://creativecommons.org/licenses/by/4.0/).

1. Introduction

Ninety percent of pancreatic cancer cases are attributed to pancreatic ductal adenocarcinoma (PDAC), characterized by its aggressive and lethal nature [1]. With a 5-year survival rate of approximately 8%, PDAC stands as the sole cancer exhibiting an escalating mortality rate for both men and women. Nowadays, surgery remains the most radical treatment option for PDAC, but only 15–20% of cases are deemed surgically resectable at the time of diagnosis [2]. PDAC is in fact commonly referred to as the "silent killer" owing to its characteristic late diagnosis, with only 7% of cases identified at an early stage due to the absence of specific early symptoms; unfortunately, the tumor becomes apparent only after infiltrating surrounding tissues or metastasizing [3]. Given the tight correlation between survival rates of PDAC patients and disease stage, an early detection of the neoplasm becomes a pressing necessity to significantly enhance treatment effectiveness [4]. To date, there are no screening or surveillance programs for the early diagnosis of PDAC. The primary tool for determining the localization, extent, and clinical staging of the mass relies on imaging techniques. In the last few decades, there has been a significant increase in emphasis on circulating biomarkers as early warning systems for assessing disease risk. They have become a potent and cost-effective tool in cancer management [5]. However, in the case of PDAC, compared to other solid neoplasms, the clinical utilization and subsequent benefits of biomarkers remain considerably limited. Current guidelines regarding PDAC suggest Carbohydrate Antigen 19.9 (CA19.9) as the gold-standard circulating biomarker. However, CA19.9 has several limitations, since altered levels have been observed in patients with non-cancerous conditions such as biliary obstructions, chronic pancreatitis, and non-malignant jaundice. Additionally, not all PDAC patients exhibit elevated CA19.9 levels, particularly in Lewis-negative individuals, meaning its diagnostic accuracy is significantly reduced [6]. The suboptimal diagnostic performance of CA19.9 and the absence of other molecules indicating the presence of PDAC underscore the compelling need to explore new biomarkers with enhanced sensitivity and specificity. In recent years, there has been growing interest in the protein induced by vitamin K absence (PIVKA-II), an immature form of prothrombin also known as DCP (Des-gamma-Carboxy Prothrombin). Prothrombin, a vitamin K-dependent coagulation factor, is naturally synthesized by the liver under physiological conditions. In instances of vitamin K deficiency or when its action is hindered, such as with the administration of antivitamin-K drugs, PIVKA-II is released into the bloodstream [7].

Several lines of evidence underscore the connection between the absence of vitamin K and cancer. Vitamin K, an essential nutrient, has recently been explored as a potential anticancer agent, as demonstrated by its ability to inhibit the survival of certain pancreatic cancer cell lines through apoptotic mechanisms [8,9]. The deficiency of vitamin K can be detected using molecules such as PIVKA-II. This aspect has garnered significant interest within the scientific community [6].

Although PIVKA-II has proven to be a recognized tool in the diagnosis and prognosis of hepatocellular carcinoma (HCC) [10–12], it has also been observed that serum PIVKA-II alone could be a reliable biomarker for the detection of pancreatic cancer, showing superior diagnostic performance compared to other biomarkers [7,13]. In a recent report, we found overexpression of PIVKA-II in PDAC tissue and reduced circulating PIVKA-II levels after surgery in PDAC patients. The decrease in circulating PIVKA-II levels post-surgery suggests a reduction in tumor load. It can be speculated that baseline high PIVKA-II levels are a result of direct production by PDAC cells [14].

Considering these observations and the limited information available on the potential mechanisms underlying PIVKA-II's diagnostic efficiency as a biomarker for PDAC, we transitioned from clinical observations to laboratory investigations to functionally characterize the role of this protein in this specific cancer type. Thus, in this study, our primary aim was to identify an in vitro model using stabilized PDAC cell lines to comprehensively characterize PIVKA-II expression, distribution, and the molecular mechanisms leading to its release. Within our experimental model, we also explored the relationship between PIVKA-II and glucose. Recent findings highlighted that 80% of PDAC patients exhibit glucose intolerance or frank diabetes, and the variability of glucose levels in PDAC cell lines has been associated with both tumor proliferation and metastasis [15–17]. Finally, considering the frequently attributed association between hyperglycemia and poor prognosis in PDAC due to glucose-dependent alterations in the epithelial–mesenchymal transition process (EMT) [18], we aimed to analyze the kinetics of PIVKA-II expression and release in correlation with EMT.

2. Results

2.1. PIVKA-II Expression in PDAC Cell Lines

In order to study the potential role of PIVKA-II as an early biomarker of pancreatic cancer in vitro, we examined the expression of PIVKA-II protein in a panel of PDAC cell lines whose main features are summarized in Table 1.

Table 1. General characteristics of cell lines selected for this experimental study.

Cell Line	Age	Gender	Origin	Cell Type	Mutations
PANC-1 [19,20]	56	F	Primary tumor	Epithelial	KRAS, TP53, CDKN2A/p16
PaCa44 [21]	65	M	Primary tumor	Epithelial	KRAS, TP53, CDKN2A/p16
H-PAF-II [22]	44	M	Ascites	Epithelial	KRAS, TP53, CDKN2A/p16
Capan-1 [22]	40	M	Liver metastasis	Epithelial	KRAS, TP53, CDKN2A/p16 SMAD4/DPC4
HaCaT [23]	nd	nd	Keratynocyte	Epithelial	

Cultured cells were lysed, separated on SDS-PAGE, and finally analyzed by Western blotting using Moab-anti PIVKA-II and anti b-actin. Considering that there are no cell lines known in the literature to be used as a PIVKA-II positive control, we separated a high-PIVKA-II titer serum from a PDAC patient previously determined by immunometry (CLEIA) using SDS-PAGE. Nontumorigenic HaCaT cells were used as a negative control because they do not express PIVKA-II. As shown in Figure 1A, pancreatic cell line PaCa 44, PANC-1, H-Paf II, and Capan-1 express PIVKA-II, although production levels are different for each cell type, as reported by the quantitative analysis of PIVKA-II protein calculated in relation to the b-actin detected (Figure 1B). Based on this result, we elected PANC-1 cells as a reference model to study PIVKA-II protein.

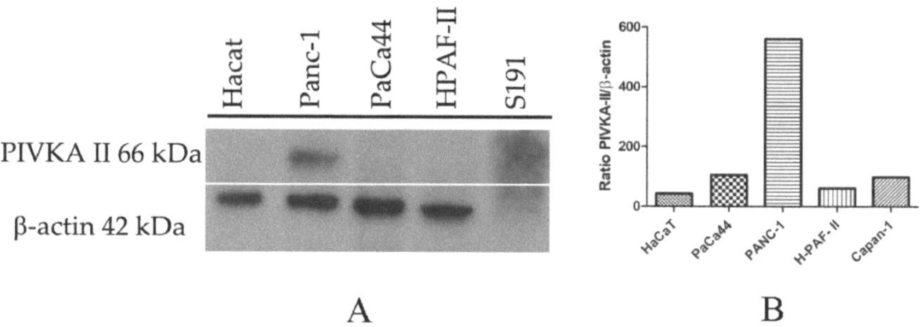

Figure 1. PIVKA-II protein expression in different cell lines. One representative experiment out of three is shown. (**A**) Western blot analysis of PIVKA-II protein in different pancreatic adenocarcinoma cell lines. (**B**) Densitometric evaluation of PIVKA-II protein in PDAC cell lines. Histograms represent the mean of the densitometric analysis of the ratio of PIVKA-II/β-actin. Densitometric analysis was performed with ImageJ software (1.47 version, NIH, Bethesda, MD, USA), which was downloaded from the NIH website (http://imagej.nih.gov, accessed on 1 August 2022) and plotted with GraphPad Prism 5.0 software.

2.2. PIVKA-II Localization in PANC-1 Cells

In order to better characterize the cellular distribution of this novel biomarker, we analyzed PIVKA II localization in PANC-1 cells by using indirect immunofluorescence (IFA) [24]. Figure 2A shows an IFA performed on PANC-1 cells (left panels) and on HaCaT cells (negative control, right panels) labeled with the monoclonal antibody (mo-ab) directed against the PIVKA-II protein (red); nuclei were counterstained with DAPI (blue).

As shown in Figure 2, PIVKA-II expression seemed to be exclusive to PANC-1 cells and undetectable in HaCaT cells, thus confirming the immunoblotting results. PIVKA-II is mainly distributed in the cell cytoplasm, with an enrichment in the perinuclear zone of certain cells, and the distribution is granular and appears to be associated with the fibrous structures of the cells. This distribution is comparable to that observed in vivo [14].

To evaluate the organization of the actin cytoskeleton in PANC-1 cells, a morphological analysis was performed by double immunofluorescence microscopy. To this end, human PANC-1 cells and HaCaT cells were stained with PIVKA II (red) and phalloidin (green), which specifically recognize the filamentous actin cytoskeleton (Figure 2B). The morphological analysis shows no co-localization between actin and PIVKA II protein.

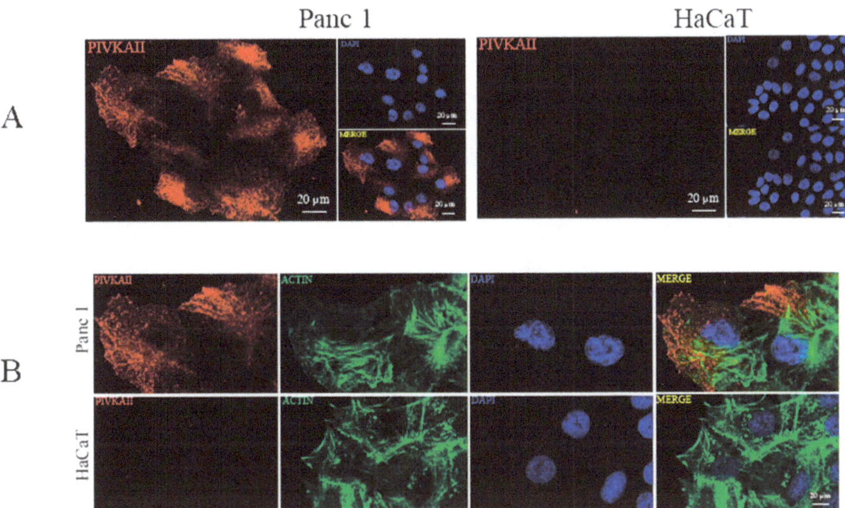

Figure 2. PIVKA-II localization in PANC-1 cells. (**A**) IFA performed on PANC-1 (left panel) and HaCaT (right panel) cell lines, showing PIVKA-II (red) and nuclei (blue). Representative images out of three are shown. (**B**) Double IFA performed on PANC-1 (upper panel) and HaCaT (lower panel) cell lines, showing PIVKA-II (red) and nuclei (blue) and actin (green). Representative images out of three are shown.

2.3. PIVKA-II Release in PANC-1 Cell Lines Is Glucose-Dependent

PIVKA-II is a free-circulating biomarker in vivo; however, nothing is yet known regarding the release mechanism of this protein. Several lines of evidence suggest that a large number (up to 80%) of pancreatic cancer patients suffer from hyperglycemia or diabetes, both characterized by elevated blood glucose levels [25]. Based on these observations, we wanted to investigate whether glucose could play a role in inducing the release of the PIVKA-II protein in vitro. To this end, we incubated PANC-1 cells in presence of increasing glucose doses (0 mM, 5 mM, 25 mM and 50 mM) for 24 h and 48 h. Following treatment, cells (Pell) and supernatants (Sup) were collected, separated by SDS-PAGE, and immunoblotted with different antibodies. As shown by the Western blotting analysis (Figure 3, left panel), we observed that PIVKA-II protein is released in the presence of 25 mM glucose following 48 h treatment, with an increase in the presence of 50 mM glucose; otherwise, this protein is retained in the absence of glucose or in the presence of very low concentrations of this sugar (5 mM) [16]. No increments in PIVKA-II production and release were observed following 24 h treatment. Ponceau staining of the immunoblot (red, lower left panel) was used as a supernatant loading control. In the same set of experiments, we also examined PANC-1 intracellular production of PIVKA-II protein in the presence of increasing glucose doses. Western blotting analysis of Figure 3B shows that the production of PIVKA-II seems to be directly related to the increase in glucose concentration in the culture medium, as also reported in the densitometric analysis presented in Figure 3C; b-actin was used as a loading control and human serum S191 was considered the positive control in the experiment. Taken together, these observations suggest that biomarker production and release in PANC-1 cells is promoted by glucose in a dose-dependent manner.

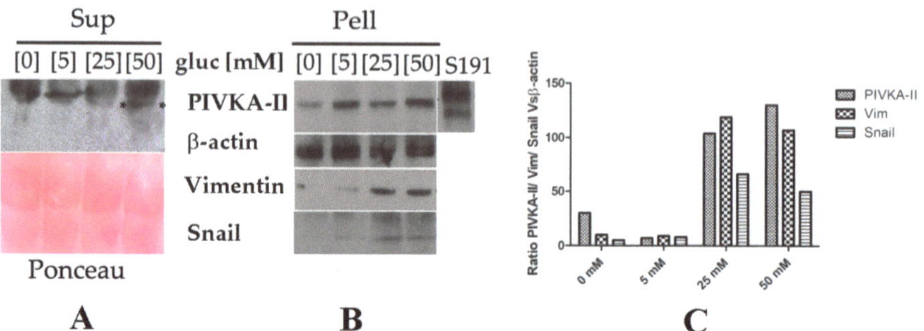

Figure 3. Expression and release of PIVKA-II in the presence of increasing doses of glucose in PANC-1 cells. One representative experiment out of three is shown. (**A**,**B**) Western blotting analysis performed on PANC-1 cells treated in the presence of increasing doses of glucose (5 mM, 25 mM, 50 mM) for 48 h. Following the treatment, Super (**A**) and Pellet (**B**) were recovered, separated on 12% SDS-PAGE, and analyzed by Western blotting with the indicated antibodies. Asterisks (*) in Figure 3A indicates released PIVKA-II protein. Correct loading of the supernatants was checked by Ponceau staining of the membrane (pink panel). b-actin was used as the lysates' loading control. (**C**) Densitometric evaluation of PIVKA-II protein in different cell lines. Histograms represent the mean of the densitometric analysis (performed with ImageJ) of the ratio of PIVKA-II–Vimentin–Snail vs. β-actin.

2.4. PIVKA-II Release in PANC-1 Cells Is Simultaneous with Epithelial–Mesenchymal Transition Activation

In order to study PIVKA-II release as a function of EMT onset, we evaluated the expression of two well-acknowledged EMT markers, Vimentin and Snail, in PANC-1 cells and in the presence of increasing doses of glucose [26]. As shown in Figure 3B, we observed the appearance of both Vimentin and Snail proteins in presence of 25 mM and 50 mM glucose. It is noteworthy that in this experiment, at the same glucose concentrations (i.e., 25 mM and 50 mM), PIVKA-II protein was released into the supernatant, thus suggesting that the biomarker in this in vitro model was released as EMT began; notably, PIVKA-II production in the PANC-1 cell line occurred before the complete activation of EMT.

3. Discussion

In developed nations, PDAC is presently ranked the fourth among the leading causes of cancer deaths; its mortality rate relative to its incidence has been a constant over the last two decades. Thus, improvements in timely PDAC diagnosis are mainly dictated by the necessity of setting up the decision-making process within a short space of time [27,28]. PDAC is characterized by an early and aggressive local invasion which, associated with a delayed clinical presentation and high metastatic potential, makes it a tumor with a poor prognosis [29]. In this scenario, circulating biomarkers, due to their availability, represent a powerful tool for early-stage diagnosis, prognosis, and follow up. Although many serum markers for the diagnosis of PDAC have been evaluated, to date, no reliable biomolecules (such as CEA, CA19.9, CA242) [30] have been identified for optimal clinical management. A notable push in this direction has been provided by recent studies focused on new biomolecules showing more reliable diagnostic performance [6,31,32]. Among these, the most promising is PIVKA-II protein, a modified prothrombin whose expression is related to vitamin K deficiency.

Recently, increasing attention has been paid to the relationship between vitamin K and malignancy [33]; several in vivo observational studies have established a relationship between vitamin K intake and cancer mortality [33,34]. An in vitro study also reported that vitamin K retains a peculiar cytotoxicity towards cancer cells through different mechanisms implicated in cell growth arrest and suppression of proliferation [33]. Taking these observations into account, all the products developed following vitamin K absence, such as PIVKA-II, acquire a new role in the management of cancer. This biomarker is commonly used for HCC diagnosis and prognosis, and recently altered values of PIVKA-II have been detected in various gastroenteric neoplasms such as gastric cancer, colon cancer, and PDAC, providing a new perspective on the diagnosis of these neoplasms [7,35]. The presence of this protein in PDAC is perhaps due to the fact that pancreas and liver tissues share a common embryological origin from the mesoderm, retaining a latent ability to trans-differentiate one into the other; it therefore seems reasonable to hypothesize that the characteristic expression in HCC could also be present in PDAC [36,37].

As a consequence of previous in vivo studies demonstrating that circulating levels of PIVKA-II were altered in patients with PDAC, here, we aimed to study the molecular aspect of PIVKA-II protein in vitro [14]. The in vitro study provided valuable information on the biological aspects of PIVKA-II; for the first time, in fact, we demonstrated that this protein is expressed in several cell lines originating from PDACs. Given the high basal level of protein expression, we chose PANC-1 cells as an in vitro model to study the properties of PIVKA-II in relation to PDAC. The peculiar cytoplasmic and granular distribution of PIVKA-II observed in vivo can be also confirmed in a PANC-1 model, thus strengthening previous reports' findings and further supporting the choice of this cellular model. However, morphological analysis of PIVKA-II distribution has currently only provided us with a preliminary dataset, and further studies will be needed to clarify how the cellular structure is altered in response to an appropriate stimulus to facilitate the intracellular transport and release of the protein [24,38].

Currently, one of the hallmarks of PDAC is the cellular metabolism reprogramming promoted by mutations in the KRAS oncogene [39,40]. It is well documented that cancer cells use a large amount of glucose, which is processed to produce lactate even in the presence of oxygen, a process described as the Warburg effect [41]. Cancer cells are known to have markedly increased glycolytic flux even in the presence of oxygen and normal mitochondrial function. The main role of glycolytic flux in carbon metabolism is not limited to the adenosine triphosphate (ATP) production but is also pivotal for providing biomass for the anabolic processes that support cell proliferation. In 2011, Han et al. reported that proliferation of pancreatic cell lines was affected by different concentrations of glucose in a concentration-dependent manner [42]. Thus, in light of these observations, we investigated the effect of glucose on the PIVKA-II protein in our model. In our study, we applied the same experimental conditions of Han et al. [42] and we observed that increasing doses of glucose promote PIVKA-II cellular production in a dose-dependent manner. Similar experiments have also been conducted on PaCa44 cell lines but unfortunately, we did not observe any effects on PIVKA-II production or release. We can speculate that one possible explanation is related to the fact that while PANC-1 cells are responsive to glucose treatment [42], the other cell lines could activate PIVKA-II production and release through different stimuli.

Since there is no information in the literature on whether PIVKA-II could be considered an early or late disease marker, we studied its expression in relation to two established EMT-related molecules in order to evaluate if PIVKA-II release could be associated with EMT-dictated tumor progression. It has been demonstrated that high glucose levels could promote pancreatic cancer proliferation and invasion as well as EMT and metastasis [18]. EMT, the hallmark of cancer metastasis, is a complex developmental program in which epithelial cells lose many of their characteristics and acquire a mesenchymal phenotype that permits the invasion of surrounding tissues, distant metastasis, metabolic reprogramming, resistance to chemotherapy, and immune system suppression [26,43,44]. EMT is generally associated with a poor prognosis since the activation of this mechanism confers characteristic aggressiveness to the tumor [45,46], and in pancreatic cancer progression, it has been demonstrated that tumor seeding of distant organs occurs before and concomitantly with tumor formation at the primary site [47]. Biomarkers are widely used in EMT studies to characterize the state in which cells are found, and some of them are already associated with this process, such as growth factors (TFG-β and Wnts), transcription factors (SNAIL and TWIST), adhesion molecules (cadherins), and molecules present in the cytoskeleton (Vimentin) [48,49]. In our study, using the EMT biomarkers Snail and Vimentin, we observed that while PIVKA-II production take place independently of EMT onset, the release of this protein occurs concomitantly with the beginning of glucose-induced EMT. Taken together, these experimental data suggest that PIVKA-II may represent an early signal of cancer progression and can be considered a novel valuable tool for timely PDAC diagnosis. The overall information deriving from our study underlines the importance, in the field of biomedical research, of identifying preclinical experimental models that are useful both for characterizing specific cellular mechanisms involved in the progression of PDAC and for evaluating the effects of possible targeted therapeutic strategies. It must also be considered that in recent years, tumor markers have not only been used as disease indicators but are often molecules that actively participate in tumor progression [50].

This study does have some limitations. The primary limitation of this study is the use of a single cell line and the omission of investigating other models of carcinogenesis activation, such as a cytokine cocktail and hypoxia. Moreover, it will be important to make a comparison with results from other studies in this particular field. Notwithstanding, the results obtained in this research are to be considered preliminary; they offer promising new perspectives for establishing a new effective PDAC biomarker for early diagnosis.

Appropriate biomarkers are crucial for the screening, early diagnosis, treatment, and prognosis of PDAC. We believe that PIVKA-II could be a useful tool for PDAC screening in selected populations; particularly, it could be a valid aid for people at increased risk of developing PDAC, i.e., patients with diabetes or glucose intolerance.

The overall information derived from our study emphasizes the importance, in the field of biomedical research, of identifying preclinical experimental models that are useful both for characterizing specific cellular mechanisms involved in the progression of PDAC and for evaluating the effects of possible targeted therapeutic strategies.

4. Materials and Methods

4.1. Cell Culture and Treatments

The human PDAC-derived cell lines PANC-1 [19], PaCa44 [20], HPAF II [22] and Capan-1 [21] were cultured in RPMI 1640, 10% fetal calf serum (FCS) (Aurogene, Rome, Italy), Lglutamine (2 mM), streptomycin (100 mg/mL), and penicillin (100 U/mL) in 5% CO_2 at 37 °C. Human keratinocytes (HaCaTs) are a spontaneously immortalized non-tumorigenic human keratinocyte line [23] and were maintained in D-Mem, 10% heat-inactivated fetal bovine serum (Aurogene), 2 mg of L-glutammin (Aurogene), and penicillin/streptomycin (100 unit of penicillin, 100 mg/mL streptomycin) before being incubated at 37 °C in and 5% CO_2. To study the effect of glucose concentration, cells were grown at 70% confluence on six wells washed with PBS and starved for 5 h at 37 °C in and 5% CO_2. Following the starvation, complete medium (RPMI) was replaced in the presence of glucose concentrations varying from 5.0 to 50 mM for 12 h, 24 h, or 48 h.

4.2. Indirect Immunofluorescence (IFA)

Untreated or treated cells were seeded on sterilized coverslips and grown at 70% confluence than washed (PBS 1x), air-dried, fixed, and permeabilized as described elsewhere [24]. The following primary antibodies were used: mouse monoclonal anti-PIVKA-II (Biorbyt-Durham, NC, USA, 1:1000) and FITC-phalloidin (Sigma-Aldrich, St. Louis, MO, USA) (1:50). Sheep anti-mouse IgG-Cy3 (SAM-Cy3, Jackson-Ely, UK; 1:2000) antibodies were used as secondary antibodies. Nuclei were stained with DAPI (1:5000 in PBS, Sigma) 1 min RT. Immunofluorescence was analyzed by using an Axio Observer Z1 inverted microscope equipped with an ApoTome.2 System (Carl Zeiss Inc., Ober Kochen, Germany). Digital images were acquired with an AxioCam MRm high-resolution digital camera (Zeiss) and processed with the AxioVision 4.8.2 software (Zeiss). ApoTome optical sectioning images of fluorescent cells were recorded under 40/0.75 objective (Zeiss).

4.3. Western-Blot Analysis

Treated or untreated cells were washed in PBS 1X and lysed in a RIPA buffer 1x (150 mM NaCl, 1% NP-40, 50 mM TrisHCl, pH 8, 0.5% deoxycholic acid, 0.1% SDS, protease and phosphatase inhibitors) on ice for 30 min, as described elsewhere [24]. Protein concentration was measured by using a BCA protein assay kit (Sigma 71285-M) and 15 µg of protein was subjected to electrophoresis on 10% TGX FastCast (TGX FastCast Acrylamide Kits-Bio Rad, San Francicsco, CA, USA), according to the manufacturer's instruction. The gels were transferred to nitrocellulose membranes (Bio-Rad, Hercules, CA, USA) for 45 min in Tris-glycine buffer, and the membranes were incubated in blocking solution (1 × PBS, 0.1% Tween20 and 3% of BSA (SERVA Electrophoresis GmbH, Heidelberg, Germany) containing the specific antibodies and developed using ECL Blotting Substrate (Advansta, San Jose, CA, USA). Concerning supernatants' analysis, 2 mL out of 20 mL cell culture medium was loaded and separated by SDS-PAGE 10% (TGX FastCast-Kits-Bio Rad, San Francicsco, CA, USA) then immunoblotted on nitrocellulose membranes, as described elsewhere [24]. Membranes were then probed with anti PIVKA-II (Biorbyt 1:1000), anti-β-actin (Santa Cruz 1:1000, Santa Cruz, CA, USA), anti-Snail (Cell Signaling 1:100, Beverly, MA, USA), anti-Vimentin (Santa Cruz 1:200), polyclonal anti-mouse IgG-HRP (Bethyl, 1:10,000, Montgomery, TX, USA), and anti-rabbit IgG-HRP (Bethyl, 1:20,000). Detection was performed using Western Bright (Advansta, Menlo Park, CA, USA).

4.4. Densitometric Analysis

Quantification of protein bands was performed by densitometric analysis using Image J software (1.47 version, NIH, Bethesda, MD, USA), which was downloaded from the NIH website (http://imagej.nih.gov, accessed on 1 August 2022). The densitometric analysis was performed using GraphPad Prism 5.0 software (GraphPad Software Inc., La Jolla, CA, USA).

Author Contributions: Study design: A.F. and E.A. Experiments performed: A.F. and A.P. Analysis and interpretation data: A.F., E.A. and P.M. Densitometric analysis: A.F. and A.P. Immunofluorescence acquisition: A.P. and P.M. Manuscript preparation: A.F., S.T., P.M., A.A. and E.A. All authors have read and agreed to the published version of the manuscript.

Funding: This work was funded by University of Rome la Sapienza, Ateneo 2021 (RP12117A768C8030) and Ateneo 2022 (AR22218167FCBC4B).

Institutional Review Board Statement: The study was conducted in accordance with the Declaration of Helsinki and approved by the Institutional Review Board (or Ethics Committee) of Sapienza University of Rome (protocol code ME-3-PIvka) for studies involving humans.

Informed Consent Statement: Informed consent was obtained from all subjects involved in the study.

Data Availability Statement: Data are contained within the article.

Acknowledgments: We are thankful to Valentina Viggiani, Giuseppina Gennarini. Francesca Cortese, and Barbara Colaprisca for their technical assistance and to D. Kanton for providing language help.

Conflicts of Interest: All authors have disclosed any financial or personal relationship with organizations that could potentially be perceived as influencing the described research.

References

1. Sarantis, P.; Koustas, E.; Papadimitropoulou, A.; Papavassiliou, A.G.; Karamouzis, M.V. Pancreatic ductal adenocarcinoma: Treatment hurdles, tumor microenvironment and immunotherapy. *World J. Gastrointest. Oncol.* **2020**, *12*, 173–181. [CrossRef]
2. Sakin, A.; Sahin, S.; Sakin, A.; Atci, M.M.; Arici, S.; Yasar, N.; Demir, C.; Geredeli, C.; Cihan, S. Factors affecting survival in operated pancreatic cancer: Does tumor localization have a significant effect on treatment outcomes? *N. Clin. Istanb.* **2020**, *7*, 487–493. [CrossRef]

3.   Rawla, P.; Sunkara, T.; Gaduputi, V. Epidemiology of pancreatic cancer: Global trends, etiology and risk factors. *World J. Oncol.* **2019**, *10*, 10–27. [CrossRef]
4.   Pereira, S.P.; Oldfield, L.; Ney, A.; Hart, P.A.; Keane, M.G.; Pandol, S.J.; Li, D.; Greenhalf, W.; Jeon, C.Y.; Koay, E.J.; et al. Early detection of pancreatic cancer. The lancet. *Gastroenterol. Hepatol.* **2020**, *5*, 698–710.
5.   Bodaghi, A.; Fattahi, N.; Ramazani, A. Biomarkers: Promising and valuable tools towards diagnosis, prognosis and treatment of COVID-19 and other diseases. *Heliyon* **2023**, *9*, e13323. [CrossRef]
6.   Yang, Y.; Li, G.; Zhang, Y.; Cui, Y.; Liu, J. Protein Induced by Vitamin K Absence II: A Potential Biomarker to Differentiate Pancreatic Ductal Adenocarcinoma from Pancreatic Benign Lesions and Predict Vascular Invasion. *J. Clin. Med.* **2023**, *12*, 2769. [CrossRef]
7.   Tartaglione, S.; Pecorella, I.; Zarrillo, S.R.; Granato, T.; Viggiani, V.; Manganaro, L.; Marchese, C.; Angeloni, A.; Anastasi, E. Protein Induced by Vitamin K Absence II (PIVKA-II) as a potential serological biomarker in pancreatic cancer: A pilot study. *Biochem. Med.* **2019**, *29*, 020707. [CrossRef]
8.   Dahlberg, S.; Ede, J.; Schött, U. Vitamin K and cancer. *Scand. J. Clin. Lab. Investig.* **2017**, *77*, 555–567. [CrossRef]
9.   Gul, S.; Maqbool, M.F.; Maryam, A.; Khan, M.; Shakir, H.A.; Irfan, M.; Ara, C.; Li, Y.; Ma, T. Vitamin K: A novel cancer chemosensitizer. *Biotechnol. Appl. Biochem.* **2022**, *69*, 2641–2657. [CrossRef]
10.   Tian, S.; Chen, Y.; Zhang, Y.; Xu, X. Clinical value of serum AFP and PIVKA-II for diagnosis, treatment and prognosis of hepatocellular carcinoma. *J. Clin. Lab. Anal.* **2023**, *37*, e24823. [CrossRef]
11.   Takahashi, Y.; Inoue, T.; Fukusato, T. Protein induced by vitamin K absence or antagonist II-producing gastric cancer. *World J. Gastrointest. Pathophysiol.* **2010**, *1*, 129–136. [CrossRef] [PubMed]
12.   Yang, Y.; Li, G.; Lu, Z.; Liu, Y.; Kong, J.; Liu, J. Progression of Prothrombin Induced by Vitamin K Absence-II in Hepatocellular Carcinoma. *Front. Oncol.* **2021**, *11*, 726213. [CrossRef]
13.   Kemik, A.S.; Kemik, O.; Purisa, S.; Tuzun, S. Serum des-gamma-carboxyprothrombin in patients with pancreatic head adenocarcinoma. *Bratisl. Lek. Listy* **2011**, *112*, 552–554.
14.   Tartaglione, S.; Mancini, P.; Viggiani, V.; Chirletti, P.; Angeloni, A.; Anastasi, E. PIVKA-II: A biomarker for diagnosing and monitoring patients with pancreatic adenocarcinoma. *PLoS ONE* **2021**, *16*, e0251656. [CrossRef]
15.   Wang, F.; Herrington, M.; Larsson, J.; Permert, J. The relationship between diabetes and pancreatic cancer. *Mol. Cancer* **2003**, *2*, 4. [CrossRef]
16.   Jian, Z.; Cheng, T.; Zhang, Z.; Raulefs, S.; Shi, K.; Steiger, K.; Maeritz, N.; Kleigrewe, K.; Hofmann, T.; Benitz, S.; et al. Glycemic Variability Promotes Both Local Invasion and Metastatic Colonization by Pancreatic Ductal Adenocarcinoma. *Cell Mol. Gastroenterol. Hepatol.* **2018**, *6*, 429–449. [CrossRef]
17.   Ying, H.; Kimmelman, A.C.; Lyssiotis, C.A.; Hua, S.; Chu, G.C.; Fletcher-Sananikone, E.; Locasale, J.W.; Son, J.; Zhang, H.; Coloff, J.L.; et al. Oncogenic Kras maintains pancreatic tumors through regulation of anabolic glucose metabolism. *Cell* **2012**, *149*, 656–670. [CrossRef]
18.   Li, W.; Liu, H.; Qian, W.; Cheng, L.; Yan, B.; Han, L.; Xu, Q.; Ma, Q.; Ma, J. Hyperglycemia aggravates microenvironment hypoxia and promotes the metastatic ability of pancreatic cancer. *Comput. Struct. Biotechnol. J.* **2018**, *16*, 479–487. [CrossRef]
19.   Ungefroren, H.; Thürling, I.; Färber, B.; Kowalke, T.; Fischer, T.; De Assis, L.V.M.; Braun, R.; Castven, D.; Oster, H.; Konukiewitz, B.; et al. The Quasimesenchymal Pancreatic Ductal Epithelial Cell Line PANC-1—A Useful Model to Study Clonal Heterogeneity and EMT Subtype Shifting. *Cancers* **2022**, *14*, 2057. [CrossRef]
20.   Watanabe, M.; Sheriff, S.; Lewis, K.B.; Cho, J.; Tinch, S.L.; Balasubramaniam, A.; Kennedy, M.A. Metabolic Profiling Comparison of Human Pancreatic Ductal Epithelial Cells and Three Pancreatic Cancer Cell Lines using NMR Based Metabonomics. *J. Mol. Biomark. Diagn.* **2012**, *3*, S3-002. [CrossRef] [PubMed]
21.   Dalla Pozza, E.; Manfredi, M.; Brandi, J.; Buzzi, A.; Conte, E.; Pacchiana, R.; Cecconi, D.; Marengo, E.; Donadelli, M. Trichostatin A alters cytoskeleton and energy metabolism of pancreatic adenocarcinoma cells: An in depth proteomic study. *J. Cell Biochem.* **2018**, *119*, 2696–2707. [CrossRef] [PubMed]
22.   Deer, E.L.; González-Hernández, J.; Coursen, J.D.; Shea, J.E.; Ngatia, J.; Scaife, C.L.; Firpo, M.A.; Mulvihill, S.J. Phenotype and genotype of pancreatic cancer cell lines. *Pancreas* **2010**, *39*, 425–435. [CrossRef] [PubMed]
23.   Boukamp, P.; Petrussevska, R.T.; Breitkreutz, D.; Hornung, J.; Markham, A.; Fusenig, N.E. Normal keratinization in a spontaneously immortalized aneuploid human keratinocyte cell line. *J. Cell Biol.* **1988**, *106*, 761–771. [CrossRef] [PubMed]
24.   Gonnella, R.; Dimarco, M.; Farina, G.A.; Santarelli, R.; Valia, S.; Faggioni, A.; Angeloni, A.; Cirone, M.; Farina, A. BFRF1 protein is involved in EBV-mediated autophagy manipulation. *Microbes Infect.* **2020**, *22*, 585–591. [CrossRef] [PubMed]
25.   De Souza, A.; Irfan, K.; Masud, F.; Saif, M.W. Diabetes Type 2 and Pancreatic Cancer: A History Unfolding. *JOP J. Pancreas* **2016**, *17*, 144–148.
26.   Ribatti, D.; Tamma, R.; Annese, T. Epithelial-Mesenchymal Transition in Cancer: A Historical Overview. *Transl. Oncol.* **2020**, *13*, 100773. [CrossRef]
27.   Miller, K.D.; Goding Sauer, A.; Ortiz, A.P.; Fedewa, S.A.; Pinheiro, P.S.; Tortolero-Luna, G.; Martinez-Tyson, D.; Jemal, A.; Siegel, R.L. Cancer Statistics for Hispanics/Latinos, 2018. *CA A Cancer J. Clin.* **2018**, *68*, 425–445. [CrossRef]
28.   Siegel, R.L.; Miller, K.D.; Fuchs, H.E.; Jemal, A. Cancer statistics, 2022. *CA Cancer J Clin.* **2022**, *72*, 7–33. [CrossRef]
29.   Huang, J.; Lok, V.; Ngai, C.H.; Zhang, L.; Yuan, J.; Lao, X.Q.; Ng, K.; Chong, C.; Zheng, Z.J.; Wong, M.C.S. Worldwide Burden of, Risk Factors for, and Trends in Pancreatic Cancer. *Gastroenterology* **2021**, *160*, 744–754. [CrossRef]

30. Kane, L.E.; Mellotte, G.S.; Mylod, E.; O'Brien, R.M.; O'Connell, F.; Buckley, C.E.; Arlow, J.; Nguyen, K.; Mockler, D.; Meade, A.D.; et al. Diagnostic Accuracy of Blood-based Biomarkers for Pancreatic Cancer: A Systematic Review and Meta-analysis. *Cancer Res. Commun.* **2022**, *2*, 1229–1243. [CrossRef]

31. Yang, J.; Xu, R.; Wang, C.; Qiu, J.; Ren, B.; You, L. Early screening and diagnosis strategies of pancreatic cancer: A comprehensive review. *Cancer Commun.* **2021**, *41*, 1257–1274. [CrossRef] [PubMed]

32. Matsumura, K.; Hayashi, H.; Uemura, N.; Zhao, L.; Higashi, T.; Yamao, T.; Kitamura, F.; Nakao, Y.; Yusa, T.; Itoyama, R.; et al. Prognostic Impact of Coagulation Activity in Patients Undergoing Curative Resection for Pancreatic Ductal Adenocarcinoma. *In Vivo Sep.* **2020**, *34*, 2845–2850. [CrossRef] [PubMed]

33. Welsh, J.; Bak, M.J.; Narvaez, C.J. New insights into vitamin K biology with relevance to cancer. *Trends Mol. Med.* **2022**, *28*, 864–881. [CrossRef] [PubMed]

34. Markowska, A.; Antoszczak, M.; Markowska, J.; Huczyński, A. Role of Vitamin K in Selected Malignant Neoplasms in Women. *Nutrients* **2022**, *14*, 3401. [CrossRef] [PubMed]

35. Caviglia, G.P.; Ribaldone, D.G.; Abate, M.L.; Ciancio, A.; Pellicano, R.; Smedile, A.; Saracco, G.M. Performance of protein induced by vitamin K absence or antagonist-II assessed by chemiluminescence enzyme immunoassay for hepatocellular carcinoma detection: A meta-analysis. *Scand. J. Gastroenterol.* **2018**, *53*, 734–740. [CrossRef]

36. Yi, F.; Liu, G.H.; Izpisua Belmonte, J.C. Rejuvenating liver and pancreas through cell transdifferentiation. *Cell Res.* **2012**, *22*, 616–619. [CrossRef]

37. Gordillo, M.; Evans, T.; Gouon-Evans, V. Orchestrating liver development. *Development* **2015**, *142*, 2094–2108. [CrossRef] [PubMed]

38. Shankar, J.; Nabi, I.R. Actin Cytoskeleton Regulation of Epithelial Mesenchymal Transition in Metastatic Cancer Cells. *PLoS ONE* **2015**, *10*, e0119954. [CrossRef]

39. Qin, C.; Yang, G.; Yang, J.; Ren, B.; Wang, H.; Chen, G.; Zhao, F.; You, L.; Wang, W.; Zhao, Y. Metabolism of pancreatic cancer: Paving the way to better anticancer strategies. *Mol. Cancer* **2020**, *19*, 50. [CrossRef]

40. Perera, R.M.; Bardeesy, N. Pancreatic Cancer Metabolism: Breaking It Down to Build It Back Up. *Cancer Discov.* **2015**, *5*, 1247–1261. [CrossRef]

41. Liberti, M.V.; Locasale, J.W. The Warburg Effect: How Does it Benefit Cancer Cells? *Trends Biochem. Sci.* **2016**, *41*, 211–218, Erratum in: *Trends Biochem. Sci.* **2016**, *41*, 287. [CrossRef]

42. Han, L.; Ma, Q.; Li, J.; Liu, H.; Li, W.; Ma, G.; Xu, Q.; Zhou, S.; Wu, E. High glucose promotes pancreatic cancer cell proliferation via the induction of EGF expression and transactivation of EGFR. *PLoS ONE.* **2011**, *6*, e27074. [CrossRef]

43. Pastushenko, I.; Blanpain, C. EMT Transition States during Tumor Progression and Metastasis. *Trends Cell Biol.* **2019**, *29*, 212–226. [CrossRef] [PubMed]

44. Polyak, K.; Weinberg, R. Transitions between epithelial and mesenchymal states: Acquisition of malignant and stem cell traits. *Nat. Rev. Cancer* **2009**, *9*, 265–273. [CrossRef] [PubMed]

45. Palamaris, K.; Felekouras, E.; Sakellariou, S. Epithelial to Mesenchymal Transition: Key Regulator of Pancreatic Ductal Adenocarcinoma Progression and Chemoresistance. *Cancers* **2021**, *13*, 5532. [CrossRef]

46. Hu, X.; Chen, W. Role of epithelial-mesenchymal transition in chemoresistance in pancreatic ductal adenocarcinoma. *World J. Clin. Cases* **2021**, *9*, 4998–5006. [CrossRef] [PubMed]

47. Rhim, A.D.; Mirek, E.T.; Aiello, N.M.; Maitra, A.; Bailey, J.M.; McAllister, F.; Reichert, M.; Beatty, G.L.; Rustgi, A.K.; Vonderheide, R.H.; et al. EMT and dissemination precede pancreatic tumor formation. *Cell* **2012**, *148*, 349–361. [CrossRef]

48. Thompson, E.W.; Newgreen, D.F.; Tarin, D. Carcinoma invasion and metastasis: A role for epithelial-mesenchymal transition? *Cancer Res.* **2005**, *65*, 5991–5995. [CrossRef]

49. Arko-Boham, B.; Lomotey, J.T.; Tetteh, E.N.; Tagoe, E.A.; Aryee, N.A.; Owusu, E.A.; Okai, I.; Blay, R.M.; Clegg-Lamptey, J.N. Higher serum concentrations of vimentin and DAKP1 are associated with aggressive breast tumour phenotypes in Ghanaian women. *Biomark. Res.* **2017**, *5*, 21. [CrossRef]

50. Anastasi, E.; Farina, A.; Granato, T.; Colaiacovo, F.; Pucci, B.; Tartaglione, S.; Angeloni, A. Recent Insight about HE4 Role in Ovarian Cancer Oncogenesis. *Int. J. Mol. Sci.* **2023**, *24*, 10479. [CrossRef]

Disclaimer/Publisher's Note: The statements, opinions and data contained in all publications are solely those of the individual author(s) and contributor(s) and not of MDPI and/or the editor(s). MDPI and/or the editor(s) disclaim responsibility for any injury to people or property resulting from any ideas, methods, instructions or products referred to in the content.

International Journal of
Molecular Sciences

MDPI

Article

First-in-Class Humanized Antibody against Alternatively Spliced Tissue Factor Augments Anti-Metastatic Efficacy of Chemotherapy in a Preclinical Model of Pancreatic Ductal Adenocarcinoma

Clayton S. Lewis [1], Charles Backman [1], Sabahat Ahsan [1], Ashley Cliff [2], Arthi Hariharan [2], Jen Jen Yeh [2,3], Xiang Zhang [4], Changchun Xie [5], Davendra P. S. Sohal [1] and Vladimir Y. Bogdanov [1,*]

[1] Division of Hematology/Oncology, Department of Internal Medicine, College of Medicine, University of Cincinnati, Cincinnati, OH 45267, USA; clayton.lewis@uc.edu (C.S.L.); backmaca@mail.uc.edu (C.B.); ahsansh@mail.uc.edu (S.A.); sohalda@uc.edu (D.P.S.S.)

[2] Lineberger Comprehensive Cancer Center, University of North Carolina at Chapel Hill, Chapel Hill, NC 27599, USA; ashley_cliff@med.unc.edu (A.C.); arthih@unc.edu (A.H.); jen_jen_yeh@med.unc.edu (J.J.Y.)

[3] Departments of Surgery and Pharmacology, University of North Carolina at Chapel Hill, Chapel Hill, NC 27599, USA

[4] Division of Environmental Genetics and Molecular Toxicology, Department of Environmental and Public Health Sciences, College of Medicine, University of Cincinnati, Cincinnati, OH 45267, USA; xiang.zhang@uc.edu

[5] Division of Biostatistics and Bioinformatics, Department of Environmental and Public Health Sciences, College of Medicine, University of Cincinnati, Cincinnati, OH 45267, USA; changchun.xie@uc.edu

* Correspondence: vladimir.bogdanov@uc.edu; Tel.: +1-513-558-6276

Citation: Lewis, C.S.; Backman, C.; Ahsan, S.; Cliff, A.; Hariharan, A.; Yeh, J.J.; Zhang, X.; Xie, C.; Sohal, D.P.S.; Bogdanov, V.Y. First-in-Class Humanized Antibody against Alternatively Spliced Tissue Factor Augments Anti-Metastatic Efficacy of Chemotherapy in a Preclinical Model of Pancreatic Ductal Adenocarcinoma. *Int. J. Mol. Sci.* **2024**, *25*, 2580. https://doi.org/10.3390/ijms25052580

Academic Editors: Donatella Delle Cave and Claudio Luchini

Received: 1 January 2024
Revised: 5 February 2024
Accepted: 13 February 2024
Published: 23 February 2024

Copyright: © 2024 by the authors. Licensee MDPI, Basel, Switzerland. This article is an open access article distributed under the terms and conditions of the Creative Commons Attribution (CC BY) license (https://creativecommons.org/licenses/by/4.0/).

Abstract: Alternatively spliced tissue factor (asTF) promotes the progression of pancreatic ductal adenocarcinoma (PDAC) by activating β1-integrins on PDAC cell surfaces. hRabMab1, a first-in-class humanized inhibitory anti-asTF antibody we recently developed, can suppress PDAC primary tumor growth as a single agent. Whether hRabMab1 has the potential to suppress metastases in PDAC is unknown. Following in vivo screening of three asTF-proficient human PDAC cell lines, we chose to make use of KRAS G12V-mutant human PDAC cell line PaCa-44, which yields aggressive primary orthotopic tumors with spontaneous spread to PDAC-relevant anatomical sites, along with concomitant severe leukocytosis. The experimental design featured orthotopic tumors formed by luciferase labeled PaCa-44 cells; administration of hRabMab1 alone or in combination with gemcitabine/paclitaxel (gem/PTX); and the assessment of the treatment outcomes on the primary tumor tissue as well as systemic spread. When administered alone, hRabMab1 exhibited poor penetration of tumor tissue; however, hRabMab1 was abundant in tumor tissue when co-administered with gem/PTX, which resulted in a significant decrease in tumor cell proliferation; leukocyte infiltration; and neovascularization. Gem/PTX alone reduced primary tumor volume, but not metastatic spread; only the combination of hRabMab1 and gem/PTX significantly reduced metastatic spread. RNA-seq analysis of primary tumors showed that the addition of hRabMab1 to gem/PTX enhanced the downregulation of tubulin binding and microtubule motor activity. In the liver, hRabMab1 reduced liver metastasis as a single agent. Only the combination of hRabMab1 and gem/PTX eliminated tumor cell-induced leukocytosis. We here demonstrate for the first time that hRabMab1 may help suppress metastasis in PDAC. hRabMab1's ability to improve the efficacy of chemotherapy is significant and warrants further investigation.

Keywords: pancreatic ductal adenocarcinoma (PDAC); alternatively spliced tissue factor (asTF); humanized monoclonal antibody

1. Introduction

PDAC is associated with high rates of venous thromboembolism (VTE); one of the key contributors to this morbidity is tissue factor (TF, also known as CD142, thromboplastin, coagulation factor III) [1]. The much-studied, plasma membrane-bound form of TF protein termed full-length (fl)TF, is the obligatory cofactor of the plasma serine protease fVIIa and triggers blood clotting either upon tissue damage, or aberrant expression in cells that come in contact with circulating blood; in PDAC, cancer cell-associated and extracellular vesicle-bound flTF both contribute to VTE [2]. Hypoxia synergizes with such oncogenic drivers as KRAS to induce TF (*F3*) gene expression via the amplification of PI3K-Akt and p38-NFkB signaling pathways, both of which are prominent in KRAS-mutant cancers including PDAC [3]; HIF-1α can also induce TF in cancer cells indirectly, via the upregulation of VEGF expression [4]. In addition to cancer cells, stromal cells such as monocytes/macrophages, fibroblasts, and microvascular endothelial cells express TF in cancer lesions [5]. Aside from causing thrombosis in PDAC, high TF expression was long known to correlate with PDAC's histological grade [6]; in 1999, it was reported that TF can promote PDAC growth and tumor cell invasion in vivo [7]. Nitori and colleagues suggested that TF may have prognostic significance in PDAC: "high TF" patients presented with larger tumors and more advanced metastatic disease with TF prominently expressed at the invasive front of the primary tumor [8]. More recently, the Flick laboratory reported that the flTF/fVIIa complex can contribute to metastatic seeding and immune evasion by cleaving protease-activated receptors on PDAC cell surfaces [9], and this year, Zhang et al. reported that TF overexpression can promote resistance to the newest class of KRAS-G12C inhibitors [10].

Another layer of complexity to TF function involves activity that is not protease dependent, but rather integrin mediated; it is largely executed by TF's minimally coagulant alternatively spliced form, asTF. Unlike flTF, asTF lacks a transmembrane domain and can, thus, be secreted as a free protein; a splicing-dependent shift in asTF's open reading frame creates a unique 40 amino acid C-terminal epitope in asTF, which makes it possible to develop asTF-specific antibodies [11]. asTF binds a subset of β1 integrins in close proximity to the "knee" region, causing a conformational change that amplifies integrin-linked outside-in signaling. When bound to integrins on benign cells, e.g., endothelium, asTF promotes cell migration and the expression of leukocyte adhesion molecules, yet not cell proliferation [12]; however, when bound to integrins on malignant cells, asTF fuels both proliferation as well as systemic spread [13,14]. Given asTF's cell-agonist properties, along with its dispensability for normal hemostasis, asTF is an attractive therapeutic target.

In 2021, we reported the results of the first study that evaluated the in vivo efficacy of an asTF-specific, inhibitory, humanized antibody termed hRabMab1 [15]. We found that hRabMab1 was able to suppress the growth of pre-formed, orthotopically grown PDAC tumors (KRAS G12D-mutant cell line Pt45.P1) when administered intravenously as a single agent. hRabMab1 exhibited a favorable pharmacokinetic (PK) profile in mice with no toxicity detected at the dose of 18 mg/kg; examination of PDAC tumor tissue post-hRabMab1 treatment showed the reduction of cancer cell proliferation and decreased monocyte/macrophage infiltration of the lesions. In this study, we assessed hRabMab1's ability to suppress the progression of experimental PDAC in a model featuring a more aggressive asTF-proficient human PDAC cell line—KRAS G12V-mutant PaCa-44 cells [16]—alone and in combination with a standard-of-care regimen, gemcitabine/paclitaxel (gem/PTX).

2. Results

2.1. Orthotopic Implantation of PaCa44 Cells Yields Stroma-Rich Primary Tumors with Spontaneous Metastases

Recent studies have demonstrated that circulating tumor cells (CTCs) arising endogenously from solid primary tumors undergo a multi-step metastatic process that is not recapitulated by cell lines grown in vitro limiting the value of tumor cell-seeding approaches such as tail vein injections and hemi-splenic injections for studying PDAC metastatic seeding in the lung and the liver, respectively [17,18]. Given our desire to determine the anti-metastatic potential of hRabMab1, we first sought to ascertain the metastatic potential of three asTF-proficient human PDAC cell lines grown orthotopically in SCID mice. The expression profile of asTF-target integrins and the ability of each cell line to yield spontaneous metastases in an orthotopic setting are shown in Figure 1.

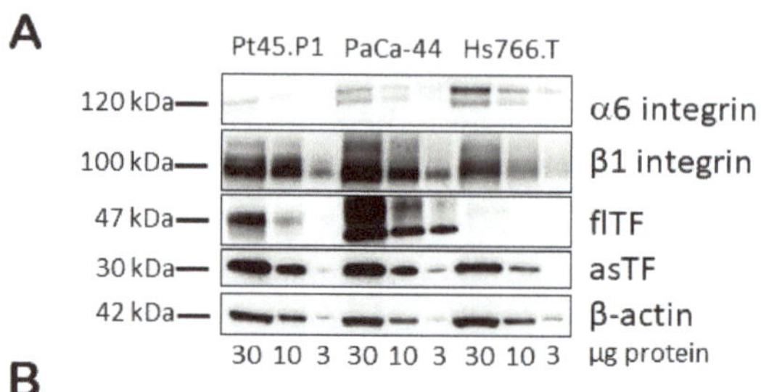

Cell line	Primary tumor growth	Spont. lung mets	Spont. liver mets	Spont. abdominal mets
Pt45.P1	∨ ∨	X	∨	∨ ∨
PaCa-44	∨ ∨ ∨	∨ ∨ ∨	∨ ∨ ∨	∨ ∨ ∨
Hs766.T	∨	X	X	∨

Figure 1. (**A**) Expression of TF protein variants and α6β1 integrins in Pt45.P1, PaCa-44, and Hs766.T cells. (**B**) In vivo properties of Pt45.P1, PaCa-44, and Hs766.T cells were evaluated (2 in vivo studies per cell line); ∨: positive outcome; X: negative outcome.

asTF-proficient, KRAS G12V-mutant cell line PaCa-44 yields reproducible-size, aggressive primary orthotopic tumors that spread spontaneously to PDAC-relevant anatomical sites (Figure 2); other useful features of the PaCa-44 model comprise its 100% penetrance of metastases to the site of surgical incision (wound closure area in the abdominal wall), and severe leukocytosis.

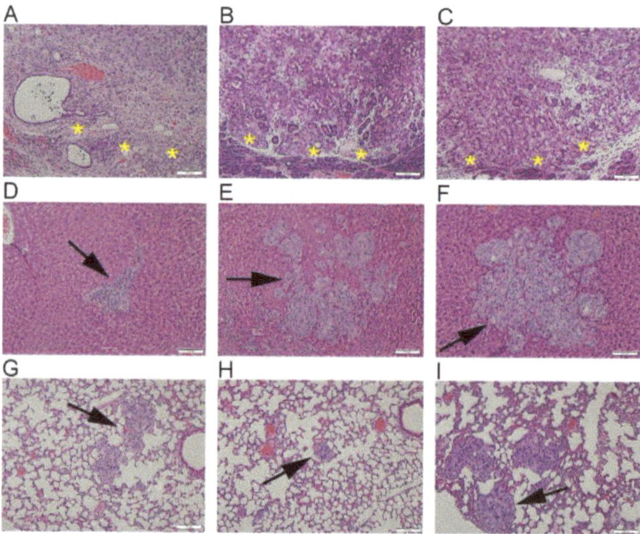

Figure 2. Representative images of PaCa-44 primary tumor leading edge (demarked by yellow asterisks, (**A–C**)), liver metastases (arrows, (**D–F**)), and lung metastases (arrows, (**G–I**)) from 3 different mice; animal-specific tissues arranged in vertical columns. Hematoxylin and eosin stain; original magnification: 20×; 100 µm scale bar shown in bottom right of each micrograph.

2.2. Effects of hRabMab1 in the Primary Tumor Tissue

We determined that sequential administration of gem/PTX (50 mg/kg/3 mg/kg, respectively) in line with Wolfe et al. [19] resulted in a significant reduction in primary volume (~80%, $p = 0.002$, vehicle vs. gem/PTX). To study the effect of hRabMab1 in this model, 5×10^5 luciferase-labeled PaCa-44 cells were implanted into the pancreata of NOD.scid mice ($n = 40$); 10 days post-implantation, mice were randomized into 4 cohorts: vehicle; hRabMab1 (IV at 18 mg/kg); gem/PTX (50 mg/kg and 3 mg/kg, respectively); and the combination of hRabMab1 with gem/PTX. Post-mortem analysis of primary tumors revealed that hRabMab1 did not significantly impact the primary tumor volume when used alone or in combination with gem/PTX (Figure 3).

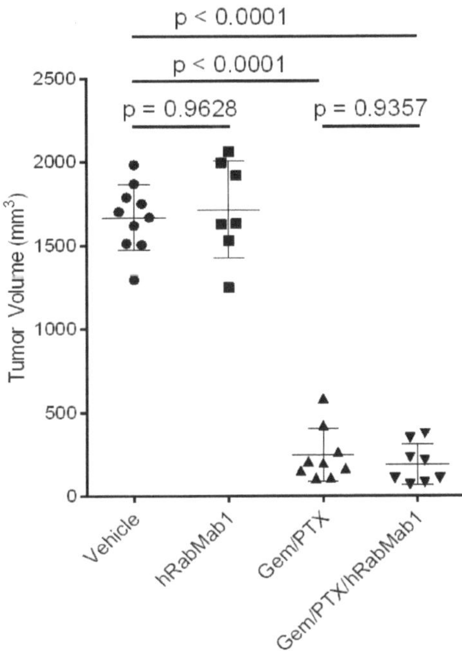

Figure 3. Tumor volume in four experimental cohorts as indicated; 1-way ANOVA with Tukey's multiple comparison test was used to assess significance.

Immunohistochemical (IHC) analysis revealed that, unlike in thin-capsule forming Pt45.P1 tumors, which are penetrable by hRabMab1 as a single agent [15], intratumoral hIgG was not detectable in well-encapsulated PaCa-44 tumors in mice that received hRabMab1 as a single agent. However, when hRabMab1 was combined with gem/PTX, intratumoral hIgG was found to be abundant throughout the tumor tissue (Figure 4A). Ki67+ signal was significantly suppressed by the addition of hRabMab1 to gem/PTX; there were also significantly fewer neutrophils, monocytes, and microvessels in hRabMab1+gem/PTX tumors compared to gem/PTX tumors (Figure 4B–E and Supplementary Materials Figure S1).

RNA-seq analysis of the tumor tissue followed by gene-set enrichment of differentially expressed genes (three tumors per cohort: control; gem/PTX; and hRabMab1+gem/PTX, Figure 5) revealed that the addition of hRabMab1 to gem/PTX downregulated tubulin binding and microtubule motor activity. Genes involved in neovascularization were upregulated in response to gem/PTX and the addition of hRabMab1 to gem/PTX weakened this compensatory effect, which is consistent with our IHC data (Figure 4E).

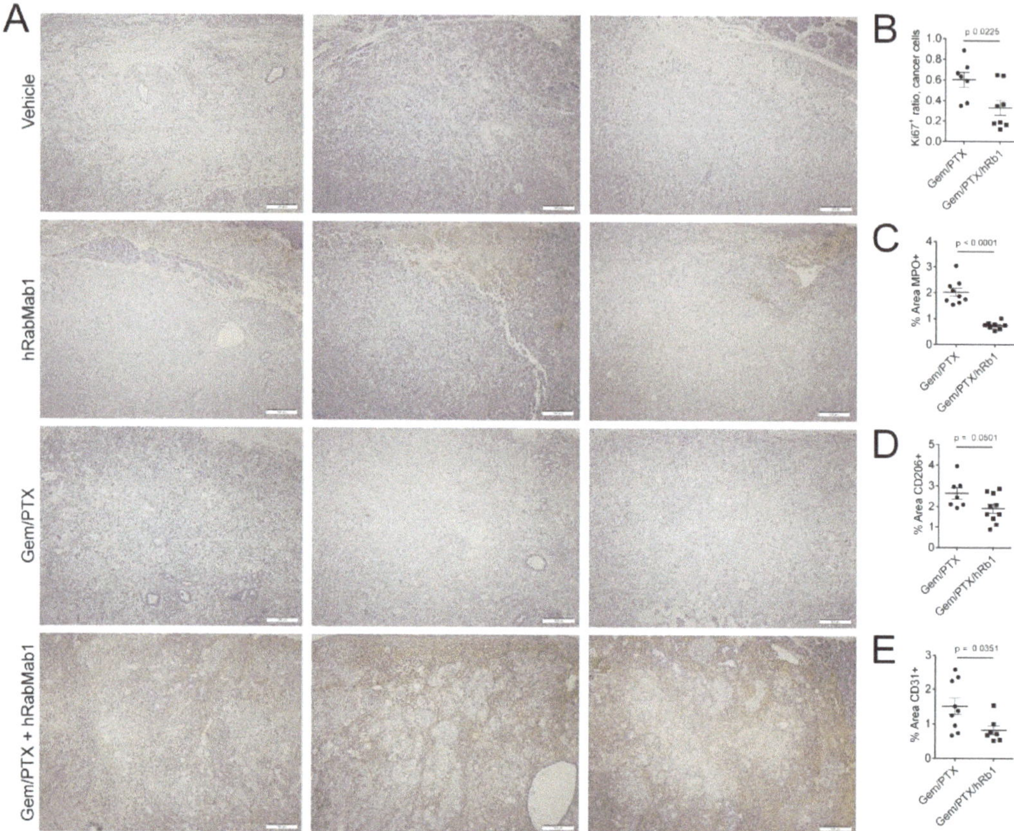

Figure 4. (**A**) IHC of PaCa-44 primary tumor tissue (3 representative specimens per cohort as indicated) stained for human IgG; original magnification: $20\times$; 100 μm scale bar shown in bottom right of each micrograph. (**B–E**) Quantification of Ki67 positivity; neutrophil infiltration; monocyte/macrophage infiltration; and neovascularization, respectively; please see Section 4 for details.

2.3. Systemic Effects of hRabMab1

As assessed by quantitative luciferase imaging, cumulative metastatic spread (liver, lung, abdominal cavity) was not significantly reduced by gem/PTX alone; however, the addition of hRabMab1 to gem/PTX significantly reduced whole-body metastatic spread ($p = 0.0415$, vehicle vs. hRabMab1+gem/PTX, Figure 6).

In the liver, hRabMab1 significantly reduced metastatic burden as a single agent to a comparable degree to that achieved by gem/PTX (hRabMab1: $p = 0.0089$ vs. vehicle; gem/PTX: $p = 0.0008$ vs. vehicle; Supplementary Materials Figure S2). At the surgical incision site, where PaCa-44 metastases routinely engraft, the following results were obtained: vehicle, 100% penetrance (10/10); hRabMab1, 100% penetrance (8/8); gem/PTX, 44% penetrance (4/9); hRabMab1+gem/PTX, 0% penetrance (0/8; $p = 0.0378$ for gem/PTX vs. hRabMab1+gem/PTX, Chi-square test with 0.1 replacing 0). Only the combination of hRabMab1 with gem/PTX was able to eliminate neutrophil-driven leukocytosis in our model (Figure 7 and Supplementary Materials Figure S3); the levels of circulating neutrophils, as well as monocytes, were significantly lower in the hRabMab1+gem/PTX cohort when compared to the gem/PTX cohort ($p = 0.035$ and $p = 0.051$, respectively).

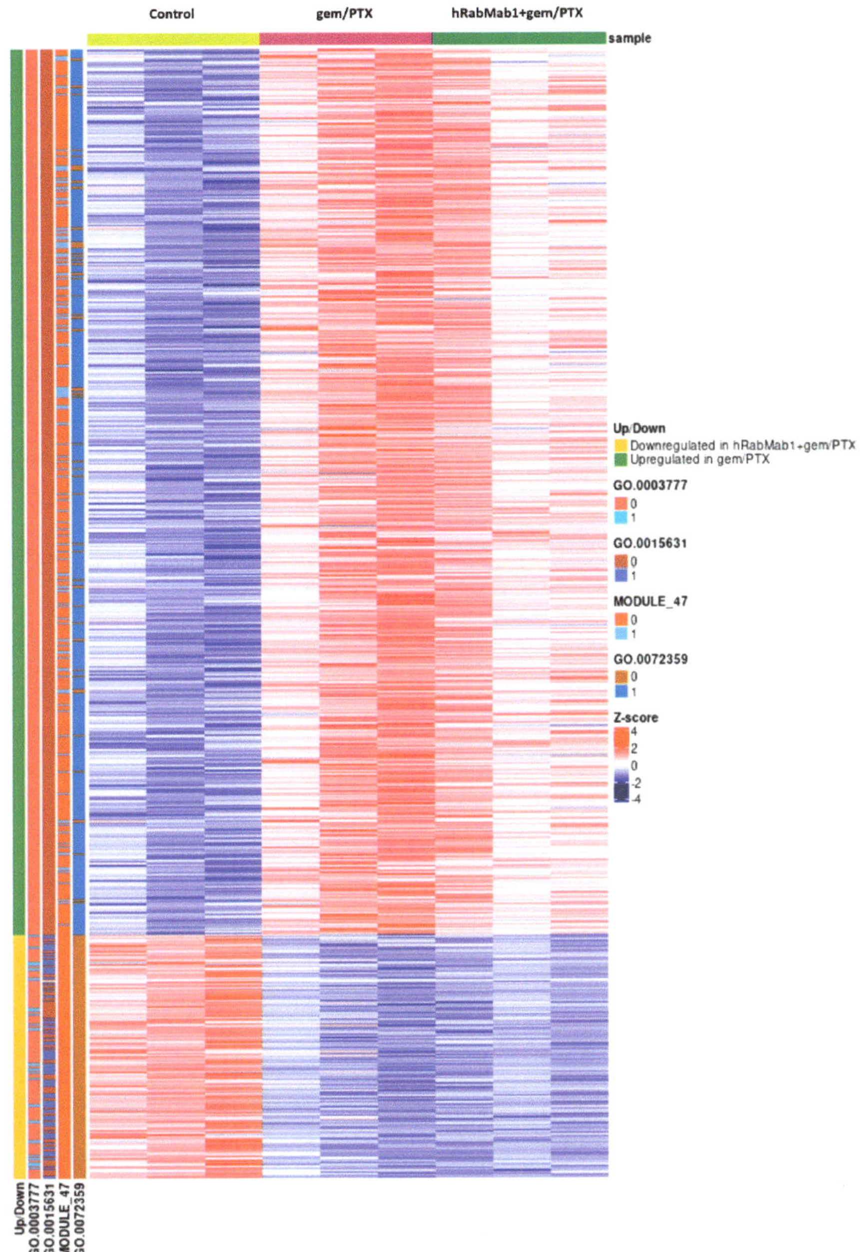

Figure 5. Heatmap of z-score normalized, variance stabilizing transformed differentially expressed gene counts. Up/Down trackbar indicates whether the gene is upregulated in gem/PTX samples, or downregulated in hRabMab1+gem/PTX samples. The trackbars for Gene Ontology terms indicate the presence (1) or absence (0) of the gene in the set. GO:0003777: microtubule motor activity; GO.0015631: tubulin binding; GO:0072359: circulatory system development; MODULE_47: ECM and collagens.

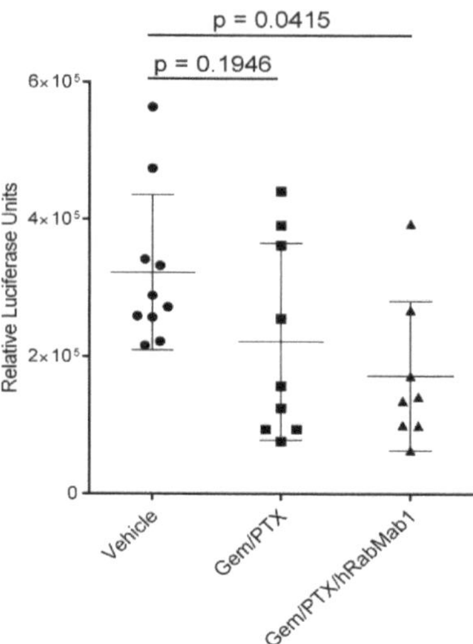

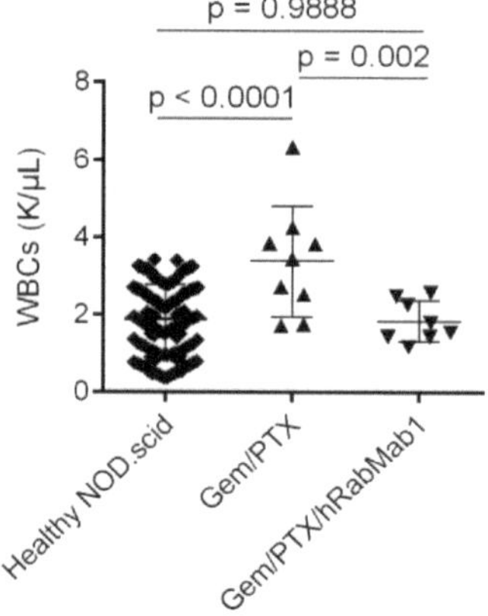

Figure 6. Cumulative metastatic spread assessed by quantitative luciferase imaging in experimental cohorts as indicated; 1-way ANOVA with Tukey's multiple comparison test was used to assess significance.

Figure 7. White blood cell (WBC) counts in NOD.scid mice: historical NOD.scid reference data and the experimental cohorts as indicated; 1-way ANOVA with Tukey's multiple comparison test was used to assess significance.

Neutrophils and monocytes recruited from the circulation promote tumor progression in PDAC [20,21]. To address the potential physiological significance of neutrophil count normalization by hRabMab1+gem/PTX in the PaCa-44 model, we performed correlation analysis between circulating neutrophil counts and tumor volumes across all 4 cohorts; a highly significant positive correlation between neutrophil counts and tumor volumes was identified (R = 0.628, p = 0.00002), mirroring findings reported in human patients [22]. No differences in body weight of mice were observed between gem/PTX and hRabMab1+gem/PTX (not shown).

3. Discussion

In this study, we evaluated the effects of hRabMab1, administered intravenously as a single agent and in combination with gem/PTX, on primary tumor growth and systemic spread of orthotopically implanted human PDAC cells. Our main findings are as follows:

(i) hRabMab1 was able to suppress liver metastases as a single agent while improving the efficacy of gem/PTX in suppressing metastases to other anatomical sites; (ii) hRabMab1 was able to penetrate primary tumor tissue when co-administered with gem/PTX, which led to the suppression of cancer cell proliferative potential, as well as leukocyte infiltration of primary tumors; (iii) the combination of hRabMab1+gem/PTX normalized WBC counts in tumor-bearing mice. These results agree with and expand on our published findings pointing to hRabMab1's potential to stem the growth of human PDAC cells in vivo [15,23]; hRabMab1's potential to increase the anti-metastatic efficacy of gem/PTX with no additional toxicity is particularly significant from a clinical perspective. The observations we describe here pose a number of new questions about the biologic function(s) of hRabMab1, e.g., what are the mechanisms underlying its ability to penetrate primary tumor tissue when co-administered with gem/PTX? Are anti-metastatic effects of hRabMab1 largely due to its ability to suppress the growth of PDAC cells already homed to distal sites, or is there also an effect on CTC intravasation, extravasation, and/or homing capacity? What causes the normalization of WBC counts in the hRabMab1+gem/PTX cohort? With regard to the tumor-tissue penetrance of hRabMab1, the most likely explanation is that the gem/PTX-elicited disruption of the fibrous capsule facilitated hRabMab1's diffusion throughout the tumor ECM. With regard to the anti-metastatic properties of hRabMab1, these effects are most likely exerted on PDAC cells at various steps of metastatic dissemination. Integrins largely mediate a CTC's capacity for motility and metastatic colonization, as well as their anchorage-independent survival, and ECM-β1 integrin interactions contribute to chemotherapy resistance of orthotopically grown PDAC tumors [24–26]. Thus, when considering hRabMab1's perceived mode of action, i.e., the diminution of asTF-induced integrin activation, our data indicate hRabMab1 likely disrupts these integrin-driven metastatic processes. We note that the suppression of liver metastases by hRabMab1 alone may be due to the highly vascularized nature of the liver, which may have facilitated hRabMab1's access to liver metastases. Further studies using additional primary tumor-driven endogenous metastatic models, as well as tail vein injections and hemi-splenic injections, will address whether hRabMab1's suppresses PDAC cell seeding in the lung and the liver, respectively. Likewise, future studies are planned to feature tumor harvesting at regular intervals so that we can better assess the longitudinal dynamics of hRabMab1's effects on primary tumor growth. Lastly, the normalization of WBC counts in the hRabMab1+gem/PTX cohort likely reflects a lowered systemic response due to a decreased cancer cell burden in these mice. It cannot be fully excluded, however, that the combination of hRabMab1+gem/PTX was more toxic than gem/PTX alone; that being said, such a scenario is unlikely given that there were no significant differences in BW between gem/PTX and hRabMab1+gem/PTX cohorts.

asTF is a soluble TF variant that has no role in normal blood clotting. flTF and asTF both play a role in cancer progression and interact with β1 integrins, yet with different consequences. In non-malignant cells, flTF keeps β1 integrins in an inactive state whereas in cancer, flTF activates $\alpha\beta$ subsets and triggers protease-activated receptor 1 activation and thrombin generation that, collectively, promote cancer progression [23]. Earlier findings from other groups that the depletion of flTF can reduce tumor growth and thrombosis in murine models led to the exploration of targeting "total TF" in human clinical trials. The most well-characterized TF-targeting therapy, an antibody-drug conjugate (ADC) tisotumab vedotin (TF epitope unspecified), has shown promise with a phase II objective response rate of 15.6%; however, nearly 50% of patients had a treatment-emergent serious adverse event and nearly 70% experienced epistaxis [27–29]. While (fl)TF is still being actively pursued clinically, we posit that asTF is the preferred TF form to target in PDAC and other solid tumors due to both a low risk of bleeding complications, as well as superior selectivity for cancer cells and tissues; further, hRabMab1 is not an ADC and is, thus, not likely to cause tissue toxicity. As mentioned in the Introduction, "total TF" was recently implicated in the development of resistance to KRAS-G12C inhibitors [10]; as such, asTF-targeting via hRabMab1 may hold future promise in tackling this phenomenon in PDAC and other cancers driven by mutant KRAS.

Targeting asTF also comprises a novel way to impede integrin-linked, cancer-promoting signaling. By disrupting asTF-integrin interactions, hRabMab1 inhibits outside-in integrin signaling without the limitations associated with direct pharmacological inhibition of integrins: the key adverse issue of direct integrin inhibitors being their paradoxical ability to induce a conformational change in the integrin dimer, leading to a high-affinity ligand-binding state [30]. Indirectly inhibiting integrin-linked outside-in signaling cascades, however, has been one of the few methodologies that have provided additional benefit to mainline gemcitabine chemotherapy in the treatment of PDAC. When combined with gemcitabine and pembrolizumab, defactinib, a small molecule inhibitor of focal adhesion kinase (FAK), one of the kinases downstream of asTF/integrins in PDAC cells [14], showed preliminary efficacy without added toxicity [31]. We note that the use of antibody inhibitors of checkpoint proteins in combination with gemcitabine, with or without PTX, has largely yielded no additional benefit in the clinic in the treatment of PDAC, highlighting the importance of defactinib in this therapeutic regimen. Indeed, most monoclonal antibodies (mAbs) have failed in the clinic for the treatment of PDAC, largely due to the robust nature of PDAC tumor capsules, poor tumor vascularity, and the choice of therapeutic targets. Excitingly, we have recently seen that mAbs do have a place in the clinic for the treatment of PDAC, as the combination of pamrevlumab, a mAb targeting connective tissue growth factor (CTGF), with gem/PTX for the neoadjuvant treatment of locally advanced PDAC enabled study participants to advance to successful surgical resection in 8 of 24 cases (33%) compared to 1 of 13 (8%) patients who received gem/PTX alone [32]. Interestingly, CTGF elicits cellular mitogenic responses through an $\alpha v\beta 3$-dependent mechanism in microvascular endothelial cells in vitro [33]. Thus, targeting integrin-linked signaling appears to hold promise as a new way to treat PDAC; we note that inhibiting entities upstream of integrin-linked signaling, such as asTF, is conceptually more robust compared to downstream targeting. We show here that when combined with gem/PTX, hRabMab1 both effectively penetrate tumor tissue and suppresses metastases.

In conclusion, we here show for the first time that asTF-inhibitory humanized antibody hRabMab1 holds promise to enhance the anti-metastatic effects of chemotherapy in PDAC. The main limitation of our study comprises the use of a single PDAC cell line PaCa-44; we note, however, that this cell line yields primary tumors that spontaneously metastasize to relevant anatomical sites, which augments the likely biological significance of the obtained results. Moreover, we previously showed the cancer cell-suppressing effect of inhibiting asTF in models that used other asTF-proficient PDAC cell lines. Future studies will explore hRabMab1's ability to suppress primary tumor growth and experimental metastases in models featuring additional asTF-proficient PDAC cell lines and patient-derived xenografts.

4. Materials and Methods

4.1. PDAC Cell Lines and Culture Conditions; Western Blotting

Human PDAC cell lines Pt45.P1 (a kind gift of Prof. Holger Kalthoff), PaCa-44 (a kind gift of Prof. Stephan Haas), and HS766.T (ATCC) were cultured in DMEM supplemented with 10% fetal calf serum and antibiotics. Lysates were prepared with RIPA buffer (50 mM Tris HCl, pH 8.0; 150 mM NaCl, 0.5% *w/v* sodium deoxycholate; 0.1% SDS; 1% NP-40) containing 5mM EDTA along with Halt protease and phosphatase inhibitor cocktail (Thermo Fisher Scientific, Waltham, MA, USA, ref. 1861281) and loaded into 10% TGX gels (BioRad Hercules, CA, USA). Protein was transferred onto PVDF membranes and probed with antibodies specific for human $\alpha 6$ integrin (Cell Signaling Technology, Danvers, MA, USA, #3750, 1:1000), $\beta 1$ integrin (Cell Signaling Technology, Danvers, MA, USA, #34971, 1:1000), flTF (clone TF9-10H10, Invitrogen/Thermo Fisher Scientific, Waltham, MA, USA, 1:500), asTF (custom rabbit polyclonal, ref. 14, 2 μg/mL), and beta-actin (Cell Signaling Technology, Danvers, MA, USA, #3700, 1:1000).

4.2. In Vivo Studies

Orthotopic tumor implantation: On the day of surgery, PaCa-44 cells were detached from tissue culture plates with 0.25% trypsin. Trypsin was neutralized with DMEM containing 10% FBS. Cells were washed 2× with DMEM before preparing a final suspension of 2.5×10^7 cells/mL. Then, 20 μL of PaCa-44 cell suspension (containing 5×10^5 cells) was injected into the pancreata proximal to the duodenum of NOD.scid mice (The Jackson Laboratory, Bar Harbor, ME, USA, 001303). Ten days post-implantation, mice were randomized into four cohorts: vehicle; hRabMab1 (IV at 18 mg/kg); gem/PTX (50 mg/kg IP and 3 mg/kg IV, respectively); and hRabMab1+gem/PTX. In the vehicle and hRabMab1 cohorts, mice were sacrificed when tumor volume reached 1500 mm^3; in the gem/PTX and hRabMab1+gem/PTX cohorts, mice were sacrificed on day 40 post-implantation. In vivo imaging was carried out weekly using the IVIS Spectrum System (Xenogen Corporation, Alameda, CA, USA). Blood counts were determined at sacrifice using a HEMAVET automated hematology analyzer (Drew Scientific, Miami Lakes, FL, USA); historical reference values for white blood cell count data for healthy NOD.scid mice were retrieved from Charles River hematology records [34]. Tumor volumes were derived using the formula $V = (W(2) \times L)/2$ for caliper measurements. A portion of each tumor material was flash frozen, as well as fixed in 10% formalin and embedded in paraffin for RNA-seq and IHC analyses, respectively.

4.3. Immunohistochemistry

Formalin-fixed tissues were embedded in paraffin and sectioned into 5 μm sections. Sections were deparaffinized and rehydrated into PBS. Antigen retrieval was carried out whenever indicated by the antibody manufacturer; native peroxidase activity was squelched using 0.4% hydrogen peroxide. Blocking was carried out using 5% bovine serum albumin in PBS and sections were incubated with the following antibodies at the manufacturer-indicated dilutions: CD31 (Novus Biologicals, Centennial, CO, USA, AF3628, 10 μg/mL), CD206 (Cell Signaling Technology, Danvers, MA, USA, #24595, 1:200), Ki67 (Novus Biologicals, Centennial, CO, USA, NB110-89717, 1:250), Myeloperoxidase (Abcam Waltham, MA, USA, AB 300650, 1:1000), and goat anti-human IgG biotinylated antibody (Vector Laboratories, Newark, CA, USA, BA-3000-1.5, 1:500). Species-specific, HRP-conjugated anti-antibody polymers and DAB+ reagent (both—Cell Signaling) were used to visualize unlabeled primary antibody binding and HRP-streptavidin reagent (SA-5704, Vector Laboratories) was used to visualize anti-human IgG antibody; all sections were counterstained with hematoxylin. Representative images (n = 6 per tissue specimen) were captured using an Olympus BX51 (Center Valley, PA, USA) equipped with Olympus DP72 digital camera and used for statistical analyses. Staining intensity and/or positive staining events were analyzed using ImageJ.

4.4. RNA-seq

Using an RNeasy kit (Qiagen, Germantown, MD, USA), total RNA was isolated from frozen tissue specimens representing median tumor volumes from each experimental cohort. Directional polyA RNA-seq was performed by the Genomics, Epigenomics and Sequencing Core at the University of Cincinnati, using established protocols. The quality of total RNA was QC analyzed by Bioanalyzer (Agilent, Santa Clara, CA, USA). To enrich polyA RNA for library preparation, NEBNext Poly(A) mRNA Magnetic Isolation Module (New England BioLabs, Ipswich, MA, USA) was used with 1 μg good quality total RNA as input. Next, NEBNext Ultra II Directional RNA Library Prep kit (New England BioLabs) was used for library preparation under PCR (cycle number: 8). After library QC and Qubit quantification (ThermoFisher, Waltham, MA, USA), the normalized libraries were sequenced using NextSeq 2000 Sequencer (Illumina, San Diego, CA, USA) under the setting of PE 2x61 bp to generate an average of 42.3 M reads. Once the sequencing was completed, FASTQ files for downstream data analysis were generated and transferred/shared via BaseSpace Sequence Hub (Illumina). A quality control check on the FASTQ files was performed using FASTQC (https://www.bioinformatics.babraham.ac.uk/projects/fastqc/, accessed 21 December 2023) and MultiQC [35] to verify data quality. The FASTQ files were processed with STAR (v. 2.7.7a) [36] with the Gencode GRCh38 as an index to determine alignment. Salmon [37] quant (version 1.9.0), with default parameters and the RefSeq CRCm39 index, was used to obtain counts. Differential gene expression analysis was performed with unnormalized counts and default parameters using the DESeq2 package (Ver 1.34.0) between the following groups of samples: vehicle vs. hRabMab1, vehicle vs. gem/PTX, vehicle vs. hRabMab1+gem/PTX [38,39]. Wald test was used to test the null hypothesis of no differential expression across the two sample groups along with Benjamini–Hochberg correction to adjust for multiple testing. Functional gene set enrichment was performed with lists of genes that had significant differential expression (adjusted p-value < 0.05) using the ToppFun application in the ToppGene suite (https://toppgene.cchmc.org/) [40].

4.5. Statistics

Continuous variables were summarized as means $\pm$ SD or median and inter-quartile ranges (IQR). Two-tailed t test and 1-way ANOVA with Tukey's multiple comparison test were used to test the difference between two or more cohorts, respectively (GraphPad Prism v.6.0); p values ≤ 0.05 were deemed significant. Categorical variables were summarized as counts and percentages. The Chi-square test was used to test the association.

Supplementary Materials: The following supporting information can be downloaded at: https://www.mdpi.com/article/10.3390/ijms25052580/s1.

Author Contributions: Conceptualization, C.S.L., D.P.S.S. and V.Y.B.; methodology, C.S.L., A.C., A.H., J.J.Y., X.Z., C.X. and V.Y.B.; validation, C.S.L. and V.Y.B.; formal analysis, C.S.L., A.C., A.H., X.Z., C.X. and V.Y.B.; investigation, in vitro and/or in vivo experiments: C.S.L., C.B., S.A. and V.Y.B.; resources, J.J.Y. and V.Y.B.; data curation, C.S.L., A.C., A.H. and X.Z.; writing—original draft preparation, C.S.L. and V.Y.B.; writing—review and editing, C.S.L., A.C., A.H., X.Z., C.X. and V.Y.B.; supervision, J.J.Y. and V.Y.B.; project administration, V.Y.B.; funding acquisition, D.P.S.S. and V.Y.B. All authors have read and agreed to the published version of the manuscript.

Funding: This research was funded by the 2021 Catalyst Award, Dr. Ralph and Marian Falk Medical Research Trust (no grant number assigned), Principal Investigators, V.Y.B. and D.P.S.S.; and 2023 Pilot Project Award Program, University of Cincinnati Cancer Center (no grant number assigned), Principal Investigator, V.Y.B. C.B. is a recipient of 2023 Fellowship in the Medical Student Summer Research Program (MSSRP), grant number 5T35DK060444-20.

Studies Involving Animals: The animal study protocol was approved by University of Cincinnati's Institutional Animal Care and Use Committee (IACUC); protocol # 21-09-16-01, most recent approval date: 29 September 2023.

Data Availability Statement: The datasets presented in this study can be found in the online repository; accession number: GSE252286.

Conflicts of Interest: The authors declare no conflicts of interest.

References

1. Hisada, Y.; Mackman, N. Cancer-associated pathways and biomarkers of venous thrombosis. *Blood* **2017**, *130*, 1499–1506. [CrossRef]
2. Wang, J.-G.; Geddings, J.E.; Aleman, M.M.; Cardenas, J.C.; Chantrathammachart, P.; Williams, J.C.; Kirchhofer, D.; Bogdanov, V.Y.; Bach, R.R.; Rak, J.; et al. Tumor-derived tissue factor activates coagulation and enhances thrombosis in a mouse xenograft model of human pancreatic cancer. *Blood* **2012**, *119*, 5543–5552. [CrossRef]
3. Van Den Berg, Y.W.; Osanto, S.; Reitsma, P.H.; Versteeg, H.H. The relationship between tissue factor and cancer progression: Insights from bench and bedside. *Blood* **2012**, *119*, 924–932. [CrossRef]
4. Sun, L.; Liu, Y.; Lin, S.; Shang, J.; Liu, J.; Li, J.; Yuan, S.; Zhang, L. Early growth response gene-1 and hypoxia-inducible factor-1α affect tumor metastasis via regulation of tissue factor. *Acta Oncol.* **2013**, *52*, 842–851. [CrossRef]
5. Milsom, C.; Yu, J.; May, L.; Magnus, N.; Rak, J. Diverse roles of tissue factor-expressing cell subsets in tumor progression. *Semin. Thromb. Hemost.* **2008**, *34*, 170–181. [CrossRef]

6. Kakkar, A.K.; Lemoine, N.R.; Scully, M.F.; Tebbutt, S.; Williamson, R.C.N. Tissue factor expression correlates with histological grade in human pancreatic cancer. *Br. J. Surg.* **1995**, *82*, 1101–1104. [CrossRef]
7. Kakkar, A.K.; Chinswangwatanakul, V.; Lemoine, N.R.; Tebbutt, S.; Williamson, R.C.N. Role of tissue factor expression on tumour cell invasion and growth of experimental pancreatic adenocarcinoma. *Br. J. Surg.* **1999**, *86*, 890–894. [CrossRef]
8. Nitori, N.; Ino, Y.; Nakanishi, Y.; Yamada, T.; Honda, K.; Yanagihara, K.; Kosuge, T.; Kanai, Y.; Kitajima, M.; Hirohashi, S. Prognostic Significance of Tissue Factor in Pancreatic Ductal Adenocarcinoma. *Clin. Cancer Res.* **2005**, *11*, 2531–2539. [CrossRef]
9. Yang, Y.; Stang, A.; Schweickert, P.G.; Lanman, N.A.; Paul, E.N.; Monia, B.P.; Revenko, A.S.; Palumbo, J.S.; Mullins, E.S.; Flick, M.J. Thrombin Signaling Promotes Pancreatic Adenocarcinoma through PAR-1-Dependent Immune Evasion. *Cancer Res.* **2019**, *79*, 3417–3430. [CrossRef] [PubMed]
10. Zhang, Y.; Liu, L.; Pei, J.; Ren, Z.; Deng, Y.; Yu, K. Tissue factor overexpression promotes resistance to KRAS-G12C inhibition in non-small cell lung cancer. *Oncogene* **2024**. [CrossRef]
11. Bogdanov, V.Y.; Balasubramanian, V.; Hathcock, J.; Vele, O.; Lieb, M.; Nemerson, Y. Alternatively spliced human tissue factor: A circulating, soluble, thrombogenic protein. *Nat. Med.* **2003**, *9*, 458–462. [CrossRef]
12. Berg, Y.W.V.D.; Hengel, L.G.V.D.; Myers, H.R.; Ayachi, O.; Jordanova, E.; Ruf, W.; Spek, C.A.; Reitsma, P.H.; Bogdanov, V.Y.; Versteeg, H.H. Alternatively spliced tissue factor induces angiogenesis through integrin ligation. *Proc. Natl. Acad. Sci. USA* **2009**, *106*, 19497–19502. [CrossRef]
13. Kocatürk, B.; Berg, Y.W.V.D.; Tieken, C.; Mieog, J.S.D.; de Kruijf, E.M.; Engels, C.C.; van der Ent, M.A.; Kuppen, P.J.; Van de Velde, C.J.; Ruf, W.; et al. Alternatively spliced tissue factor promotes breast cancer growth in a β1 integrin-dependent manner. *Proc. Natl. Acad. Sci. USA* **2013**, *110*, 11517–11522. [CrossRef]
14. Unruh, D.; Turner, K.; Srinivasan, R.; Kocatürk, B.; Qi, X.; Chu, Z.; Aronow, B.J.; Plas, D.R.; Gallo, C.A.; Kalthoff, H.; et al. Alternatively spliced tissue factor contributes to tumor spread and activation of coagulation in pancreatic ductal adenocarcinoma. *Int. J. Cancer* **2014**, *134*, 9–20. [CrossRef]
15. Lewis, C.S.; Karve, A.; Matiash, K.; Stone, T.; Li, J.; Wang, J.K.; Versteeg, H.H.; Aronow, B.J.; Ahmad, S.A.; Desai, P.B.; et al. A First-In-Class, Humanized Antibody Targeting Alternatively Spliced Tissue Factor: Preclinical Evaluation in an Orthotopic Model of Pancreatic Ductal Adenocarcinoma. *Front. Oncol.* **2021**, *11*, 691685. [CrossRef]
16. Moore, P.S.; Sipos, B.; Orlandini, S.; Sorio, C.; Real, F.X.; Lemoine, N.R.; Gress, T.; Bassi, C.; Klöppel, G.; Kalthoff, H.; et al. Genetic profile of 22 pancreatic carcinoma cell lines. *Virchows Arch.* **2001**, *439*, 798–802. [CrossRef]
17. Hamza, B.; Miller, A.B.; Meier, L.; Stockslager, M.; Ng, S.R.; King, E.M.; Lin, L.; DeGouveia, K.L.; Mulugeta, N.; Calistri, N.L.; et al. Measuring kinetics and metastatic propensity of CTCs by blood exchange between mice. *Nat. Commun.* **2021**, *12*, 5680. [CrossRef]
18. Hebert, J.D.; Neal, J.W.; Winslow, M.M. Dissecting metastasis using preclinical models and methods. *Nat. Rev. Cancer* **2023**, *23*, 391–407. [CrossRef]
19. Wolfe, A.R.; Robb, R.; Hegazi, A.; Abushahin, L.; Yang, L.; Shyu, D.L.; Trevino, J.G.; Cruz-Monserrate, Z.; Jacob, J.R.; Williams, T.M.; et al. Altered gemcitabine and nab-paclitaxel scheduling improves therapeutic efficacy compared with standard concurrent treatment in preclinical models of pancreatic cancer. *Clin. Cancer Res.* **2021**, *27*, 554–565. [CrossRef]
20. Bellomo, G.; Rainer, C.; Quaranta, V.; Astuti, Y.; Raymant, M.; Boyd, E.; Stafferton, R.; Campbell, F.; Ghaneh, P.; Schmid, M.C.; et al. Chemotherapy-induced infiltration of neutrophils promotes pancreatic cancer metastasis via Gas6/AXL signalling axis. *Gut* **2022**, *71*, 2284–2299. [CrossRef]
21. Nywening, T.M.; Belt, B.A.; Cullinan, D.R.; Panni, R.Z.; Han, B.J.; Sanford, D.E.; Jacobs, R.C.; Ye, J.; Patel, A.A.; Gillanders, W.E.; et al. Targeting both tumour-associated CXCR2+ neutrophils and CCR2+ macrophages disrupts myeloid recruitment and improves chemotherapeutic responses in pancreatic ductal adenocarcinoma. *Gut* **2018**, *67*, 1112–1123. [CrossRef]
22. Holub, K.; Conill, C. Unveiling the mechanisms of immune evasion in pancreatic cancer: May it be a systemic inflammation responsible for dismal survival? *Clin. Transl. Oncol.* **2020**, *22*, 81–90. [CrossRef]
23. Unruh, D.; Ünlü, B.; Lewis, C.S.; Qi, X.; Chu, Z.; Sturm, R.; Keil, R.; Ahmad, S.A.; Sovershaev, T.; Adam, M.; et al. Antibody-based targeting of alternatively spliced tissue factor: A new approach to impede the primary growth and spread of pancreatic ductal adenocarcinoma. *Oncotarget* **2016**, *7*, 25264–25275. [CrossRef]
24. Hamidi, H.; Ivaska, J. Every step of the way: Integrins in cancer progression and metastasis. *Nat. Rev. Cancer* **2018**, *18*, 533–548. [CrossRef]
25. Brannon, A.; Drouillard, D.; Steele, N.; Schoettle, S.; Abel, E.V.; Crawford, H.C.; Pasca di Magliano, M. Beta 1 integrin signaling mediates pancreatic ductal adenocarcinoma resistance to MEK inhibition. *Sci. Rep.* **2020**, *10*, 11133. [CrossRef]
26. Görte, J.; Danen, E.; Cordes, N. Therapy-Naive and Radioresistant 3-Dimensional Pancreatic Cancer Cell Cultures Are Effectively Radiosensitized by β1 Integrin Targeting. *Int. J. Radiat. Oncol. Biol. Phys.* **2022**, *112*, 487–498. [CrossRef]
27. De Bono, J.S.; Concin, N.; Hong, D.S.; Thistlethwaite, F.C.; Machiels, J.P.; Arkenau, H.T.; Plummer, R.; Hugh Jones, R.; Nielsen, D.; Lassen, U.; et al. Tisotumab vedotin in patients with advanced or metastatic solid tumours (InnovaTV 201): A first-in-human, multicentre, phase 1–2 trial. *Lancet Oncol.* **2019**, *20*, 383–393. [CrossRef] [PubMed]
28. Coleman, R.L.; Lorusso, D.; Gennigens, C.; González-Martín, A.; Randall, L.; Cibula, D.; Lund, B.; Pignata, S.; Harris, J.R.; Bhatia, S.; et al. Efficacy and safety of tisotumab vedotin in previously treated recurrent or metastatic cervical cancer (innovaTV 204/GOG-3023/ENGOT-cx6): A multicentre, open-label, single-arm, phase 2 study. *Lancet Oncol.* **2021**, *22*, 609–619. [CrossRef]

29. Yonemori, K.; Kuboki, Y.; Hasegawa, K.; Iwata, T.; Kato, H.; Takehara, K.; Hirashima, Y.; Kato, H.; Passey, C.; Buchbjerg, J.K.; et al. Tisotumab vedotin in Japanese patients with recurrent/metastatic cervical cancer: Results from the innovaTV 206 study. *Cancer Sci.* **2022**, *113*, 2788–2797. [CrossRef] [PubMed]
30. Li, J.; Fukase, Y.; Shang, Y.; Zou, W.; Muñoz-Félix, J.M.; Buitrago, L.; Zhang, Y.; Hara, R.; Okamoto, R.; Coller, B.S.; et al. Novel Pure αvβ3 Integrin Antagonists That Do Not Induce Receptor Extension, Prime the Receptor, or Enhance Angiogenesis at Low Concentrations. *ACS Pharmacol. Transl. Sci.* **2019**, *2*, 387–401. [CrossRef]
31. Wang-Gillam, A.; Lim, K.H.; McWilliams, R.; Suresh, R.; Lockhart, A.C.; Brown, A.; Breden, M.; Belle, J.I.; Herndon, J.; DeNardo, D.G.; et al. Defactinib, Pembrolizumab, and Gemcitabine in Patients with Advanced Treatment Refractory Pancreatic Cancer: A Phase I Dose Escalation and Expansion Study. *Clin. Cancer Res.* **2022**, *28*, 5254–5262. [CrossRef]
32. Picozzi, V.; Alseidi, A.; Winter, J.; Pishvaian, M.; Mody, K.; Glaspy, J.; Larson, T.; Matrana, M.; Carney, M.; Porter, S.; et al. Gemcitabine/nab-paclitaxel with pamrevlumab: A novel drug combination and trial design for the treatment of locally advanced pancreatic cancer. *ESMO Open* **2020**, *5*, e000668. [CrossRef]
33. Babic, A.M.; Chen, C.-C.; Lau, L.F. Fisp12/mouse connective tissue growth factor mediates endothelial cell adhesion and migration through integrin alphavbeta3, promotes endothelial cell survival, and induces angiogenesis in vivo. *Mol. Cell Biol.* **1999**, *19*, 2958–2966. [CrossRef] [PubMed]
34. Charles River Laboraties: Biochemistry and Hematology for NOD SCID Mouse Colonies in North American for January 2011–December 2011. Available online: https://www.criver.com/products-services/find-model/nod-scid-mouse?region=36112011 (accessed on 6 June 2023).
35. Ewels, P.; Magnusson, M.; Lundin, S.; Käller, M. MultiQC: Summarize analysis results for multiple tools and samples in a single report. *Bioinformatics* **2016**, *32*, 3047–3048. [CrossRef]
36. Dobin, A.; Davis, C.A.; Schlesinger, F.; Drenkow, J.; Zaleski, C.; Jha, S.; Batut, P.; Chaisson, M.; Gingeras, T.R. STAR: Ultrafast universal RNA-seq aligner. *Bioinformatics* **2013**, *29*, 15–21. [CrossRef] [PubMed]
37. Patro, R.; Duggal, G.; Love, M.I.; Irizarry, R.A.; Kingsford, C. Salmon provides fast and bias-aware quantification of transcript expression. *Nat. Methods* **2017**, *14*, 417–419. [CrossRef]
38. Love, M.I.; Huber, W.; Anders, S. Moderated estimation of fold change and dispersion for RNA-seq data with DESeq2. *Genome Biol.* **2014**, *15*, 550. [CrossRef]
39. Zhu, A.; Ibrahim, J.G.; Love, M.I. Heavy-tailed prior distributions for sequence count data: Removing the noise and preserving large differences. *Bioinformatics* **2019**, *35*, 2084–2092. [CrossRef]
40. Chen, J.; Bardes, E.E.; Aronow, B.J.; Jegga, A.G. ToppGene Suite for gene list enrichment analysis and candidate gene prioritization. *Nucleic Acids Res.* **2009**, *37*, W305–W311. [CrossRef] [PubMed]

Disclaimer/Publisher's Note: The statements, opinions and data contained in all publications are solely those of the individual author(s) and contributor(s) and not of MDPI and/or the editor(s). MDPI and/or the editor(s) disclaim responsibility for any injury to people or property resulting from any ideas, methods, instructions or products referred to in the content.

International Journal of
Molecular Sciences

MDPI

Article

Synergistic Anticancer Activity of Plumbagin and Xanthohumol Combination on Pancreatic Cancer Models

Ranjith Palanisamy [1], Nimnaka Indrajith Kahingalage [1], David Archibald [2], Ilaria Casari [1] and Marco Falasca [1,3,*]

[1] Metabolic Signalling Group, Curtin Medical School, Curtin Health Innovation Research Institute, Curtin University, Perth 6102, Australia; ranjith.palanisamy@telethonkids.org.au (R.P.); indrajithnimnaka7@gmail.com (N.I.K.); ilacasari@gmail.com (I.C.)

[2] Backreef Oil Pty Ltd., Perth 6015, Australia; david.archibald@westnet.com.au

[3] Department of Medicine and Surgery, University of Parma, Via Volturno 39, 43125 Parma, Italy

[*] Correspondence: marco.falasca@unipr.it

Abstract: Among diverse cancers, pancreatic cancer is one of the most aggressive types due to inadequate diagnostic options and treatments available. Therefore, there is a necessity to use combination chemotherapy options to overcome the chemoresistance of pancreatic cancer cells. Plumbagin and xanthohumol, natural compounds isolated from the Plumbaginaceae family and *Humulus lupulus*, respectively, have been used to treat various cancers. In this study, we investigated the anticancer effects of a combination of plumbagin and xanthohumol on pancreatic cancer models, as well as the underlying mechanism. We have screened in vitro numerous plant-derived extracts and compounds and tested in vivo the most effective combination, plumbagin and xanthohumol, using a transgenic model of pancreatic cancer KPC (KrasLSL.G12D/+; p53R172H/+; PdxCretg/+). A significant synergistic anticancer activity of plumbagin and xanthohumol combinations on different pancreatic cancer cell lines was found. The combination treatment of plumbagin and xanthohumol influences the levels of B-cell lymphoma (BCL2), which are known to be associated with apoptosis in both cell lysates and tissues. More importantly, the survival of a transgenic mouse model of pancreatic cancer KPC treated with a combination of plumbagin and xanthohumol was significantly increased, and the effect on BCL2 levels has been confirmed. These results provide a foundation for a potential new treatment for pancreatic cancer based on plumbagin and xanthohumol combinations.

Keywords: pancreatic cancer; plumbagin; xanthohumol; drug combinations; phytochemicals

Citation: Palanisamy, R.; Indrajith Kahingalage, N.; Archibald, D.; Casari, I.; Falasca, M. Synergistic Anticancer Activity of Plumbagin and Xanthohumol Combination on Pancreatic Cancer Models. *Int. J. Mol. Sci.* **2024**, *25*, 2340. https://doi.org/10.3390/ijms25042340

Academic Editors: Claudio Luchini and Donatella Delle Cave

Received: 23 January 2024
Revised: 12 February 2024
Accepted: 12 February 2024
Published: 16 February 2024

Copyright: © 2024 by the authors. Licensee MDPI, Basel, Switzerland. This article is an open access article distributed under the terms and conditions of the Creative Commons Attribution (CC BY) license (https://creativecommons.org/licenses/by/4.0/).

1. Introduction

Pancreatic cancer (PC) is one of the most hostile and intractable types of cancer in the world with an incidence-to-mortality ratio of close to one. It has an extremely poor prognosis resulting in a five-year survival rate of approximately 12% [1,2]. Pancreatic cancer is estimated to become the second most common cause of cancer-related death in 2040 [3]. Surgery is the only possible treatment for primary-stage or resectable pancreatic tumours [4] and, even in this case, cancer relapses and metastasis development are a major concern [5]. Presently, chemotherapy is the only feasible option available to inoperable patients, and gemcitabine combined with albumin-bound paclitaxel is the standard first-line treatment [6]. For cancer metastatic patients who can tolerate it, FOLFIRINOX, a combination agent, is the treatment of preference [7]. However, the lack of effectiveness due to the high chemoresistance of PC, high toxicity, and the wide variety of side effects of chemotherapy demand a search for new lines of treatment [4,8]. The benefits of a diet rich in fruit and vegetables and its role in preventing numerous types of cancer have become common knowledge in the last few decades [9]. Consequently, many researchers are currently focusing on natural plant extracts and their isolated compounds as a treatment for pancreatic and other cancer types [10]. Dietary-derived anti-cancer agents are divided into blocking agents and suppressing agents, depending on their effect as chemopreventive agents throughout the various stages of the carcinogenic process [11,12]. Examples of phytochemicals contained in food and vegetables that battle cancer include cancer-blocking agents such as sulforaphane, the natural compound found in broccoli, and suppressing agents such as 6–gingerol, found in ginger, and capsaicin, found in the chilli pepper [13,14]. Some natural plant extracts have been identified as effective anticancer agents and have reached the clinical trial stage [15,16]. Polyphenols, and in particular flavonoids, are biologically active molecules present in many food plants. As well as having antioxidant activity, they are known to down-regulate the production of antiapoptotic proteins in the apoptotic cascade of the caspases [17]. The flavonoid xanthohumol (XH) has been found to inhibit cell proliferation and to sensitise cells to chemotherapy in in vitro studies on colorectal cancer cells [18,19]. Interestingly, XH has been found to have antitumour activity in different cancer cell lines including pancreatic cancer cell lines [20–22]. Similarly, plumbagin (PL), a plant-derived naphthoquinone, has been shown to have anticancer activity in vitro and in vivo [22,23].

The objectives of this study were to test the potential anticancer activity of different plant extracts and plant-derived bioactive compounds in pancreatic cancer. The most promising compounds were selected and tested in combination on a panel of different pancreatic human cell lines. Studies on the possible mechanisms of action of these compounds were also undertaken. As protein signalling is a mechanism in the progression of cancer, various protein signalling pathways were examined to determine which protein pathways were affected by the extracts or compounds. Two plant-based agents, PL and XH, were selected after the initial screening thanks to their enhanced action on cancer cell growth and proliferation. In the present study, we present data showing the anticancer activity of PL and XH combinations used as a treatment in vitro and substantiated in a transgenic mouse model of pancreatic cancer.

2. Results

2.1. In Vitro Screening

A variety of plant extracts and compounds were tested on a panel of human pancreatic (HPAF-II, AsPC-1) and mouse pancreatic (mT4-2D and KPC) cancer cell lines. The IC_{50} of each different agent was calculated after dose-response experiments were performed (Table 1). The most interesting compound studied was PL, which showed IC_{50} values on human pancreatic cancer cell lines HPAF-II and AsPC-1 of 1.33 and 0.98 µg/mL, respectively (Table 1). PL was even more effective when tested on human AsPC-1 tumourspheres, showing increased activity with a lower IC_{50} compared to parental AsPC-1 cancer cells of 0.53 µg/mL. When tested on two transgenic pancreatic mouse-derived cancer cell lines, mT4-2D and KPC, cell lines derived from KPC organoid cultures and KPC mice, respectively, IC_{50} values of 0.4 and 0.58 µg/mL, respectively, were observed (Table 1). Subsequently, PL was tested on the human ovarian cancer cell line A2780, the human prostate cancer cell line PC3, and the breast adenocarcinoma syngeneic cell line 4T1, obtaining IC_{50} values of 0.56, 0.93, and 0.74 µg/mL, respectively (Table 1). In addition, IC_{50} of XH on HPAF-II and AsPC-1 were recorded as 3.89 µg/mL and 6.8 µg/mL, respectively. An IC_{50} of 9.15 µg/mL was recorded treating mT4-2D cells with XH.

Other cancer cell lines, A2780, PC3, and 4T1, expressed IC_{50} values of 0.3478 µg/mL, 4.968 µg/mL, and 1.26 µg/mL, respectively, after treatments with XH. As observed with PL, when XH was tested on Human AsPC-1 tumour spheres, it showed an increased activity with a lower IC_{50} compared to parental AsPC-1 cancer cells of 1.3 µg/mL. Amongst the extracts, ginger and hops were demonstrated to be the most active. Sulforaphane, 6-shogaol, and 6-gingerol showed an interesting effect, with sulforaphane being by far the most effective agent. On the contrary, broccoli seed and broccoli seed combined with chilli extracts did show very little activity on all cell lines tested, with the IC_{50} values of these extracts being quite high. Peppermint and marjoram demonstrated to have modest activity with an IC_{50} above 40, and pepper performed slightly better (IC_{50} above 20 µg/mL). The active compound of ginger, zingerone, has only mild cytotoxicity on ASPC-1 cells (IC_{50}

37.4 μg/mL); piperine and capsaicin were fairly active on those cells (IC_{50} 13.2 μg/mL and 15.3 μg/mL, respectively), but capsaicin was more active on HPAFII cells (IC_{50} 6.1 μg/mL).

Table 1. IC_{50} values for each extract/compound determined with the cell viability assay after 72 h incubation in different cancer cell line (tumour source in brackets). IC_{50} was determined by a dose-response curve generated from at least five different inhibitor concentrations after three repeats using the GraphPad Prism software.

Extract/ Compound (Values in μg/mL)	Pancreas HPAF-II (H)	Pancreas ASPC-1 (H)	Tumour Type Ovary A2780 (H)	Prostate PC3 (H)	Pancreas mt4-2D (M)	Breast 4T1 (M)
Broccoli seed	184.2	237.1	228.5			
Broccoli seed and Chilli	207.3	245.7	250			592
Ginger	6.9	14.3	14.73	11.66		5.14
Pepper	22.42		28.33			
Peppermint		45.92				
Marjoram		41.22				
Hop extract	9.193	6.3	2.68	9.783		14.26
6-shogaol	4.22	4.94				
Zingerone		37.4				
Piperine		13.2				
Capsaicin	6.1	15.3				
Sulforaphane	1.46	1.32				
Xanthohumol	3.89	6.8	0.3478	4.968	9.15	1.26
6-gingerol	8.5	10.5				
Plumbagin	1.33	0.98	0.56	0.93	0.4	0.74

μg/mL = microgram per millilitre; Tumour source: H—human, M—mouse.

2.2. In Vitro Combination Studies

After having screened the extracts and compounds as single agents, the most effective ones were selected and tested again in various combinations to assess their anticancer potential and possible synergistic effects. The MTT assay was used to examine the potential synergistic effect of extract and compound combinations which were tested on HPAF-II, Mt4-2D, and KPC pancreatic cancer cell lines. The combination of PL/XH was found to be the most active. According to the Chou–Talalay method [24], a Combination Index (CI) value of less than 1 can be considered as synergism, while CI = 1 represents an additive effect and CI > 1 is antagonistic. Interestingly, we observed synergistic effects (CI < 1) in every concentration of XH combined with a concentration close to IC_{50} of PL on Mt4-2D cells. Effective synergism was observed when XH concentrations were combined with a concentration close to IC_{25} of PL on HPAF-II, Mt4-2D, and KPC cells (Figure 1). All of these results were consistent with the MTT assay reading obtained after combination treatments (Figure 1). The CompuSyn analysis gave a Combination Index (CI) consistent with MTT assay readings, showing a synergistic effect at low concentrations on HPAF-II (PL 2.5 and 5 μM, XH 0.25, 0.5, and 1 μg/mL) (Figure 1). Additionally, a synergistic effect was observed when the Mt4-2D cell line was treated with low concentrations of both PL and XH (PL 1 and 2.5 μM, XH 0.25, 0.5, 1, 2.5, and 5 μg/mL) (Figure 1). On the KPC cell line, the synergistic effect was observed using concentrations of 1 and 1.5 μM for PL and a higher concentration of 5, 10, and 20 μg/mL for XH (Figure 1). Hence, it can be clearly seen that PL and XH combination treatments are more effective than individual treatments in our in vitro studies.

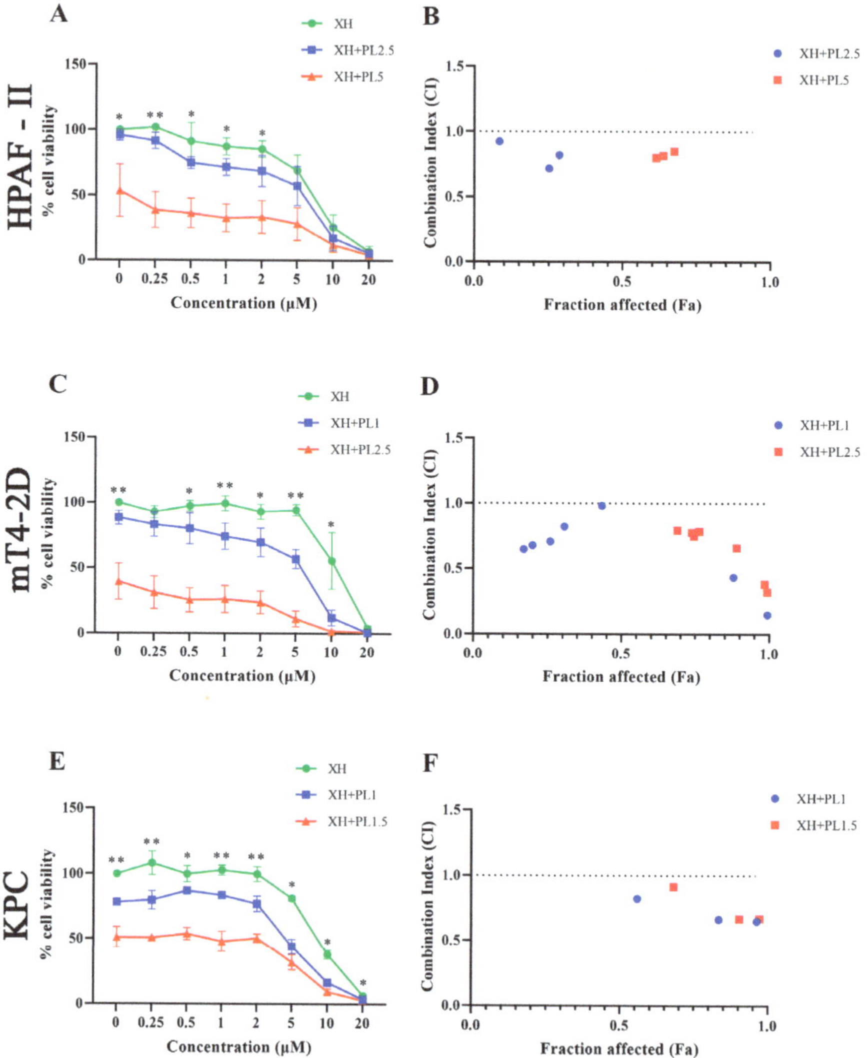

Figure 1. Synergistic effects of XH and PL on PDAC cell lines. (**A,C,E**) The viability of the cells treated with XH combined with PL was detected with dose-response MTT assays. KPC, Mt4-2D, and HPAF-II cells were grown and treated with XH or PL alone, or XH combined with PL, at two different concentrations for 72 h. Each experiment was repeated three times independently. Data are shown as means ± S.E and were analysed with one-way ANOVA. * p 0.05, ** $p < 0.01$ vs. 0 μg/mL. (**B,D,F**) Combination Index (CI)-effect plot. The CI values were calculated using the CompuSyn program to determine the combined effects of XH and PL in PDAC cell lines. The combinations were synergistic when CI values were < 1. The data are the mean values from three independent experiments. XH—Xanthohumol; PL—Plumbagin.

2.3. Plumbagin and Xanthohumol Target BCL2 on HPAF-II and mT4-2D Pancreatic Cancer Cell Lines

To gain an insight into the mechanism of action of the PL/XH combination, western blot analysis was performed to investigate signalling pathways involved in cancer development. HPAF-II and mT4-2D pancreatic cells were treated for 24 h with PL and XH alone and in two different combination concentrations of PL and XH (PL 1, 2.5, and

5 µM, XH at 2.5 µg/mL), as shown in Figure 2. Western blot analysis shows that combined concentrations of PL and XH in both cell lines decreased the level of the BCL2 (B-cell lymphoma 2, *BCL2*) protein compared to both the DMSO control and individual concentrations (Figure 2A–D), with the combination PL at 5 µM plus XH at 2.5 µg/mL being the most effective in HPAF-II cells (Figure 2B), while the combination treatments PL 1 and 2.5 µM plus XH at 2.5 µg/mL on mT4-2D cells reduced the level of BCL2 in a statistically significant way (Figure 2D). CXCR4 (C-X-C Motif Chemokine Receptor 4, *CXCR4*) and phospho-Akt 473 (AKT serine/threonine kinase 1, *AKT1*) levels were also tested in HPAF and mT4-2D cells upon 24 h treatment with a combination of PL and XH, but the treatments did not give consistent results (Supplementary Figure S1).

HPAF-II cell line

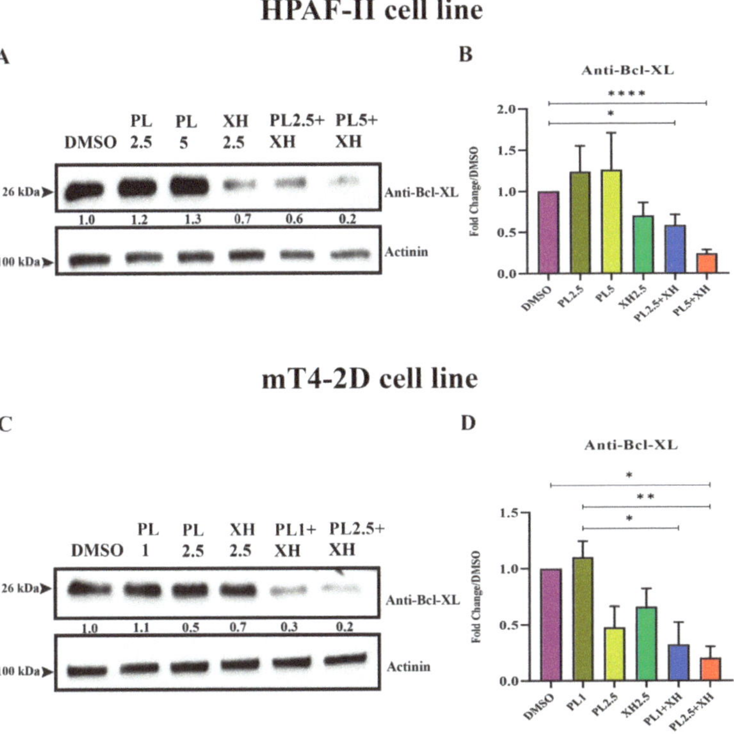

Figure 2. (**A**) Western blot image showing HPAF-II pancreatic cancer cell line after treatment with DMSO, PL (2.5 and 5 µM), XH (2.5 µg/mL), and combinations of XH 2.5 µg/mL with PL 2.5 and 5 µM concentrations. (**B**) Quantification of protein bands (BCL2) in HPAF-II cells analysed using the Image Lab 5.1 software. (**C**) Western blot image showing mT4-2D pancreatic cancer cell line after treatment with DMSO, PL (1 and 2.5 µM), XH (2.5 µg/mL), and combinations of XH 2.5 µg/mL with PL 1 and 2.5 µM concentrations. (**D**) Quantification of protein bands (BCL2) in mT4-2D cells analysed using the Image Lab 5.1 software. All experiments are presented as mean $\pm$ SEM of 3 independent experiments. Unpaired two-tailed student's *t*-test and GraphPad Prism version 9.4.1 were used for statistical analysis, * $p < 0.01$, ** $p < 0.001$, **** $p < 0.0001$.

2.4. In Vivo Studies

To assess the anticancer activity of PL in vivo, a KPC transgenic mouse model of pancreatic cancer was employed. This model is characterised by KrasLSL.G12D/+, p53R172H/+, and PdxCretg/+ mutations, and it is designed to spontaneously develop PDAC in a way that strongly mimics the human disease, both histologically and pathologically [25]. PL was combined with XH to evaluate the effects of this combination treatment in the KPC mice model. Once tumour development was established by palpation, mice were randomly divided into a control group and three treatment groups, composed of 10–14 mice. The four groups were then treated daily by intraperitoneal injection with vehicle only, 2 mg/kg of PL, 40 mg/kg of XH, and a combination of 2 mg/kg of PL and 40 mg/kg of XH. The treatment with PL resulted in increased survival of the mice belonging to the PL group (p = 0.0628) (Figure 3). No difference in survival was observed in mice treated with XH alone. Interestingly, a statistically significant increase (p = 0.0300) in the survival of mice treated with the PL/XH combination was observed (Figure 3). The weight of the pancreas extracted from all the mice also showed a significant reduction in the XH/PL combination compared to the control (Supplementary Figure S2).

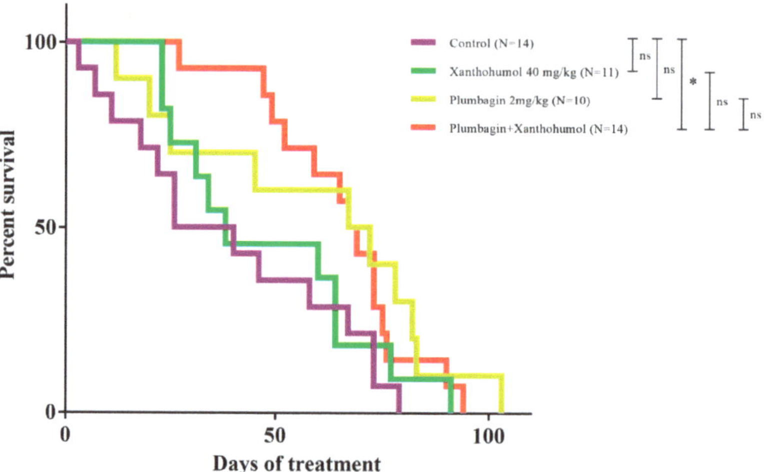

Figure 3. Kaplan–Meier survival curve of KPC mice treated with vehicle (n = 14), xanthohumol (n = 11), plumbagin (n = 10), and xanthohumol plus plumbagin (n = 14). Logrank (Mantel–Cox) test p = 0.0845, Logrank test for trend p = 0.0228, Gehan–Breslow–Wilcoxon test p = 0.0484. The xanthohumol plus plumbagin was the only treatment group to show a significant difference in the survival of KPC mice when compared to the control. Logrank test (Mantel–Cox) p = 0.0300, Gehan–Breslow–Wilcoxon test p = 0.0086; * p < 0.05. Curves indicate days after the start of each treatment.

Having established the existence of an anticancer activity (prolonged median survival) of the PL/XH combination in KPC mice, we proceeded to compare it to the standard chemotherapy treatment for PDAC, a gemcitabine plus albumin-bound paclitaxel (nab-paclitaxel) combination. A group of nine mice was administered intravenously with 50 mg/kg of gemcitabine plus 6.25 mg/kg of nab-paclitaxel (on days 1, 8, and 15 of each 28-day treatment cycle), while the nine mice in the control group received only vehicle. The ratio and schedule of the two drugs used are the same used in the treatment of pancreatic cancer patients. Interestingly, only three mice out of nine (33%) responded to the gemcitabine plus nab-paclitaxel combination, and the overall chemotherapy treatment was not statistically significant (Figure 4). In addition, the PL/XH combination had better median survival than the chemotherapy treatment and a twofold increase compared to the control group (Supplementary Figure S3). To test the effect of different treatments on their weight, the KPC mice were weighed every day and at the end of the experiment. The recorded data are presented in Supplementary Figure S4, which shows that all treatments did not lead to any significant weight loss. Weight loss was noted in the KPC mice when the administration of PL or the XH/PL combination was started, but the weight loss on average was around 5% among the mice in those treatment groups. The weights plateaued out around 2 weeks into treatment and the combination group gained some weight along the period compared to the plumbagin group.

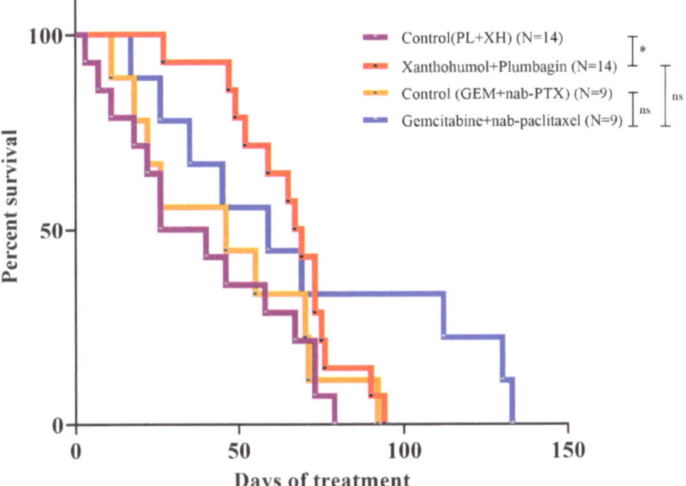

Figure 4. Kaplan–Meier survival curve showing KPC mice survival of xanthohumol plus plumbagin treatment compared against gemcitabine and nab-paclitaxel combination with their respective controls. Curves indicate days after the start of each treatment. Logrank (Mantel–Cox) test $p = 0.0626$, Logrank test for trend $p = 0.0639$, Gehan–Breslow–Wilcoxon test $p = 0.0535$; * $p < 0.05$. The curves control (XH/PL) and xanthohumol plus plumbagin are the same as presented in Figure 3.

2.5. Plumbagin and Xanthohumol Target BCL2 and pSTAT3 on Mice Tissue

Frozen pancreatic tissues of mice treated with XH alone, PL alone, and a PL/XH combination were analysed by western blot to further corroborate the anticancer effect of this combination on proteins involved in the development of PDAC. A statistically significant reduction of the expression of the BCL2 protein in the mice group treated with PL/XH was found, compared to the expression of the vehicle-treated mice group (Figure 5 and Supplementary Figure S5). Similarly, pSTAT3 Y705 (Signal Transducer and Activator of Transcription 3, *STAT3*) levels in mice tissues were significantly reduced by both the treatment with single agents and the combination, while no significant decrease was observed for total STAT3 (Figure 5). The levels of CXCR4 could not be detected in mice tissues.

A

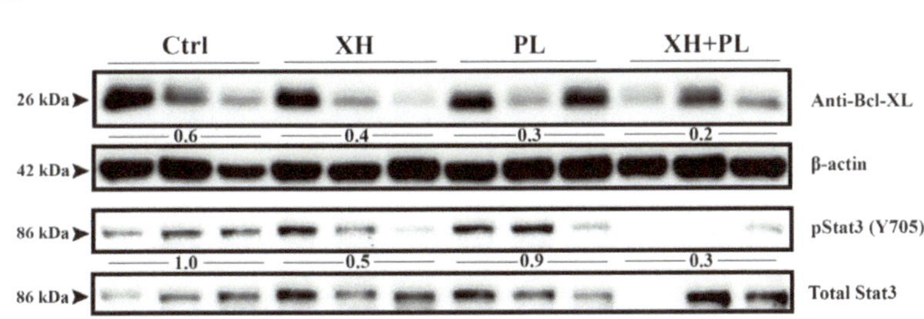

Figure 5. *Cont.*

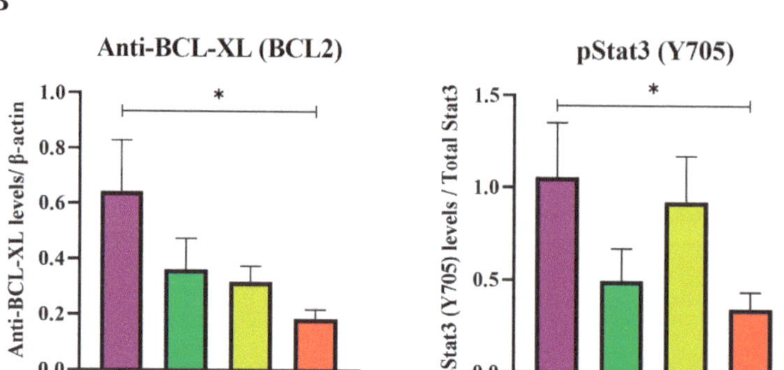

Figure 5. (A) Representative western blot of KPC mice tissues treated with vehicle (Control), xanthohumol (XH), plumbagin (PL), and a xanthohumol and plumbagin combination (XH/PL) [six mice were used per each treatment group] and probed for Anti-BCL-XL (BCL-2) and pStat3 Y705. β-actin was used as the loading control for BCL-2 and Total Stat3 for pStat3 (Y705). (B) Averaged quantification of Anti-BCL-XL and pStat3 (Y705) western blot signals normalised to their loading controls of the corresponding treatment groups. Data are means ± S.E. and an unpaired two-tailed student's *t*-test was used for statistical analysis, * $p < 0.01$.

3. Discussion

The potential benefits of using natural plant extracts and plant-isolated compounds and their role in diminishing the risk of cancer have been widely reported [26–28]. In this study, after screening in vitro different plant extracts and isolated active compounds, we have focused on the most promising agents, PL, the active molecule from *Plumbago indica*, and XH, found in the inflorescences of *Humulus lupulus* (Figure 6).

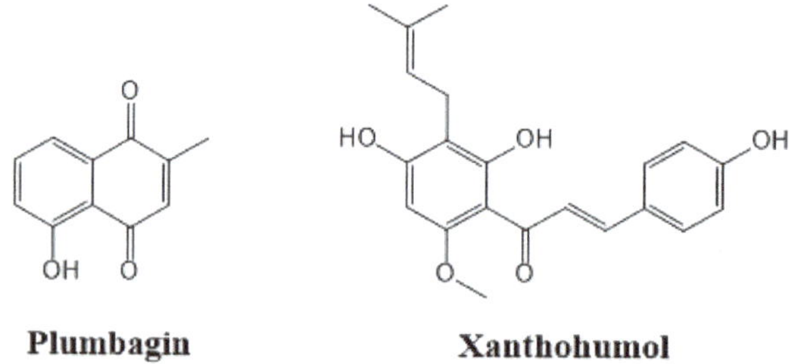

Plumbagin   **Xanthohumol**

Figure 6. Chemical structures of plumbagin and xanthohumol.

When we tested PL and XH on pancreatic cancer models to examine the anticancer activity, the combination showed convincing results on different human pancreatic cancer cell lines. (Table 1). Interestingly, PL indicated a higher activity and the second lowest IC$_{50}$ 0.53 µg/mL in AsPC-1 tumourspheres compared to the other cancer cell lines tested (Table 1). Pancreatic tumourspheres are cancer stem-like cells that are identified as a population of slow-cycling cells inside the tumour with self-renewal abilities [29]. Furthermore, these cells' distinctive features are an increased level of tumourigenicity and a higher chemo-resistance [29]. Therefore, the above activity of PL on AsPC-1 tumourspheres is a promising discovery in this study. We subsequently tested the PL/XH combination on transgenic mouse pancreatic cancer cell lines (KPC and mT4-2D) and observed an increased anticancer activity on mT4-2D cells, with the lowest IC$_{50}$ value (0.4 µg/mL) obtained from PL-treated cancer cell lines in the study (Table 1). Given the good results shown by PL in pancreatic cancer cell lines, we decided to test PL in cancer cell lines from the following other tumour types: human ovarian (A2780), human prostate (PC3), and breast adenocarcinoma (4T1) cell lines. In these cell lines, treatment with PL also produced low IC_{50} values, demonstrating good anticancer effects (Table 1). In addition, the anticancer activities of PL on prostate, pancreatic, breast, and lung cancer cell lines are also reported in the literature [30–32]. The IC_{50} values of XH on the human pancreatic cell lines HPAF-II and AsPC-1 were found to be 3.89 μg/mL and 6.8 μg/mL, respectively, showing XH to be relatively less effective compared with the IC_{50} values of PL. Furthermore, XH was found to have a stronger anticancer effect against the A2780 ovarian cancer cell lines, with an IC_{50} value of 0.3478 μg/mL (Table 1).

Chemo-resistance is becoming one of the main difficulties in cancer treatment [8,33]. Therefore, effective combination therapies are more reliable for the treatment of cancer. In this study, our focus was to perform combination treatments on pancreatic cancer cell lines to find a good and effective synergistic combination to progress to a transgenic model of pancreatic cancer. We observed a good synergistic effect after combining PL concentrations close to IC_{25} and IC_{50} with different concentrations of XH on both pancreatic cancer cell lines HPAF-II and mT4-2D that were tested. There is no data available in the literature for the synergism activity of PL with XH on any type of cancer.

More importantly, we showed that the PL and XH combination strikingly reduces tumour progression in an established and clinically relevant model of PDAC. To determine the mechanism of action of PL and XH treatments individually and in combination on pancreatic cancer cell lines, western blot analysis was performed on pancreatic cancer cell lines HPAF-II and mT4-2D. Results indicate that PL and XH both individually and in combination treatments target BCL2 protein on both cell lines. Moreover, based on the western blot images and quantification graphs, it is clearly demonstrated that both PL and XH combination treatments tested decreased the level of BCL2 protein compared to the control and compared to PL and XH treatments as single agents in both cell lines (Figure 2). BCL2 is known to be associated with apoptosis [34]. A clinical study showed that PL was able to regulate the intrinsic mitochondrial apoptotic pathway to induce cell apoptosis [35]. Furthermore, it was identified that the degradation of IκBα could be inhibited by PL and then suppressed the NF-κB activation and its translocation to the nucleus for function, which caused the downregulation of the anti-apoptosis protein BCL2 expression on various cancer cells [32,36,37]. The higher effect on BCL2 of the combination of PL and XH could also be attributed to a similar mechanism. Overexpression of CXCR4 in cancer cells is known to lead to tumour growth, invasion, angiogenesis, metastasis, relapse, and therapeutic resistance to cancer [38]. Therefore, CXCR4 becomes an important therapeutic target in the treatment of cancer [39]. Some studies identified PL as a major key player in downregulating the expression of the CXCR4 chemokine receptor in various types of cancers [40]. Consequently, identifying that PL and XH combinations could be able to downregulate CXCR4 more than PL alone can be considered a promising finding. Interestingly, a recent study that investigated the immunological consequences of inhibiting CXCR4 in PDAC patients who have historically resisted immunotherapy found that using continuous administration for one week of a small-molecule inhibitor of CXCR4 promotes an integrated immune response in metastatic lesions from these patients [41]. In addition, recent data also found that PL exerts a potent antitumour activity through the induction of anticancer immune response in non-small cell lung cancer models [23]. Therefore, it would be interesting to investigate in future studies the potential of the PL and XH combination in immunotherapy.

Our main data are obtained using a transgenic mouse model of pancreatic cancer (KPC) that is based upon the pancreatic-specific expression of endogenous mutant *Kras* and *Trp53* alleles. This model develops primary pancreatic tumours that precisely recapitulate the clinical, histopathologic, pharmacokinetic, and molecular features of human disease. In our study, we investigated the antitumour efficacy and the molecular mechanism of action of PL and XH in KPC mice survival. In addition, we have compared the activity of the combination to the gemcitabine and nab-paclitaxel combination, a treatment option for people with advanced pancreatic cancer. Our data show that the combination of gemcitabine and nab-paclitaxel gives a good response only in a portion of animals (as reported in clinical use) wherein it is not effective in most of them, whereas the PL/XH combination shows a good response in all animals. In addition, both PL and XH are used in clinical trials and therefore there is an immediate translational potential. Indeed, our search of available databases (ClinicalTrails.gov, accessed on 11 February 2024) found different ongoing clinical studies related to XH (NCT06225258, NCT05711212, NCT04590508, NCT03735420) and a synthetic form of PL (PCUR-101: NCT04677855, NCT03137758).

4. Materials and Methods

4.1. Extracts and Compounds Preparation

Plant extracts (ginger, chilli, pepper, peppermint, marjoram) were obtained from Flavex, Germany. The broccoli seed extract was obtained from CS Health (Louisville, KY, USA). The XH fraction of the hop extract was obtained from Hopsteiner (Au in der Hallertau, Germany). The active compounds (plumbagin, 6-gingerol, sulforaphane, 6-shogoal, capsaicin, zingerone, and piperine) were purchased from Sigma Aldrich (St. Louis, MO, USA). The pre-prepared powdered extracts were weighed and dissolved in dimethyl sulfoxide (DMSO) at the desired concentration (50 mg of solute per ml of DMSO for broccoli, broccoli and chilli, pepper, 10 mg/mL for ginger, peppermint, marjoram, hop extract, and XH). The mixture was filtered to obtain a clear solution that contained the components of interest. The active compound extracts, 6-gingerol, sulforaphane, 6-shogoal, capsaicin, zingerone, piperine, and PL, were dissolved in DMSO at a concentration of 10 mM; small 10 µL aliquots of the extracts and active compounds were prepared and frozen until needed. Each extract and compound was prepared specifically for the treatment, and after optimising concentrations, each treatment was undertaken 3 times on each cell line to achieve accurate results. Depending on the results obtained, the initial concentrations were modified, and combinations of compounds were tested on the cell lines. Subsequently, cells were treated with the most active extracts and compounds and protein analysis was undertaken by the western blotting technique.

4.2. Cell Culture

The mT4-2D cell line was a kind gift from Professor David Tuveson (CSHL) and was maintained in complete Dulbecco's Modified Eagle's Medium (DMEM) (Sigma, St. Louis, MO, USA) containing 10% foetal bovine serum (FBS) (Gibco, Waltham, MA, USA), 100 U/mL penicillin, 100 mg/mL streptomycin sulphate, and 2 mM L-glutamine [42]. The KPC (KrasLSL.G12D/+; p53R172H/+; PdxCretg/+) primary cell line, which we established as previously described [29], was cultured in standard cell culture vessels with DMEM supplemented with 10% FBS and added 1% Penicillin/Streptomycin. All other cancer cell lines were acquired from ATCC. HPAF-II was cultured in MEM medium (Sigma); AsPC-1, A2780, PC3, and 4T1 were cultured in RPMI 1640 medium (Sigma).

All the culture media, except for the cancer stem-like cell medium, were supplemented with 1% penicillin-streptomycin-glutamine (PSG) and 10% of FBS. Cancer stem-like cells were cultured in DMEM/F-12 medium (Sigma) supplemented as previously described and treated according to the protocol [29]. Cells were maintained in a humidified chamber (Nuaire DH auto flow CO_2 air-jacketed incubator) at 37 °C with 5% CO_2.

4.3. Cell Viability Assay

Cell viability assays were performed using a haemocytometer as described previously [43]. After seeding the cells for 24 h in 12 well plates, cells were treated with different plant extracts and pure compounds in a dose-response manner. After three independent repeats, results were examined and the IC_{50} for each extract and compound was calculated using the GraphPad Prism software 9.4.1.

4.4. MTT Assay

MTT assays were conducted using combination treatments of plant extracts and their compounds to examine the synergistic or antagonistic effect of the combinations. Thiazoyl Blue Tetrazolium Bromide (MTT) (Sigma) was prepared at 5 mg/mL in 1X PBS and was used for this assay. Approximately 5000 cells per well were seeded in 96 well tissue culture plates and treated with the combination treatments on the following day. After 72 h from the treatment, MTT (5 mg /mL, 10 µL per well) combined with 90 µL of media was added after the treatment was removed. Then, the plates were incubated at 37 °C for 2–4 h. After that, MTT was removed, and plates were allowed to dry for another 2 h. When fully dried, 70 µL of DMSO were added into each well and mixed gently using a shaker. At last, plates were read on a multimode plate reader (PerkinElmer, Waltham, MA, USA) at 570 nm. All data obtained were analysed using the CompuSyn software 1.0 by the Chou–Talalay method [24].

4.5. Determination of Combination Index and Dose Reduction Index

To study the effects of a combination treatment using PL and XH, cell viability was determined using the MTT assay after incubating cells with various concentrations of PL and XH in combination for 72 h. The Combination Index (CI) was calculated using the CompuSyn software 1.0 and following the classic isobologram equation: CI = D1/Dx1 + D2/Dx2, where Dx1 and Dx2 are the individual doses of PL and XH required to inhibit a given level of viability (x), and D1 and D2 are the doses of PL and XH required to inhibit the same level of viability (x) in combination. The CI values are then used to plot the CI versus Fraction Affected (FA) using GraphPad Prism software 9.4.1. The CI values help determine if the combination has an antagonistic effect (CI > 1) or a synergistic effect (CI < 1).

4.6. Western Blot Analysis

Western blot analysis was conducted to examine the expression patterns of cell apoptosis and cancer metastasis-regulating proteins (BCL2 and CXCR4) on pancreatic cancer cell lines treated with plant extracts and compounds. Tissues of mice treated with XH, PL, and combinations of the two were homogenised in RIPA buffer using pestle and mortar, sonicated and probed for BLC2, pERK, total ERK, pAkt 473, P-Akt 308, pStat3, and total Stat3 with western blot analysis, which was conducted according to the standard procedures as described previously [43]. All primary antibodies were used at 1:1000 dilution and incubated overnight at 4 °C with primary antibodies from Abcam: Anti-Bcl-XL (BCL2) (ab32370), Novus Biologicals CXCR4 (NB100-74396), Cell Signaling Technology®, Phospho-ERK (#4370), total ERK (#4695), Phospho-Akt 473 (#4060), Phospho-Akt (T308) (#4056), Phospho-Stat3 (Y705) (#9138), total Stat3 (#9139), and Actinin (#6487), the latter used as a loading control. The following day, membranes were incubated with HRP-conjugated secondary antibodies (Cell Signaling Technologies, Danvers, MA, USA) at a 1:40,000 dilution in 0.75% BSA in TBS/0.05% Tween-20 buffer for 1 h at room temperature. The signal was detected using the chemiluminescent detection reagent Amersham ECL PrimeWestern Blotting Detection Reagent (GE Healthcare Life Sciences, Chicago, IL, USA) imaged using ChemiDoc XRS+ (BioRad, Hercules, CA, USA), and quantified using Image Lab 5.1 software.

4.7. Animal Studies

All animal experiments were carried out in accordance with the National Health and Medical Research Council guidelines for the care and use of Laboratory animals (NHMRC, the Australian Code of Practice for the Care and Use of Animals for scientific purposes). All animal studies comply with the ARRIVE guidelines. The Curtin University Animal Ethics Committee approved procedures on KPC mice (AEC_2019_22). All animals were kept at 21 °C in ventilated cages cleaned weekly, with a 12 h light/12 h dark cycle and provided with water and food ad libitum. The sample size was estimated based on the power calculations performed previously in our group [44,45]. KRASWT/G12D,

P53WT/R172H, PDX-1CRE+/+ (KPC) transgenic mice (both male and female), and control mice were maintained and genotyped (C57BL/6 genetic background) by the Animal Resources Centre (Murdoch, Western Australia) according to the original protocol [25]. Following the anticipated 80-day period, which was previously determined as the expected initiation of tumour development [44,45], KPC mice underwent daily palpation for the evaluation of tumour growth. After the tumours reached palpable size, animals were treated with PL (2 mg/kg), XH (40 mg/kg), and with the combination of two treatments or vehicle daily by intraperitoneal injections. Mice were assigned to the four arms (Vehicle, XH, PL, XH/PL) by simple randomisation by writing treatments on a piece of paper, folding, mixing, and then picking one by one. The animals underwent daily monitoring, and euthanasia was performed upon the detection of apparent signs of pain and distress, including substantial weight loss (exceeding 20% of the initial body weight), dehydration, ascites formation, breathing difficulties attributable to lymphoma progression, or signs of pain. Euthanasia was carried out by snipping the main cardiac vein, followed by organ perfusion through the heart. Mouse survival was depicted using a Kaplan–Meier curve and assessed using a log-rank test.

4.8. Statistics

The results are all presented as mean ± SEM and are representative of at least three independent experiments. Statistical analyses were carried out using GraphPad Prism V9.4.1 software. Statistical significances were determined by performing two-tailed *t*-tests and one-way analysis of variance (ANOVA). *p*-value less than 0.05 is considered statistically significant. Kaplan–Meier curve was used to describe survival rates and its significance is quantified by log rank (Mantel–Cox) test.

5. Conclusions

In conclusion, we provide evidence of the pharmacological potential of the PL and XH combination that can block PDAC progression in vivo without discernible toxicity. Indeed, PL and XH individually displayed good anticancer activity in in vitro and in vivo studies and, interestingly, when used in combination, they demonstrated a much stronger effect compared to individual treatments, showing synergistic activity. Even more striking, a pancreatic cancer animal model treated with a combination of PL and XH resulted in a statistically significant increase in the survival of mice receiving this treatment compared to the controls. Western blot analysis of pancreatic cancer cell lines proved that the combination treatment affected the levels of BCL2, known to be associated with apoptosis and cancer metastasis, and in tissues from treated mice, a decrease in BCL2 and phospho-STAT3 (Y705) was found. Therefore, the combination of PL and XH demonstrated to be a promising candidate to be further validated as a potential anti-cancer treatment in clinical studies in the future.

Supplementary Materials: The following supporting information can be downloaded at: https://www.mdpi.com/article/10.3390/ijms25042340/s1.

Author Contributions: R.P.: conceptualisation, investigation, data curation, formal analysis, methodology, writing original draft, writing review, and editing. N.I.K.: conceptualisation, investigation, data curation, formal analysis, methodology, and writing original draft. I.C.: investigation, data curation, formal analysis, methodology, writing original draft, writing review, and editing. D.A.: study design and data analysis, methodology, writing review, and editing. M.F.: conceptualisation, investigation, data curation, formal analysis, methodology, writing original draft, writing review, and editing; project administration, funding acquisition, and supervision. All authors have read and agreed to the published version of the manuscript.

Funding: This work was supported by Backreef Oil Pty Limited.

Institutional Review Board Statement: This study has been conducted in accordance with ethical standards and according to the Declaration of Helsinki and the national and international guidelines and has been approved by the authors' institutional review board. The Curtin University Animal Ethics Committee approved procedures on KPC mice (AEC_2019_22).

Informed Consent Statement: Not applicable.

Data Availability Statement: The datasets generated during the current study are available from the corresponding author upon reasonable request.

Acknowledgments: The authors would like to acknowledge the support of PanKind, The Australian Pancreatic Cancer Foundation (www.pankind.org.au accessed on 14 February 2024) and the Keith & Ann Vaughan Pancreatic Cancer Fund.

Conflicts of Interest: David Archibald is on the board of Backreef Oil Pty Ltd. All other authors declare they have no competing interests.

References

1. Falasca, M.; Kim, M.; Casari, I. Pancreatic cancer: Current research and future directions. *Biochim. Biophys. Acta* **2016**, *1865*, 123–132. [CrossRef]
2. Mizrahi, J.D.; Surana, R.; Valle, J.W.; Shroff, R.T. Pancreatic cancer. *Lancet* **2020**, *395*, 2008–2020. [CrossRef]
3. Rahib, L.; Wehner, M.R.; Matrisian, L.M.; Nead, K.T. Estimated Projection of US Cancer Incidence and Death to 2040. *JAMA Netw. Open* **2021**, *4*, e214708. [CrossRef]
4. Adamska, A.; Domenichini, A.; Falasca, M. Pancreatic Ductal Adenocarcinoma: Current and Evolving Therapies. *Int. J. Mol. Sci.* **2017**, *18*, 1338. [CrossRef] [PubMed]
5. Springfeld, C.; Ferrone, C.R.; Katz, M.H.G.; Philip, P.A.; Hong, T.S.; Hackert, T.; Büchler, M.W.; Neoptolemos, J. Neoadjuvant therapy for pancreatic cancer. *Nat. Rev. Clin. Oncol.* **2023**, *20*, 318–337. [CrossRef] [PubMed]
6. Wood, L.D.; Canto, M.I.; Jaffee, E.M.; Simeone, D.M. Pancreatic Cancer: Pathogenesis, Screening, Diagnosis, and Treatment. *Gastroenterology* **2022**, *163*, 386–402.e1. [CrossRef] [PubMed]
7. Conroy, T.; Castan, F.; Lopez, A.; Turpin, A.; Ben Abdelghani, M.; Wei, A.C.; Mitry, E.; Biagi, J.J.; Evesque, L.; Artru, P.; et al. Five-Year Outcomes of FOLFIRINOX vs. Gemcitabine as Adjuvant Therapy for Pancreatic Cancer: A Randomized Clinical Trial. *JAMA Oncol.* **2022**, *8*, 1571–1578. [CrossRef] [PubMed]
8. Adamska, A.; Elaskalani, O.; Emmanouilidi, A.; Kim, M.; Abdol Razak, N.B.; Metharom, P.; Falasca, M. Molecular and cellular mechanisms of chemoresistance in pancreatic cancer. *Adv. Biol. Regul.* **2018**, *68*, 77–87. [CrossRef] [PubMed]
9. Key, T.J.; Schatzkin, A.; Willett, W.C.; Allen, N.E.; Spencer, E.A.; Travis, R.C. Diet, nutrition and the prevention of cancer. *Public Health Nutr.* **2004**, *7*, 187–200. [CrossRef]
10. Agrawal, M.Y.; Gaikwad, S.; Srivastava, S.; Srivastava, S.K. Research Trend and Detailed Insights into the Molecular Mechanisms of Food Bioactive Compounds against Cancer: A Comprehensive Review with Special Emphasis on Probiotics. *Cancers* **2022**, *14*, 5482. [CrossRef] [PubMed]
11. Steward, W.P.; Brown, K. Cancer chemoprevention: A rapidly evolving field. *Br. J. Cancer* **2013**, *109*, 1–7. [CrossRef] [PubMed]
12. Wattenberg, L.W. Chemoprevention of cancer. *Cancer Res.* **1985**, *45*, 1–8. [CrossRef] [PubMed]
13. Falasca, M.; Casari, I. Cancer chemoprevention by nuts: Evidence and promises. *Front. Biosci.* **2012**, *4*, 109–120. [CrossRef]
14. Noonan, D.M.; Benelli, R.; Albini, A. Angiogenesis and cancer prevention: A vision. *Recent Results Cancer Res.* **2007**, *174*, 219–224. [PubMed]
15. Falasca, M.; Casari, I.; Maffucci, T. Cancer chemoprevention with nuts. *J. Natl. Cancer Inst.* **2014**, *106*, dju238. [CrossRef] [PubMed]
16. Turrini, E.; Ferruzzi, L.; Fimognari, C. Natural compounds to overcome cancer chemoresistance: Toxicological and clinical issues. *Expert Opin. Drug Metab. Toxicol.* **2014**, *10*, 1677–1690. [CrossRef] [PubMed]
17. Clere, N.; Faure, S.; Martinez, M.C.; Andriantsitohaina, R. Anticancer properties of flavonoids: Roles in various stages of carcinogenesis. *Cardiovasc. Hematol. Agents Med. Chem.* **2011**, *9*, 62–77. [CrossRef]
18. Scagliarini, A.; Mathey, A.; Aires, V.; Delmas, D. Xanthohumol, a Prenylated Flavonoid from Hops, Induces DNA Damages in Colorectal Cancer Cells and Sensitizes SW480 Cells to the SN38 Chemotherapeutic Agent. *Cells* **2020**, *9*, 932. [CrossRef]
19. Turdo, A.; Glaviano, A.; Pepe, G.; Calapà, F.; Raimondo, S.; Fiori, M.E.; Carbone, D.; Basilicata, M.G.; Di Sarno, V.; Ostacolo, C.; et al. Nobiletin and Xanthohumol Sensitize Colorectal Cancer Stem Cells to Standard Chemotherapy. *Cancers* **2021**, *13*, 3927. [CrossRef]
20. Kunnimalaiyaan, S.; Trevino, J.; Tsai, S.; Gamblin, T.C.; Kunnimalaiyaan, M. Xanthohumol-Mediated Suppression of Notch1 Signaling Is Associated with Antitumor Activity in Human Pancreatic Cancer Cells. *Mol. Cancer Ther.* **2015**, *14*, 1395–1403. [CrossRef]
21. Jiang, W.; Zhao, S.; Xu, L.; Lu, Y.; Lu, Z.; Chen, C.; Ni, J.; Wan, R.; Yang, L. The inhibitory effects of xanthohumol, a prenylated chalcone derived from hops, on cell growth and tumorigenesis in human pancreatic cancer. *Biomed. Pharmacother.* **2015**, *73*, 40–47. [CrossRef]
22. Jiang, C.H.; Sun, T.L.; Xiang, D.X.; Wei, S.S.; Li, W.Q. Anticancer Activity and Mechanism of Xanthohumol: A Prenylated Flavonoid from Hops (*Humulus lupulus* L.). *Front. Pharmacol.* **2018**, *9*, 530. [CrossRef]

23. Jiang, Z.B.; Xu, C.; Wang, W.; Zhang, Y.Z.; Huang, J.M.; Xie, Y.J.; Wang, Q.-Q.; Fan, X.-X.; Yao, X.-J.; Xie, C.; et al. Plumbagin suppresses non-small cell lung cancer progression through downregulating ARF1 and by elevating CD8$^+$ T cells. *Pharmacol. Res.* **2021**, *169*, 105656. [CrossRef]

24. Chou, T.C. Drug combination studies and their synergy quantification using the Chou-Talalay method. *Cancer Res.* **2010**, *70*, 440–446. [CrossRef]

25. Hingorani, S.R.; Wang, L.; Multani, A.S.; Combs, C.; Deramaudt, T.B.; Hruban, R.H.; Rustgi, A.K.; Chang, S.; Tuveson, D.A. *Trp53R172H* and *KrasG12D* cooperate to promote chromosomal instability and widely metastatic pancreatic ductal adenocarcinoma in mice. *Cancer Cell.* **2005**, *7*, 469–483. [CrossRef] [PubMed]

26. Lin, S.R.; Chang, C.H.; Hsu, C.F.; Tsai, M.J.; Cheng, H.; Leong, M.K.; Sung, P.J.; Chen, J.C.; Weng, C.F. Natural compounds as potential adjuvants to cancer therapy: Preclinical evidence. *Br. J. Pharmacol.* **2020**, *177*, 1409–1423. [CrossRef] [PubMed]

27. Braicu, C.; Mehterov, N.; Vladimirov, B.; Sarafian, V.; Nabavi, S.M.; Atanasov, A.G.; Berindan-Neagoe, I. Nutrigenomics in cancer: Revisiting the effects of natural compounds. *Semin. Cancer Biol.* **2017**, *46*, 84–106. [CrossRef]

28. Swetha, M.; Keerthana, C.K.; Rayginia, T.P.; Anto, R.J. Cancer Chemoprevention: A Strategic Approach Using Phytochemicals. *Front. Pharmacol.* **2021**, *12*, 809308.

29. Domenichini, A.; Edmands, J.S.; Adamska, A.; Begicevic, R.R.; Paternoster, S.; Falasca, M. Pancreatic cancer tumorspheres are cancer stem-like cells with increased chemoresistance and reduced metabolic potential. *Adv. Biol. Regul.* **2019**, *72*, 63–77. [CrossRef]

30. Aziz, M.H.; Dreckschmidt, N.E.; Verma, A.K. Plumbagin, a medicinal plant-derived naphthoquinone, is a novel inhibitor of the growth and invasion of hormone-refractory prostate cancer. *Cancer Res.* **2008**, *68*, 9024–9032. [CrossRef] [PubMed]

31. Gomathinayagam, R.; Sowmyalakshmi, S.; Mardhatillah, F.; Kumar, R.; Akbarsha, M.A.; Damodaran, C. Anticancer mechanism of plumbagin, a natural compound, on non-small cell lung cancer cells. *Anticancer Res.* **2008**, *28*, 785–792. [PubMed]

32. Hafeez, B.B.; Jamal, M.S.; Fischer, J.W.; Mustafa, A.; Verma, A.K. Plumbagin, a plant derived natural agent inhibits the growth of pancreatic cancer cells in in vitro and in vivo via targeting EGFR, Stat3 and NF-κB signaling pathways. *Int. J. Cancer* **2012**, *131*, 2175–2186. [CrossRef]

33. Robey, R.W.; Pluchino, K.M.; Hall, M.D.; Fojo, A.T.; Bates, S.E.; Gottesman, M.M. Revisiting the role of ABC transporters in multidrug-resistant cancer. *Nat. Rev. Cancer* **2018**, *18*, 452–464. [CrossRef] [PubMed]

34. Green, D.R. The Mitochondrial Pathway of Apoptosis Part II: The BCL-2 Protein Family. *Cold Spring Harb. Perspect. Biol.* **2022**, *14*, a041046. [CrossRef] [PubMed]

35. Pandey, K.; Tripathi, S.K.; Panda, M.; Biswal, B.K. Prooxidative activity of plumbagin induces apoptosis in human pancreatic ductal adenocarcinoma cells via intrinsic apoptotic pathway. *Toxicol. Vitr.* **2020**, *65*, 104788. [CrossRef]

36. Sandur, S.K.; Ichikawa, H.; Sethi, G.; Ahn, K.S.; Aggarwal, B.B. Plumbagin (5-hydroxy-2-methyl-1,4-naphthoquinone) suppresses NF-kappaB activation and NF-kappaB-regulated gene products through modulation of p65 and IkappaBalpha kinase activation, leading to potentiation of apoptosis induced by cytokine and chemotherapeutic agents. *J. Biol. Chem.* **2006**, *281*, 17023–17033.

37. Ahmad, A.; Banerjee, S.; Wang, Z.; Kong, D.; Sarkar, F.H. Plumbagin-induced apoptosis of human breast cancer cells is mediated by inactivation of NF-kappaB and Bcl-2. *J. Cell Biochem.* **2008**, *105*, 1461–1471. [CrossRef]

38. Schottelius, M.; Herrmann, K.; Lapa, C. In Vivo Targeting of CXCR4-New Horizons. *Cancers* **2021**, *13*, 5920. [CrossRef]

39. Chatterjee, S.; Behnam Azad, B.; Nimmagadda, S. The intricate role of CXCR4 in cancer. *Adv. Cancer Res.* **2014**, *124*, 31–82.

40. Manu, K.A.; Shanmugam, M.K.; Rajendran, P.; Li, F.; Ramachandran, L.; Hay, H.S.; Kannaiyan, R.; Swamy, S.N.; Vali, S.; Kapoor, S.; et al. Plumbagin inhibits invasion and migration of breast and gastric cancer cells by downregulating the expression of chemokine receptor CXCR4. *Mol. Cancer* **2011**, *10*, 107. [CrossRef]

41. Biasci, D.; Smoragiewicz, M.; Connell, C.M.; Wang, Z.; Gao, Y.; Thaventhiran, J.E.D.; Basu, B.; Magiera, L.; Johnson, T.I.; Bax, L.; et al. CXCR4 inhibition in human pancreatic and colorectal cancers induces an integrated immune response. *Proc. Natl. Acad. Sci. USA* **2020**, *117*, 28960–28970. [CrossRef] [PubMed]

42. Sacks, D.; Baxter, B.; Campbell, B.C.V.; Carpenter, J.S.; Cognard, C.; Dippel, D.; Eesa, M.; Fischer, U.; Hausegger, K.; Hirsch, J.A.; et al. Multisociety Consensus Quality Improvement Revised Consensus Statement for Endovascular Therapy of Acute Ischemic Stroke. *Int. J. Stroke* **2018**, *13*, 612–632. [CrossRef] [PubMed]

43. Adamska, A.; Ferro, R.; Lattanzio, R.; Capone, E.; Domenichini, A.; Damiani, V.; Chiorino, G.; Akkaya, B.G.; Linton, K.J.; De Laurenzi, V.; et al. ABCC3 is a novel target for the treatment of pancreatic cancer. *Adv. Biol. Regul.* **2019**, *73*, 100634. [CrossRef] [PubMed]

44. Ferro, R.; Adamska, A.; Lattanzio, R.; Mavrommati, I.; Edling, C.E.; Arifin, S.A.; Fyffe, C.A.; Sala, G.; Sacchetto, L.; Chiorino, G.; et al. GPR55 signalling promotes proliferation of pancreatic cancer cells and tumour growth in mice, and its inhibition increases effects of gemcitabine. *Oncogene* **2018**, *37*, 6368–6382. [CrossRef]

45. Adamska, A.; Domenichini, A.; Capone, E.; Damiani, V.; Akkaya, B.G.; Linton, K.J.; Di Sebastiano, P.; Chen, X.; Keeton, A.B.; Ramirez-Alcantara, V.; et al. Pharmacological inhibition of ABCC3 slows tumour progression in animal models of pancreatic cancer. *J. Exp. Clin. Cancer Res.* **2019**, *38*, 312. [CrossRef]

Disclaimer/Publisher's Note: The statements, opinions and data contained in all publications are solely those of the individual author(s) and contributor(s) and not of MDPI and/or the editor(s). MDPI and/or the editor(s) disclaim responsibility for any injury to people or property resulting from any ideas, methods, instructions or products referred to in the content.

International Journal of
Molecular Sciences

Article

Bile Microbiome Signatures Associated with Pancreatic Ductal Adenocarcinoma Compared to Benign Disease: A UK Pilot Study

Nabeel Merali [1,2,3,*], Tarak Chouari [2,3], Julien Terroire [4,5], Maria-Danae Jessel [3], Daniel S. K. Liu [6], James-Halle Smith [7], Tyler Wooldridge [3], Tony Dhillon [3], José I. Jiménez [8], Jonathan Krell [6], Keith J. Roberts [7], Timothy A. Rockall [1], Eirini Velliou [9], Shivan Sivakumar [10], Elisa Giovannetti [11,12], Ayse Demirkan [4,5], Nicola E. Annels [3] and Adam E. Frampton [1,2,3,6]

1. Minimal Access Therapy Training Unit (MATTU), Royal Surrey County Hospital NHS Foundation Trust, Egerton Road, Guildford GU2 7XX, UK
2. Department of Hepato-Pancreato-Biliary (HPB) Surgery, Royal Surrey County Hospital NHS Foundation Trust, Egerton Road, Guildford GU2 7XX, UK
3. Section of Oncology, Department of Clinical and Experimental Medicine, Faculty of Health and Medical Science, University of Surrey, Guildford GU2 7WG, UK
4. Surrey Institute for People-Centred AI, University of Surrey, Guildford GU2 7XH, UK
5. Section of Statistical Multi-Omics, Department of Clinical and Experimental Medicine, Faculty of Health and Medical Science, University of Surrey, Guildford GU2 7WG, UK
6. Division of Cancer, Department of Surgery and Cancer, Imperial College London, Hammersmith Hospital Campus, London W12 0NN, UK
7. Hepatobiliary and Pancreatic Surgery Unit, Queen Elizabeth Hospital Birmingham, College of Medical and Dental Sciences, University of Birmingham, Birmingham B15 2TH, UK
8. Department of Life Sciences, Imperial College London, London SW7 2AZ, UK
9. Centre for 3D Models of Health and Disease, Division of Surgery and Interventional Science, University College London (UCL), London W1W 7TY, UK
10. Oncology Department, Institute of Immunology and Immunotherapy, Birmingham Medical School, University of Birmingham, Birmingham B15 2TT, UK
11. Department of Medical Oncology, VU University Medical Center, Cancer Center Amsterdam, 1081 HV Amsterdam, The Netherlands
12. Cancer Pharmacology Lab, AIRC Start Up Unit, Fondazione Pisana per la Scienza, San Giuliano Terme PI, 56017 Pisa, Italy
* Correspondence: n.merali@nhs.net

Citation: Merali, N.; Chouari, T.; Terroire, J.; Jessel, M.-D.; Liu, D.S.K.; Smith, J.-H.; Wooldridge, T.; Dhillon, T.; Jiménez, J.I.; Krell, J.; et al. Bile Microbiome Signatures Associated with Pancreatic Ductal Adenocarcinoma Compared to Benign Disease: A UK Pilot Study. *Int. J. Mol. Sci.* **2023**, 24, 16888. https://doi.org/10.3390/ijms242316888

Academic Editors: Donatella Delle Cave and Claudio Luchini

Received: 24 October 2023
Revised: 18 November 2023
Accepted: 24 November 2023
Published: 28 November 2023

Copyright: © 2023 by the authors. Licensee MDPI, Basel, Switzerland. This article is an open access article distributed under the terms and conditions of the Creative Commons Attribution (CC BY) license (https://creativecommons.org/licenses/by/4.0/).

Abstract: Pancreatic ductal adenocarcinoma (PDAC) has a very poor survival. The intra-tumoural microbiome can influence pancreatic tumourigenesis and chemoresistance and, therefore, patient survival. The role played by bile microbiota in PDAC is unknown. We aimed to define bile microbiome signatures that can effectively distinguish malignant from benign tumours in patients presenting with obstructive jaundice caused by benign and malignant pancreaticobiliary disease. Prospective bile samples were obtained from 31 patients who underwent either Endoscopic Retrograde Cholangiopancreatography (ERCP) or Percutaneous Transhepatic Cholangiogram (PTC). Variable regions (V3–V4) of the 16S rRNA genes of microorganisms present in the samples were amplified by Polymerase Chain Reaction (PCR) and sequenced. The cohort consisted of 12 PDAC, 10 choledocholithiasis, seven gallstone pancreatitis and two primary sclerosing cholangitis patients. Using the 16S rRNA method, we identified a total of 135 genera from 29 individuals (12 PDAC and 17 benign). The bile microbial beta diversity significantly differed between patients with PDAC vs. benign disease (Permanova p = 0.0173). The separation of PDAC from benign samples is clearly seen through unsupervised clustering of Aitchison distance. We found three genera to be of significantly lower abundance among PDAC samples vs. benign, adjusting for false discovery rate (FDR). These were *Escherichia* (FDR = 0.002) and two unclassified genera, one from *Proteobacteria* (FDR = 0.002) and one from *Enterobacteriaceae* (FDR = 0.011). In the same samples, the genus *Streptococcus* (FDR = 0.033) was found to be of increased abundance in the PDAC group. We show that patients with obstructive jaundice caused by PDAC have an altered microbiome composition in the bile compared to those with benign disease. These bile-based microbes could be developed into potential diagnostic and prognostic biomarkers for PDAC and warrant further investigation.

Keywords: pancreatic cancer; microbiome; 16S rRNA gene; bile; biomarker

1. Introduction

Pancreatic cancer (pancreatic ductal adenocarcinoma, PDAC) is a devastating disease. It is projected that by 2030, PDAC will become the 2nd leading cause of cancer-related death [1]. The incidence and mortality rates of PDAC are increasing. Poor outcomes are partly due to late diagnosis and these patients have either inoperable local disease or incurable metastatic disease. As a result, most patients are ineligible for surgery, and systemic treatments are not sufficient. Even after potentially curative surgical resection, the recurrence rates are very high. Optimal surgery and adjuvant chemotherapy results in a median disease-free survival (mDFS) of 13.9 months (range 12.1–16.6) with gemcitabine and capecitabine [2]; and 21.6 months (range 17.7–27.6) with FOLFIRINOX (FOLFIRI-NOX = Folic acid, Fluorouracil, Irinotecan and Oxaliplatin) [3]. Indeed, despite advances in surgical and oncological treatments, 5-year overall-survival (OS) is only 6% [4].

The tumour microbiome is gaining more interest recently in terms of prognosis and response to therapy. We now know the pancreas is not necessarily a sterile organ and can be infected by the gut microbiome refluxing into the pancreatic duct by the upper gastrointestinal tract [5,6]. Studies have shown that the pancreatic intra-tumoural microbiome can influence tumourigenesis, chemoresistance and the patients' immune response to the cancer [5,7–12]. Furthermore, specific microbes, such as *Gammaproteobacteria*, can inactivate gemcitabine chemotherapy leading to worse survival in PDAC mouse models [8]. Riquelme et al. disclosed the intra-tumoral microbiome composition of PDAC patients and identified a specific intra-tumoral microbiome signature (*Pseudoxanthomonas-Streptomyces-Saccharopolyspora-Bacillus clausii*) predicting the long-term survivorship of PDAC [9].

Assessing the influence of the microbiota in human physiology has revolutionised our understanding of medicine. Nejman et al. found that intra-tumoral microbiome composition is diverse and cancer type-specific [13]. The presence of bacteria in the pancreas can stimulate resident leukocytes to produce Interleukin-1β (IL-1β), which produces proangiogenic factors in the tumour microenvironment (TME) (e.g., Vascular Endothelial Growth Factor (VEGF), Tissue Necrosis Factor (TNF)) [14]. Das et al. demonstrate that tumour-derived IL-1β is required for the establishment of the immunosuppressive pancreatic TME [15]. This weakens the host immune defence system by the activation of inflammatory pathway mediators; Toll-like receptors (TLRs) and microorganism-associated molecular patterns (MAMPs) that leads to bacterial trans-location and chronic inflammation. IL-6 is an important proinflammatory cytokine that leads to tumour progression. A recent study identified *Helicobacter pylori* (*H. pylori*), can alter the expression of IL-6 by microRNA regulation [16] and can induce contact between leukocytes and other microorganisms [17]. Pushalkar et al. showed the depletion of the gut microbiome led to a reduction of Myeloid-derived suppressor cells (MDSC) infiltration and reprogramming of Tumour-associated macrophages (TAMs) toward a tumour-protective M1-like phenotype. Therefore, ablation of the gut microbiome highlighted T Helper 1 Cells 1 (Th1) polarization of Cluster of Differentiation 4 (CD4+) T cells and enhanced the cytotoxic phenotype of CD8+ T cells [5]. Certain microbes can cause genotoxic effects (i.e., colibactin) that damage the host DNA and activates IL-23-producing myeloid cells that promote tumour growth [18]. The microbiome can also modulate innate and adaptive immune responses to further contribute to the formation of the immunosuppressive TME found in PDAC [19].

The bile is potentially a rich source of novel biomarkers for PDAC and Biliary tract cancers (BTC) due to its intimate proximity to the malignant lesion. The bile duct was once considered a sterile environment. However, it is now well-regarded that microbiota exists within the bile duct in both benign and malignant diseases of the hepato-pancreato-biliary system, as well as other diseases of the alimentary canal [20–27]. Over the last two decades, several studies have explored the bile microbiota in the context of benign biliary tract disease [21,28,29]. Yet there is a paucity of studies investigating the bile microbiome in the context of PDAC [22,23,26,30].

A recent study has shown that the bile does have a distinct microbial fingerprint in PDAC, as compared to other pancreatic biliary diseases [31]. Furthermore, alteration of the bile microbiome can have a direct effect on the pancreatic cell survival [28]. Therefore, investigating the intra-tumoral microbiome through the role played by the bile microbiota in biliary cancers is the next frontier in clinical cancer treatment. Given the high fatality rate and the silent progression of early disease, identifying risk factors for the prevention and early detection of biliary tract cancers is critical. Therefore, the aim of this work was to define differentiating bile microbial signatures in patients presenting with obstructive jaundice caused by PDAC and benign pancreaticobiliary disease.

2. Results

Patient Characteristics

A total of 31 patients were enrolled in the study, corresponding to PDAC ($n = 12$) and benign cases ($n = 19$). Unfortunately, two benign samples were excluded due to low read counts for analysis as we did not have enough bile volume and insufficient DNA quality. Therefore, reliable data was available and analysed from only 12 PDAC and 17 benign cases (10 cases of common bile duct stones, six cases of gallstone pancreatitis, and one patient with primary sclerosing cholangitis). Table S1 shows a summary of the patient cohort.

All the PDAC cases had tumours in the head of the pancreas and were stented. The common aetiology found at Endoscopic Retrograde Cholangiopancreatography (ERCP) in the benign cases were for Common Bile Duct (CBD) stones and benign inflammatory strictures. There were an equal number of cholangitic patients in each group with similar median C-Reactive Protein (CRP) values at the time of ERCP, with no significance between the groups. The PDAC group presented with worsening jaundice, identified with statistically significant bilirubin and Carbohydrate Antigen 19-9 (CA19-9) tumour markers. Only two benign cases had a course of antibiotics within the previous month for the management of a bile leak.

Using 16S rRNA gene analysis, we identified a total of 135 genera from 29 individuals (12 PDAC and 17 benign) and their relative abundances are shown in Figure 1.

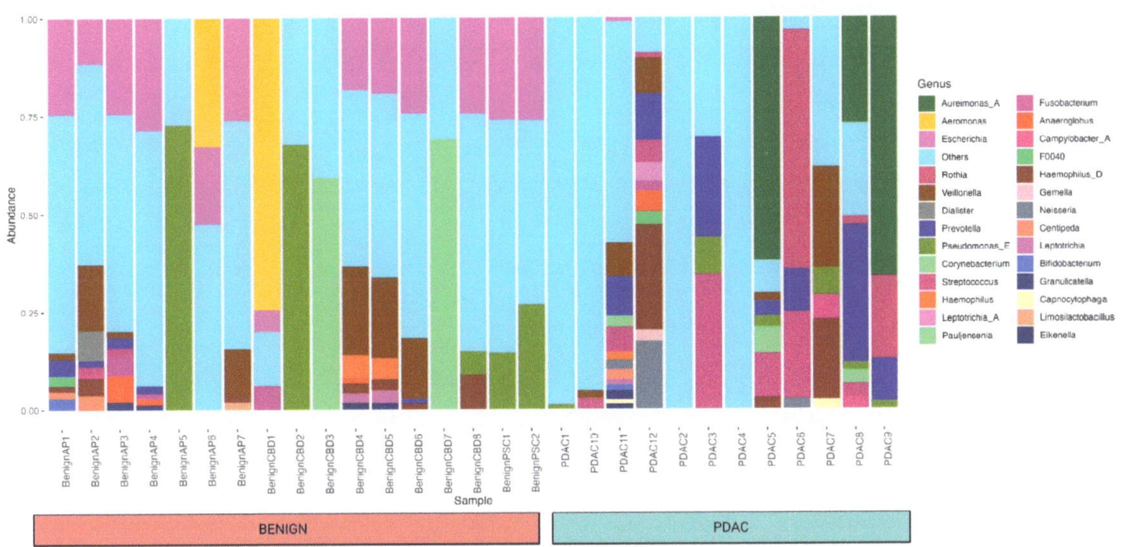

Figure 1. Bar plot showing the relative abundance of different bacteria within each sample and cohort at the genus level. A total of 135 genera from 29 individuals (12 PDAC and 17 benign) were identified, and a relative abundance of 10% and above was included in the figure. Forty-one different taxa that were not mapped to any bacteria at the genus level were clustered under "others" for the simplicity of this figure.

We only used the taxa identified at the genera level for our research. As seen in the bar plot, this occurred for a few samples that had all taxa identified as "known" at the genus level. Whereas 41 different taxa were not mapped to any bacteria at the genus level and clustered under "others". These taxa were included as separate entities in the differential abundance analysis.

We compared the alpha and beta diversity of the bacterial communities per group (PDAC vs. Benign). Alpha diversity did not significantly differ between sample groups (Figure 2). In regard to beta diversity, we used Aitchison distances as the measure of inter-sample differences in the compositions of gut metagenomes, which revealed a significant difference in average microbiome composition between bile from individuals with PDAC compared to individuals with benign samples by PERMANOVA ($p = 0.0173$) (Figure 3).

We then performed unsupervised clustering of the PDAC and benign groups' metagenomes based on Canberra distances of CLR-transformed abundance counts, as shown in Figure 4. The first cluster identified consists of 16 samples, 12 of which are PDAC, whereas no PDAC samples were assigned to the second cluster of 13 samples.

We next tested the differences in the relative abundance of microbial communities between PDAC and benign samples, using Maaslin2 default parameters. We found three genera to be of significantly lower abundance among PDAC samples compared to benign after adjusting for false discovery rate (FDR). These were *Escherichia* (FDR = 0.002), an unclassified genus from *Proteobacteria* (FDR = 0.002) and an unclassified genus from *Enterobacteriaceae* (FDR = 0.011). In the same samples, the genus *Streptococcus* (FDR = 0.033) had increased abundance in the PDAC group. This has been summarised in Table S2. Our data is compatible with Minimum information about a marker gene sequence (MIMARKS) and minimum information about any (x) sequence (MIxS) specifications [32] and is summarised as a checklist in Table S3.

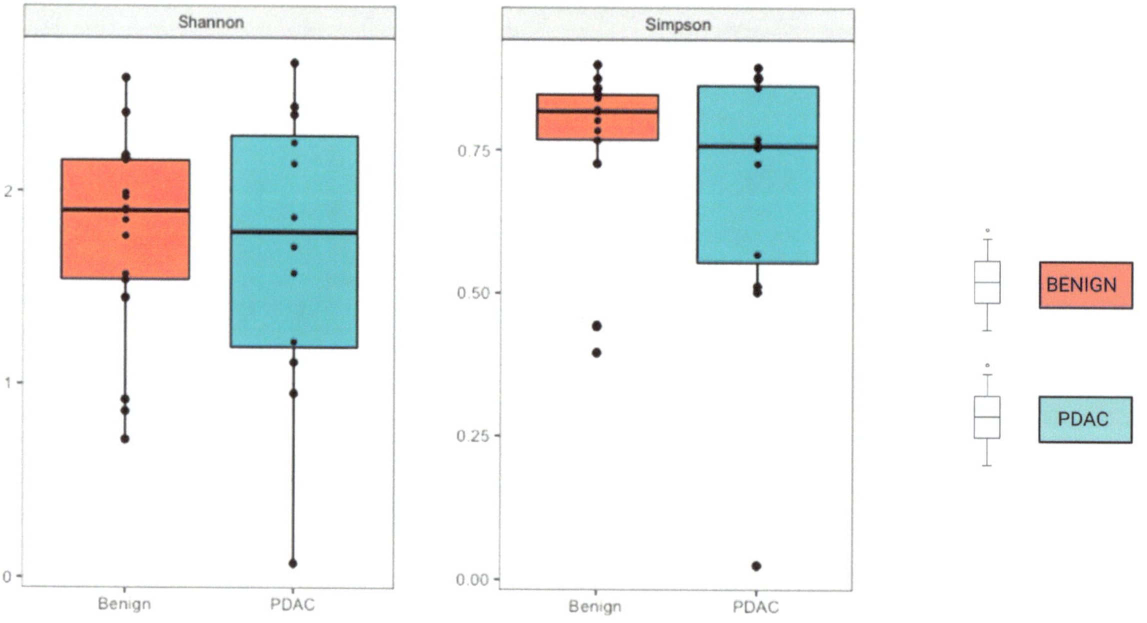

Figure 2. No difference was observed in Alpha diversity using the Shannon and Simpson index. Mean alpha diversity is higher among the benign samples, with *p*-values of 0.31 for Shannon and 0.3 for Simpson indices.

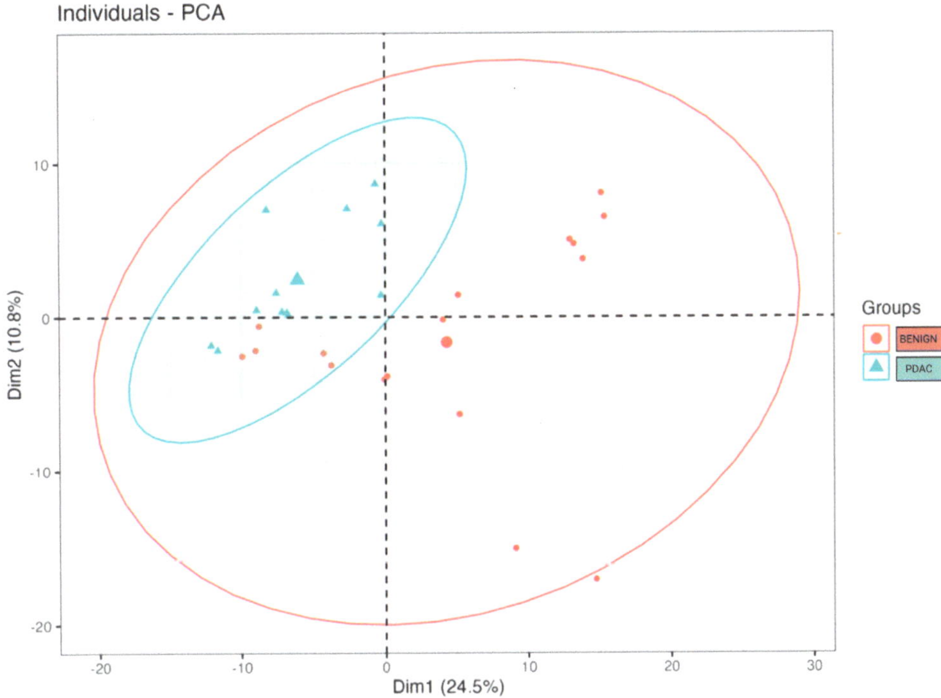

Figure 3. Microbial beta diversity significantly differs between bile from individuals with PDAC vs. bile from individuals with benign samples (p = 0.0173). The clusters were visualised by plotting the first two components that explain up to 35.3% of the variation in the sample space. The large blue triangle and the large red dot represent the centroids of the PDAC and benign sample groups.

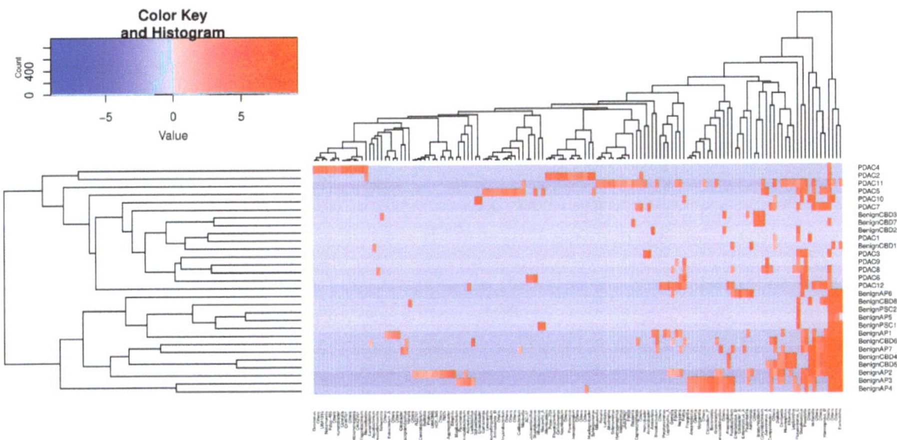

Figure 4. Unsupervised clustering of selected genera abundances separates PDAC from benign samples. The phylogenic tree clustering learned from 135 genera is shown in the *x*-axis. The *y*-axis shows the individual samples, which are separated into two major clusters, as shown by the dendrogram on the left. The colour key code is a graphic representation of centre-log ratio (CLR) abundance that uses all taxon read counts within a sample divided by this geometric mean, and the log fold changes in this ratio between samples are compared. If the abundance of a bacteria is lower than the mean CLR value, then it will be negative (blue), and if it is higher, then it will be positive (red).

3. Discussion

PDAC is an aggressive cancer with a high risk of invasion and metastasis. Furthermore, they are resistant to most cytotoxic agents and are often diagnosed at advanced stages. In the absence of an obvious mass lesion on cross-sectional imaging, determining the benign or malignant nature of a biliary stricture is important and can be even more challenging [33]. Evaluation of indeterminate strictures typically involves cytological and histological assessment. Biliary brush cytology and intraductal biopsies that are routinely performed during ERCP to assess malignant-appearing biliary strictures are limited by relatively low sensitivity [34]. It is critical to establish new diagnostic, prognostic, and therapeutic biomarkers which can complement the cytological and histological assessment of such strictures as well as any therapeutic strategies. One potential avenue of study is bile biomarkers. Bile is a potentially rich source of novel biomarkers for PDAC due to its intimate proximity to pancreatic parenchyma, which can be readily acquired via ERCP. This may prove valuable in the assessment of the underlying aetiology of biliary strictures. Emerging studies have revealed the role of the microbiome as a causative, prognostic, and predictive factor in various cancers and their treatment, including but not limited to PDAC [35]. Therefore, investigating the bile microbiota in PDAC may be the next frontier in diagnostics, prognostication, and management strategies.

Using a targeted amplicon sequencing approach for 16S rRNA gene to investigate the bile microbiome in PDAC, we have demonstrated that patients with obstructive jaundice secondary to PDAC, have an altered microbiome in the bile, compared to those with benign disease. We have identified four statistically significant microbes that are associated with PDAC. This study confirms the growing body of evidence that high microbial diversity is present within the biliary milieu of patients with benign and malignant pancreaticobiliary conditions [22–24,26,36–38].

For example, previous studies have evaluated the oral, gut, bile and/or intra-tumoural microbiota in relation to PDAC, and have found links between *Escherichia*, *Proteobacteria*, *Enterobacteriaceae* and *Streptococcus* [22,25,30,39,40]. Nagata et al. found an enrichment of *Streptococcus* spp. in the gut microbiome of PDAC patients [39]. A further study by Chen et al. also found *Streptococcus* as one of the gut pathogenic genera that exhibited a significant increase in abundance in patients with pancreatic cancer [25]. Previous in vitro and in vivo work has showcased *Streptococcus* can modify the biological effects of bile on PDAC cancer cell survival [28]. Our results also suggest an increased relative abundance of the genus *Streptococcus* in the bile of PDAC patients. The work described above may support the hypothesis that retrograde translocation of certain gut microbiome constituents into the CBD may have implications on cancer cell survival in PDAC. A larger study correlating the relationship between gut and bile microbiome analysis and clinical outcomes should be considered.

It should be noted that other work has drawn some conflicting results. For example, Li et al. recently investigated the microbiome differences among 53 patients with benign and malignant hepato-pancreato-biliary tract diseases. They found specific microbial bile markers for various malignant and benign disease states. *Streptococcus* was actually identified as a marker for distal cholangiocarcinoma (dCCA) and not PDAC. In vitro work has also shown *Streptococcus* has a pathognomic role in disease progression in Primary Sclerosing cholangitis to biliary dysplasia [27]. Interestingly, in PDAC, they found 24 microbial biomarkers at a genus level, none of which are in keeping with the 4 markers found in our study. The 3 most abundant markers for pancreatic cancer included *Pseudomonas*, *Chloroplast* and *Acinetobacter*, compared to other etiologies [26].

Our work demonstrates a low relative abundance of Escherichia and Enterobacteriaceae at a genus level. In a similar vein, work has shown that patients diagnosed with PDAC were associated with more bactibilia and *Escherichia* spp. was a negative predictor of PDAC [40]. Yet other work has found somewhat contradictory findings. For example, *Escherichia coli* [41] and *Escherichia-Shigella* [41] were found to be abundant in the biliary microbes of PDAC patients versus their benign counterparts [28]. Maekawa et al. investigated the presence of bacteria in pancreatic juice samples (taken post-operatively from drainage tubes) and found that *Enterobacter* and *Enterococcus* spp. were the major microbes in patients with PDAC [30]. Poudel et al. explored ERCP-derived bile microbial signatures in 46 patients with either PDAC, Cholangiocarcinoma (CCA), gallbladder cancer or benign biliary tract pathology. They demonstrated a distinct bile microbiome signature capable of differentiating all malignant pancreatic-biliary disease from benign disease samples. In fact, they identified a predominance of genera *Dickeya*, *Eubacterium hallii group*, *Bacteroides*, *Faecalibacterium*, *Escherichia-Shigella* and *Ruminococcus 1*, in bile samples from pancreato-biliary malignancies as compared to benign disease [22]. This study also highlighted a distinct dysbiosis not only between pancreaticobiliary cancers and benign disease but between different malignancies of the pancreaticobiliary system. Unfortunately, they have not compared subgroup bile microbiome profiling of PDAC versus benign disease. The studies described above have drawn some similar conclusions to our work yet other contradictory findings.

The conflicting findings of such studies [26] may be due to several reasons. We must consider that there are nuances to the bile microbiome we do not yet understand relating to environmental, host and tumour factors. Studies have previously shown certain clinical variables may be associated with significant changes in specific microbiota abundance found in bile, whilst other factors are of no significance [24]. Unfortunately, there is heterogeneity in terms of available clinical information and exclusion criteria in the studies described above. For example, Li et al. excluded patients with other systemic diseases, previous neoplastic disease or those receiving proton-pump inhibitors/antibiotics/prebiotics within one month [26]. Kirishima et al. did not specify if patients received antimicrobials whilst some patients had received chemotherapy [23]. These factors logically have implications on the microbiome and results observed. Patients in our study were treatment naïve. The fact patients received anti-neoplastic therapies in some of the above studies suggests their cohorts were at different stages in the patient journey once the diagnosis had already been confirmed [42]. Thus, our study may be more applicable to the initial diagnostic role of bile microbiome analysis in jaundiced patients. Furthermore, it is not entirely clear how the stage of disease or use of systemic therapies implicates the bile microbiome. Thus, making it challenging to draw any robust comparisons between our study and others. However, it is likely that heterogeneity in cohorts can explain the conflicting findings. This extends to our study as well, we have included patients with Stage I-IIA and III disease (Supplementary Materials, Table S1).

It should also be noted that previous work investigating ERCP-derived bile samples in PDAC has not clarified if patients were undergoing their first ERCP or had prior instrumentation of the CBD [22,26]. Such information is of particular importance when we consider the growing body of evidence suggesting the CBD and PDAC TME become colonized through a retrograde fashion from the duodenum [5,12,43] or after prior instrumentation. If indeed patients underwent a prior ERCP, it may in part explain some contradictory findings noted between our work and others. Furthermore, other work has used bile samples from surgically removed gallbladders [23]. It is not clear if the gallbladder-specific microbiome compartment correlates with the CBD-specific compartment. Other work has used pancreatic juice fluid sampled from surgical drains in the post operative period [30]. Again, such factors in part explain discrepancies noted between studies and between our study and others.

Nonetheless, all studies begin to fill the knowledge gap associated with the PDAC-associated bile microbiome and add value as a resource for future studies to build on. Our study has demonstrated a significant inter-sample difference in the average microbiome composition of bile in PDAC versus benign disease. However, fundamental questions remain on how we can generalise the findings of our study and contextualise it with other studies in this field. It is clear a degree of standardisation in terms of both study design and available demographic information on host, tumour and environmental factors is required in future studies to contextualise any findings observed. Emphasising the need for a further larger, comprehensive study into the four significant bacteria that we have identified in relation to PDAC as well as others identified in other studies.

Other work has alluded to the role of bile microbiome analysis in prognostication. For example, in PDAC or CCA, the relative abundance of certain microbiota correlated with prognosis after adjusting for clinicopathological variables [23]. Their results showed no common microbe correlated with a poor prognosis between tumour types. This may suggest different microbiome shifts at play within the disease-specific microenvironment, with implications on prognosis. This may have implications for clinical decision-making in the future if validated in larger studies. Follow-up with the collection of relevant clinicopathological variables in our study cohort may provide valuable insights into the correlation between bile microbiome and outcomes.

Whilst other work has demonstrated systemic therapies can alter the biliary microbiome with subsequent clinical implications. For example, S. Nadeem et al. assessed the impact of neoadjuvant chemotherapy (NAC) on the biliary microbiome in 168 patients with PDAC. Concluding that patients who received NAC exhibited significantly increased growth of Gram-negative anaerobic bacteria ($p = 0.043$), stating that perioperative antibiotic prophylaxis should be tailored to cover Gram-negative organisms and *enterococci* [42]. A direct pathological role for the bile microbiome has yet to be established. However, a previous study attributed the reduced abundance of oral *Streptococcus mitis* in PDAC to the protective role it plays against cariogenic bacterial adhesion [44]. This may result in a loss of colonisation by *Streptococcus* spp., which is thought to contribute to aggressive periodontitis [45], a risk factor for PDAC. Other work has proposed a bacterial-induced carcinogenesis model for the PDAC [46]. Whilst pre-clinical work suggests an alteration in the bile microbiome from biliary stenting has direct implications on pancreatic cell survival [28]. Further work is required to understand the effects other microbes in the bile (or PDAC TME) may have on these 4 genera (and vice versa) as this may help create a comprehensive picture of how the microbiome impacts PDAC carcinogenesis. Metagenomic assessment of the bile microbiome may shed further light on our functional understanding of the bile microbiome in PDAC carcinogenesis.

Precautions were taken to avoid intestinal milieu contamination during ERCP collection, we cannot rule that bacteria originating from the duodenum were included in the bile, since separated milieus were not screened. Likewise, direct contamination from the endoscope route leading to the introduction of bacteria from patients' oral and oesophago-gastric route cannot be ruled out. Furthermore, we only obtained one bile sample per patient, additional sampling in future study may further minimise the impact of contamination on findings.

Secondly, this is a single-centre research study with a small sample size, which should be expanded in the future. First, because of the non-randomized nature of the study, our study provides room for the traditional confounders of selection bias. In our study, bile was sampled at a diagnostic stage, where bile signatures correlating with diagnosis may be reflective of their potential future clinical role in diagnostics whilst also providing insights into bile microbial changes and carcinogenesis, prior to any systemic therapy or disease progression which may alter microbiome compositions. Of course, an understanding of the linear changes of the bile microbiome with duration of disease, antimicrobial/antineoplastic therapies received as well as the stage/extent of the disease is required to further contextualise this. Unfortunately, we remain at a primitive stage in our understanding of the bile microbiome in both pancreatic and biliary tract malignancy with a scarcity of studies exploring this topic. Unfortunately, a rate-limiting step in our understanding of the above, is the knowledge gap in understanding what a healthy bile microbiome entail. This is ethically challenging to ascertain and will likely prove to be a major hurdle in our understanding of the bile microbiome moving forward. Molinero et al. have tried to circumvent this hurdle by evaluating bile from liver donors without a history of biliary or hepatic disorders. They found an abundance of sequences belonging to the family *Propionibacteriaceae* in healthy controls compared to patients with cholelithiasis who had an abundance of sequences belonging to the families *Bacteroids*, *Prevotellaceae* and *Veillonellaceae* [21].

4. Materials and Methods

4.1. Patient Enrolment

A prospective, non-randomised study in which 31 patients undergoing their first endoscopic retrograde cholangio-pancreatography (ERCP) at the time of obstructive jaundice for benign and malignant pancreatico-biliary disease/strictures were recruited and assessed for their microbial signatures in biliary fluid (Figure 5).

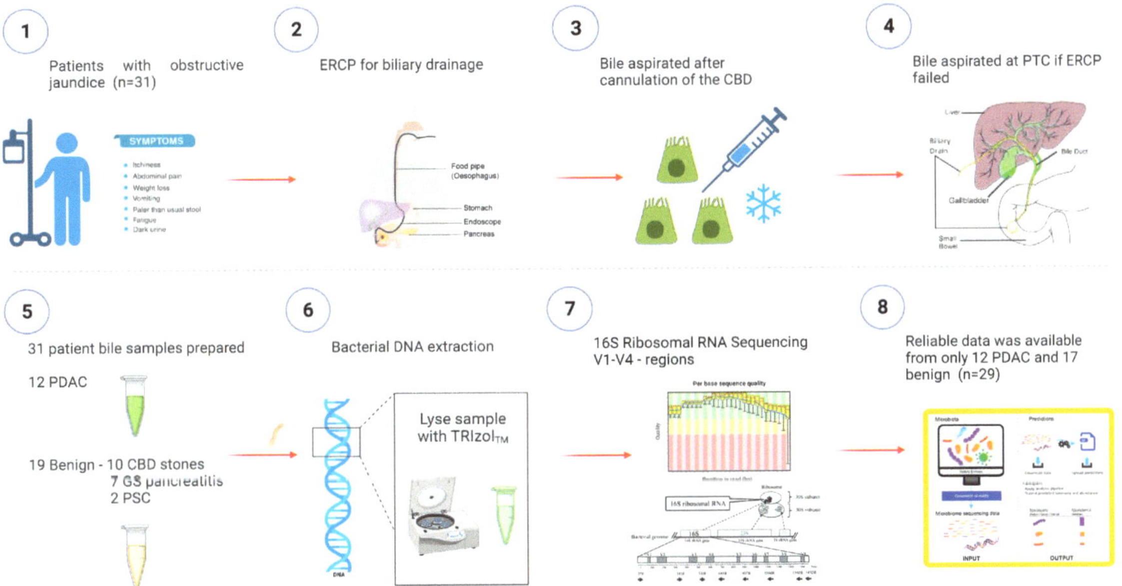

Figure 5. Study Flowchart outlining aspiration of bile at the time of obstructive jaundice at ERCP and subsequent sequencing of the bile microbiome. Prospective bile samples were obtained from 31 patients who underwent either ERCP or PTC. The cohort consisted of 12 PDAC, 10 choledocholithiasis, seven gallstone pancreatitis and two primary sclerosing cholangitis patients. Using the 16S rRNA method, we identified a total of 135 genera from 29 individuals (12 PDAC and 17 benign).

Patient biospecimens and clinical information were obtained from participants enrolled in microRNAs as BILE-based biomarkers in the Pancreaticobiliary Cancers (MIRABILE) research project. The pathological diagnoses were performed by NHS pathologists. Clear notice and signed written informed consent was obtained from all participants of this research (ICHTB HTA ethics license 12275/REC Wales approval: 17/WA/0161).

The patients were included according to the following inclusion criteria: (1) Age $\geq$ 18 years; (2) WHO performance status 0, 1 or 2; (3) willing and mentally able to provide written informed consent; (4) presented with obstructive jaundice and an indeterminate biliary stricture; (5) known benign or malignant pancreaticobiliary disease and undergoing their first ERCP or percutaneous transhepatic biliary drainage (PTBD). The exclusion criteria were as follows: (1) No clinical or image data suggestive of pancreaticobiliary disease and no need for endoscopic intervention or investigation; (2) pregnant women; (3) patients undergoing ERCP post-bariatric surgery, hepatico-jejunostomy or Bilroth II surgery; (4) any infectious disease such as hepatitis or HIV (human immunodeficiency virus). Side-viewing endoscopes, which were strictly sterile before the operation, were used to keep their working channel sterile. New plugs for the working channel were applied to every patient. To avoid contact and contamination with the duodenal mucosa upon ERCP, once the endoscope canal exit was positioned to the biliary duct entry, a biliary catheter was used for bile duct canalization and 5 mL of bile was aspirated in the lower third of CBD through the sphincterotome and into a sterile syringe. We obtained the bile before contrast injection, brushing and stenting in all subjects enrolled in our study. If the ERCP procedure was abandoned or technically not possible and the patient underwent PTBD, then bile will be aspirated from the drain bag after the procedure. After aspiration, bile was then frozen in our tissue bank at $-80\,^{\circ}$C until DNA extraction.

4.2. DNA Extraction, Amplification, and Sequencing

The Invitrogen™ TRIzol® LS Reagent, method [47] was employed for the extraction of total genomic DNA. The bile was vortexed and centrifuged at $300\times g$ for 10 min at 4 °C to remove cells and debris. The supernatant was then transferred to a sterile tube and centrifuged at $16,000\times g$ for 10 min at 4 °C to remove further debris. An amount of 1 mL of filtered bile was transferred into a sterile 5 mL container, and 3 mL Invitrogen™ TRIzol® LS reagent was added (3:1), followed by a brief vortex. The mixture was pipetted up and down to homogenise the content. An amount of 800 µL of chloroform was added into the mixture (4 mL) and shaken vigorously, incubated for 10 min at room temperature, followed by a centrifuge at $12,000\times g$ for 15 min at 4 °C. The organic and interphase layers were transferred into a 15 mL sterile falcon tube, and 1.2 mL of 100% ethanol (0.3 mL per 0.75 mL of Invitrogen™ TRIzol™ LS reagent) was added to the mixture. The tube was capped and was centrifuged several times until the DNA pellet was resuspended in 2 mL of 0.1 M sodium citrate in 10% ethanol, pH 8.5 (1 mL, per 0.75 mL of Invitrogen™ TRIzol™ LS Reagent) and transferred to a 2 mL eppendorf tube. The DNA pellet was resuspended in 2 mL of 75% ethanol (1.5–2 mL per 0.75 mL of Invitrogen™ TRIzol™ LS reagent used) and incubated for 20 min at room temperature. Finally, air-drying the DNA pellet for 5 min and washing the DNA was performed by resuspending the pellet in 20 µL Invitrogen™ TE Buffer (10 mM Tris-HCl (pH 8.0) and 0.1 mM EDTA).

DNA concentration and purity were checked on 1% agarose gels, and sterile water was used to dilute the DNA samples. DNA concentrations were determined using a Thermo Scientific™ NanoDrop™ 2000/2000c Spectrophotometer, Dover, DE, USA.

4.3. Library Preparation & Sequencing (V3–V4)

The variable regions, i.e., the V3–V4 area of the bacterial 16S ribosomal RNA (rRNA) gene, was amplified by polymerase chain reaction (PCR) using 515 forward (GGTGCCAGC MGCCGCGGTAA) and 806 reverse (GACTACHVG GGTWTCTAAT) primers Samples were prepared following the protocol in [48], using KAPA HiFi Polymerase to amplify variable region 3-4 of the 16S rRNA gene. Samples underwent 30 cycles of PCR. The libraries were sequenced at Diversigen (New Brighton, MN, USA) on an Illumina MiSeq using paired-end 2×250 reads with the MiSeq Reagent Kit V3 (Illumina, 600 cycle kit, San Diego, CA, USA). PCR controls included one negative and three positives (mock community, *E. coli* isolate and manufactured control for GC bias). All controls passed in the project.

4.4. Amplicon Sequence Variant (ASV) Picking (V4/V3V4)

Cutadapt was used to remove adaptors and primers from sequencing reads. The reads were quality checked by DADA2 (v1.25.2) R package [27214047], using default parameters of filterAndTrim function on forward and reverse reads; these specifically allowed for no ambiguous nucleotides ("maxN = 0"), truncated reads at the first instance of quality score < 2 ("truncQ = 2") and allowed for maximum expected errors of less than 2 ("maxEE = c(2,2)") where expected errors are calculated from the nominal definition of the quality score: $EE = \text{sum}(10^{(-Q/10)})$ by DADA2.

The error rates were calculated by the built-in machine learning function "learnErros" in DADA2. Then, forward and reverse reads were merged using the "mergePairs" function. Finally, chimeric reads were removed by "removeBimeraDenovo" function. As a result, two samples (samples no. 33 and 36) which had less than 1000 non-chimeric reads were removed. Next, by using the "assignTaxonomy" function of DADA2, the reads were mapped to the DADA2 formatted GTDB database ("GTDB_bac120_arc122_ssu_r202_fullTaxo.fa.gz"). The resulting taxa table, sequence table and associated meta-data were merged into the phyloseq object of 767 ASVs identified in a set of 29 individuals by phyloseq (1.42.0) package [23630581]. Eventually, 135 genera with minimum abundance greater than 0 were identified (Table S2—abundance counts) and 31 of them were observed >10% of the samples. Alpha diversity measures were calculated using the "diversity" function from the vegan R package [2.6–4]. Shannon and Simpson diversity per group were plotted using the "plot richness" function from Phyloseq. The diversity indices were compared by linear regression between benign and PDAC by adjusting for the number of non-chimeric reads for each sample. The principal components were extracted from central log-ratio-transformed (CLR) read counts ($n + 1$) using the "prcomp" function in R. PERMANOVA tests were calculated by adonis2 from vegan R package, while PERMADISP was calculated using betadisper (vegan), both adjusted by number nonchimeric reads per sample and 9999 using permutations. Linear regression analyses to test the association between PDAC and genera abundance were performed by Maaslin2 for common genera with abundance > 10 and by adjusting for non-chimeric reads per sample, as shown in Table S5. The Benjamini–Hochberg correction was used to correct for multiple testing. Patient characteristics were analysed on GraphPad Prism version 9.5.1. Independent *t*-test, Mann–Whitney U test and Chi-squared test were used to calculate the *p*-values. The full code scripts used in the analysis are made available in the Supplementary Materials.

5. Conclusions

Given the close relationship between microbiota and cancer, microbiome-targeted therapies are believed to be the next frontier in clinical cancer treatment. It also has a role in diagnostics and prognostication. Thus, with a greater understanding and definition of the bacterial microbiome in PDAC, there lies promise to develop novel biomarkers and therapeutic strategies. This study has demonstrated that patients with obstructive jaundice caused by PDAC have an altered microbiome in the bile compared to those with benign disease. We have identified four microbes that are associated with PDAC, and the genus *Streptococcus* (FDR = 0.033) was found to be of increased abundance in the PDAC group. Identification of specific bacteria in the bile may potentially enable the detection and stratification of PDAC. Patients undergoing biliary drainage could have bile analysed and "their microbial signature" targeted prior to surgery or neoadjuvant chemotherapy in order to optimise survival outcomes. Therefore, our study provides new insights into the link between the bile microbiome and PDAC. The results are promising and warrant a future larger study with metagenomic sequencing to investigate the function of these bacteria.

Supplementary Materials: The following supporting information can be downloaded at: https://www.mdpi.com/article/10.3390/ijms242316888/s1. The full scripts used in the analysis are available in the Supplementary Materials.

Author Contributions: Conceptualization, N.M.; methodology, N.M., S.S., E.G. and N.E.A.; software, J.T. and A.D.; validation, T.D., J.I.J., J.K., K.J.R., T.A.R., E.V., S.S., E.G., A.D., N.E.A. and A.E.F.; formal analysis, N.M., J.T. and A.D.; investigation, N.M., T.C. and D.S.K.L.; resources, N.M., J.-H.S. and T.W.; data curation, N.M., J.T. and A.D.; writing—original draft preparation, N.M., T.C. and M.-D.J.; writing—review and editing, T.D., J.I.J., J.K., K.J.R., T.A.R., E.V., S.S., E.G., A.D., N.E.A. and A.E.F.; visualization, N.M.; supervision, T.A.R., E.V., S.S., E.G., A.D., N.E.A. and A.E.F.; funding acquisition, N.M., J.I.J., J.K., T.A.R. and A.E.F. All authors have read and agreed to the published version of the manuscript.

Funding: This research was funded by the CRUK Development Fund Imperial Centre grant number PSM295_LCIB, Mason Medical Research Trust grant number RC3597 and Royal College of Surgeons of England Research Fellowship. The APC was funded by the University of Surrey.

Institutional Review Board Statement: Clear notice and signed written informed consent were obtained from all participants of this research (ICHTB HTA ethics license 12275/REC Wales approval: 17/WA/0161).

Informed Consent Statement: Informed consent was obtained from all subjects involved in the study.

Data Availability Statement: All information is included in the manuscript or Supplementary Materials. The bioinformatic code is available in the supporting files. 16S V3/V4 sequencing data is deposited and available on NCBI BioProject Accession Number: PRJNA1018343.

Acknowledgments: J.J., J.K. and A.E.F. acknowledge the support received from the Development fund of the CRUK Imperial Centre. N.M., T.A.R. and A.E.F. acknowledge the support received from Royal College of Surgeons of England Research Fellowship. N.M. and A.E.F. acknowledge the support received from the Mason Medical Research Fellowship.

Conflicts of Interest: The authors declare no conflict of interest.

References

1. McGuigan, A.; Kelly, P.; Turkington, R.C.; Jones, C.; Coleman, H.G.; McCain, R.S. Pancreatic cancer: A review of clinical diagnosis, epidemiology, treatment and outcomes. *World J. Gastroenterol.* **2018**, *24*, 4846–4861. [CrossRef]
2. Neoptolemos, J.P.; Palmer, D.H.; Ghaneh, P.; Psarelli, E.E.; Valle, J.W.; Halloran, C.M.; Faluyi, O.; O'Reilly, D.A.; Cunningham, D.; Wadsley, J.; et al. Comparison of adjuvant gemcitabine and capecitabine with gemcitabine monotherapy in patients with resected pancreatic cancer (ESPAC-4): A multicentre, open-label, randomised, phase 3 trial. *Lancet* **2017**, *389*, 1011–1024. [CrossRef]
3. Conroy, T.; Hammel, P.; Hebbar, M.; Ben Abdelghani, M.; Wei, A.C.; Raoul, J.-L.; Choné, L.; Francois, E.; Artru, P.; Biagi, J.J.; et al. FOLFIRINOX or Gemcitabine as Adjuvant Therapy for Pancreatic Cancer. *N. Engl. J. Med.* **2018**, *379*, 2395–2406. [CrossRef] [PubMed]
4. Kamisawa, T.; Wood, L.D.; Itoi, T.; Takaori, K. Pancreatic cancer. *Lancet* **2016**, *388*, 73–85. [CrossRef] [PubMed]
5. Pushalkar, S.; Hundeyin, M.; Daley, D.; Zambirinis, C.P.; Kurz, E.; Mishra, A.; Mohan, N.; Aykut, B.; Usyk, M.; Torres, L.E.; et al. The Pancreatic Cancer Microbiome Promotes Oncogenesis by Induction of Innate and Adaptive Immune Suppression. *Cancer Discov.* **2018**, *8*, 403–416. [CrossRef] [PubMed]
6. Akshintala, V.S.; Talukdar, R.; Singh, V.K.; Goggins, M. The Gut Microbiome in Pancreatic Disease. *Clin. Gastroenterol. Hepatol.* **2018**, *17*, 290–295. [CrossRef]
7. Guan, S.-W.; Lin, Q.; Yu, H.-B. Intratumour microbiome of pancreatic cancer. *World J. Gastrointest. Oncol.* **2023**, *15*, 713–730. [CrossRef]
8. Geller, L.T.; Barzily-Rokni, M.; Danino, T.; Jonas, O.H.; Shental, N.; Nejman, D.; Gavert, N.; Zwang, Y.; Cooper, Z.A.; Shee, K.; et al. Potential role of intratumor bacteria in mediating tumor resistance to the chemotherapeutic drug gemcitabine. *Science* **2017**, *357*, 1156–1160. [CrossRef]
9. Riquelme, E.; Zhang, Y.; Zhang, L.; Montiel, M.; Zoltan, M.; Dong, W.; Quesada, P.; Sahin, I.; Chandra, V.; Lucas, A.S.; et al. Tumor Microbiome Diversity and Composition Influence Pancreatic Cancer Outcomes. *Cell* **2019**, *178*, 795–806.e12. [CrossRef]
10. Guo, W.; Zhang, Y.; Guo, S.; Mei, Z.; Liao, H.; Dong, H.; Wu, K.; Ye, H.; Zhang, Y.; Zhu, Y.; et al. Tumor microbiome contributes to an aggressive phenotype in the basal-like subtype of pancreatic cancer. *Commun. Biol.* **2021**, *4*, 1019. [CrossRef] [PubMed]
11. Chakladar, J.; Kuo, S.Z.; Castaneda, G.; Li, W.T.; Gnanasekar, A.; Yu, M.A.; Chang, E.Y.; Wang, X.Q.; Ongkeko, W.M. The Pancreatic Microbiome Is Associated with Carcinogenesis and Worse Prognosis in Males and Smokers. *Cancers* **2020**, *12*, 2672. [CrossRef]
12. Aykut, B.; Pushalkar, S.; Chen, R.; Li, Q.; Abengozar, R.; Kim, J.I.; Shadaloey, S.A.; Wu, D.; Preiss, P.; Verma, N.; et al. The fungal mycobiome promotes pancreatic oncogenesis via activation of MBL. *Nature* **2019**, *574*, 264–267. [CrossRef]
13. Nejman, D.; Livyatan, I.; Fuks, G.; Gavert, N.; Zwang, Y.; Geller, L.T.; Rotter-Maskowitz, A.; Weiser, R.; Mallel, G.; Gigi, E.; et al. The human tumor microbiome is composed of tumor type–specific intracellular bacteria. *Science* **2020**, *368*, 973–980. [CrossRef]
14. Carmi, Y.; Dotan, S.; Rider, P.; Kaplanov, I.; White, M.R.; Baron, R.; Abutbul, S.; Huszar, M.; Dinarello, C.A.; Apte, R.N.; et al. The Role of IL-1β in the Early Tumor Cell–Induced Angiogenic Response. *J. Immunol.* **2013**, *190*, 3500–3509. [CrossRef] [PubMed]
15. Das, S.; Shapiro, B.; Vucic, E.A.; Vogt, S.; Bar-Sagi, D. Tumor Cell–Derived IL1β Promotes Desmoplasia and Immune Suppression in Pancreatic Cancer. *Cancer Res* **2020**, *80*, 1088–1101. [CrossRef] [PubMed]
16. Vainer, N.; Dehlendorff, C.; Johansen, J.S. Systematic literature review of IL-6 as a biomarker or treatment target in patients with gastric, bile duct, pancreatic and colorectal cancer. *Oncotarget* **2018**, *9*, 29820–29841. [CrossRef]
17. Uciechowski, P.; Dempke, W.C. Interleukin-6: A Masterplayer in the Cytokine Network. *Oncology* **2020**, *98*, 131–137. [CrossRef] [PubMed]
18. Parekh, P.J.; A Balart, L.; A Johnson, D. The Influence of the Gut Microbiome on Obesity, Metabolic Syndrome and Gastrointestinal Disease. *Clin. Transl. Gastroenterol.* **2015**, *6*, e91. [CrossRef]
19. Miyabayashi, K.; Ijichi, H.; Fujishiro, M. The Role of the Microbiome in Pancreatic Cancer. *Cancers* **2022**, *14*, 4479. [CrossRef] [PubMed]
20. Hsu, C.L.; Schnabl, B. The gut–liver axis and gut microbiota in health and liver disease. *Nat. Rev. Microbiol.* **2023**, *21*, 719–733. [CrossRef] [PubMed]
21. Molinero, N.; Ruiz, L.; Milani, C.; Gutiérrez-Díaz, I.; Sánchez, B.; Mangifesta, M.; Segura, J.; Cambero, I.; Campelo, A.B.; García-Bernardo, C.M.; et al. The human gallbladder microbiome is related to the physiological state and the biliary metabolic profile. *Microbiome* **2019**, *7*, 100. [CrossRef] [PubMed]
22. Poudel, S.K.; Padmanabhan, R.; Dave, H.; Guinta, K.; Stevens, T.; Sanaka, M.R.; Chahal, P.; Sohal, D.P.S.; Khorana, A.A.; Eng, C. Microbiomic profiles of bile in patients with benign and malignant pancreaticobiliary disease. *PLoS ONE* **2023**, *18*, e0283021. [CrossRef] [PubMed]
23. Kirishima, M.; Yokoyama, S.; Matsuo, K.; Hamada, T.; Shimokawa, M.; Akahane, T.; Sugimoto, T.; Tsurumaru, H.; Ishibashi, M.; Mataki, Y.; et al. Gallbladder microbiota composition is associated with pancreaticobiliary and gallbladder cancer prognosis. *BMC Microbiol.* **2022**, *22*, 147. [CrossRef]
24. Avilés-Jiménez, F.; Guitron, A.; Segura-López, F.; Méndez-Tenorio, A.; Iwai, S.; Hernández-Guerrero, A.; Torres, J. Microbiota studies in the bile duct strongly suggest a role for *Helicobacter pylori* in extrahepatic cholangiocarcinoma. *Clin. Microbiol. Infect.* **2015**, *22*, 178.e11–178.e22. [CrossRef] [PubMed]
25. Chen, T.; Li, X.; Li, G.; Liu, Y.; Huang, X.; Ma, W.; Qian, C.; Guo, J.; Wang, S.; Qin, Q.; et al. Alterations of commensal microbiota are associated with pancreatic cancer. *Int. J. Biol. Markers* **2023**, *38*, 89–98. [CrossRef]
26. Li, Z.; Chu, J.; Su, F.; Ding, X.; Zhang, Y.; Dou, L.; Liu, Y.; Ke, Y.; Liu, X.; Liu, Y.; et al. Characteristics of bile microbiota in cholelithiasis, perihilar cholangiocarcinoma, distal cholangiocarcinoma, and pancreatic cancer. *Am. J. Transl. Res.* **2022**, *14*, 2962–2971.
27. Pereira, P.; Aho, V.; Arola, J.; Boyd, S.; Jokelainen, K.; Paulin, L.; Auvinen, P.; Färkkilä, M. Bile microbiota in primary sclerosing cholangitis: Impact on disease progression and development of biliary dysplasia. *PLoS ONE* **2017**, *12*, e0182924. [CrossRef]

28. Shrader, H.R.; Miller, A.M.; Tomanek-Chalkley, A.; McCarthy, A.; Coleman, K.L.; Ear, P.H.; Mangalam, A.K.; Salem, A.K.; Chan, C.H.F. Effect of bacterial contamination in bile on pancreatic cancer cell survival. *Surgery* **2021**, *169*, 617–622. [CrossRef]

29. Sabino, J.; Vieira-Silva, S.; Machiels, K.; Joossens, M.; Falony, G.; Ballet, V.; Ferrante, M.; Van Assche, G.; Van der Merwe, S.; Vermeire, S.; et al. Primary sclerosing cholangitis is characterised by intestinal dysbiosis independent from IBD. *Gut* **2016**, *65*, 1681–1689. [CrossRef]

30. Maekawa, T.; Fukaya, R.; Takamatsu, S.; Itoyama, S.; Fukuoka, T.; Yamada, M.; Hata, T.; Nagaoka, S.; Kawamoto, K.; Eguchi, H.; et al. Possible involvement of Enterococcus infection in the pathogenesis of chronic pancreatitis and cancer. *Biochem. Biophys. Res. Commun.* **2018**, *506*, 962–969. [CrossRef]

31. Wheatley, R.C.; Kilgour, E.; Jacobs, T.; Lamarca, A.; Hubner, R.A.; Valle, J.W.; McNamara, M.G. Potential influence of the microbiome environment in patients with biliary tract cancer and implications for therapy. *Br. J. Cancer* **2021**, *126*, 693–705. [CrossRef]

32. Yilmaz, P.; Kottmann, R.; Field, D.; Knight, R.; Cole, J.R.; Amaral-Zettler, L.; A Gilbert, J.; Karsch-Mizrachi, I.; Johnston, A.; Cochrane, G.; et al. Minimum information about a marker gene sequence (MIMARKS) and minimum information about any (x) sequence (MIxS) specifications. *Nat. Biotechnol.* **2011**, *29*, 415–420. [CrossRef]

33. Jamieson, N.B.; Denley, S.M.; Logue, J.; MacKenzie, D.J.; Foulis, A.K.; Dickson, E.J.; Imrie, C.W.; Carter, R.; McKay, C.J.; McMillan, D.C. A Prospective Comparison of the Prognostic Value of Tumor- and Patient-Related Factors in Patients Undergoing Potentially Curative Surgery for Pancreatic Ductal Adenocarcinoma. *Ann. Surg. Oncol.* **2011**, *18*, 2318–2328. [CrossRef]

34. Farrell, R.J.; Jain, A.K.; Brandwein, S.L.; Wang, H.; Chuttani, R.; Pleskow, D.K. The combination of stricture dilation, endoscopic needle aspiration, and biliary brushings significantly improves diagnostic yield from malignant bile duct strictures. *Gastrointest. Endosc.* **2001**, *54*, 587–594. [CrossRef]

35. Merali, N.; Chouari, T.; Kayani, K.; Rayner, C.J.; Jiménez, J.I.; Krell, J.; Giovannetti, E.; Bagwan, I.; Relph, K.; Rockall, T.A.; et al. A Comprehensive Review of the Current and Future Role of the Microbiome in Pancreatic Ductal Adenocarcinoma. *Cancers* **2022**, *14*, 1020. [CrossRef] [PubMed]

36. Wei, M.-Y.; Shi, S.; Liang, C.; Meng, Q.C.; Hua, J.; Zhang, Y.-Y.; Liu, J.; Bo, Z.; Xu, J.; Yu, X.J. The microbiota and microbiome in pancreatic cancer: More influential than expected. *Mol. Cancer* **2019**, *18*, 97. [CrossRef] [PubMed]

37. Segura-López, F.K.; Avilés-Jiménez, F.; Güitrón-Cantú, A.; Valdéz-Salazar, H.A.; León-Carballo, S.; Guerrero-Pérez, L.; Fox, J.G.; Torres, J. Infection with *Helicobacter bilis* but not *Helicobacter hepaticus* was Associated with Extrahepatic Cholangiocarcinoma. *Helicobacter* **2015**, *20*, 223–230. [CrossRef]

38. Sepich-Poore, G.D.; Zitvogel, L.; Straussman, R.; Hasty, J.; Wargo, J.A.; Knight, R. The microbiome and human cancer. *Science* **2021**, *371*, eabc4552. [CrossRef] [PubMed]

39. Nagata, N.; Nishijima, S.; Kojima, Y.; Hisada, Y.; Imbe, K.; Miyoshi-Akiyama, T.; Suda, W.; Kimura, M.; Aoki, R.; Sekine, K.; et al. Metagenomic Identification of Microbial Signatures Predicting Pancreatic Cancer From a Multinational Study. *Gastroenterology* **2022**, *163*, 222–238. [CrossRef] [PubMed]

40. Serra, N.; Di Carlo, P.; Gulotta, G.; d'Arpa, F.; Giammanco, A.; Colomba, C.; Melfa, G.; Fasciana, T.; Sergi, C. Bactibilia in women affected with diseases of the biliary tract and pancreas. A STROBE guidelines-adherent cross-sectional study in Southern Italy. *J. Med. Microbiol.* **2018**, *67*, 1090–1095. [CrossRef] [PubMed]

41. Di Carlo, P.; Serra, N.; D'Arpa, F.; Agrusa, A.; Gulotta, G.; Fasciana, T.; Rodolico, V.; Giammanco, A.; Sergi, C. The microbiota of the bilio-pancreatic system: A cohort, STROBE-compliant study. *Infect. Drug Resist.* **2019**, *12*, 1513–1527. [CrossRef]

42. Nadeem, S.O.; Jajja, M.R.; Maxwell, D.W.; Pouch, S.M.; Sarmiento, J.M. Neoadjuvant chemotherapy for pancreatic cancer and changes in the biliary microbiome. *Am. J. Surg.* **2020**, *222*, 3–7. [CrossRef] [PubMed]

43. Langheinrich, M.; Wirtz, S.; Kneis, B.; Gittler, M.M.; Tyc, O.; Schierwagen, R.; Brunner, M.; Krautz, C.; Weber, G.F.; Pilarsky, C.; et al. Microbiome Patterns in Matched Bile, Duodenal, Pancreatic Tumor Tissue, Drainage, and Stool Samples: Association with Preoperative Stenting and Postoperative Pancreatic Fistula Development. *J. Clin. Med.* **2020**, *9*, 2785. [CrossRef] [PubMed]

44. Van Hoogmoed, C.G.; van der Mei, H.C.; Busscher, H.J. The Influence of Biosurfactants Released by *S. mitis* BMS on the Adhesion of Pioneer Strains and Cariogenic Bacteria. *Biofouling* **2004**, *20*, 261–267. [CrossRef] [PubMed]

45. Stingu, C.-S.; Eschrich, K.; Rodloff, A.C.; Schaumann, R.; Jentsch, H. Periodontitis is associated with a loss of colonization by Streptococcus sanguinis. *J. Med. Microbiol.* **2008**, *57*, 495–499. [CrossRef]

46. Arteta, A.A.; Sánchez-Jiménez, M.; Dávila, D.F.; Palacios, O.G.; Cardona-Castro, N. Biliary Tract Carcinogenesis Model Based on Bile Metaproteomics. *Front. Oncol.* **2020**, *10*, 1032. [CrossRef]

47. Rio, D.C.; Ares, M., Jr.; Hannon, G.J.; Nilsen, T.W. Purification of RNA Using TRIzol (TRI Reagent). *Cold Spring Harb. Protoc.* **2010**, *2010*, pdb-prot5439. [CrossRef]

48. Gohl, D.M.; Vangay, P.; Garbe, J.; MacLean, A.; Hauge, A.; Becker, A.; Gould, T.J.; Clayton, J.B.; Johnson, T.J.; Hunter, R.; et al. Systematic improvement of amplicon marker gene methods for increased accuracy in microbiome studies. *Nat. Biotechnol.* **2016**, *34*, 942–949. [CrossRef]

Disclaimer/Publisher's Note: The statements, opinions and data contained in all publications are solely those of the individual author(s) and contributor(s) and not of MDPI and/or the editor(s). MDPI and/or the editor(s) disclaim responsibility for any injury to people or property resulting from any ideas, methods, instructions or products referred to in the content.

International Journal of
Molecular Sciences

MDPI

Article

IK Channel-Independent Effects of Clotrimazole and Senicapoc on Cancer Cells Viability and Migration

Paolo Zuccolini [†], Raffaella Barbieri, Francesca Sbrana, Cristiana Picco, Paola Gavazzo and Michael Pusch *

Biophysics Institute, National Research Council, 16149 Genova, Italy; paolo.zuccolini@ibf.cnr.it (P.Z.); raffaella.barbieri@ibf.cnr.it (R.B.); francesca.sbrana@ibf.cnr.it (F.S.); cristiana.picco@ibf.cnr.it (C.P.); paola.gavazzo@ibf.cnr.it (P.G.)
* Correspondence: michael.pusch@ibf.cnr.it
[†] Present address: Membrane Transport Biophysics Section, National Institutes of Neurological Disorders and Stroke, Bethesda, MD 20892, USA.

Abstract: Many studies highlighted the importance of the IK channel for the proliferation and the migration of different types of cancer cells, showing how IK blockers could slow down cancer growth. Based on these data, we wanted to characterize the effects of IK blockers on melanoma metastatic cells and to understand if such effects were exclusively IK-dependent. For this purpose, we employed two different blockers, namely clotrimazole and senicapoc, and two cell lines: metastatic melanoma WM266-4 and pancreatic cancer Panc-1, which is reported to have little or no IK expression. Clotrimazole and senicapoc induced a decrease in viability and the migration of both WM266-4 and Panc-1 cells irrespective of IK expression levels. Patch-clamp experiments on WM266-4 cells revealed Ca^{2+}-dependent, IK-like, clotrimazole- and senicapoc-sensitive currents, which could not be detected in Panc-1 cells. Neither clotrimazole nor senicapoc altered the intracellular Ca^{2+} concentration. These results suggest that the effects of IK blockers on cancer cells are not strictly dependent on a robust presence of the channel in the plasma membrane, but they might be due to off-target effects on other cellular targets or to the blockade of IK channels localized in intracellular organelles.

Keywords: IK; KCa3.1; *KCNN4*; cancer; melanoma; pancreatic duct adenocarcinoma (PDAC); blockers; clotrimazole; senicapoc

Citation: Zuccolini, P.; Barbieri, R.; Sbrana, F.; Picco, C.; Gavazzo, P.; Pusch, M. IK Channel-Independent Effects of Clotrimazole and Senicapoc on Cancer Cells Viability and Migration. *Int. J. Mol. Sci.* **2023**, *24*, 16285. https://doi.org/10.3390/ijms242216285

Academic Editor: Marco Falasca

Received: 16 August 2023
Revised: 10 November 2023
Accepted: 10 November 2023
Published: 14 November 2023

Copyright: © 2023 by the authors. Licensee MDPI, Basel, Switzerland. This article is an open access article distributed under the terms and conditions of the Creative Commons Attribution (CC BY) license (https://creativecommons.org/licenses/by/4.0/).

1. Introduction

In recent years, ion channels have emerged as potential targets for cancer treatment [1–4]. This should not surprise, considering the multitude of physiological cellular processes in which they take part [5]. To present a few examples: changes in the membrane potential are important for the regulation of the cell cycle [6,7]; Cl^-, K^+ and Ca^{2+} channels are involved in apoptosis [8]; and the regulation of cell volume requires a class of specialized volume-sensitive channels [9,10]. Among the different ion channels that populate the membranes of cancer cells, many studies have focused on the Ca^{2+}-gated K^+ channel KCa3.1 (commonly known as IK), which is encoded by the gene *KCNN4* [11]. IK was first discovered in the 1950s by Gárdos, who observed that intracellular Ca^{2+} can enhance the K^+ permeability of human erythrocytes [12]. The channel was later cloned and heterologously expressed in different cell types, allowing a detailed characterization of its biophysical properties and pharmacology [13–15]. IK opens in response to an increase in $[Ca^{2+}]_i$. The gating mechanism became clear with the cryo-EM structure solved by Lee and MacKinnon [16]. Briefly, IK is a homo-tetrameric protein displaying an architecture resembling that of non-swapped 6-TM K^+ channels [16] with three characteristic cytosolic helices at the C-terminal end of S6 [16]. Pore opening is regulated by intracellular Ca^{2+} binding to calmodulin (CaM) [17], which is constitutively associated with the C-terminal cytosolic helices of the channel [16]. Most of the known IK blockers, including the antimycotic-derived ones, inhibit IK by directly binding the channel pore module just below the selectivity filter [16,18] (see Table 1). Such molecules have been proposed as potential cures for medical conditions in which the IK channel is directly involved.

It has been known since the late 1990s that the IK channel is blocked by the scorpion toxin charybdotoxin and by clotrimazole, which is a member of the imidazole antimycotics family [13,15,18–20]. Clotrimazole blocks the channel very efficiently but is also an inhibitor of cytochrome P450 (CYP) enzymes from different species (see for example [21–25]). In order to obtain a more selective molecule, Wulff and colleagues used a rational design strategy to develop a clotrimazole analog lacking the imidazole ring, which is strictly required for cytochrome P450 inhibition, obtaining the molecule known as TRAM-34 (IC_{50} = 20 nM) [26]. However, TRAM-34 displays a few shortcomings, which include a low oral bioavailability, even after enteric coating, and a short half-life [27,28]. Moreover, it has been reported that TRAM-34 can inhibit human and rat CYP isoforms although at relatively high concentrations [29]. Three years later, senicapoc, another IK inhibitor, was developed [30]. Senicapoc, when compared to TRAM-34, has a longer half-life, is more orally bioavailable in humans, has a lower IC_{50} (11 nM), displays an increased metabolic stability and no effects on CYP enzymes have been reported [18,27,30–32].

Table 1. Names, 2D structures and references of the compounds mentioned in the introduction.

Name	2D structure	Reference
Clotrimazole	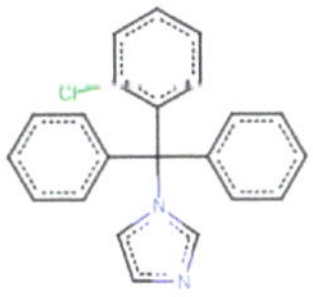	Reference [20]
Tram-34	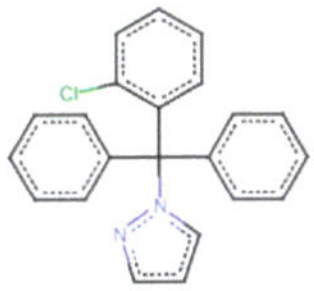	Reference [26]
Senicapoc	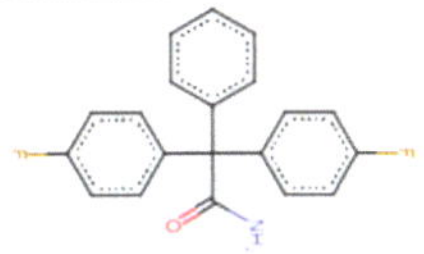	Reference [30]

As previously mentioned, IK expression has been reported to be altered in different types of cancer cells. Epigenetic dysregulation of the *KCNN4* gene, leading to a high-level expression of IK, has been correlated with the aggressiveness of lung cancer [33], and channel upregulation has been observed also in glioblastoma cells [34]. Moreover, channel expression and activity turned out to be important for the progression of the cell cycle, as observed for example in breast cancer and endometrial cancer cells [35,36]. It is indeed well known that potassium channels are involved in the regulation of the cell cycle of healthy and tumor cells [6]. In this scenario, different groups have employed IK blockers in the attempt to arrest the cell cycle of cancer cells and to reduce their migration. It was reported that 20 to 30 µM TRAM-34 can slow down the proliferation and the migration speed of lung cancer cells and that senicapoc, when administered in vivo, reduced the tumor mass in mice [33]. Experiments on intrahepatic cholangiocarcinoma cells showed that 40 µM TRAM-34 induced a reduction of ~50% on cell proliferation after 72 h and decreased invasiveness and migration; also in this case, senicapoc was able to reduce the tumor mass in vivo [37]. TRAM-34, at concentrations up to 40 µM, can arrest the cell cycle and therefore the growth of endometrial cancer cells; the same effect could be observed, at lower concentrations, with the less IK-specific clotrimazole [35]. The latter has been reported to block (at a concentration of 20 µM) breast cancer cells in the G1 phase [36] and to decrease the proliferation of pancreatic cancer cells [38,39]. Also in non-small cell lung cancer cells [40], cervical cancer cells [41], and triple-negative breast cancer cells [42], IK block reduced proliferation. It should be noted that the concentrations of the various IK blockers used in these cellular and in vivo studies are in the tens of μM range, which is much higher than the reported EC50 concentrations required to achieve channel block in patch-clamp experiments. The reason is that clotrimazole, TRAM-34 and senicapoc bind to serum proteins [31,40,43,44], such that their effective concentration is much lower in the presence of serum, which is commonly not present in the patch-clamp experiments.

In recent years, our lab has been studying ion channels expressed in different types of cancers [45–47]. In particular, we are interested in understanding the importance of these proteins in the viability, proliferation and migration of melanoma and pancreatic duct cancer cells.

In the present work, we characterized the impact of IK channel blockers on the proliferation and the migration of the metastatic melanoma cells WM266-4 [48]. A deeper knowledge of melanoma metastasis is indeed crucial for the battle against this type of cancer, as its outcome becomes worse when the tumor starts to invade other tissues [49]. The 5-year survival rate drops from 93% for stage IIIA to 32% for stage IIID [50], and for patients with stage IV metastatic melanoma, the median survival is less than one year [51].

Thus, understanding whether IK channel blockers act on cancer cells also through IK-independent mechanisms and whether a clear correlation between their effects and IK expression levels exists could help elucidate the molecular mechanisms underlying melanoma development. Channel modulators can affect cells in multiple ways. For example, we recently found that activators of the BK channel can change the intracellular Ca^{2+} concentration of melanoma cells and that the VRAC blocker DCPIB directly activates the BK channel as an off-target effect [46,52]. For this reason, in the present work, we also tried to address general questions regarding IK blockers action on cancer cells: are the obtained results solely due to channel blockade? Is there a correlation between the presence in the membrane of functional channels and the observed effects of IK blockers? To this purpose, we employed two different blockers and two different cell lines. We chose clotrimazole, known for having side effects, and senicapoc, one of the most IK-selective blockers available. The two compounds were tested in parallel on two cell lines: WM266-4 and the pancreatic duct adenocarcinoma (PDAC) line Panc-1, which is reported to express little or no IK protein and to lack IK-like currents [38,39,53] and therefore is regarded as a reference cell line. To our surprise, migration and proliferation were decreased in both cell lines regardless of IK expression levels. To obtain more insights into this unexpected phenomenon, we performed electrophysiology and calcium imaging experiments. The results suggest that indeed, the effect of IK blockers on cancer cells might not strictly depend on a robust presence of functional channels in the membrane.

2. Results

2.1. IK Blockers Reduce Viability of Both WM266-4 and Panc-1 Regardless of Expression Levels

In order to confirm the data reported in the literature regarding the low expression of IK in Panc-1 cells, we performed RT-qPCR experiments (Figure 1A). Such an analysis showed that the channel relative expression is much higher in WM266-4 cells compared to Panc-1, validating our idea of using these two cell lines to test for a correlation between IK blockers effects and channel expression. We therefore proceeded to test the impact of clotrimazole and senicapoc on cell viability using the MTT assays. Based on previous studies, cells were seeded in a multi-well plate and treated with 30 μM clotrimazole or 30 μM senicapoc or the corresponding volume of the solvent DMSO for 72 h. In the literature, it is reported that the concentration of serum in the medium can affect the impact of IK blockers on cancer cell proliferation. Indeed, it was shown that a decrease in the proliferation of PDAC cells could be obtained also at lower concentrations of IK inhibitors as long as the serum concentration was 1% [38]. We preferred to use higher concentrations of the compounds rather than modify such a critical factor for cell growth as the serum in the medium.

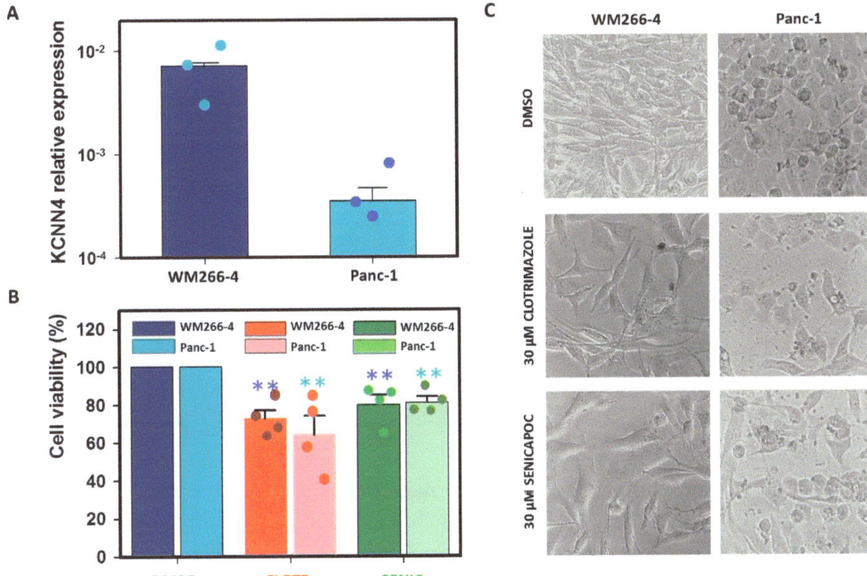

Figure 1. IK blockers affect WM266−4 and Panc−1 viability regardless of IK channel expression. (**A**) Relative *KCNN4* expression from RT-qPCR showing the difference in *KCNN4* mRNA levels between WM266-4 and Panc-1 (N = 3). (**B**) Results from MTT viability assays after 72 h of exposure to 30 μM clotrimazole, 30 μM senicapoc or the corresponding amount of DMSO (N = 4 for all). Data are reported as absorbance (570 nm) ratio drug-/DMSO-treated cells (color code reported in the legend). (**C**) Exemplary pictures from experiments in (**B**): WM266-4 and Panc-1 cells treated for 72 h with 30 μM clotrimazole, 30 μM senicapoc or the corresponding amount of DMSO. Significance level is indicated by two asterisks ($p < 0.01$).

In agreement with its lack of specificity for IK, clotrimazole caused a reduction in viable cells with respect to the DMSO-treated controls in both WM266-4 and Panc-1 cells (Figure 1B,C; $-27.4 \pm 4.6\%$ for WM266-4 and $-36 \pm 10\%$ for Panc-1). To our big surprise, also the presumably IK-selective senicapoc induced a decrease in cell viability in both cell lines to a similar extent (Figure 1B,C; $-20.2 \pm 5.1\%$ for WM266-4 and $-19 \pm 3.1\%$ for Panc-1 with respect to DMSO-treated cells), in contrast with the results of channel expression analysis, which showed that the *KCNN4* mRNA levels are higher in WM266-4 compared to Panc-1.

2.2. Clotrimazole and Senicapoc Affect the Migration of WM266-4 and Panc-1

The MTT assay results suggested that IK channel expression is not proportional to the impact of clotrimazole and senicapoc on cell growth, so we went further by testing if such a correlation could be present regarding cell migration. To evaluate cell migration, we performed trans-well migration assays and wound (scratch)-healing assays on both WM266-4 and Panc-1, which is our reference for poor IK expression. To perform trans-well migration, cells were seeded on the upper side of the cell-permeable membranes (8 μm pores, see methods) in serum-free medium containing 30 μM clotrimazole or 30 μM senicapoc or the corresponding concentration of DMSO, while on the other side of the membrane, a complete medium served as a chemo-attractor to stimulate migration. Migrated cells were counted after 24 h (see Supplementary Figure S1A). In Figure 2A, we report the obtained data as the migration rate: the number of drug-exposed migrated cells compared to those which migrated after treatment with DMSO. The reduction in the migration induced by clotrimazole was similar between the two cell lines. For WM266-4, migrated cells were 73.9 ± 16% of those migrated in DMSO, while for Panc-1, they were 73 ± 8%. However, the IK-selective senicapoc impacted the migration of both cell lines as well: WM266-4 migrated cells were 59.4 ± 7.3% of those migrated in DMSO, while the ratio for Panc-1 was 74.2 ± 5%.

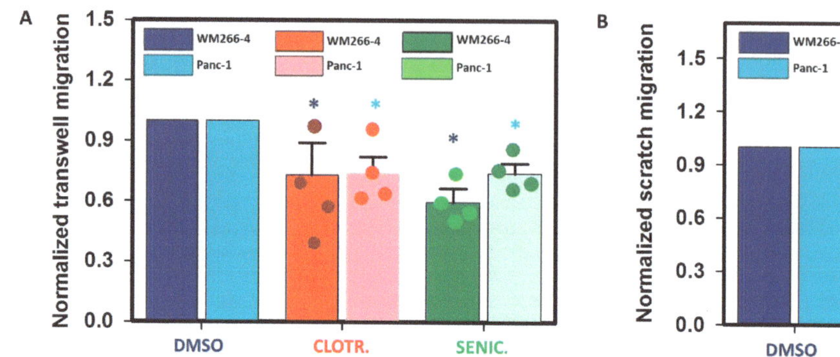
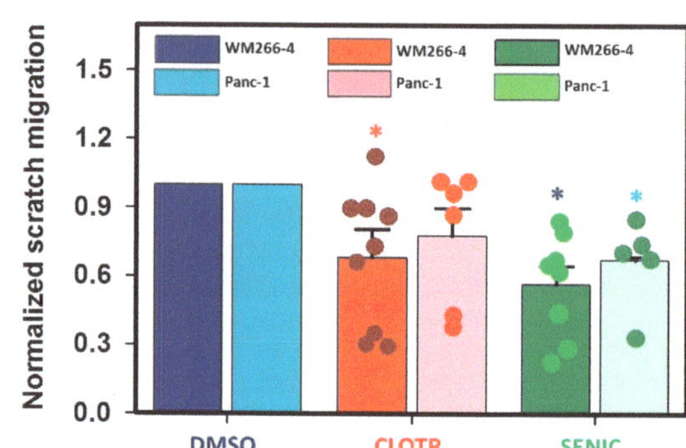

Figure 2. Clotrimazole and senicapoc decrease the migration of WM266-4 and Panc-1. (**A**) Migration rate from trans-well migration assays of cells exposed to 30 µM clotrimazole or 30 µM senicapoc with respect to DMSO-treated cells (N = 4 for all). Different cell lines and treatments are color coded as reported in the legend. (**B**) Relative increases of cell-covered areas at t = 24 h with respect to t = 0 from the same petri dish for WM266-4 (DMSO N = 9, clotrimazole N = 9, senicapoc N = 8) and Panc-1 (DMSO N = 6, clotrimazole N = 6, senicapoc N = 5). Data are significantly different from control for WM266-4, clotrimazole ($p = 0.0128$), WM266-4, senicapoc ($p = 0.001$), and Panc-1, senicapoc ($p = 0.0246$). Significance level is indicated by an asterisk ($p < 0.05$).

To further evaluate the impact of the employed compounds on WM266-4 and Panc-1 migration ability, we performed also scratch-healing assays. Cell migration from the scratch edges was monitored with a holographic microscope (see Supplementary Figure S1B). Figure 2B shows the movement of scratch edges as increases of cell-covered areas after 24 h long treatment with 30 µM clotrimazole or 30 µM senicapoc with respect to control DMSO-treated cells. In both cell lines, we observed a high variability in wound edges after 24 h long exposure to 30 µM clotrimazole. From experiments with senicapoc, we obtained more uniform datasets which showed that wound healing was significantly lower than that of DMSO-treated cells for both WM266-4 and Panc-1. Similarly to trans-well migration, WM226-4 cells were more affected by senicapoc than Panc-1 cells, which were nevertheless sensitive to the molecule. This indicated that the reduction in WM266-4 cell migration is possibly caused by a combination of IK blockage and an IK-independent effect, the latter being responsible for the Panc-1 migration decrease.

2.3. Characterization of WM266-4 and Panc-1 Whole-Cell Currents in High Intracellular Ca^{2+}

Our results indicated that the IK-selective compound senicapoc can impact the growth and the migration of WM266-4 cells, but also of Panc-1, in which IK is poorly expressed. Since the presence of mRNA coding for a given channel does not always correlate with the presence of the correspondent conductance in the membrane, we performed whole-cell patch-clamp experiments to assess the presence of Ca^{2+}-activated currents with properties of IK channels in WM266-4 and Panc-1 cells. In WM266-4, we observed that 1 µM Ca^{2+} in the recording pipette triggered a voltage-independent current, which could not be detected when an intracellular solution with nominally 0 mM free Ca^{2+} was employed (Figure 3A–D). Currents reverted from negative to positive values at negative potentials as can be seen from the traces reported in Figure 3C,D, where pre- and post-pulse voltages correspond to the cell resting potential (around −50 mV). We observed a high variability in current amplitude among different cells, and some cells did not even display any Ca^{2+}-activated current (out of 33 patched WM266-4 cells, 12 had currents at 100 mV smaller than 0.5 nA, 13 had currents between 0.5 and 1.5 nA, and 8 had currents larger than 1.5 nA; see Figure 3B). The Ca^{2+}-elicited currents were strongly reduced by senicapoc (Figure 3C,E) and clotrimazole (Figure 3D,F) already at a concentration of 1 μM. Figure 3G,H report the degree of block by the two compounds of the Ca^{2+}-evoked currents at the voltages corresponding to their maximal amplitude.

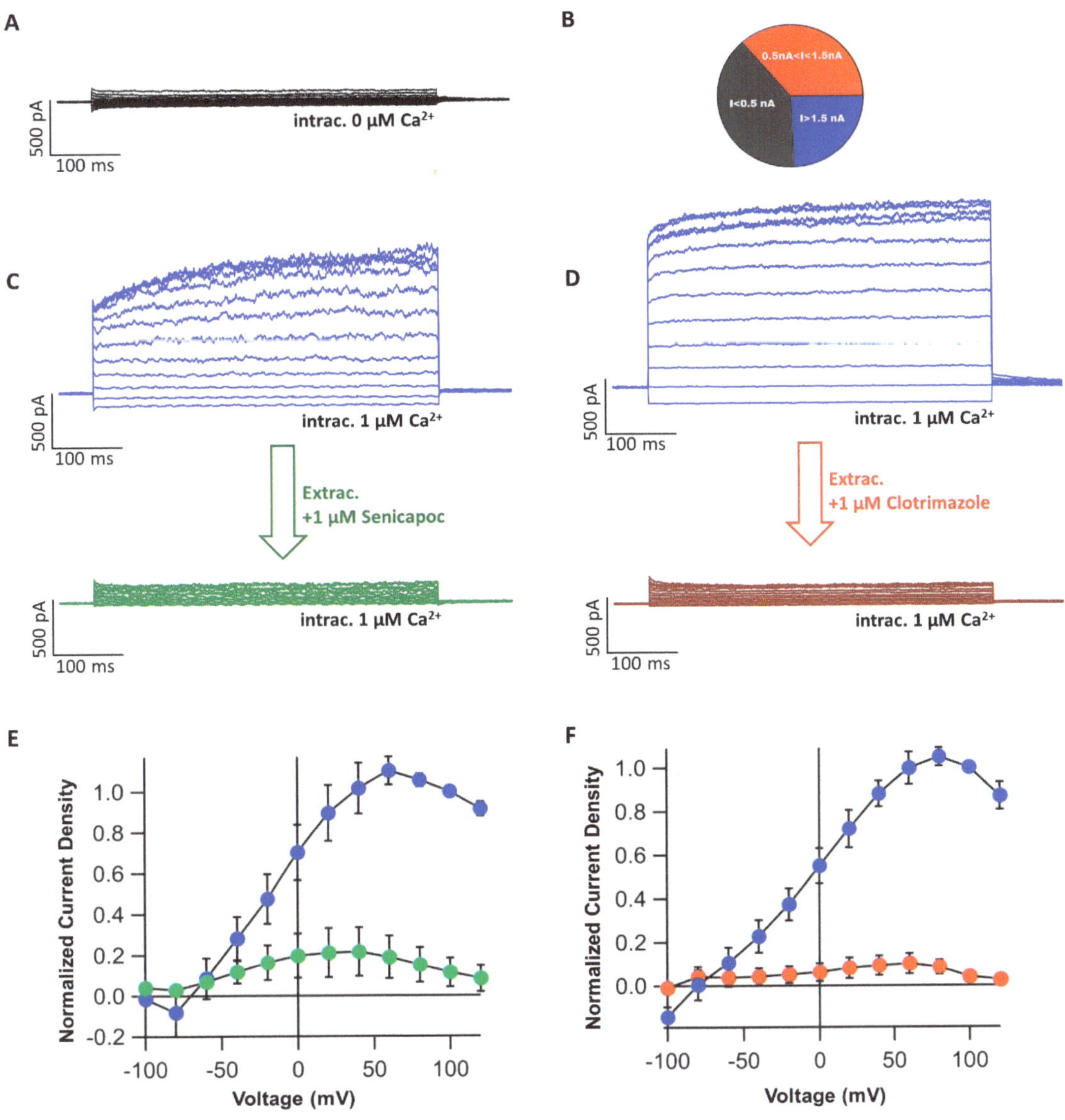

Figure 3. *Cont.*

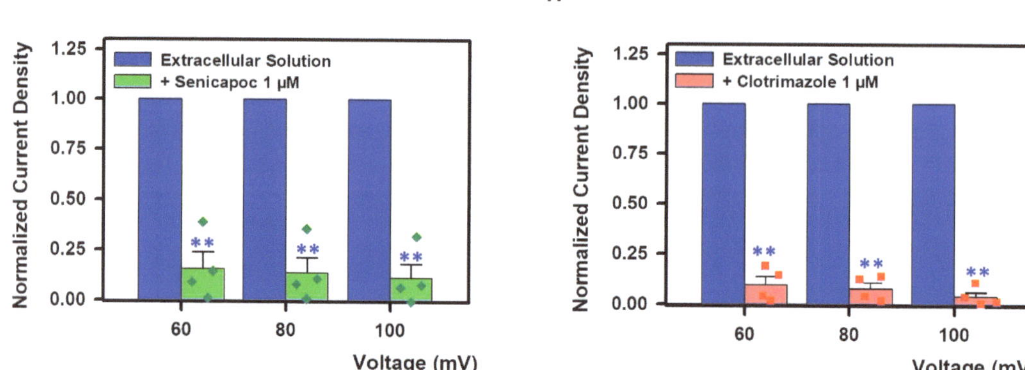

Figure 3. Ca^{2+}−evoked whole-cell currents of WM266−4 cells. (**A**) Exemplary current traces of WM266-4 whole-cell currents when the patch pipette was filled with the Ca^{2+}-free intracellular solution. (**B**) Distribution of current amplitudes in WM266-4 cells measured at 100 mV with 1 µM Ca^{2+} in the pipette solution. (**C**) Exemplary WM266-4 current traces from recordings with 1 µM Ca^{2+} in the intracellular solution; cells were perfused with standard bath solution (top) or with the same solution + 1 µM senicapoc (bottom). (**D**) Exemplary Panc-1 current traces from recordings with 1 µM Ca^{2+} in the intracellular solution; cells were perfused with standard bath solution (top) or with the same solution + 1µM clotrimazole (bottom). (**E**) Average normalized IVs of WM266-4 cells before (blue circles) and after (green circles) application of 1 µM senicapoc (N = 4 cells). Currents are normalized to the value at 100 mV. (**F**) Average normalized IVs of WM266-4 cells before (blue circles) and after (red circles) application of 1 µM clotrimazole (N = 4). Currents are normalized to the value at 100 mV. (**G**) Background-subtracted currents (background was calculated from the mean of 4 cells measured in Ca^{2+}-free conditions) in the presence/absence of senicapoc normalized to the currents measured in standard bath solution at the same voltage in the same cells (mean ± SE, N = 4). (**H**) Background-subtracted currents in the presence/absence of clotrimazole normalized for the currents measured in standard bath solution at the same voltage in the same cells (mean ± SE, N = 4). Significance level is indicated by two asterisks ($p < 0.01$).

As previously mentioned, we observed a relatively large variability in the amplitude of the Ca^{2+}-dependent currents, while the currents recorded in the absence of intracellular Ca^{2+} were small and more comparable among different cells. For this reason, the apparent degree of block is also variable between different cells, as background conductances influence the results more strongly in cells that express less IK current. To obtain a more accurate quantification of the inhibitory effect, we subtracted background currents from the currents recorded in the presence/absence of the blockers. Both senicapoc and clotrimazole strongly reduced the currents induced by 1 µM Ca^{2+}. The features and the pharmacology of the above-described currents strongly suggested that they were carried by IK channels.

Since IK blockers had a significant impact also on viability and migration properties of Panc-1 cells, we needed to verify if these cells showed IK-like currents in the membrane as elsewhere reported [53]. We therefore applied 1 µM senicapoc (Figure 4A) or 1 µM clotrimazole (Figure 4B) to Panc-1 cells measured in the whole cell configuration with 1 µM intracellular Ca^{2+}. Before drug application, large BK currents were seen in all patched cells [45,52], and none of the drugs had an appreciable effect (Figure 4A–C). Conversely, the subsequent application of 1 µM paxilline blocked most of the currents as expected (Figure 4A–C). Overall, these experiments exclude a significant presence of functional IK channels in Panc-1 cells.

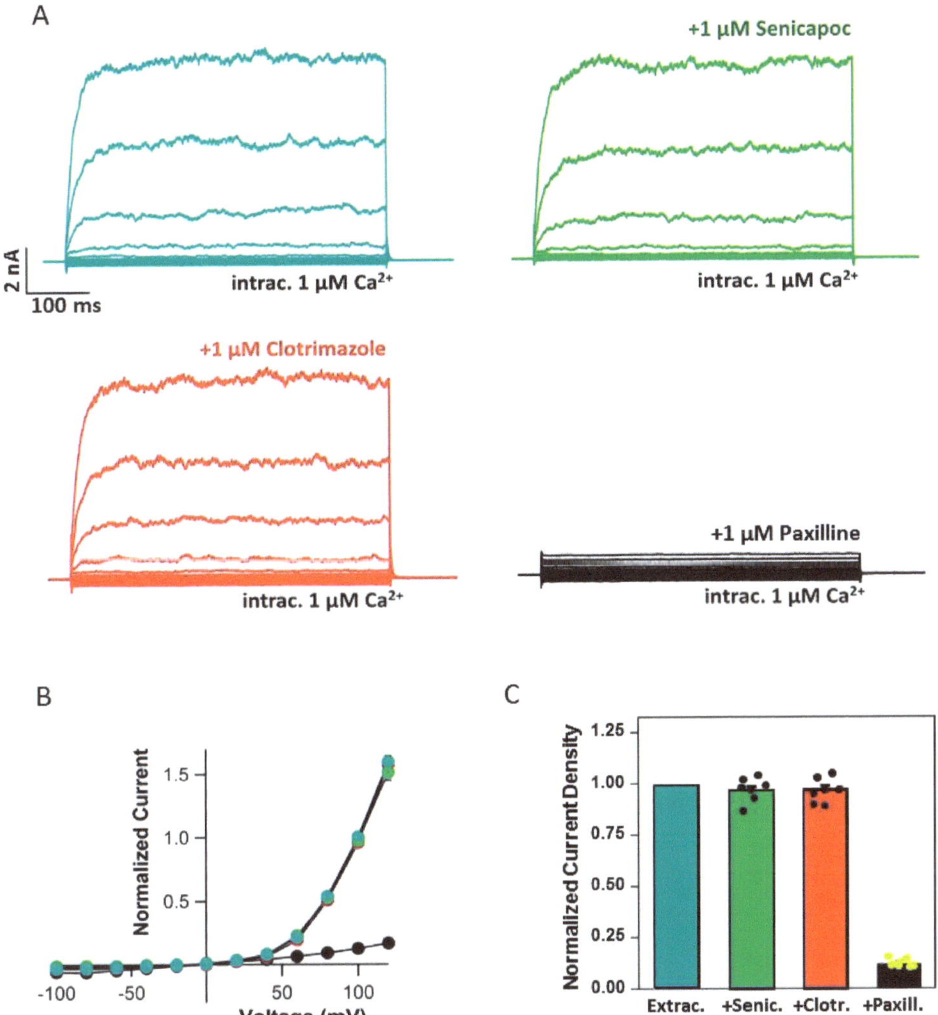

Figure 4. Panc−1 cells lack functional IK channels. (**A**) Current traces measured from a typical Panc-1 cell with 1 µM Ca^{2+} in the patch pipette in standard extracellular solution (top left), during perfusion with 1 µM senicapoc (top right), with 1 µM clotrimazole (bottom left) or 1 µM paxilline (bottom right). (**B**) Average normalized current voltage relationship in control conditions (turquoise symbols), in 1 µM senicapoc (green symbols, N = 7), 1 µM clotrimazole (red symbols, N = 7) and 1 µM paxilline (black symbols, N = 7) (currents are normalized to those measured in control conditions at 100 mV; error bars indicate SEM). (**C**) Average current density normalized to that measured in control conditions at 100 mV in the indicated conditions.

2.4. IK Blockers Do Not Affect the Intracellular Ca²⁺ Concentration

We have previously reported that K^+ channel modulators can alter the intracellular calcium concentration ($[Ca^{2+}]_i$) of cancer cells [46]. It was also observed by other authors that the IK channel can regulate calcium entry: in prostate cancer cells, its activation induces calcium uptake through TRP channels by increasing the driving force for calcium [54]. To check if our results could be explained by $[Ca^{2+}]_i$ alterations, we performed Ca^{2+}-imaging experiments on Panc-1 and WM266-4 cells testing whether clotrimazole or senicapoc can have some effects on $[Ca^{2+}]_i$. $[Ca^{2+}]_i$ was monitored over time with the fura-2 fluorescent probe before and after the application of 30 µM clotrimazole or 30 µM senicapoc. In Figure 5A,C the mean $[Ca^{2+}]_i$ values from these experiments are reported. Neither WM266-4 cells nor Panc-1 showed a significant variation of $[Ca^{2+}]_i$ after the application of the two molecules; rather, there were only small fluctuations that were never larger than 15 nM. The average $\Delta[Ca^{2+}]_i$ values (with respect to the standard bath solution) after the perfusion of clotrimazole and senicapoc on WM266-4 cells were only 4 ± 1.3 nM and 5 ± 2.5 nM, respectively (Figure 5B, N = 32 cells). Such small variations in $[Ca^{2+}]_i$ were observed also for Panc-1: $\Delta[Ca^{2+}]_i$ values were 12 ± 5 nM for clotrimazole and 5 ± 2 nM for senicapoc (Figure 5D, N = 32 cells). The lack of significant variations in $[Ca^{2+}]_i$ caused by the two compounds indicated that they induced neither the activation of a Ca^{2+} membrane conductance nor the release of Ca^{2+} from intracellular stores.

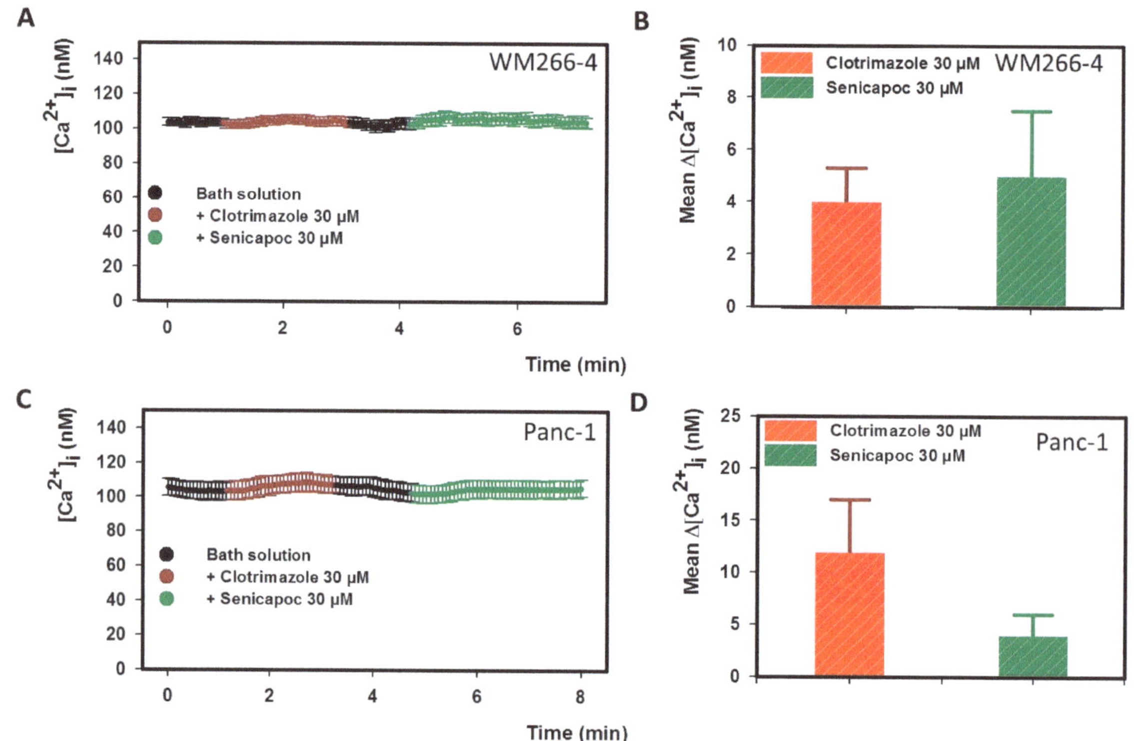

Figure 5. Senicapoc and clotrimazole do not alter the intracellular Ca^{2+} concentration. (**A**) $[Ca^{2+}]_i$ over time from 32 WM266-4 cells (color code in the legend). (**B**) Mean $\Delta[Ca^{2+}]_i$ from (**A**) with respect to bath solution (color-coded as in (**A**), N = 32). (**C**) Mean $[Ca^{2+}]_i$ over time from 32 Panc-1 cells (color-coded as in (**A**)). (**D**) mean $\Delta[Ca^{2+}]_i$ from (**C**) with respect to bath solution (color-coded as in (**C**), N = 32).

2.5. Clotrimazole and Senicapoc Do Not Alter the F-Actin Organization

Another hypothesis we formulated to explain our unexpected results was that IK blocker treatments could affect cellular cytoskeleton organization. This idea emerged since the invasive migration of WM266-4 and Panc-1 was altered by the exposure to the tested molecules. Thus, in order to evaluate the effect of clotrimazole and senicapoc on Panc-1 and WM266-4 actin organization, we utilized phalloidin staining, which was able to bind the filamentous actin (F-actin). Phalloidin staining did not appear to modify both WM266-4 and Panc-1 cells after 24 h of treatment (Figure 6). However, changes in the F-actin organization could be detected after 72 h of treatment with 30 µM of clotrimazole or senicapoc (Supplementary Figure S2). These observations suggested that exposure to clotrimazole or senicapoc did not significantly alter the F-actin organization of Panc-1 and WM266-4 cells after 24 h.

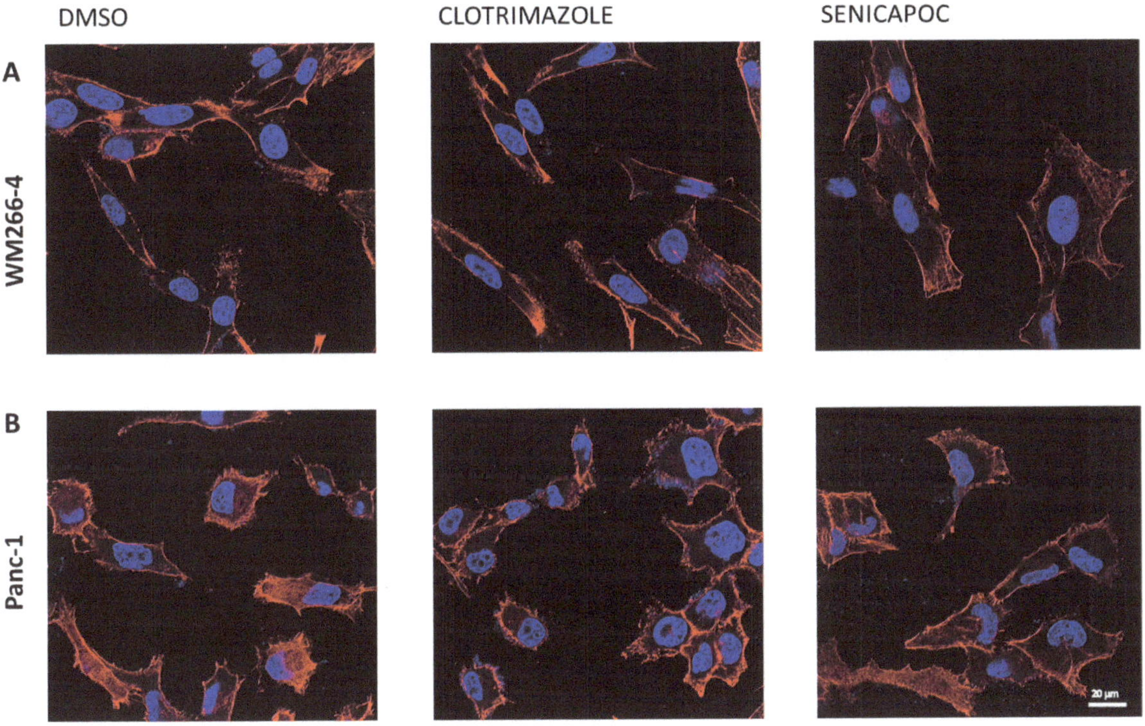

DMSO CLOTRIMAZOLE SENICAPOC

A

WM266-4

B

Panc-1

20 µm

Figure 6. Effect of IK blockers on F-actin organization. WM266-4 (**A**) and Panc-1 (**B**) cells incubated for 24 h in vehicle alone or with 30 µM clotrimazole, 30 µM senicapoc or the corresponding volume of DMSO, which have subsequently been labeled with phalloidin (red) and DAPI (blue) and processed for fluorescence microscopy (see Section 4).

2.6. Effect on Cell Viability of BA6b9, an Allosteric IK Blocker

Our data indicate that the clotrimazole-derived compound senicapoc exerts IK-independent effects on PDAC cells. Since clotrimazole and senicapoc share the same inhibition mechanism (they act on the channel pore), we decided to test if IK inhibitors with a different block mechanism were able to induce a decrease in Panc-1 viability similar to what we observed with clotrimazole and senicapoc. To this purpose, we employed an allosteric blocker called BA6b9, which acts on the CaM-PIP2-binding domain at the interface of the proximal carboxyl terminus and the linker S4–S5 [55]. We first tested whether the compound was able to inhibit IK currents in WM266-4 cells. As shown in Figure 7A,B, 20 µM BA6b9 inhibited around 60% of currents induced by 1 µM intracellular Ca^{2+}, while 60 µM BA6b9 inhibited around 80%, with the residual currents being at least partially unspecific leak. The degree of inhibition is line what has been reported by Burg et al. [55]. We next repeated the viability assays using 30 µM, 60 µM and 100 µM BA6b9 employing the same experimental conditions as in Figure 1B. Interestingly, the compound did not induce any significant alteration in Panc-1 viability (Figure 7C), while we could only observe a slight but significant decrease in WM266-4 viable cells (Figure 7C).

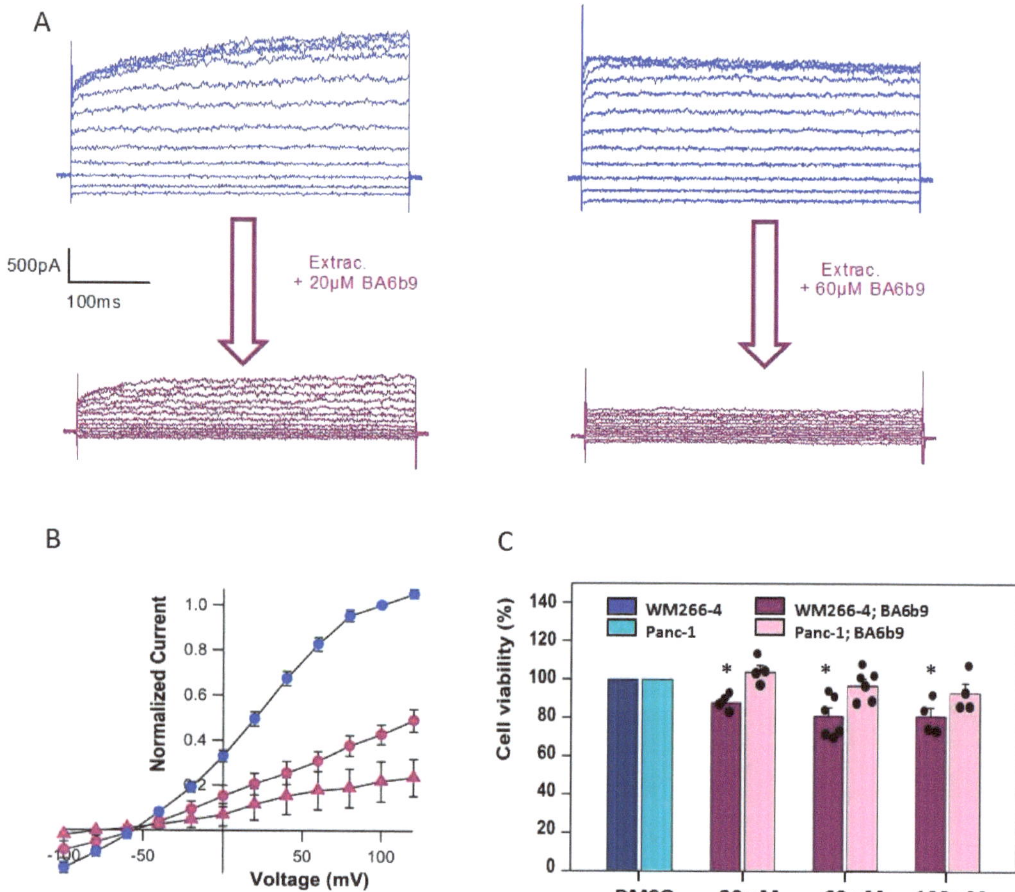

Figure 7. Effect of BA6b9 on IK−currents in WM266−4 cells and on viability in WM266−4 and Panc−1 cells. (**A**) Example currents measured from WM266-4 cells with 1 µM Ca^{2+} in the patch pipette in standard extracellular solution (top) and during perfusion with 20 µM (left) or 60 µM (right) BA69B. (**B**) Average normalized current voltage relationship in control conditions (blue symbols), in 20 µM BA6b9 (magenta circles), and 60 µM BA6b9 (magenta triangles) (currents are normalized to those measured in control conditions at 100 mV; N = 4 each, error bars indicate SEM). (**C**) Results from MTT viability assays after 72 h of exposure to DMSO (control) and the indicated concentrations of BA6b9 (N ≥ 4). Data are reported as absorbance (570 nm) ratio drug-/DMSO-treated cells (color-code reported in the legend. Significance level is indicated by an asterisk ($p < 0.05$).

3. Discussion

Like other K^+ channels, IK was reported to have important roles in the proliferation and the migration of cancer cells; accordingly, clotrimazole, TRAM-34 and senicapoc were able to reduce cancer cells' growth and migration ability as well as induce a decrease in tumor mass when administered in vivo [33–39]. It was also observed that the expression of IK varies throughout the cell cycle and seems to be important for its correct progression [36]. In these studies, the importance of IK for cancer cell tumorigenic processes has usually been highlighted by comparing aggressive IK-overexpressing cancer cells with blocker-exposed cells of the same line, siRNA IK-knocked-down cells or healthy cells from the same tissue of the tumor of interest.

Here, we chose a different but complementary approach to study the impact of IK blockers on the growth/migration of cancer cells, in particular on metastatic melanoma cells WM266-4. We selected clotrimazole and senicapoc as the two inhibitors: the former is already known to have side effects besides blocking the IK channel, while the latter is reported to be IK-selective. The idea of comparing the effects of these two compounds was to test if we could distinguish an IK-dependent from an IK-independent component in case the two added up. Moreover, we decided to employ Panc-1 cells from primary pancreatic cancer as a term of comparison, since they poorly express the IK channel. The latter idea turned out to be quite fruitful, as the most insightful results were obtained from the comparison of the two cell lines.

To estimate cell growth, we seeded the same number of Panc-1 and WM266-4 cells and performed MTT viability assays after a 72 h long treatment with clotrimazole, senicapoc or the corresponding amount of their solvent DMSO. The unspecific clotrimazole caused a decrease in viability in both WM266-4 and Panc-1 cells. This result is compatible with previous observations that clotrimazole inhibits the activity of cytochrome P450, a large family of heme-containing oxidases, which play essential roles in endogenous signaling and metabolic pathways. However, intriguingly, also senicapoc was able to induce a reduction in cell viability for both cell lines. This decrease was similar for WM266-4 and Panc-1, so it did not reflect the difference in IK expression assessed by RT-qPCR.

The ability to migrate through the 8 μM wide pores is a good way to estimate the ability of cancer cells to migrate and invade other tissues, therefore, we performed trans well migration assays. We found that clotrimazole affected the migration ability of both cell lines, which was probably as a result of its lack of target specificity. Surprisingly, as observed for cell viability, also the IK-specific compound senicapoc affected both cell lines: the molecule induced a decrease in cell migration in both Panc-1 and WM266-4. Similar results were obtained with wound-healing assays. The latter were performed to observe the combination of the molecules' actions on proliferation and migration at the same time, and for this reason, we did not employ proliferation inhibitors.

The data collected suggested that growth and migration might be hampered in WM266-4 cells via channel blockade by senicapoc but, in addition, also by an IK-independent effect visible in the control Panc-1. Therefore, our data seemed to diverge from the most accepted hypothesis about the action mechanism of IK blockers on cancer cells: that is, via inhibiting IK-mediated K^+ conductances. We therefore performed patch-clamp experiments to evaluate the presence/absence of IK conductances in WM266-4/Panc-1 cells plasma membranes. In WM266-4 cells, we could measure Ca^{2+}-triggered, voltage-independent currents whose features and pharmacology suggested that they were mediated by IK channels. The current density varied between different cells, which is compatible with the fact that IK over-expression in cancer cells is not constitutive but occurs only in certain phases of the cell cycle [35,36]. Upon exposure to senicapoc, the Ca^{2+}-triggered currents dramatically dropped toward the level of the background, Ca^{2+}-independent, ones. This suggested that IK was indeed the main mediator of the recorded currents. It excludes also the significant presence of other calcium-activated K channels, like for example BK, which is in accordance with what has been published earlier [53]. Currents like those recorded in WM266-4 could not be observed in Panc-1 cells when measured in the whole-cell configuration with Ca^{2+}-enriched pipette solution. This confirms our hypothesis that the reduction in migration and proliferation observed in drug-exposed Panc-1 is not the direct result of IK blockade. It should, however, be kept in mind that the electrophysiological recordings only reveal IK channels localized in the plasma membrane. Thus, we cannot exclude putative effects on IK channels localized to intracellular membranes.

As outlined in the Introduction, the activation of IK can be expected to lead to an increase in $[Ca^{2+}]_i$ due to an increase in the electrochemical driving force. Conversely, the inhibition of IK might lead to a decrease in $[Ca^{2+}]_i$. However, in addition to these driving-force mediated effects, we know from our previous experience that K^+ channel modulators can alter the intracellular calcium concentration of cancer cells in more unspecific ways [37,46,52]. For example, it could be that the two molecules could induce calcium uptake from the extracellular environment directly by opening Ca^{2+} channels. Another hypothesis was that the employed inhibitors were able to induce the release of Ca^{2+} from intracellular stores. In both cases, we would expect to see a large raise in $[Ca^{2+}]_i$ after acute exposure to clotrimazole and senicapoc. Calcium imaging experiments on WM266-4 and Panc-1 cells did not show any significant increase in $[Ca^{2+}]_I$ after the perfusion of the two compounds, excluding that they can have secondary effects similar to those observed, for example, for BK modulators [46].

Regarding the decrease in the trans-well motility of both WM266-4 and Panc-1, we reasoned that the compounds might affect the cytoskeleton organization of these cells. To test this hypothesis, cells were fixed and stained with phalloidin after exposure to clotrimazole or senicapoc. No changes in F-actin organization could be detected after 24 h treatment with senicapoc or clotrimazole, suggesting that blocker effects on migration are not correlated with large alterations of the F-actin organization. Thus, it is difficult to distinguish whether the rather marked effects on actin organization and cell size and shape seen after 72 h treatment are caused by a direct action on F-actin organization, are indirect consequences of other cellular alterations, or are simply linked with the loss of cell viability. Similar changes in F-actin organization have been reported for other ion channel modulators [56]; however, also in this case, it was difficult to distinguish between direct and indirect effects.

Taken together, the results we collected from all the experiments supported our hypothesis that clotrimazole and senicapoc can affect the carcinogenic behavior of PDAC cells independently of the presence of IK conductances in the plasma membrane. To test whether this is a specific shortcoming of clotrimazole-derived compounds, we tested the viability of WM266-4 and Panc-1 cells after 72 h of treatment with another molecule, namely BA6b9, which is an allosteric blocker of the IK channel. BA6b9, instead of binding the pore module of IK like antimycotic-derived compounds, hampers crucial interactions between S4–S5 linker, CaM and PIP2 [55]. We confirmed that BA6b9 inhibited IK currents in WM266-4 cells, necessitating however larger concentrations compared to clotrimazole and senicapoc, which is in agreement with the literature [55]. When we repeated MTT assays with BA6b9, we observed that Panc-1 cells were not affected by the exposure to this drug even at concentrations up to 100 μM. These experiments suggest that the reduction in viability observed in Panc-1 cells after treatment with senicapoc was due only to interactions of the molecule with secondary targets. Such secondary targets do not seem to be shared with BA6b9. Regarding WM266-4, we observed a slight viability decrease in BA6b9-exposed cells, but we believe that further studies will be required to determine if this drug can be used as a tool to reduce cancer cells viability and migration by targeting the IK channel. We used BA6b9 to test our hypothesis about the promiscuous behavior of senicapoc, but a deeper and systematic characterization of the impact of this compound on neoplastic cells can be an interesting topic for a whole new project.

The importance of IK channel expression for cancer progression has been suggested in a number of studies: knocking down the *KCNN4* gene can reduce carcinogenic behavior, and treatment with IK blockers had been reported to have an outcome comparable to that of knocking down the IK gene [37,38]. Therefore, we believe that IK blockade might reduce cancer cell growth. Nevertheless, in the present work, we discovered that relatively high concentrations of IK blockers can affect the proliferation and the migration also of cancer cells that do not display IK conductance in the plasma membrane. This suggests that IK blockade of plasma membrane-localized IK channels might not be the only mechanism by which senicapoc and other compounds exert their action on cancer cells. Another possible mechanism could be related to the presence of IK channels on the membrane of intracellular organelles like mitochondria [53], although this pathway is strongly dependent on the chemical nature of the drug. This hypothesis could be further analyzed in future studies. We believe that our findings should be taken into consideration when considering IK blockers (existing or to be developed in the future) as tools to slow down cancer growth and metastasis formation.

4. Materials and Methods

4.1. Cell Culture

Melanoma cell line WM266-4 (RRID:CVCL_2765) was cultured in MEM medium (Thermo Fisher Scientific, Waltham, MA, USA) supplemented with 10% FBS, 2 mM L-glutamine, 100 U/mL penicillin, 100 µg/mL streptomycin, and 1% non-essential amino acids (Sigma-Aldrich, St. Louis, MO, USA). The PDAC line Panc-1 (RRID:CVCL_0480) was grown in high-glucose DMEM (Thermo Fisher Scientific) enriched with 10% FBS, 100 U/mL penicillin, 100 µg/mL streptomycin, and 4 mM L-glutamine (Sigma-Aldrich). Cells were grown at 37 °C in a 5% CO2/95% air atmosphere (ESCO Lifesciences Group, Singapore). To split the cells into new flasks or to seed them in petri dishes and multi-well plates, they were washed with PBS (Euroclone, Pero, Italy) and then detached from the flask with 1 mL of trypsin–EDTA solution (Sigma-Aldrich).

4.2. RT-qPCR

Total RNA was extracted from WM266-4 or Panc-1 cells grown until sub-confluency in 25 cm^2 flasks, using the PureLink RNA mini kit (Ambion Inc.-Austin, TX, USA), and 1 µg was reverse transcribed using the Super ScriptIV VILO cDNA synthesis kit (Thermo-Fisher Scientific, Milan, Italy). The obtained cDNAs were used as a template for RT-qPCR performed in the thermal cycler CFX Connect from Bio-Rad (Hercules, CA, USA). Gene expression was assessed by SYBR Green quantitative real time PCR using the SsoAdvanced Universal SYBR Green Supermix. The thermal protocol consisted of a denaturation step at 95 °C for 3 min, which was followed by 39 two-step cycles composed of a denaturation step at 95 °C for 10 sec and of annealing/extension at 55 °C for 30 sec. No template control (NTC) and no reverse-transcription control (NAC) were included to avoid false positives. Expression levels of the *KCNN4* target gene (encoding KCa3.1) were assessed in triplicate and then normalized to the expression of the housekeeping gene Actin. The following *KCNN4* primers were used: forward gctgcgtctctacctggtg; reverse cgatgctgcggtaggaag. Results were visualized with BIO-RAD software Bio-Rad CFX Manager-RRID:SCR_017251; Bio-Rad). We refer to the PCR cycle at which amplification fluorescence exceeds the background signal as the quantification cycle (Cq). Data are reported in the figures as relative expression with respect to the housekeeping gene GAPDH. This value is $2^{-\Delta Cq}$, where ΔCq is the difference between the Cq of the housekeeping gene and that of the target gene.

4.3. Materials

Clotrimazole, senicapoc and paxilline were purchased from Merck (Milan, Italy)and dissolved in DMSO to prepare stock solutions according to the information provided from the manufacturer. BA6b9 was kindly provided by Bernard Attali (Tel Aviv University). The compounds were added to solutions or mediums at the desired concentration prior to experiments. The final concentration of DMSO never exceeded 0.1%. MTT was purchased from Promega (Milan, Italy).

4.4. Cell Viability Assay (MTT)

WM266-4 and Panc-1 cells were seeded into 96-well plates (5×10^3 cells per well). The following day, the wells medium was replaced with fresh medium containing the indicated amount of clotrimazole and senicapoc or the corresponding volume of DMSO. Incubation with IK blockers lasted 72 h, after which 20 µL of 3-(4,5-dimethylthiazol-2-yl)-2,5-diphenyltetrazolium bromide solution (MTT, Promega) was added to each well. Cells were incubated with MTT for 2 h at 37 °C. Relative cell viability was derived from the absorbance ratio (at 570 nm) between drug- and DMSO-exposed cells (plate reader from

Molecular Devices, San Jose, CA, USA). For each experiment, each condition was tested in triplicate. Data are reported as cell viability normalized to the control condition (DMSO).

4.5. Trans-Well Migration Assay

WM266-4 and Panc-1 cells were suspended in serum-free medium containing the indicated final concentration of IK blockers or the corresponding volume of DMSO and then seeded into the upper side of cell culture inserts (8×10^3 cells/insert) with cell-permeable membranes (8 μm pores, Sarstedt, Nümbrecht, Germany). The inserts were placed in a 24-well plate, in which the wells contained 600 μL of complete medium. In this way, the presence of serum in the medium below the membrane acts as a chemo-attractant. After 24 h, the cells were fixed with cold methanol and stained with crystal violet (Sigma-Aldrich). Unmigrated cells were scraped away from the upper layer of the membrane with a cotton swab. Pictures of the membrane bottom layers were taken with the help of a microscope (Nikon, Tokyo, Japan). Migrated cells were counted with the software FIJI [57] (version 1.53c, https://imagej.net/ij/, accessed on 15 August 2023). For each experiment, each condition was tested in duplicate. Data are reported as migrated cells with respect to the control condition (DMSO).

4.6. Scratch-Healing Assays

Panc-1 and WM266-4 cells were seeded in 35 mm petri dishes (2.5×10^5 cells/petri). Then, 24 h after seeding, cells reached 100% confluency, and a wound in the cells layer was created with a 200 μL pipette tip. The cells' medium was then removed and substituted with new medium enriched with 30 μM clotrimazole, 30 μM senicapoc, or the corresponding volume of DMSO. Cell migration was monitored in time under a holographic microscope Holomonitor M4 live cell imaging system (Phase Holographic Imaging, PHI AB, Lund, Sweden). Data analysis was performed using HStudio (PHI AB, Lund, Sweden). Pictures were taken after 24 h. Data are reported as increase in cell-covered areas after 24 h with respect to the control condition (DMSO).

4.7. Patch-Clamp Experiments

Cells were seeded in 35 mm dishes 24 h before the experiments. The standard intracellular pipette solution contained the following (in mM): 140 K-Asp, 4.3 $CaCl_2$, 2.06 $MgCl_2$, 5 EGTA, and 10 HEPES, and a pH of 7.2 was reached with KOH. The free calcium in this solution was calculated to be 1 μM using the Maxchelator program (Stanford University). Alternatively, a Ca^{2+}-free pipette solution was used, in which $CaCl_2$ was omitted and EGTA was increased to 10 mM. The extracellular solution contained the following (in mM): 150 Na-Asp, 5 KCl, 2 $CaCl_2$, 1 $MgCl_2$, 10 Glucose, and 10 HEPES, and a pH of 7.4 was reached with NaOH. The standard voltage-clamp protocol consisted of 500 ms long voltage steps, ranging from -100 to 120 mV in steps of 20 mV. The holding potential was set to the observed cell's resting potential. To monitor the response of cells to different stimuli over time, we used a 'time course protocol', which administers to the cells a pulse of +100 mV lasting 50 ms. All currents were measured at 20 °C, using an Axon amplifier (Molecular Devices-San Jose, CA, USA) and filtered with a low-pass filter at 10 kHz. We used the acquisition software GePulse (freely available at http://users.ge.ibf.cnr.it/pusch/programs-mik.htm, accessed on 25 February 2022). Currents were digitized with a National Instruments DAQ interface (Austin, TX, USA). Current traces were further analyzed with the freely available Ana analysis program (http://users.ge.ibf.cnr.it/pusch/programs-mik.htm, accessed on 13 August 2021).

4.8. Calcium Imaging Experiments

Measurements of cytosolic calcium ($[Ca^{2+}]_i$) were performed using the fluorescent indicator fura-2 AM. Before the experiments, cells were incubated for 45 min at 37 °C with 5 μM fura-2 AM (Sigma-Aldrich) dissolved in the same extracellular solutions of patch-clamp experiments, adding 0.1% pluronic acid in order to improve dye uptake. Coverslips were then transferred on the stage of an inverted Nikon TE200 fluorescence microscope. Cells were excited at 340 and 380 nm at 0.5 Hz with a dual excitation fluorometric Ca^{2+} imaging system (Hamamatsu, Sunayama-Cho, Japan). Fluorescence emission was measured at 510 nm and was acquired with a digital CCD camera (Hamamatsu C4742-95-12ER). To monitor $[Ca^{2+}]_i$, the fluorescence ratio F340/F380 was used. Monochromator settings, chopper frequency, and data acquisition were controlled by a dedicated software (Aqua Cosmos/Ratio, version U7501-01, Hamamatsu, Japan). $[Ca^{2+}]_i$ was calculated according to Grynkiewicz et al. [58]. We used a dissociation constant value for the Ca^{2+}/fura-2 complex of 140 nM.

4.9. Phalloidin Staining

Cells were treated with IK blockers for 72 h, washed with PBS and fixed in 4% formaldehyde for 15 min at RT. After permeabilization with 0.1% Triton X-100 in PBS for 5 min and washing, they were incubated in rhodamine–phalloidin solution (1:100 in PBS) and DAPI for 1 h and washed with PBS three times. Then, each sample was examined by confocal microscopy using a Leica STELLARIS 8 Falcon τSTED (Leica Microsystems, Mannheim, Germany) inverted confocal/STED microscope. The fluorescence image (1024 × 1024 × 16 bit) acquisition was performed using an HC PL APO CS oil immersion objective 100× (1.40 NA).

4.10. Data Analysis

Data are reported as mean ± SE. When bar charts are depicted, also individual data points are superimposed. When data are normalized, it is stated in the text or figure legends. Differences between data groups reported in the same graph were checked for statistical significance by means of a paired-sample *t*-test or ANOVA followed by Tukey tests for mean comparison (>2 groups, normal distribution). For statistical analysis, we used OriginLab (Northampton, MA, USA). Figures were prepared using Sigmaplot (Spss Inc.-Chicago, IL, USA). The chosen significance thresholds of 0.05 and 0.01 are indicated by an asterisk (*) and double asterisk (**), respectively. Cells to be patched, imaged or used for any purpose were chosen randomly.

5. Study Limitations

In the present study, no genetic knock-out or knock-down of the *KCNN4* gene has been performed, which would provide more direct evidence of *KCNN4* independent effects of KCa3.1 inhibitors on cancer cell viability and migration.

Supplementary Materials: The supporting information can be downloaded at https://www.mdpi.com/article/10.3390/ijms242216285/s1.

Author Contributions: P.Z., P.G. and M.P. initiated the study. P.Z., R.B., F.S., C.P. and P.G. performed experiments. P.Z., R.B., F.S., C.P., P.G. and M.P. performed data analysis and wrote the manuscript. P.Z. and M.P. revised the final version of the manuscript for intellectual contents. All authors have read and agreed to the published version of the manuscript.

Funding: This research was funded by grants to MP from Fondazione AIRC per la Ricerca sul Cancro (grant # IG 21558), the Italian Research Ministry (PRIN 20174TB8KW), Fondazione Telethon (grant # GMR22T1029) and Fondazione Telethon/Cariplo (grant # GJC22008).

Institutional Review Board Statement: Not applicable.

Informed Consent Statement: Not applicable.

Data Availability Statement: The data presented in this study are available on request from the corresponding author.

Acknowledgments: We thank Francesca Quartino and Alessandro Barbin for technical assistance. We gratefully acknowledge the Nanoscale Biophysics Group at DIFI-LAB, Department of Physics, University of Genova, and in particular Elena Gatta, for their support and assistance in the use of Stellaris 8, Falcon τ-STED, Leica Microsystems Confocal microscope. We thank Bernard Attali (Tel Aviv University) for a kind gift of BA6b9.

Conflicts of Interest: The authors declare no conflict of interest.

References

1. Anderson, K.J.; Cormier, R.T.; Scott, P.M. Role of ion channels in gastrointestinal cancer. *World J. Gastroenterol.* **2019**, *25*, 5732–5772. [CrossRef]
2. Lastraioli, E.; Iorio, J.; Arcangeli, A. Ion channel expression as promising cancer biomarker. *Biochim. Biophys. Acta* **2015**, *1848*, 2685–2702. [CrossRef]
3. Huang, X.; Jan, L.Y. Targeting potassium channels in cancer. *J. Cell Biol.* **2014**, *206*, 151–162. [CrossRef]
4. D'Amico, M.; Gasparoli, L.; Arcangeli, A. Potassium channels: Novel emerging biomarkers and targets for therapy in cancer. *Recent Pat. Anti-Cancer Drug Discov.* **2013**, *8*, 53–65. [CrossRef]
5. Abdul Kadir, L.; Stacey, M.; Barrett-Jolley, R. Emerging Roles of the Membrane Potential: Action Beyond the Action Potential. *Front. Physiol.* **2018**, *9*, 1661. [CrossRef] [PubMed]
6. Blackiston, D.J.; McLaughlin, K.A.; Levin, M. Bioelectric controls of cell proliferation: Ion channels, membrane voltage and the cell cycle. *Cell Cycle* **2009**, *8*, 3527–3536. [CrossRef] [PubMed]
7. Cone, C.D., Jr.; Cone, C.M. Induction of mitosis in mature neurons in central nervous system by sustained depolarization. *Science* **1976**, *192*, 155–158. [CrossRef]
8. Lang, F.; Föller, M.; Lang, K.S.; Lang, P.A.; Ritter, M.; Gulbins, E.; Vereninov, A.; Huber, S.M. Ion channels in cell proliferation and apoptotic cell death. *J. Membr. Biol.* **2005**, *205*, 147–157. [CrossRef]
9. Bertelli, S.; Remigante, A.; Zuccolini, P.; Barbieri, R.; Ferrera, L.; Picco, C.; Gavazzo, P.; Pusch, M. Mechanisms of Activation of LRRC8 Volume Regulated Anion Channels. *Cell. Physiol. Biochem.* **2021**, *55*, 41–56. [CrossRef]
10. Strange, K.; Yamada, T.; Denton, J.S. A 30-year journey from volume-regulated anion currents to molecular structure of the LRRC8 channel. *J. Gen. Physiol.* **2019**, *151*, 100–117. [CrossRef]
11. Mohr, C.J.; Steudel, F.A.; Gross, D.; Ruth, P.; Lo, W.Y.; Hoppe, R.; Schroth, W.; Brauch, H.; Huber, S.M.; Lukowski, R. Cancer-Associated Intermediate Conductance Ca^{2+}-Activated K^+ Channel $K_{Ca}3.1$. *Cancers* **2019**, *11*, 109. [CrossRef]
12. Gardos, G. The function of calcium in the potassium permeability of human erythrocytes. *Biochim. Biophys. Acta* **1958**, *30*, 653–654. [CrossRef] [PubMed]
13. Jensen, B.S.; Strobaek, D.; Christophersen, P.; Jorgensen, T.D.; Hansen, C.; Silahtaroglu, A.; Olesen, S.P.; Ahring, P.K. Characterization of the cloned human intermediate-conductance Ca^{2+}-activated K^+ channel. *Am. J. Physiol.* **1998**, *275*, C848–C856. [CrossRef] [PubMed]
14. Joiner, W.J.; Wang, L.Y.; Tang, M.D.; Kaczmarek, L.K. hSK4, a member of a novel subfamily of calcium-activated potassium channels. *Proc. Natl. Acad. Sci. USA* **1997**, *94*, 11013–11018. [CrossRef] [PubMed]
15. Ishii, T.M.; Silvia, C.; Hirschberg, B.; Bond, C.T.; Adelman, J.P.; Maylie, J. A human intermediate conductance calcium-activated potassium channel. *Proc. Natl. Acad. Sci. USA* **1997**, *94*, 11651–11656. [CrossRef]
16. Lee, C.H.; MacKinnon, R. Activation mechanism of a human SK-calmodulin channel complex elucidated by cryo-EM structures. *Science* **2018**, *360*, 508–513. [CrossRef]
17. Fanger, C.M.; Ghanshani, S.; Logsdon, N.J.; Rauer, H.; Kalman, K.; Zhou, J.; Beckingham, K.; Chandy, K.G.; Cahalan, M.D.; Aiyar, J. Calmodulin mediates calcium-dependent activation of the intermediate conductance K_{Ca} channel, *IKCa1*. *J. Biol. Chem.* **1999**, *274*, 5746–5754. [CrossRef]
18. Brown, B.M.; Shim, H.; Christophersen, P.; Wulff, H. Pharmacology of Small- and Intermediate-Conductance Calcium-Activated Potassium Channels. *Annu. Rev. Pharmacol. Toxicol.* **2020**, *60*, 219–240. [CrossRef]
19. Brugnara, C.; de Franceschi, L.; Alper, S.L. Inhibition of Ca(2+)-dependent K^+ transport and cell dehydration in sickle erythrocytes by clotrimazole and other imidazole derivatives. *J. Clin. Investig.* **1993**, *92*, 520–526. [CrossRef]
20. Alvarez, J.; Montero, M.; Garcia-Sancho, J. High affinity inhibition of Ca(2+)-dependent K^+ channels by cytochrome P-450 inhibitors. *J. Biol. Chem.* **1992**, *267*, 11789–11793. [CrossRef]
21. Pihlaja, T.L.M.; Niemissalo, S.M.; Sikanen, T.M. Cytochrome P450 Inhibition by Antimicrobials and Their Mixtures in Rainbow Trout Liver Microsomes In Vitro. *Environ. Toxicol. Chem.* **2022**, *41*, 663–676. [CrossRef]
22. Guengerich, F.P.; McCarty, K.D.; Chapman, J.G. Kinetics of cytochrome P450 3A4 inhibition by heterocyclic drugs defines a general sequential multistep binding process. *J. Biol. Chem.* **2021**, *296*, 100223. [CrossRef]
23. Suzuki, S.; Kurata, N.; Nishimura, Y.; Yasuhara, H.; Satoh, T. Effects of imidazole antimycotics on the liver microsomal cytochrome P450 isoforms in rats: Comparison of in vitro and ex vivo studies. *Eur. J. Drug Metab. Pharmacokinet.* **2000**, *25*, 121–126. [CrossRef]
24. Ayub, M.; Levell, M.J. Structure-activity relationships of the inhibition of human placental aromatase by imidazole drugs including ketoconazole. *J. Steroid Biochem.* **1988**, *31*, 65–72. [CrossRef]
25. Sheets, J.J.; Mason, J.I.; Wise, C.A.; Estabrook, R.W. Inhibition of rat liver microsomal cytochrome P-450 steroid hydroxylase reactions by imidazole antimycotic agents. *Biochem. Pharmacol.* **1986**, *35*, 487–491. [CrossRef]
26. Wulff, H.; Miller, M.J.; Hansel, W.; Grissmer, S.; Cahalan, M.D.; Chandy, K.G. Design of a potent and selective inhibitor of the intermediate-conductance Ca^{2+}-activated K^+ channel, *IKCa1*: A potential immunosuppressant. *Proc. Natl. Acad. Sci. USA* **2000**, *97*, 8151–8156. [CrossRef]
27. Wulff, H.; Castle, N.A. Therapeutic potential of $K_{Ca}3.1$ blockers: Recent advances and promising trends. *Expert Rev. Clin. Pharmacol.* **2010**, *3*, 385–396. [CrossRef] [PubMed]

28. Al-Ghananeem, A.M.; Abbassi, M.; Shrestha, S.; Raman, G.; Wulff, H.; Pereira, L.; Ansari, A. Formulation-based approach to support early drug discovery and development efforts: A case study with enteric microencapsulation dosage form development for a triarylmethane derivative TRAM-34; a novel potential immunosuppressant. *Drug Dev. Ind. Pharm.* **2010**, *36*, 563–569. [CrossRef] [PubMed]

29. Agarwal, J.J.; Zhu, Y.; Zhang, Q.Y.; Mongin, A.A.; Hough, L.B. TRAM-34, a putatively selective blocker of intermediate-conductance, calcium-activated potassium channels, inhibits cytochrome P450 activity. *PLoS ONE* **2013**, *8*, e63028. [CrossRef] [PubMed]

30. Stocker, J.W.; De Franceschi, L.; McNaughton-Smith, G.A.; Corrocher, R.; Beuzard, Y.; Brugnara, C. ICA-17043, a novel Gardos channel blocker, prevents sickled red blood cell dehydration in vitro and in vivo in SAD mice. *Blood* **2003**, *101*, 2412–2418. [CrossRef] [PubMed]

31. Staal, R.G.W.; Weinstein, J.R.; Nattini, M.; Cajina, M.; Chandresana, G.; Moller, T. Senicapoc: Repurposing a Drug to Target Microglia K(Ca)3.1 in Stroke. *Neurochem. Res.* **2017**, *42*, 2639–2645. [CrossRef]

32. Ataga, K.I.; Orringer, E.P.; Styles, L.; Vichinsky, E.P.; Swerdlow, P.; Davis, G.A.; Desimone, P.A.; Stocker, J.W. Dose-escalation study of ICA-17043 in patients with sickle cell disease. *Pharmacotherapy* **2006**, *26*, 1557–1564. [CrossRef]

33. Bulk, E.; Ay, A.S.; Hammadi, M.; Ouadid-Ahidouch, H.; Schelhaas, S.; Hascher, A.; Rohde, C.; Thoennissen, N.H.; Wiewrodt, R.; Schmidt, E.; et al. Epigenetic dysregulation of $K_{Ca}3.1$ channels induces poor prognosis in lung cancer. *Int. J. Cancer* **2015**, *137*, 1306–1317. [CrossRef] [PubMed]

34. Catacuzzeno, L.; Fioretti, B.; Franciolini, F. Expression and Role of the Intermediate-Conductance Calcium-Activated Potassium Channel $K_{Ca}3.1$ in Glioblastoma. *J. Signal Transduct.* **2012**, *2012*, 421564. [CrossRef] [PubMed]

35. Wang, Z.H.; Shen, B.; Yao, H.L.; Jia, Y.C.; Ren, J.; Feng, Y.J.; Wang, Y.Z. Blockage of intermediate-conductance-Ca^{2+}-activated K^+ channels inhibits progression of human endometrial cancer. *Oncogene* **2007**, *26*, 5107–5114. [CrossRef] [PubMed]

36. Ouadid-Ahidouch, H.; Roudbaraki, M.; Delcourt, P.; Ahidouch, A.; Joury, N.; Prevarskaya, N. Functional and molecular identification of intermediate-conductance Ca^{2+}-activated K^+ channels in breast cancer cells: Association with cell cycle progression. *Am. J. Physiol. Cell Physiol.* **2004**, *287*, C125–C134. [CrossRef]

37. Song, P.; Du, Y.; Song, W.; Chen, H.; Xuan, Z.; Zhao, L.; Chen, J.; Chen, J.; Guo, D.; Jin, C.; et al. $K_{Ca}3.1$ as an Effective Target for Inhibition of Growth and Progression of Intrahepatic Cholangiocarcinoma. *J. Cancer* **2017**, *8*, 1568–1578. [CrossRef]

38. Bonito, B.; Sauter, D.R.; Schwab, A.; Djamgoz, M.B.; Novak, I. $K_{Ca}3.1$ (IK) modulates pancreatic cancer cell migration, invasion and proliferation: Anomalous effects on TRAM-34. *Pflugers Arch.* **2016**, *468*, 1865–1875. [CrossRef]

39. Jager, H.; Dreker, T.; Buck, A.; Giehl, K.; Gress, T.; Grissmer, S. Blockage of intermediate-conductance Ca^{2+}-activated K^+ channels inhibit human pancreatic cancer cell growth in vitro. *Mol. Pharmacol.* **2004**, *65*, 630–638. [CrossRef]

40. Glaser, F.; Hundehege, P.; Bulk, E.; Todesca, L.M.; Schimmelpfennig, S.; Nass, E.; Budde, T.; Meuth, S.G.; Schwab, A. K_{Ca} channel blockers increase effectiveness of the EGF receptor TK inhibitor erlotinib in non-small cell lung cancer cells (A549). *Sci. Rep.* **2021**, *11*, 18330. [CrossRef]

41. Liu, L.; Zhan, P.; Nie, D.; Fan, L.; Lin, H.; Gao, L.; Mao, X. Intermediate-Conductance-Ca2-Activated K Channel *IKCa1* Is Upregulated and Promotes Cell Proliferation in Cervical Cancer. *Med. Sci. Monit. Basic Res.* **2017**, *23*, 45–57. [CrossRef] [PubMed]

42. Zhang, P.; Yang, X.; Yin, Q.; Yi, J.; Shen, W.; Zhao, L.; Zhu, Z.; Liu, J. Inhibition of SK4 Potassium Channels Suppresses Cell Proliferation, Migration and the Epithelial-Mesenchymal Transition in Triple-Negative Breast Cancer Cells. *PLoS ONE* **2016**, *11*, e0154471. [CrossRef] [PubMed]

43. Tubman, V.N.; Mejia, P.; Shmukler, B.E.; Bei, A.K.; Alper, S.L.; Mitchell, J.R.; Brugnara, C.; Duraisingh, M.T. The Clinically Tested Gardos Channel Inhibitor Senicapoc Exhibits Antimalarial Activity. *Antimicrob. Agents Chemother.* **2016**, *60*, 613–616. [CrossRef] [PubMed]

44. Rosenkranzkw, H.; Pütter, J. The binding of Clotrimazole to the proteins of human serum. *Eur. J. Drug Metab. Pharmacokinet.* **1976**, *1*, 73–76. [CrossRef]

45. Zuccolini, P.; Gavazzo, P.; Pusch, M. BK Channel in the Physiology and in the Cancer of Pancreatic Duct: Impact and Reliability of BK Openers. *Front. Pharmacol.* **2022**, *13*, 906608. [CrossRef]

46. Remigante, A.; Zuccolini, P.; Barbieri, R.; Ferrera, L.; Morabito, R.; Gavazzo, P.; Pusch, M.; Picco, C. NS-11021 Modulates Cancer-Associated Processes Independently of BK Channels in Melanoma and Pancreatic Duct Adenocarcinoma Cell Lines. *Cancers* **2021**, *13*, 6144. [CrossRef]

47. Ferrera, L.; Barbieri, R.; Picco, C.; Zuccolini, P.; Remigante, A.; Bertelli, S.; Fumagalli, M.R.; Zifarelli, G.; La Porta, C.A.M.; Gavazzo, P.; et al. TRPM2 oxidation activates two distinct potassium channels in melanoma cells through Intracellular calcium Increase. *Int. J. Mol. Sci.* **2021**, *22*, 8359. [CrossRef]

48. Herlyn, M.; Balaban, G.; Bennicelli, J.; Guerry, D.t.; Halaban, R.; Herlyn, D.; Elder, D.E.; Maul, G.G.; Steplewski, Z.; Nowell, P.C.; et al. Primary melanoma cells of the vertical growth phase: Similarities to metastatic cells. *J. Natl. Cancer Inst.* **1985**, *74*, 283–289.

49. Cassano, R.; Cuconato, M.; Calviello, G.; Serini, S.; Trombino, S. Recent Advances in Nanotechnology for the Treatment of Melanoma. *Molecules* **2021**, *26*, 785. [CrossRef]

50. van Willigen, W.W.; Bloemendal, M.; Boers-Sonderen, M.J.; de Groot, J.W.B.; Koornstra, R.H.T.; van der Veldt, A.A.M.; Haanen, J.; Boudewijns, S.; Schreibelt, G.; Gerritsen, W.R.; et al. Response and survival of metastatic melanoma patients treated with immune checkpoint inhibition for recurrent disease on adjuvant dendritic cell vaccination. *Oncoimmunology* **2020**, *9*, 1738814. [CrossRef]

51. Garbe, C.; Eigentler, T.K.; Keilholz, U.; Hauschild, A.; Kirkwood, J.M. Systematic review of medical treatment in melanoma: Current status and future prospects. *Oncologist* **2011**, *16*, 5–24. [CrossRef] [PubMed]

52. Zuccolini, P.; Ferrera, L.; Remigante, A.; Picco, C.; Barbieri, R.; Bertelli, S.; Moran, O.; Gavazzo, P.; Pusch, M. The VRAC blocker DCPIB directly gates the BK channels and increases intracellular Ca^{2+} in melanoma and pancreatic duct adenocarcinoma cell lines. *Br. J. Pharmacol.* **2022**, *179*, 3452–3469. [CrossRef]

53. Kovalenko, I.; Glasauer, A.; Schockel, L.; Sauter, D.R.; Ehrmann, A.; Sohler, F.; Hagebarth, A.; Novak, I.; Christian, S. Identification of $K_{Ca}^{3.1}$ channel as a novel regulator of oxidative phosphorylation in a subset of pancreatic carcinoma cell lines. *PLoS ONE* **2016**, *11*, e0160658. [CrossRef] [PubMed]

54. Lallet-Daher, H.; Roudbaraki, M.; Bavencoffe, A.; Mariot, P.; Gackiere, F.; Bidaux, G.; Urbain, R.; Gosset, P.; Delcourt, P.; Fleurisse, L.; et al. Intermediate-conductance Ca^{2+}-activated K^+ channels (IKCa1) regulate human prostate cancer cell proliferation through a close control of calcium entry. *Oncogene* **2009**, *28*, 1792–1806. [CrossRef] [PubMed]

55. Burg, S.; Shapiro, S.; Peretz, A.; Haimov, E.; Redko, B.; Yeheskel, A.; Simhaev, L.; Engel, H.; Raveh, A.; Ben-Bassat, A.; et al. Allosteric inhibitors targeting the calmodulin-PIP2 interface of SK4 K^+ channels for atrial fibrillation treatment. *Proc. Natl. Acad. Sci. USA* **2022**, *119*, e2202926119. [CrossRef]

56. Manoli, S.; Coppola, S.; Duranti, C.; Lulli, M.; Magni, L.; Kuppalu, N.; Nielsen, N.; Schmidt, T.; Schwab, A.; Becchetti, A.; et al. The Activity of Kv 11.1 Potassium Channel Modulates F-Actin Organization During Cell Migration of Pancreatic Ductal Adenocarcinoma Cells. *Cancers* **2019**, *11*, 135. [CrossRef]

57. Schindelin, J.; Arganda-Carreras, I.; Frise, E.; Kaynig, V.; Longair, M.; Pietzsch, T.; Preibisch, S.; Rueden, C.; Saalfeld, S.; Schmid, B.; et al. Fiji: An open-source platform for biological-image analysis. *Nat. Methods* **2012**, *9*, 676–682. [CrossRef]

58. Grynkiewicz, G.; Poenie, M.; Tsien, R.Y. A new generation of Ca^{2+} indicators with greatly improved fluorescence properties. *J. Biol. Chem.* **1985**, *260*, 3440–3450. [CrossRef]

Disclaimer/Publisher's Note: The statements, opinions and data contained in all publications are solely those of the individual author(s) and contributor(s) and not of MDPI and/or the editor(s). MDPI and/or the editor(s) disclaim responsibility for any injury to people or property resulting from any ideas, methods, instructions or products referred to in the content.

International Journal of
Molecular Sciences

Article

Targeting AHR Increases Pancreatic Cancer Cell Sensitivity to Gemcitabine through the ELAVL1-DCK Pathway

Darius Stukas [1,*], Aldona Jasukaitiene [1], Arenida Bartkeviciene [1], Jason Matthews [2,3], Toivo Maimets [4], Indrek Teino [4], Kristaps Jaudzems [5], Antanas Gulbinas [1] and Zilvinas Dambrauskas [1]

[1] Surgical Gastroenterology Laboratory, Institute for Digestive Research, Lithuanian University of Health Sciences, Eiveniu 4, 50103 Kaunas, Lithuania; aldona.jasukaitiene@lsmu.lt (A.J.); arenida.bartkeviciene@lsmu.lt (A.B.); antanas.gulbinas@lsmu.lt (A.G.); zilvinas.dambrauskas@lsmu.lt (Z.D.)

[2] Department of Nutrition, Institute of Basic Medical Sciences, University of Oslo, 1046 Blindern, 0317 Oslo, Norway; jason.matthews@medisin.uio.no

[3] Department of Pharmacology and Toxicology, University of Toronto, Toronto, ON M5S 1A8, Canada

[4] Institute of Molecular and Cell Biology, University of Tartu, Riia 23, 51010 Tartu, Estonia; toivo.maimets@ut.ee (T.M.); indrek.teino@ut.ee (I.T.)

[5] Latvian Institute of Organic Synthesis, Aizkraukles 21, LV-1006 Riga, Latvia; kristaps.jaudzems@osi.lv

* Correspondence: darius.stukas@lsmu.lt; Tel.: +37-060713892

Abstract: The aryl hydrocarbon receptor (AHR) is a transcription factor that is commonly upregulated in pancreatic ductal adenocarcinoma (PDAC). AHR hinders the shuttling of human antigen R (ELAVL1) from the nucleus to the cytoplasm, where it stabilises its target messenger RNAs (mRNAs) and enhances protein expression. Among these target mRNAs are those induced by gemcitabine. Increased AHR expression leads to the sequestration of ELAVL1 in the nucleus, resulting in chemoresistance. This study aimed to investigate the interaction between AHR and ELAVL1 in the pathogenesis of PDAC in vitro. *AHR* and *ELAVL1* genes were silenced by siRNA transfection. The RNA and protein were extracted for quantitative real-time polymerase chain reaction (qRT-PCR) and Western blot (WB) analysis. Direct binding between the ELAVL1 protein and *AHR* mRNA was examined through immunoprecipitation (IP) assay. Cell viability, clonogenicity, and migration assays were performed. Our study revealed that both AHR and ELAVL1 inter-regulate each other, while also having a role in cell proliferation, migration, and chemoresistance in PDAC cell lines. Notably, both proteins function through distinct mechanisms. The silencing of ELAVL1 disrupts the stability of its target mRNAs, resulting in the decreased expression of numerous cytoprotective proteins. In contrast, the silencing of *AHR* diminishes cell migration and proliferation and enhances cell sensitivity to gemcitabine through the AHR-ELAVL1-deoxycytidine kinase (DCK) molecular pathway. In conclusion, AHR and ELAVL1 interaction can form a negative feedback loop. By inhibiting AHR expression, PDAC cells become more susceptible to gemcitabine through the ELAVL1-DCK pathway.

Keywords: AHR; ELAVL1; PDAC; gemcitabine; HMOX1; DCK

Citation: Stukas, D.; Jasukaitiene, A.; Bartkeviciene, A.; Matthews, J.; Maimets, T.; Teino, I.; Jaudzems, K.; Gulbinas, A.; Dambrauskas, Z. Targeting AHR Increases Pancreatic Cancer Cell Sensitivity to Gemcitabine through the ELAVL1-DCK Pathway. *Int. J. Mol. Sci.* **2023**, 24, 13155. https://doi.org/10.3390/ijms241713155

Academic Editors: Claudio Luchini and Donatella Delle Cave

Received: 18 July 2023
Revised: 21 August 2023
Accepted: 22 August 2023
Published: 24 August 2023

Copyright: © 2023 by the authors. Licensee MDPI, Basel, Switzerland. This article is an open access article distributed under the terms and conditions of the Creative Commons Attribution (CC BY) license (https://creativecommons.org/licenses/by/4.0/).

1. Introduction

Pancreatic cancer (PC) is a devastating disease that has a less than 10% five-year survival rate [1,2], with pancreatic cancer ductal adenocarcinoma (PDAC) being the most common in pancreatic malignancies. Future projections show that the death rate from PC will increase by 1.76% annually [3] and could be one of the leading forms of cancer by mortality by 2030 [4,5]. Curative surgery followed by chemotherapy remains the recommended treatment for resectable pancreatic tumours; however, due to PC being diagnosed mainly in advanced stages, curative surgery is only available in 15–20% of cases [6]. In cases with advanced or metastatic PC, gemcitabine (GEM) remains one of the first-line drugs for chemotherapy, although other multidrug regimens are also widely used (e.g., FOLFIRINOX or FOLFOXIRI) [7]. Due to poor future projections, low diagnostic statistics, low survival, and limited treatment options, it is imperative to search for new or improve existing treatment of PC.

One of the biggest problems in PC treatment is its heterogeneity and resistance to first-line drugs. Resistance to chemotherapy can be innate (germinal genetic makeup) or acquired (mutational) [8], with the latter being common in PC. A common difficulty of PC treatment is that it becomes chemoresistant weeks after starting treatment, which could also be attributed to alterations in post-transcriptional gene regulation. The human antigen R (ELAVL1) protein is an RNA-binding protein that binds mRNAs and increases their stability. There are numerous targets of ELAVL1 which are involved in protective mechanisms (Heme oxygenase 1 [9] or Cyclooxygenase-2 [10]), proliferation, differentiation, apoptosis, and senescence of the cell [11,12]. One of the enzymes regulated by ELAVL1 is deoxycytidine kinase (DCK)—an enzyme responsible for initiating the process of GEM phosphorylation, which leads to its incorporation into the DNA and subsequently arrested cell cycle. The overexpression of ELAVL1 upregulates and silencing downregulates the expression of DCK in PC [13,14]. However, to successfully regulate the expression of DCK and the subsequent response to gemcitabine, the ELAVL1 protein has to be located in the cytoplasm of the cell. The concentration of cytoplasmic ELAVL1 is positively associated with a response to gemcitabine and survival of the patients [10]. DNA damage, such as the effect of gemcitabine, works as a stimulus for ELAVL1 to translocate from the nucleus to the cytoplasm [15]; however, even after such stimuli, ELAVL1 can be sequestered into the nucleus. One of the reasons for this sequestering of ELAVL1 could be the aryl hydrocarbon receptor (AHR) [16]. Normally, AHR is responsible for cell defence and immune system regulation [17,18]. Increases in AHR expression are seen in autoimmune diseases and various forms of cancer [19], and AHR is frequently overexpressed in pancreatic cancer [20–22]. In cases with AHR overexpression, ELAVL1 could be sequestered in the nucleus, which would increase the chemoresistance of the cells.

Therefore, we hypothesise that the modulation of AHR and ELAVL1 expression can decrease the chemoresistance of PDAC cells in vitro.

2. Results

2.1. Modulation and Relationship of AHR and ELAVL1 in PDAC Cell Lines

Human PDAC cell lines (BxPC-3, Su.86.86) were transfected with small interfering RNAs (siRNAs): siAHR or siELAVL1 for 24 h. Western blot analysis revealed that the *AHR* and *ELAVL1* genes were silenced almost completely after 24 h in both lines. In BxPC-3, siAHR decreased *AHR* mRNA to 20% and the protein was not detected; in Su.86.86, *AHR* mRNA decreased to 29% and the protein decreased to 2% compared with nontreated control cells.

When silencing *ELAVL1* in BxPC-3 cells, the *ELAVL1* mRNA levels decreased to 31%, and the ELAVL1 protein levels decreased to 38% compared with nontreated control cells. Similarly, in Su.86.86 cells, the *ELAVL1* mRNA levels decreased to 9%, and the ELAVL1 protein levels decreased to 15% compared with nontreated control cells. These results indicate successful silencing of both the target mRNA and the subsequent protein expression for ELAVL1 (see Figure 1).

Moreover, *AHR* silencing increased ELAVL1 mRNA in both cell lines (BxPC3—1.4-fold; Su.86.86—1.2-fold) (Figure 1). The same increase was seen in protein levels (BxPC-3 ELAVL1 protein increased 1.83-fold and Su.86.86 1.2-fold).

ELAVL1 silencing caused a decrease in *AHR* mRNA (in BxPC-3 *AHR* mRNA decreased significantly to 76% and in Su.86.86 to 77%) as well as in protein (in BxPC-3 AHR protein was absent and in Su.86.86 decreased to 73%).

These findings suggest a mutual influence and potential regulatory relationship between AHR and ELAVL1. As ELAVL1 primarily exerts its activity in the cytoplasm, it was important to investigate the mRNA and protein expression levels of its downstream targets, specifically DCK and HMOX1.

The silencing of *AHR* increased Heme oxygenase 1 (*HMOX1*) mRNA (4.45-fold) and protein (1.4-fold) in BxPC-3 cells; however, Su.86.86 showed no changes in either mRNA or protein of HMOX1 (Figure 2). SiAHR also increased *DCK* mRNA (1.6-fold) and protein slightly (1.1-fold) in the BxPC-3 cell line; however, it did not change *DCK* mRNA and even decreased protein levels (to 56%) in the Su.86.86 cell line.

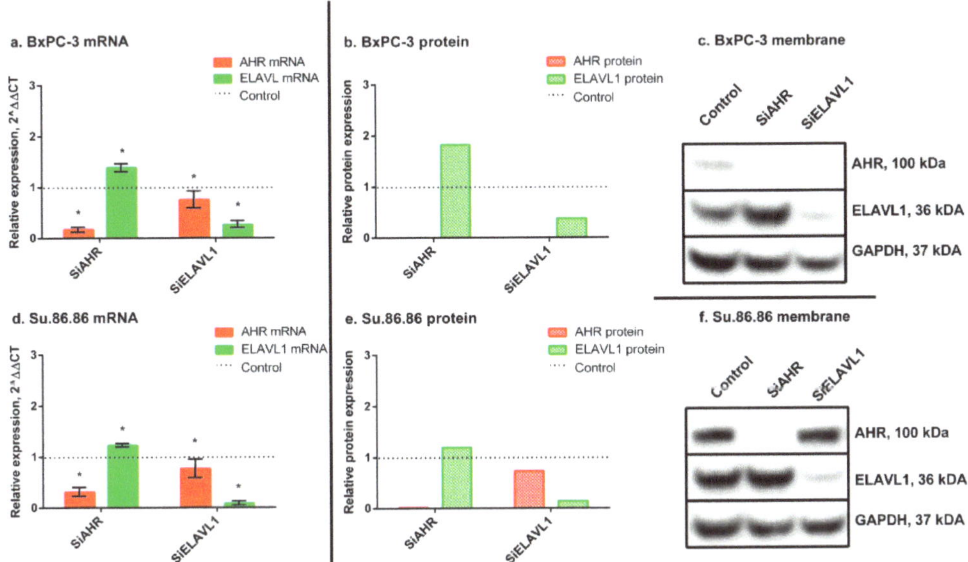

Figure 1. AHR and ELAVL1 qRT-PCR and WB analysis. mRNA and protein expression of AHR and ELAVL1 genes and proteins, after *AHR* or *ELAVL1* silencing by transfection. qRT-PCR N = 3, MEAN ± SD. * $p < 0.05$. WB N = 3; however, only 1 is shown as a representative experiment: (**a**) BxPC-3 qRT-PCR analysis, (**b**) BxPC-3 WB analysis, (**c**) membrane of BxPC-3 WB, (**d**) Su.86.86 qRT-PCR analysis, (**e**) Su.86.86 WB analysis, and (**f**) membrane of Su.86.86 WB.

The silencing of *ELAVL1* decreased *HMOX1* mRNA (BxPC-3—79%; Su.86.86—81%) and protein (BxPC-3—88%; Su.86.86—57%) in both cell lines. *DCK* mRNA did not change in either cell line; however, the DCK protein decreased in both (BxPC-3—65%; Su.86.86—57%).

These results show that ELAVL1 pathway genes and proteins somewhat correspond to changes in AHR and ELAVL1 expression levels; however, it is more noticeable for protein level and differs between cell lines.

2.2. ELAVL1-Mediated Post-Transcriptional Regulation of AHR as Demonstrated by Immunoprecipitation

Since a relationship between AHR and ELAVL1 can be seen in mRNA and protein expression levels, the IP assay was used to determine a direct link between *AHR* mRNA and the ELAVL1 protein. ELAVL1 is known to bind to various mRNAs, including *DCK* [13] and *HMOX1* [9]. The *AHR* gene has nine adenylate uridylate (AU)-rich elements (AREs) (ATTTA), which is used as a binding motif for ELAVL1 protein [23]. The qRT-PCR results of IP showed that the ELAVL1 protein binds *AHR* mRNA (Figure 3a). As a control for the assay, *GAPDH* mRNA was tested as a negative control, which the ELAVL1 protein did not bind to, and *HMOX1* mRNA as a known target for a positive control (Figure 3b). *AHR* mRNA sequence analysis, together with our experimental data, indicate that the ELAVL1 protein binds *AHR* mRNA and thereby potentially modulates AHR translation. Western blot assay was used as a control of the IP assay (Figure 3c).

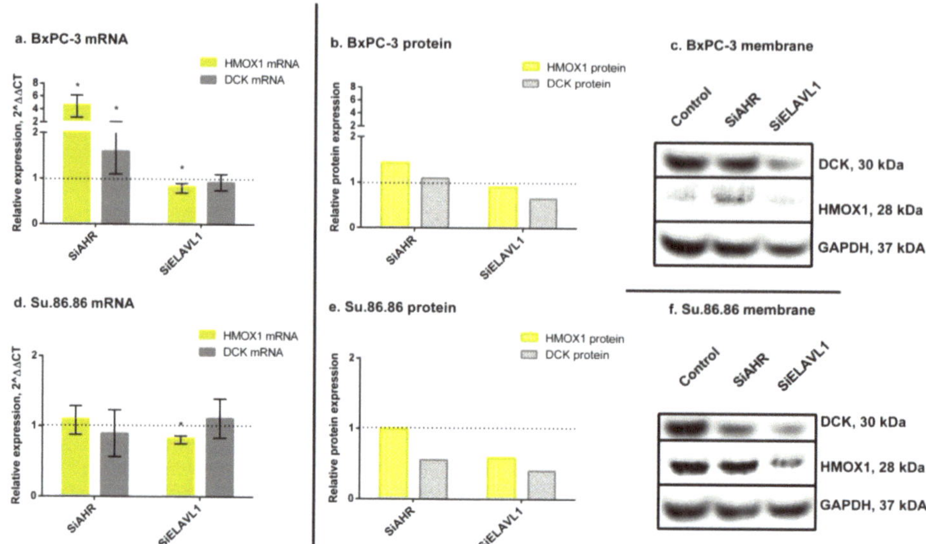

Figure 2. HMOX1 and DCK qRT-PCR and WB analysis. mRNA and protein expression of HMOX1 and DCK genes and proteins, after *AHR* or *ELAVL1* silencing. qRT-PCR N = 3, MEAN ± SD. * $p < 0.05$. WB N = 3; however, only 1 is shown as a representative experiment: (**a**) BxPC-3 qRT-PCR analysis, (**b**) BxPC-3 WB analysis, (**c**) membrane of BxPC-3 WB, (**d**) Su.86.86 qRT-PCR analysis, (**e**) Su.86.86 WB analysis, and (**f**) membrane of Su.86.86 WB.

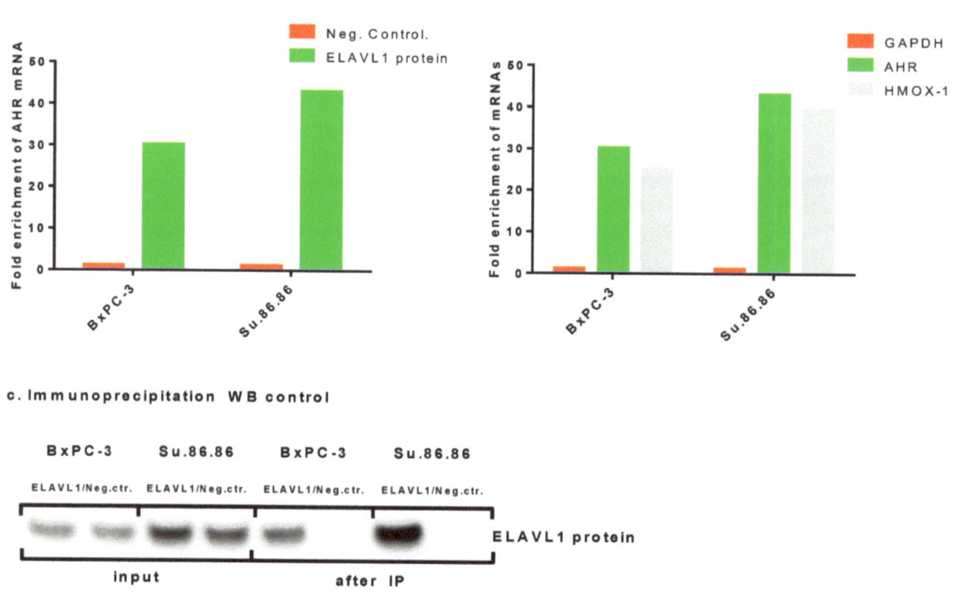

Figure 3. Immunoprecipitation of ELAVL1 protein. IP of ELAVL1 protein showing the ability of the ELAVL1 protein to bind *AHR* mRNA: N = 1 (**a**) qRT-PCR assay showing fold change in *AHR* mRNA when compared with the negative control (*GAPDH*), (**b**) qRT-PCR assay showing *GAPDH* as a negative target mRNA and *AHR* and *HMOX1* as positive mRNA targets for ELAVL1 protein, and (**c**) WB assay showing absence and presence of the ELAVL1 protein after the IP assay.

2.3. Gemcitabine IC50 Dose Determination

The metabolic activity of PDAC cells was determined by MTT assay. IC50 doses of GEM were determined separately for both cell lines: Su.86.86—36.91 $\pm$ 1.40 nM, BxPC-3—26.67 $\pm$ 1.69 nM (Figure 4a,b).

a. viability graph in nM GEM concentrations

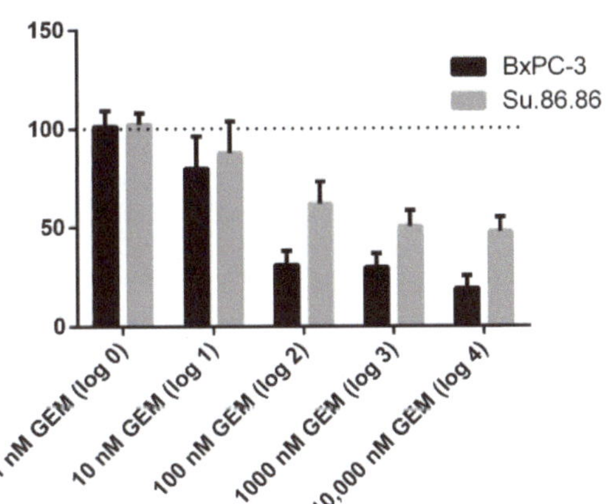

b. viability graph in log concentrations for IC50 calculation

c. calculated IC50 concentrations in nM $\pm$ SD

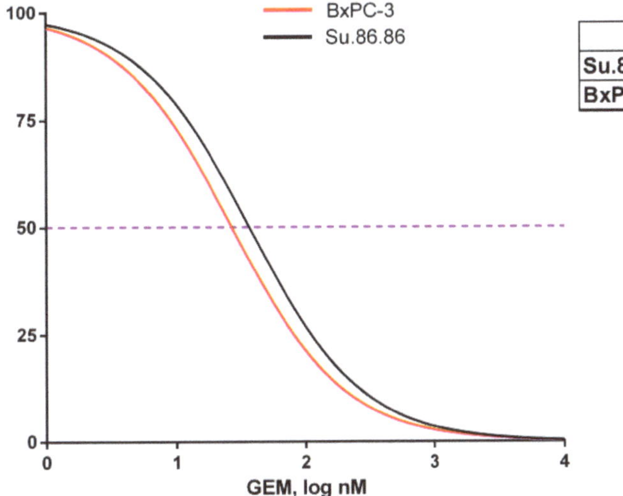

	IC50 $\pm$ SD
Su.86.86	36.91 $\pm$ 7.76
BxPC-3	26.67 $\pm$ 15.42

Figure 4. MTT assay and IC50 dose calculation of gemcitabine. IC50 doses in nM $\pm$ SD of gemcitabine after 48-h treatment: N $\geq$ 4 (**a**) viability graph in nM GEM concentration (dotted line–control (100%)), (**b**) viability graph in log concentrations for IC50 calculation (purple dotted line showing 50% viability – IC50), and (**c**) table of numerical value of IC50 doses in nM $\pm$ SD.

2.4. Cellular Localisation Changes in ELAVL1 in Response to AHR Silencing and/or Gemcitabine Treatment

Cellular localisation shifts in ELAVL1 and AHR protein after the silencing of ELAVL1 or AHR genes was elucidated by immunocytochemistry (ICC).

AHR mainly resides in the cytoplasm of both PDAC cell lines. The shift of AHR protein was not noticeable in either cell line when silencing ELAVL1 and/or treating with GEM (photos not included).

The ELAVL1 protein mainly resides in the nucleus of both PDAC cell lines (Figure 5a,e). When treating cells with GEM, the concentration of cytoplasmic ELAVL1 increases (Figure 5b,f), showing that GEM acts as a signal for ELAVL1 to shift its localisation. When silencing AHR, the concentration of cytoplasmic ELAVL1 increases (Figure 5c,g), showing that AHR can be involved in sequestering ELAVL1 in the nucleus. The joint effect of *AHR* silencing and GEM treatment increased cytoplasmic ELAVL1 concentrations even more than separately, indicating that the silencing of *AHR* could be beneficial to ELAVL1 translocation in response to GEM treatment (Figure 5d,h).

2.5. AHR and ELAVL1 Modulation Influences the Chemoresistance of PDAC Cells

GEM treatment significantly increased *AHR* mRNA (BxPC-3—4.3-fold; Su.86.86—2.4-fold) but decreased protein levels in both cell lines (BxPC-3 to 27%; Su.86.86 to 38%) (Figure 6). Treatment also increased *ELAVL1* mRNA (BxPC-3—1.34-fold; Su.86.86—1.22-fold) and protein (BxPC-3—2.1-fold; Su.86.86—1.1-fold).

When silencing *AHR* prior to GEM treatment, it increased BxPC-3 *ELAVL1* mRNA (1.46-fold) and protein (2.6-fold); however, Su.86.86 *ELAVL1* mRNA did not change, and protein decreased to 74%. Silencing *ELAVL1* prior to GEM treatment increased *AHR* mRNA in both cell lines (BxPC-3—2.2-fold; Su.86.86—1.4-fold) but greatly decreased AHR protein (BxPC-3 to 12%; Su.86.86 to 38%).

These findings highlight the differences observed between mRNA levels and protein expression levels, as well as variations across different cell lines. These differences suggest the involvement of post-transcriptional regulatory mechanisms in the cellular response to gemcitabine. To gain a deeper understanding of these distinctions, we conducted an analysis of *ELAVL1* pathway genes, specifically *HMOX1* and *DCK*.

GEM greatly increased *HMOX1* mRNA (BxPC-3—7.2-fold; Su.86.86—2.7-fold); however, an increase in protein was much less noticeable (BxPC-3—1.2-fold, Su.86.86—1.1-fold), as well as *DCK* mRNA (BxPC-3—2-fold, Su.86.86—2.3-fold) and protein (BxPC-3—2.5-fold; Su.86.86—1.3-fold) (Figure 7).

When silencing *AHR* prior to GEM treatment, *HMOX1* mRNA increased (BxPC-3—13.6-fold; Su.86.86—1.5-fold); however, the HMOX1 protein increased in BxPC-3 (1.5-fold) and decreased in Su.86.86 (to 75%). A similar pattern was also seen with *DCK* mRNA and protein levels. *DCK* mRNA increased in both cell lines (BxPC-3—2.1-fold; Su.86.86—1.3-fold); however, protein levels only increased in BxPC-3 (2.5-fold) and did not change in Su.86.86.

The silencing of *ELAVL1* prior to GEM treatment increased *HMOX1* mRNA (BxPC-3—3.1-fold; Su.86.86—1.7-fold) and protein (BxPC-3—1.1-fold; Su.86.86—1.4-fold). *DCK* mRNA was also increased (BxPC-3—2.1-fold; Su.86.86—1.4-fold), as well as protein levels (BxPC-3—2.6-fold; Su.86.86—2.1-fold).

The observed changes indicate that both AHR and ELAVL1 genes and proteins, as well as the ELAVL1 pathway genes *HMOX1* and *DCK*, respond to gemcitabine (GEM) treatment. However, notable differences exist between mRNA and protein expression levels of AHR, suggesting the involvement of post-transcriptional regulation mechanisms that affect the translation of *AHR* mRNA into protein. Furthermore, significant differences were observed between the cell lines, indicating the presence of additional mechanisms that contribute to the observed variations.

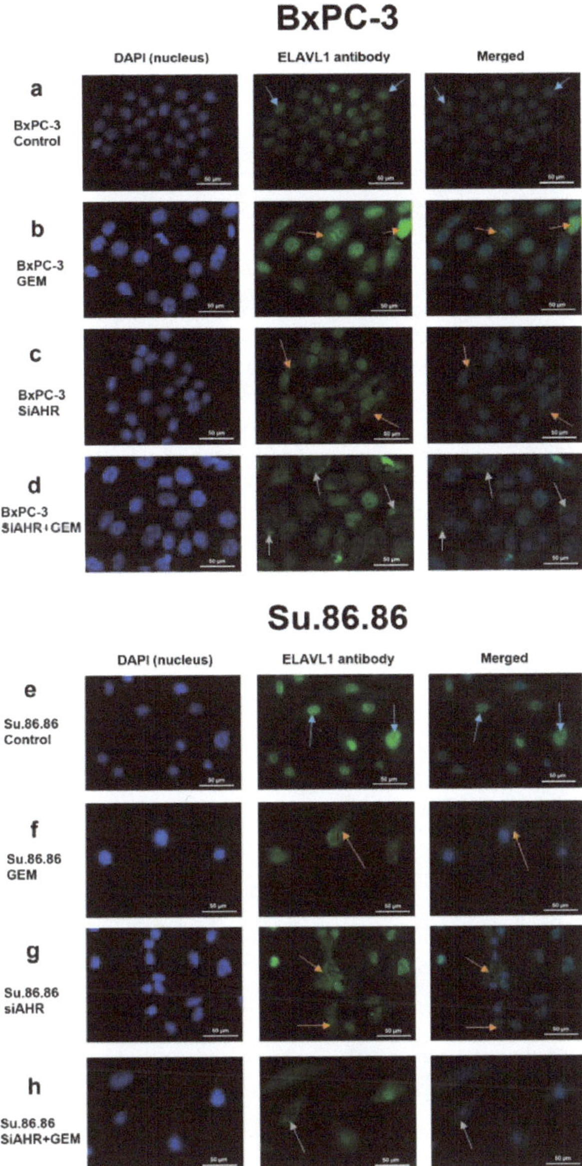

Figure 5. Example photos of PDAC lines ICC with ELAVL1 antibody. Example photos showing the ELAVL1 shift from the nucleus to the cytoplasm after the effect of SiAHR and/or GEM. ICC N = 3; however, only one is being shown as a representative experiment ($40\times$ magnification): (**a**) BxPC-3 control, (**b**) BxPC-3 GEM, (**c**) BxPC-3 siAHR, (**d**) BxPC-3 siAHR+GEM, (**e**) Su.86.86 control, (**f**) Su.86.86 GEM, (**g**) Su.86.86 siAHR, and (**h**) Su.86.86 siAHR+GEM. Blue arrows depict ELAVL1 protein being localised predominantly in the nucleus. Orange arrows depict ELAVL1 protein somewhat shifting from the nucleus to the cytoplasm. Grey arrows depict ELAVL1 being localised predominantly in the cytoplasm.

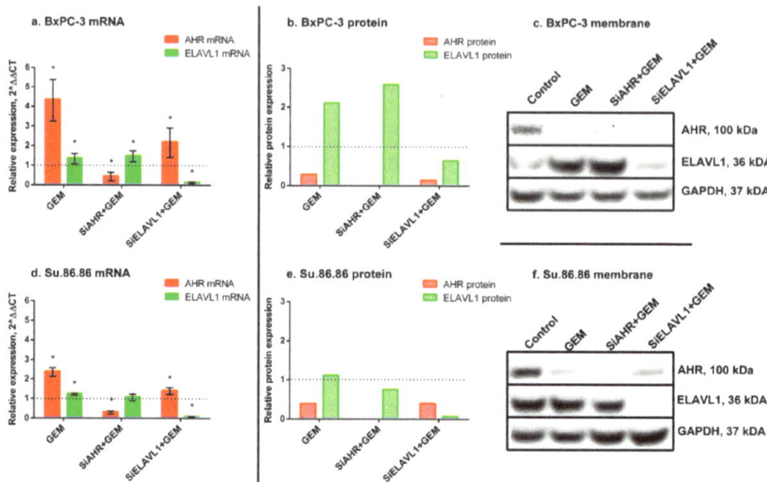

Figure 6. AHR and ELAVL1 qRT-PCR and WB analysis after GEM treatment. mRNA and protein expression of AHR and ELAVL1 genes and proteins, after *AHR* and/or *ELAVL1* silencing by transfection and treatment with IC50 GEM. qRT-PCR N = 3, MEAN ± SD. * $p < 0.05$. WB N = 3; however, only 1 is shown as a representative experiment: (**a**) BxPC-3 qRT-PCR analysis, (**b**) BxPC-3 WB analysis, (**c**) membrane of BxPC-3 WB, (**d**) Su.86.86 qRT-PCR analysis, (**e**) Su.86.86 WB analysis, and (**f**) membrane of Su.86.86 WB.

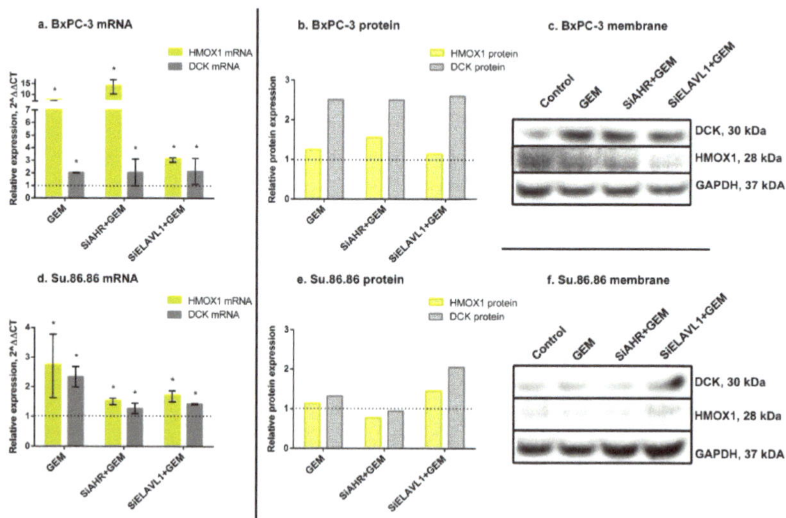

Figure 7. HMOX1 and DCK qRT-PCR and WB analysis after GEM treatment. mRNA and protein expression of HMOX1 and DCK genes and proteins, after *AHR* or *ELAVL1* silencing and treatment with GEM. qRT-PCR N = 3, MEAN ± SD. * $p < 0.05$. WB N = 3; however, only 1 is shown as a representative experiment: (**a**) BxPC-3 qRT-PCR analysis, (**b**) BxPC-3 WB analysis, (**c**) membrane of BxPC-3 WB, (**d**) Su.86.86 qRT-PCR analysis, (**e**) Su.86.86 WB analysis, and (**f**) membrane of Su.86.86 WB.

To determine the effect of *AHR* or *ELAVL1* gene silencing on cell chemoresistance, *AHR* or *ELAVL1* genes were silenced for 24 h by transfection with siAHR or siELAVL1. After silencing, the cells were treated with an IC50 dose of GEM for 48 h. The metabolic activity of all groups was compared with the control group (100%) of nontreated cells for

GEM alone or siControl for transfected cells. GEM alone significantly reduced cell viability to 57% for the BxPC-3 cell line and 48.4% for the Su.86.86 cell line. There was no significant effect on BxPC-3 or Su.86.86 cell viability after the silencing of *AHR* or *ELAVL1*. However, the silencing of *AHR* together with GEM significantly decreased the viability of both cell lines when compared with GEM alone (BxPC-3 to 25.8% and Su.86.86 to 23.1% cell viability). In BxPC-3, the silencing of *ELAVL1* together with GEM did not have a significant effect on cell viability when compared with GEM alone; however, in Su.86.86, the combined effect of siELAVL1 and GEM when compared with GEM alone significantly decreased cell viability to 23.1% (Figure 8a,b).

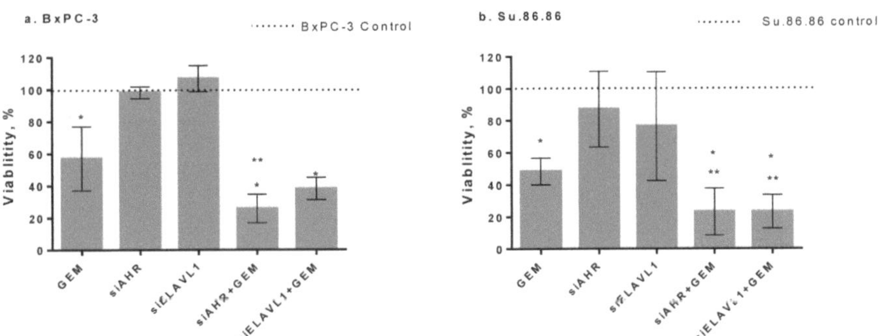

Figure 8. Cell viability analysis by MTT assay. PDAC cell viability after silencing of *AHR* or *ELAVL1* for 24 h and treatment with or without IC50 GEM for 48 h (MTT N = 4): (**a**) BxPC-3 cells and (**b**) Su.86.86 cells. N ≥ 4, MEAN ± SD * $p < 0.05$ when compared with respective control. ** $p < 0.05$ when compared with GEM alone.

The ability of cell lines to form colonies was measured by clonogenicity assay (Figure 9). The effect of gemcitabine was more detrimental to long-term than short-term (MTT) cell survival. IC50 doses (MTT) of GEM significantly decreased cell colony formation to 18.3% compared with the nontreated control in the BxPC-3 cell line and to 30.3% in the Su.86.86 cell line. The silencing of *AHR* also had a more detrimental effect on long-term than short-term survival. Both cell lines had a significant decrease in colony formation (19.3% formation in the BxPC-3 cell line and 27.7% formation in the Su.86.86 cell line compared with the nontreated control). The silencing of *ELAVL1* had a significant effect on BxPC-3 colony formation (colony formation decreased to 74.7%) but had no effect on Su.86.86 cell line colony formation. The silencing of *AHR* significantly increased the effect of GEM (BxPC-3 to 3.7% and Su.86.86 to 11.3%); however, due to severe effects on both GEM and siAHR, it is unclear whether the effect is due to increased sensitivity to GEM or the additive effect of siAHR and GEM. The silencing of *ELAVL1* did not significantly change the effects of GEM on either cell line.

Migratory ability was measured by the wound healing (scratch) assay (Figure 10). After 24 h, the BxPC-3 cell line (Figure 10a,c) showed significantly lower migratory abilities in response to GEM (41.2% wound closure compared with control cells who had 98.29% wound closure). The silencing of *AHR* significantly reduced BxPC-3 cell line migratory abilities, and 36.2% of the wound was closed after 24 h. On the other hand, the silencing of *ELAVL1* had no significant effect on the migration of BxPC-3 cells. The silencing of *AHR* or *ELAVL1* and treatment with GEM had a significant effect when compared with controls (siAHR+GEM—16.5% and siELAVL1+GEM—19.7% of the wound closed); however, neither siAHR nor siELAVL1 increased GEM effect on migratory abilities. Su.86.86 (Figure 10b,d) had a lower migratory ability overall and no significant decrease in migratory ability after any of the effects compared with the control. The silencing of *AHR* seems to have a significant effect on cell migratory abilities; however, it was noticeable only in highly migratory cells (BxPC-3), and it did not increase the effects of GEM on cell migration.

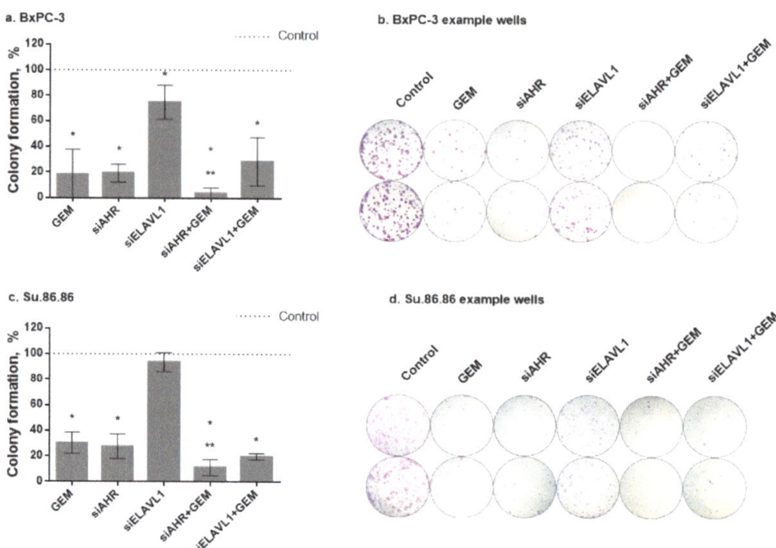

Figure 9. PDAC cell colony formation. Colony formation after silencing *AHR* or *ELAVL1* genes by transfection and/or exposure to GEM. Cells for clonogenic assay were seeded after 24 h transfection and grown for 168 h after seeding: (**a**) BxPC-3 colony formation graph, (**b**) BxPC-3 colony formation example photos, (**c**) Su.86.86 colony formation graph, and (**d**) Su.86.86 colony formation example photos. N $\geq$ 4, MEAN $\pm$ SD * $p < 0.05$ when compared with respective control. ** $p < 0.05$ when compared with GEM alone.

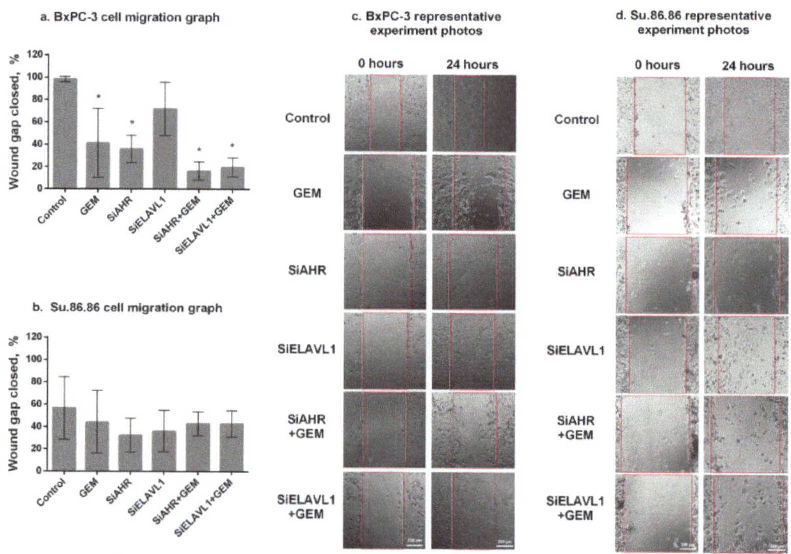

Figure 10. Cell migration by wound healing assay. Two PDAC cell lines showing migratory abilities after the silencing of *AHR* or *ELAVL1* genes and/or effect of gemcitabine: (**a**) BxPC-3 cell line migration graph, (**b**) Su.86.86 cell line migration graph, (**c**) BxPC-3 representative experiment photos (red lines show gap/wound width of photo taken at 0 h), and (**d**) Su.86.86 representative experiment photos (red lines show gap/wound width of photo taken at 0 h). N $\geq$ 3, MEAN $\pm$ SD. * $p < 0.05$ when compared with control.

3. Discussion

Chemoresistance of pancreatic cancer is a major problem limiting the success of chemotherapeutic treatments in patients. There are plenty of suggested mechanisms that cause PC resistance to chemotherapy, including common genetic mutations such as those in the *KRAS* [24,25], *MUC4* [26], and *TP53* [27] genes. MicroRNAs such as *miR-106b* [28], *miR-181b* [29], *miR-21* [30], and *miR-29a* [31] have also been implicated in the chemoresistance of PC. Even the overexpression of ribonucleotide reductase subunit M1 [32] or heat shock protein 27 [33] and many others are thought to be involved in one way or another. One of the most common proteins implicated in PC resistance to gemcitabine is the RNA-binding protein ELAVL1. Its cytoplasmic concentration is directly associated with the longer survival of PC patients [10]. AHR, which is often overexpressed in PC [20–22], is thought to be involved in blocking ELAVL1 from leaving the nucleus [17] and in turn stabilising its target mRNAs. One of the targets of ELAVL1 which is also implicated in gemcitabine resistance is DCK [34–36]. This is an enzyme that starts the activation of gemcitabine by phosphorylation; however, it is usually inactive in gemcitabine-resistant cells [37]. Therefore, in cells with upregulated AHR, ELAVL1 would be sequestered in the nucleus, which would decrease DCK protein concentrations and, in turn, contribute to gemcitabine resistance.

Our results show that *AHR* mRNA is a direct target for the ELAVL1 protein, and this interaction stabilises *AHR* mRNA, thus possibly increasing protein synthesis. AHR has been shown to sequester ELAVL1 in the nucleus [16], which in turn would block it from stabilising the mRNAs of various proteins, including *AHR*. By silencing *AHR* mRNA and in turn protein synthesis, we were able to show an increase in ELAVL1 expression as well as its localisation shift towards cytoplasm, which agrees with previous studies. However, the results of the expression of ELAVL1 pathway genes and proteins are conflicting, and both cell lines had different changes in some cases, showing that the effect of this modulation is hard to predict and can be different depending on the cell characteristics. By silencing *ELAVL1*, we showed that *AHR* mRNA and protein levels decrease, proving again that AHR can be a post-transcriptional target of ELAVL1. This complex interaction between AHR and ELAVL1 shows that they can alter the expression of each other by forming a negative feedback loop, which has never been shown before.

Due to the fact that AHR is a transcription factor and ELAVL1 is a post-transcriptional gene expression regulator, this interaction and its changes involve many different cell mechanisms involved in cytoprotection, migration, overall viability of the cell, and most importantly to PC treatment chemoresistance. Our results show that by lowering AHR or ELAVL1 expression, the cells become more susceptible to gemcitabine; however, the mechanisms most likely differ. The silencing of *AHR* greatly reduces cell migration and colony formation and might ease ELAVL1 translocation from the nucleus to the cytoplasm. However, different responses of ELAVL1 pathway genes and differences between cell lines show that there are more mechanisms involved in this regulation, warranting further investigation into the relationship between AHR and ELAVL1. Increased AHR expression can be seen in many different malignancies, for example, lymphoma [38] and leukaemia [39], as well as breast [40], kidney [41], gastrointestinal [42], and pancreatic cancers [20,21]. Since AHR is often overexpressed in cancer, it is frequently targeted in the hope of increasing the effectiveness of cancer treatment in various malignancies [20,40,43], and in this case, it could be targeted in the hopes of increasing gemcitabine efficiency in PC treatment.

The silencing of *ELAVL1* has a much lesser effect on cell migration or colony formation than *AHR*; however, our results show that it decreases *HMOX1* mRNA levels and DCK and HMOX1 protein levels. This lowers the protective mechanisms of the cell but can also increase the chemoresistance of the cell. Lowering ELAVL1 was also shown to decrease the synthesis of other protective proteins, such as COX-2 [44] or IFN-β [45], making the cell more susceptible to stress. ELAVL1 was shown to be both a positive and negative marker for various malignancies. Its interaction with various microRNAs has been shown to be a negative factor of ovarian cancer [46], prostate cancer [47], and other malignancies. It was shown to activate the *MAPK* and *JNK* signalling pathways and cause breast cancer resistance to tamoxifen [48]; its cytoplasmic concentration was shown to be a negative sign for the treatment of invasive breast cancer [49]. In pancreatic cancer, ELAVL1 was shown to regulate apoptosis through the IAP1 and IAP2 proteins [50]. It can also stop the cell cycle at the G2/M phase, allowing the cell to repair DNA damage, thus avoiding apoptosis [15]. The silencing of *ELAVL1* was also shown to increase the response to chemotherapy, although it was attributed to a decrease in cytoprotective proteins rather than a direct influence on gemcitabine activation [44]. However, in PC, more often than not, increased ELAVL1 concentrations in the cell cytoplasm are considered a positive sign, not only in terms of response to gemcitabine but also in overall patient survival [10,13,14,51].

Since ELAVL1 can be both a good and bad marker for chemotherapy resistance, it is imperative to understand how it works and find ways to utilise it in a way that would not cause any harm. Our findings suggest a mechanism (Figure 11) by which PDAC cells might be able to have increased resistance to gemcitabine through the ELAVL1-DKC pathway and highlights AHR as a target molecule to negate that resistance. Following the cellular uptake of gemcitabine, DCK starts its phosphorylation, which leads to cell cycle arrest and ideally cancer cell death (Figure 11(1–3)). ELAVL1 is a regulator of DCK and translocates to the cytoplasm following DNA damage, where it stabilises target mRNAs and increases their protein synthesis. In turn, increased DCK synthesis further strengthens the cytotoxic effect of gemcitabine (Figure 11(5,6)). However, at the same time, ELAVL1 stabilises the mRNAs of cytoprotective proteins and AHR, which in turn sequesters ELAVL1 in the nucleus, subsequently causing a negative feedback loop (Figure 11(5,7,8)). In cases of AHR overexpression in PC, ELAVL1 would be further sequestered in the nucleus, blocking it from stabilising the DCK protein and causing resistance to gemcitabine.

Overall, targeting AHR and decreasing its expression would ease the shuttling of ELAVL1 to the cytoplasm, which in turn would decrease cell resistance to gemcitabine and, at the same time, decrease cell colony formation and migration capabilities. Notably, the increase in ELAVL1 activity can increase cell cytoprotective mechanisms through the stabilisation of proteins such as HMOX1, although this effect depends on the cell characteristics according to our results, warranting further studies into the mechanisms involved.

Both AHR and ELAVL1 are involved in various molecular mechanisms. ELAVL1 is known to be involved in various cell signalling pathways, such as *MAPK* and *JNK* [48], as well as ferroptosis activation [52]. It is also involved in various systems, such as innate immune barriers in infants [53] and the overall immune system [45,53]. Similarly, AHR is also involved in various cell signalling pathways, such as epidermal growth factor family pathways [54] and xenobiotic metabolism [55]. It is also highly involved in the immune response systems [56,57]. The wide range of pathways AHR and ELAVL1 are involved with includes numerous interaction partners that could be regulated by modulating the AHR-ELAVL1 pathway, thus not limiting it to cytoprotection or gemcitabine metabolism.

Both AHR and ELAVL1 are involved in the immune system of the body. This poses a limitation to in vitro experiments due to absence of immune cells. Nonetheless, this novel mechanism provides a stepping stone for combating gemcitabine resistance in PC and should be investigated in in vivo setting. This would further elucidate the relationship of AHR and ELAVL1 in a setting with various other systems with which both AHR and ELAVL1 might interact.

There are some limitations to our study. Only lipofectamine-mediated siRNA transfection was used, which would prove to be difficult to use for patients, and so further studies with protein inhibitors would broaden the knowledge of the AHR-ELAVL1-DCK pathway. Also, only two cell lines were used, which had some different responses to AHR and ELAVL1 modulation; so, further investigation into the mechanisms causing such differences and the stratification of cancer subtypes that would benefit most from such modifications are required. Lastly, only in vitro experiments with cell cultures were performed; so, further experiments in vivo would be necessary to prove this relationship.

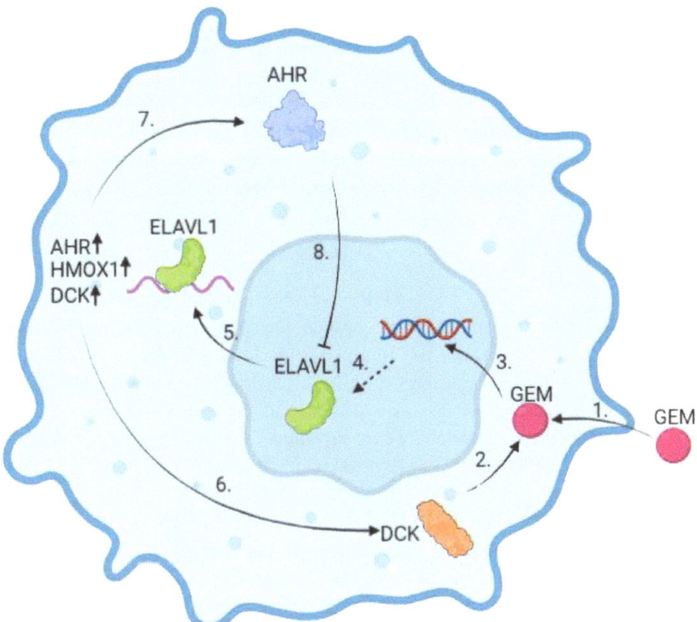

Figure 11. Suggested mechanism of ELAVL1 and AHR influence on gemcitabine effectiveness. In normal conditions, gemcitabine enters the cell through nucleoside transporters (1) where it is phosphorylated to its active state. DCK is the first enzyme to start the phosphorylation mechanism (2). Once GEM is metabolised into its active form, it integrates into the nucleus and stops further cell replication (3). This stress stimulates ELAVL1 (4) to shuttle from the nucleus to the cytoplasm (5). There, ELAVL1 stabilises its target mRNAs, thus promoting the translation of various proteins, including cytoprotective ones such as HMOX1, enzymes such as DCK (6), and the transcription factor AHR (7). Increased DCK synthesis by feedback loop subsequently promotes GEM phosphorylation. However, increased AHR synthesis blocks ELAVL1 from shuttling from the nucleus to the cytoplasm (8). This prevents the proteins of ELAVL1-targeted mRNAs from being overexpressed. (Created in Biorender.com).

4. Materials and Methods

4.1. PDAC Cell Lines and Growing Conditions

Two human pancreatic ductal adenocarcinoma cell lines, BxPc-3 and Su.86.86, were used for analysis. The BxPC-3 and Su.86.86 cell lines were a gift from the European Pancreas Centre (Heidelberg, Germany). Both cell lines were grown in RPMI medium (Gibco, Life Technologies Limited, Paisley, UK) with 10% FBS (Gibco Life Technologies Limited, Paisley, UK) and 1% penicillin/streptomycin solution (Gibco). Cells were grown in monolayers in sterile flasks/plates in an incubator, which maintains a moist temperature of 37 °C with a 5% CO_2-enriched environment.

4.2. Gemcitabine Treatment of Cells and IC50 Measurement

Cells were seeded in 96-well cell culture plates (BxPC-3—2.5×10^3 cells/well; Su.86.86 1.2×10^3 cells/well) and treated with varying concentrations of gemcitabine (GEM) (Fluorochem, Glossop, UK) ranging from 1–10,000 nM by logarithmic dilution to test the half-maximal inhibitory concentration (IC50) of GEM. Treated cells were maintained at 37 °C in a 5% CO_2 incubator for 48 h. To measure the proportion of metabolic activity, the MTT (3-(4,5-dimethylthiazol-2-yl)-2,5 diphenyltetrazolium bromide) metabolism method was used (see 'MTT metabolic activity assay'). No less than 4 replicates of the experiment were carried out.

4.3. MTT Metabolic Activity Assay

Cell viability was assessed by MTT (Invitrogen, Carlsbad, CA, USA) assay. After treatment, MTT was added to a concentration of 0.5 mg/mL, and the cells were incubated for 4 h at 37 °C. After 4 h of incubation, the medium with MTT was removed and the remaining formazan product was dissolved with DMSO (dimethyl sulfoxide) (Carl Roth GmbH, Karlsruhe, Germany) by agitation in the spectrophotometer for 60 s. The absorbance was measured with the spectrophotometer (TheSunrise (Software v7.1.), Tecan, Grodig, Austria) at a wavelength of 570 nm and reference of 620 nm.

4.4. Transfection

siELAVL1, siAHR, and negative control siRNA were purchased from Ambion (Waltham, MA, USA). Transfection was performed in 96, 6-well plates or 25 cm^2 flasks. Lipofectamine 2000 (Gibco, Life Technologies Limited, Paisley, UK) was used according to the manufacturer's instructions for all transfections with RPMI medium. All MTT experiments included two groups of control cells: untreated control and a control treated with an siRNA negative control. Four replicates of the experiment were carried out. Silencing efficiency after 24–72 h was evaluated by Western blot (WB) analysis.

4.5. RNA Extraction and Quantitative Real-Time Polymerase Chain Reaction (qRT-PCR)

Total RNA extraction was performed from cultured cells using the RNA extraction kit (Abbexa, Cambridge, UK) according to the manufacturer's protocol. Purified RNA was quantified and assessed for purity by UV spectrophotometry (NanoDrop 2000 (Software v.1.4.2), ThermoFisher Scientific, Waltham, MA, USA). cDNA was generated from 2 µg of RNA with High-Capacity cDNA Reverse Transcription Kit (Applied Biosystems, Waltham, MA, USA). The amplification of specific RNA was performed in a 20 µL reaction mixture containing 2 µL of cDNA template, 1X PCR master mix, and the primers. The PCR primers used for detection of *ELAVL1* (Hs00171039_m1), *AHR* (Hs00169233), *HMOX1* (Hs01110250_m1), and housekeeping gene *GAPDH* (Hs02758991_g1) were from Applied Biosystems. Three replicates of the experiment were carried out.

4.6. Western Blot Analysis

Whole cells were lysed using RIPA lysis buffer with protease inhibitors (Roche, Basel, Switzerland) and centrifuged at 10,000× g for 10 min at 4 °C. The supernatants were assayed for protein concentration with BCA protein assay kit (Thermo Scientific, Waltham, MA, USA). Protein samples were heated at 97 °C for 5 min before loading, and 25–50 µg of the samples was subjected to 4–12% sodium dodecyl sulfate-polyacrylamide gel electrophoresis (SDS-PAGE), then transferred to poly-vinylidene fluoride (PVDF) membranes 40 min 30 V. Next, membranes were blocked with a 5% skimmed milk blocking buffer for 60 min at room temperature. Membranes were then incubated for 1.5 h at room temperature or overnight at 4°C with primary antibodies. The following primary antibodies were used: 1:100 mouse monoclonal anti-ELAVL1 (LsBio, Lynnwood, WA, USA; Ls-C7451); 1:1000 mouse monoclonal anti-AHR (ThermoFisher Scientific, Waltham, MA, USA; MA1-514); 1:2000 rabbit monoclonal anti-HO-1 (Abcam; ab68477); 1:3000 mouse monoclonal anti-GAPDH (ThermoFisher Scientific; AM4300); and 1:2000 mouse monoclonal anti-DCK (ThermoFisher Scientific; MA5-25502). The membranes were washed with 1X Tris-Buffered Saline, 0.1% Tween 20 Detergent (TBST) antibody washing buffer or antibody wash buffer (Invitrogen, Carlsbad, CA, USA) and incubated in the appropriate peroxidase-conjugated secondary antibody solution (Invitrogen, Carlsbad, CA, USA) for 30 min or in horseradish-peroxidase-conjugated secondary antibody solution (LSbio) for 1 h. After that, membranes were washed again with TBST antibody washing buffer or antibody wash buffer (Invitrogen, Carlsbad, CA, USA) and incubated with chemiluminescence substrate (Invitrogen, Carlsbad, CA, USA) or West Pico Stable peroxidase buffer + luminol enhancer (Thermo Scientific) for 5 min. Results were analysed with a documenting system (Biorad, Hercules,

CA, USA). Three replicates of the experiment were carried out; however, only one is being shown as a representative replicate.

4.7. Immunocytochemistry

Cells were cultivated on chamber slides for 96 h either with or without treatment. A mixture of 96% ethanol with 5% glacial acetic acid was used for fixation and 0.5% Triton X-100 in PBS- for permeabilisation. Cells were subsequently incubated with 1:250 primary mouse monoclonal ELAVL1 antibody (LSBio) or 1:500 primary mouse monoclonal AHR antibody (Thermo Scientific, Waltham, MA, USA) and 1:2000 secondary antibody-Alexa Fluor 488 Goat Antimouse IgG (H+L) (Invitrogen, Carlsbad, CA, USA) and washed with PBS followed by washing with nuclease-free water. Slides were then mounted with ProLong Diamond Antifade Mountant with DAPI (Invitrogen, Carlsbad, CA, USA) for cell nuclei staining and analysed with Olympus IX71 fluorescent microscope (Olympus Corporation, Tokyo, Japan). Three replicates of the experiment were carried out; however, only one is being shown as a representative replicate.

4.8. Clonogenic Assay

The colony formation of pancreatic cancer cells was evaluated using a crystal violet stain. The cells were cultivated for 96 h with or without treatment with siAHR/siELAVL1/GEM. After treatment, the cells were detached by trypsin/EDTA, counted, and seeded into 6-well culture plates at concentration of 600 cells/well. After 7 days of growth, formed colonies were fixed with 96% ethanol and stained with crystal violet stain. After staining, crystal violet was removed, and the wells were rinsed with water. Plates were dried at room temperature, the morphology of cells was observed, and colonies were counted under an Olympus IX71 phase-contrast microscope (Olympus Corporation, Tokyo, Japan). No less than 4 replicates of the experiment were carried out.

4.9. Migration Assay

The cells were cultivated for 96 h with or without treatment with siAHR/siELAVL1/GEM. After treatment, the cells were detached by trypsin/EDTA, counted, and seeded into 24-well culture plates at concentration of 2×10^5 cells/well. After 24 h, a scratch was made with a 200 µL pipette tip, and the medium was changed into a fresh medium without FBS. The scratch was observed and photographed under an Olympus IX71 phase-contrast microscope at 0 and 24 h after making the scratch. No less than 3 replicates of the experiment were carried out.

4.10. Immunoprecipitation (IP) Assay

Cells were cultivated in 150 cm^2 flasks until they reached a confluence of 80–90%. The cells were then lysed using Magna RIP (Millipore, Burlington, MA, USA) kit and the IP assay was carried out using ELAVL1 antibody (Ls-C7451, LsBio, Lynnwood, WA, USA) and GAPDH antibody (AM4300, Invitrogen, Carlsbad, CA, USA) as a negative control. The samples for WB and qRT-PCR analysis were collected before and after immunoprecipitation. WB and qRT-PCR assays were performed as described above.

4.11. Statistical Analysis

Statistical analysis was performed using GraphPad (version 6.01; GraphPad Software Inc., La Jolla, CA, USA) software. The data are presented as mean $\pm$ SD of three or more independent experiments. A nonparametric Mann–Whitney test was used for comparison between groups. Statistical significance was defined as $p < 0.05$.

5. Conclusions

Our study revealed that both AHR and ELAVL1 inter-regulate each other, as well as having a role in cell proliferation, migration, and chemoresistance in PDAC cell lines. Notably, the effect of *AHR* silencing appears to be more pronounced than that of *ELAVL1*, and both proteins act through distinct mechanisms. The silencing of *ELAVL1* disrupts the stability of its target mRNAs, resulting in the decreased expression of numerous cytoprotective proteins. In contrast, the silencing of *AHR* diminishes cell migration and proliferation and enhances cell sensitivity to gemcitabine through the AHR-ELAVL1-DCK molecular pathway.

These findings underscore the complex interplay between AHR, ELAVL1, and important cellular processes in PDAC. The differential effects and distinct molecular pathways associated with AHR and ELAVL1 modulation highlight their potential as therapeutic targets for PDAC treatment, particularly in overcoming chemoresistance and inhibiting tumour progression. Further investigations are necessary to fully elucidate the underlying mechanisms and explore the clinical implications of targeting these pathways in PDAC management.

Author Contributions: Conceptualization, A.G., Z.D. and J.M.; methodology, D.S. and A.J.; investigation, D.S. and A.B.; data curation, A.G., Z.D., J.M. and T.M.; writing—original draft preparation, D.S.; writing—review and editing, A.G., Z.D., J.M., A.J. and I.T.; visualisation, D.S. and A.J.; supervision, T.M. and K.J.; funding acquisition, Z.D., A.G., J.M., T.M. and K.J. All authors have read and agreed to the published version of the manuscript.

Funding: This research was funded by the Baltic Research Programme Grant No. S-BMT-21-11 (LT08-2-LMT-K-01-060) financed from the funds of financial mechanism of the European Economic Area (EEA) States (Iceland and Liechtenstein) and Norway.

Institutional Review Board Statement: Not applicable.

Informed Consent Statement: Not applicable.

Data Availability Statement: The data presented in this study are available on request from the corresponding author.

Conflicts of Interest: The authors declare no conflict of interest.

References

1. Hu, J.X.; Zhao, C.F.; Chen, W.B.; Liu, Q.C.; Li, Q.W.; Lin, Y.Y.; Gao, F. Pancreatic cancer: A review of epidemiology, trend, and risk factors. *World J. Gastroenterol.* **2021**, *27*, 4298. [CrossRef] [PubMed]
2. Siegel, R.L.; Miller, K.D.; Fuchs, H.E.; Jemal, A. Cancer statistics, 2021. *CA Cancer J. Clin.* **2021**, *71*, 7–33. [CrossRef]
3. Yu, J.; Yang, X.; He, W.; Ye, W. Burden of pancreatic cancer along with attributable risk factors in Europe between 1990 and 2019, and projections until 2039. *Int. J. Cancer* **2021**, *149*, 993–1001. [CrossRef]
4. Rahib, L.; Smith, B.D.; Aizenberg, R.; Rosenzweig, A.B.; Fleshman, J.M.; Matrisian, L.M. Projecting cancer incidence and deaths to 2030: The unexpected burden of thyroid, liver, and pancreas cancers in the United States. *Cancer Res.* **2014**, *74*, 2913–2921. [CrossRef] [PubMed]
5. Quante, A.S.; Ming, C.; Rottmann, M.; Engel, J.; Boeck, S.; Heinemann, V.; Westphalen, C.B.; Strauch, K. Projections of cancer incidence and cancer-related deaths in Germany by 2020 and 2030. *Cancer Med.* **2016**, *5*, 2649–2656. [CrossRef] [PubMed]
6. Vincent, A.; Herman, J.; Schulick, R.; Hruban, R.H.; Goggins, M. Pancreatic cancer. *Lancet* **2011**, *378*, 607–620. [CrossRef] [PubMed]
7. Buscail, L.; Bournet, B.; Cordelier, P. Role of oncogenic *KRAS* in the diagnosis, prognosis and treatment of pancreatic cancer. *Nat. Rev. Gastroenterol. Hepatol.* **2020**, *17*, 153–168. [CrossRef]
8. Rueff, J.; Rodrigues, A.S. Cancer drug resistance: A brief overview from a genetic viewpoint. *Cancer Drug Res.* **2016**, *1395*, 1–18.
9. Kuwano, Y.; Rabinovic, A.; Srikantan, S.; Gorospe, M.; Demple, B. Analysis of nitric oxide-stabilized mRNAs in human fibroblasts reveals HuR-dependent heme oxygenase 1 upregulation. *Mol. Cell. Biol.* **2009**, *29*, 2622–2635. [CrossRef]
10. Richards, N.G.; Rittenhouse, D.W.; Freydin, B.; Cozzitorto, J.A.; Grenda, D.; Rui, H.; Gonye, G.; Kennedy, E.P.; Yeo, C.J.; Brody, J.R.; et al. HuR status is a powerful marker for prognosis and response to gemcitabine-based chemotherapy for resected pancreatic ductal adenocarcinoma patients. *Ann. Surg.* **2010**, *252*, 499–505. [CrossRef]
11. Wang, J.; Guo, Y.; Chu, H.; Guan, Y.; Bi, J.; Wang, B. Multiple functions of the RNA-binding protein HuR in cancer progression, treatment responses and prognosis. *Int. J. Mol. Sci.* **2013**, *14*, 10015–10041. [CrossRef]
12. Zucal, C.; D'Agostino, V.; Loffredo, R.; Mantelli, B.; Thongon, N.; Lal, P.; Latorre, E.; Provenzani, A. Targeting the multifaceted HuR protein, benefits and caveats. *Curr. Drug Targets* **2015**, *16*, 499–515. [CrossRef] [PubMed]
13. Costantino, C.L.; Witkiewicz, A.K.; Kuwano, Y.; Cozzitorto, J.A.; Kennedy, E.P.; Dasgupta, A.; Keen, J.C.; Yeo, C.J.; Gorospe, M.; Brody, J.R. The role of HuR in gemcitabine efficacy in pancreatic cancer: HuR Up-regulates the expression of the gemcitabine metabolizing enzyme deoxycytidine kinase. *Cancer Res.* **2009**, *69*, 4567–4572. [CrossRef]

14. McAllister, F.; Pineda, D.M.; Jimbo, M.; Lal, S.; Burkhart, R.A.; Moughan, J.; Winter, K.A.; Abdelmohsen, K.; Gorospe, M.; Acosta, A.J.; et al. dCK expression correlates with 5-fluorouracil efficacy and HuR cytoplasmic expression in pancreatic cancer: A dual-institutional follow-up with the RTOG 9704 trial. *Cancer Biol. Ther.* **2014**, *15*, 688–698. [CrossRef] [PubMed]
15. Lal, S.; Burkhart, R.A.; Beeharry, N.; Bhattacharjee, V.; Londin, E.R.; Cozzitorto, J.A.; Romeo, C.; Jimbo, M.; Norris, Z.A.; Yeo, C.J.; et al. HuR posttranscriptionally regulates WEE1: Implications for the DNA damage response in pancreatic cancer cells. *Cancer Res.* **2014**, *74*, 1128–1140. [CrossRef] [PubMed]
16. Zago, M.; Sheridan, J.A.; Nair, P.; Rico de Souza, A.; Gallouzi, I.E.; Rousseau, S.; Marco, S.D.; Hamid, Q.; Eidelman, D.H.; Baglole, C.J. Aryl hydrocarbon receptor-dependent retention of nuclear HuR suppresses cigarette smoke-induced cyclooxygenase-2 expression independent of DNA-binding. *PLoS ONE* **2013**, *8*, e74953. [CrossRef] [PubMed]
17. Connor, K.T.; Aylward, L.L. Human response to dioxin: Aryl hydrocarbon receptor (AhR) molecular structure, function, and dose-response data for enzyme induction indicate an impaired human AhR. *J. Toxicol. Environ. Health B* **2006**, *9*, 147–171. [CrossRef]
18. Trikha, P.; Lee, D.A. The role of AhR in transcriptional regulation of immune cell development and function. *Biochim. Biophys. Acta Rev. Cancer* **2020**, *1873*, 188335. [CrossRef]
19. Wang, Z.; Snyder, M.; Kenison, J.E.; Yang, K.; Lara, B.; Lydell, E.; Bennani, K.; Novikov, O.; Federico, A.; Monti, S.; et al. How the AHR Became Important in Cancer: The Role of Chronically Active AHR in Cancer Aggression. *Int. J. Mol. Sci.* **2020**, *22*, 387. [CrossRef]
20. Koliopanos, A.; Kleeff, J.; Xiao, Y.; Safe, S.; Zimmermann, A.; Büchler, M.W.; Friess, H. Increased arylhydrocarbon receptor expression offers a potential therapeutic target for pancreatic cancer. *Oncogene* **2002**, *21*, 6059–6070. [CrossRef]
21. Masoudi, S.; Nemati, A.H.; Fazli, H.R.; Beygi, S.; Moradzadeh, M.; Pourshams, A.; Mohamadkhani, A. An increased level of aryl hydrocarbon receptor in patients with pancreatic cancer. *Middle East J. Dig. Dis.* **2019**, *11*, 38. [CrossRef]
22. Paris, A.; Tardif, N.; Galibert, M.D.; Corre, S. AhR and cancer: From gene profiling to targeted therapy. *Int. J. Mol. Sci.* **2021**, *22*, 752. [CrossRef]
23. Fallmann, J.; Sedlyarov, V.; Tanzer, A.; Kovarik, P.; Hofacker, I.L. AREsite2: An enhanced database for the comprehensive investigation of AU/GU/U-rich elements. *Nucleic Acids Res.* **2016**, *44*, D90–D95. [CrossRef]
24. Mukhopadhyay, S.; Goswami, D.; Adiseshaiah, P.P.; Burgan, W.; Yi, M.; Guerin, T.M.; Kozlov, S.V.; Nissley, D.V.; McCormick, F. Undermining Glutaminolysis Bolsters Chemotherapy While NRF2 Promotes Chemoresistance in KRAS-Driven Pancreatic Cancers. *Cancer Res.* **2020**, *80*, 1630–1643. [CrossRef]
25. Réjiba, S.; Wack, S.; Aprahamian, M.; Hajri, A. K-ras oncogene silencing strategy reduces tumour growth and enhances gemcitabine chemotherapy efficacy for pancreatic cancer treatment. *Cancer Sci.* **2007**, *98*, 1128–1136. [CrossRef]
26. Bafna, S.; Kaur, S.; Momi, N.; Batra, S.K. Pancreatic cancer cells resistance to gemcitabine: The role of MUC4 mucin. *Br. J. Cancer* **2009**, *101*, 1155–1161. [CrossRef] [PubMed]
27. Wörmann, S.M.; Song, L.; Ai, J.; Diakopoulos, K.N.; Kurkowski, M.U.; Görgülü, K.; Ruess, D.; Campbell, A.; Doglioni, C.; Jodrell, D.; et al. Loss of P53 Function Activates JAK2-STAT3 Signaling to Promote Pancreatic Tumor Growth, Stroma Modification, and Gemcitabine Resistance in Mice and Is Associated with Patient Survival. *Gastroenterology* **2016**, *151*, 180–193. [CrossRef] [PubMed]
28. Fang, Y.; Zhou, W.; Rong, Y.; Kuang, T.; Xu, X.; Wu, W.; Wang, D.; Lou, W. Exosomal *miRNA-106b* from cancer-associated fibroblast promotes gemcitabine resistance in pancreatic cancer. *Exp. Cell Res.* **2019**, *383*, 111543. [CrossRef]
29. Takiuchi, D.; Eguchi, H.; Nagano, H.; Iwagami, Y.; Tomimaru, Y.; Wada, H.; Kawamoto, K.; Kobayashi, S.; Marubashi, S.; Tanemura, M.; et al. Involvement of microRNA-181b in the gemcitabine resistance of pancreatic cancer cells. *Pancreatology* **2013**, *13*, 517–523. [CrossRef] [PubMed]
30. Giovannetti, E.; Funel, N.; Peters, G.J.; Del Chiaro, M.; Erozenci, L.A.; Vasile, E.; Leon, L.G.; Pollina, L.E.; Groen, A.; Falcone, A.; et al. *MicroRNA-21* in pancreatic cancer: Correlation with clinical outcome and pharmacologic aspects underlying its role in the modulation of gemcitabine activity. *Cancer Res.* **2010**, *70*, 4528–4538. [CrossRef] [PubMed]
31. Nagano, H.; Tomimaru, Y.; Eguchi, H.; Hama, N.; Wada, H.; Kawamoto, K.; Kobayashi, S.; Mori, M.; Doki, Y. *MicroRNA-29a* induces resistance to gemcitabine through the Wnt/β-catenin signaling pathway in pancreatic cancer cells. *Int. J. Oncol.* **2013**, *43*, 1066–1072. [CrossRef] [PubMed]
32. Yoneyama, H.; Takizawa-Hashimoto, A.; Takeuchi, O.; Watanabe, Y.; Atsuda, K.; Asanuma, F.; Yamada, Y.; Suzuki, Y. Acquired resistance to gemcitabine and cross-resistance in human pancreatic cancer clones. *Anticancer Drugs* **2015**, *26*, 90–100. [CrossRef] [PubMed]
33. Kuramitsu, Y.; Wang, Y.; Taba, K.; Suenaga, S.; Ryozawa, S.; Kaino, S.; Sakaida, I.; Nakamura, K. Heat-shock protein 27 plays the key role in gemcitabine-resistance of pancreatic cancer cells. *Anticancer Res.* **2012**, *32*, 2295–2299. [PubMed]
34. Sebastiani, V.; Ricci, F.; Rubio-Viqueira, B.; Kulesza, P.; Yeo, C.J.; Hidalgo, M.; Klein, A.; Laheru, D.; Iacobuzio-Donahue, C.A. Immunohistochemical and genetic evaluation of deoxycytidine kinase in pancreatic cancer: Relationship to molecular mechanisms of gemcitabine resistance and survival. *Clin. Cancer Res.* **2006**, *12*, 2492–2497. [CrossRef] [PubMed]
35. Funamizu, N.; Okamoto, A.; Kamata, Y.; Misawa, T.; Uwagawa, T.; Gocho, T.; Yanaga, K.; Manome, Y. Is the resistance of gemcitabine for pancreatic cancer settled only by overexpression of deoxycytidine kinase? *Oncol. Rep.* **2010**, *23*, 471–475. [CrossRef]

36. Ohhashi, S.; Ohuchida, K.; Mizumoto, K.; Fujita, H.; Egami, T.; Yu, J.; Toma, H.; Sadatomi, S.; Nagai, E.; Tanaka, M. Down-regulation of deoxycytidine kinase enhances acquired resistance to gemcitabine in pancreatic cancer. *Anticancer Res.* **2008**, *28*, 2205–2212.

37. Saiki, Y.; Yoshino, Y.; Fujimura, H.; Manabe, T.; Kudo, Y.; Shimada, M.; Mano, N.; Nakano, T.; Lee, Y.; Shimizu, S.; et al. DCK is frequently inactivated in acquired gemcitabine-resistant human cancer cells. *Biochem. Biophys. Res. Commun.* **2012**, *421*, 98–104. [CrossRef]

38. Sherr, D.H.; Monti, S. The role of the aryl hydrocarbon receptor in normal and malignant B cell development. *Semin. Immunopathol.* **2013**, *35*, 705–716. [CrossRef]

39. Hayashibara, T.; Yamada, Y.; Mori, N.; Harasawa, H.; Sugahara, K.; Miyanishi, T.; Kamihira, S.; Tomonaga, M. Possible involvement of aryl hydrocarbon receptor (AhR) in adult T-cell leukaemia (ATL) leukemogenesis: Constitutive activation of AhR in ATL. *Biochem. Biophys. Res. Commun.* **2003**, *300*, 128–134. [CrossRef]

40. Baker, J.R.; Sakoff, J.A.; McCluskey, A. The aryl hydrocarbon receptor (AhR) as a breast cancer drug target. *Med. Res. Rev.* **2020**, *40*, 972–1001. [CrossRef]

41. Ishida, M.; Mikami, S.; Kikuchi, E.; Kosaka, T.; Miyajima, A.; Nakagawa, K.; Mukai, M.; Okada, Y.; Oya, M. Activation of the aryl hydrocarbon receptor pathway enhances cancer cell invasion by upregulating the MMP expression and is associated with poor prognosis in upper urinary tract urothelial cancer. *Carcinogenesis* **2010**, *31*, 287–295. [CrossRef] [PubMed]

42. Bogoevska, V.; Wolters-Eisfeld, G.; Hofmann, B.T.; El Gammal, A.T.; Mercanoglu, B.; Gebauer, F.; Vashist, Y.K.; Bogoevski, D.; Perez, D.; Gagliani, N.; et al. HRG/HER2/HER3 signaling promotes AhR-mediated Memo-1 expression and migration in colorectal cancer. *Oncogene* **2017**, *36*, 2394–2404. [CrossRef] [PubMed]

43. Cheong, J.E.; Sun, L. Targeting the IDO1/TDO2–KYN–AhR pathway for cancer immunotherapy–challenges and opportunities. *Trends Pharmacol. Sci.* **2018**, *39*, 307–325. [CrossRef] [PubMed]

44. Jakstaite, A.; Maziukiene, A.; Silkuniene, G.; Kmieliute, K.; Gulbinas, A.; Dambrauskas, Z. HuR mediated post-transcriptional regulation as a new potential adjuvant therapeutic target in chemotherapy for pancreatic cancer. *World J. Gastroenterol.* **2015**, *21*, 13004. [CrossRef]

45. Herdy, B.; Karonitsch, T.; Vladimer, G.I.; Tan, C.S.; Stukalov, A.; Trefzer, C.; Bigenzahn, J.W.; Theil, T.; Holinka, J.; Kiener, H.P.; et al. The RNA-binding protein HuR/ELAVL1 regulates IFN-β mRNA abundance and the type I IFN response. *Eur. J. Immunol.* **2015**, *45*, 1500–1511. [CrossRef]

46. Prislei, S.; Martinelli, E.; Mariani, M.; Raspaglio, G.; Sieber, S.; Ferrandina, G.; Shahabi, S.; Scambia, G.; Ferlini, C. MiR-200c and HuR in ovarian cancer. *BMC Cancer* **2013**, *13*, 72. [CrossRef]

47. Kojima, K.; Fujita, Y.; Nozawa, Y.; Deguchi, T.; Ito, M. MiR-34a attenuates paclitaxel-resistance of hormone-refractory prostate cancer PC3 cells through direct and indirect mechanisms. *Prostate* **2010**, *70*, 1501–1512. [CrossRef]

48. Hostetter, C.; Licata, L.A.; Costantino, C.L.; Witkiewicz, A.; Yeo, C.; Brody, J.R.; Keen, J.C. Cytoplasmic accumulation of the RNA binding protein HuR is central to tamoxifen resistance in estrogen receptor positive breast cancer cells. *Cancer Biol. Ther.* **2008**, *7*, 1496–1506. [CrossRef]

49. Wang, J.; Li, D.; Wang, B.; Wu, Y. Predictive and prognostic significance of cytoplasmic expression of ELAV-like protein HuR in invasive breast cancer treated with neoadjuvant chemotherapy. *Breast Cancer Res. Treat.* **2013**, *141*, 213–224. [CrossRef]

50. Lukosiute-Urboniene, A.; Jasukaitiene, A.; Silkuniene, G.; Barauskas, V.; Gulbinas, A.; Dambrauskas, Z. Human antigen R mediated post-transcriptional regulation of inhibitors of apoptosis proteins in pancreatic cancer. *World J. Gastroenterol.* **2019**, *25*, 205. [CrossRef]

51. Maréchal, R.; Van Laethem, J.L. HuR modulates gemcitabine efficacy: New perspectives in pancreatic cancer treatment. *Expert Rev. Anticancer Ther.* **2009**, *9*, 1439–1441. [CrossRef] [PubMed]

52. Chen, H.Y.; Xiao, Z.Z.; Ling, X.; Xu, R.N.; Zhu, P.; Zheng, S.Y. ELAVL1 is transcriptionally activated by FOXC1 and promotes ferroptosis in myocardial ischemia/reperfusion injury by regulating autophagy. *Mol. Med.* **2021**, *27*, 1–14. [CrossRef] [PubMed]

53. Li, C.; Liu, X.; Huang, Z.; Zhai, Y.; Li, H.; Wu, J. Lactoferrin Alleviates Lipopolysaccharide-Induced Infantile Intestinal Immune Barrier Damage by Regulating an ELAVL1-Related Signaling Pathway. *Int. J. Mol. Sci.* **2022**, *23*, 13719. [CrossRef] [PubMed]

54. Haarmann-Stemmann, T.; Bothe, H.; Abel, J. Growth factors, cytokines and their receptors as downstream targets of arylhydrocarbon receptor (AhR) signaling pathways. *Biochem. Pharmacol.* **2009**, *77*, 508–520. [PubMed]

55. Larigot, L.; Juricek, L.; Dairou, J.; Coumoul, X. AhR signaling pathways and regulatory functions. *Biochim. Open* **2018**, *7*, 1–9. [CrossRef]

56. Stevens, E.A.; Mezrich, J.D.; Bradfield, C. A The aryl hydrocarbon receptor: A perspective on potential roles in the immune system. *Immunology* **2009**, *127*, 299–311. [CrossRef]

57. Schiering, C.; Wincent, E.; Metidji, A.; Iseppon, A.; Li, Y.; Potocnik, A.J.; Omenetti, S.; Henderson, C.J.; Wolf, C.R.; Nebert, D.W.; et al. Feedback control of AHR signalling regulates intestinal immunity. *Nature* **2017**, *542*, 242–245. [CrossRef]

Disclaimer/Publisher's Note: The statements, opinions and data contained in all publications are solely those of the individual author(s) and contributor(s) and not of MDPI and/or the editor(s). MDPI and/or the editor(s) disclaim responsibility for any injury to people or property resulting from any ideas, methods, instructions or products referred to in the content.

International Journal of
Molecular Sciences

Exosomal miRNA Biomarker Panel for Pancreatic Ductal Adenocarcinoma Detection in Patient Plasma: A Pilot Study

Amy Makler [1,2] and Waseem Asghar [1,*]

[1] Micro and Nanotechnology in Medicine, Department of Electrical Engineering and Computer Science, College of Engineering and Science, Florida Atlantic University, Boca Raton, FL 33431, USA
[2] Department of Biomedical Science, Charles E. Schmidt College of Medicine, Florida Atlantic University, Boca Raton, FL 33431, USA
* Correspondence: wasghar@fau.edu

Abstract: Pancreatic ductal adenocarcinoma (PDAC) is rapidly becoming one of the leading causes of cancer-related deaths in the United States, and with its high mortality rate, there is a pressing need to develop sensitive and robust methods for detection. Exosomal biomarker panels provide a promising avenue for PDAC screening since exosomes are highly stable and easily harvested from body fluids. PDAC-associated miRNAs packaged within these exosomes could be used as diagnostic markers. We analyzed a series of 18 candidate miRNAs via RT-qPCR to identify the differentially expressed miRNAs ($p < 0.05$, *t*-test) between plasma exosomes harvested from PDAC patients and control patients. From this analysis, we propose a four-marker panel consisting of miR-93-5p, miR-339-3p, miR-425-5p, and miR-425-3p with an area under the curve (AUC) of the receiver operator characteristic curve (ROC) of 0.885 with a sensitivity of 80% and a specificity of 94.7%, which is comparable to the CA19-9 standard PDAC marker diagnostic.

Keywords: biomarker panel; diagnostics; exosomes; microRNA; pancreatic ducal adenocarcinoma; plasma

Citation: Makler, A.; Asghar, W. Exosomal miRNA Biomarker Panel for Pancreatic Ductal Adenocarcinoma Detection in Patient Plasma: A Pilot Study. *Int. J. Mol. Sci.* **2023**, *24*, 5081. https://doi.org/10.3390/ijms24065081

Academic Editors: Claudio Luchini and Donatella Delle Cave

Received: 24 January 2023
Revised: 1 March 2023
Accepted: 1 March 2023
Published: 7 March 2023

Copyright: © 2023 by the authors. Licensee MDPI, Basel, Switzerland. This article is an open access article distributed under the terms and conditions of the Creative Commons Attribution (CC BY) license (https://creativecommons.org/licenses/by/4.0/).

1. Introduction

Pancreatic ductal adenocarcinoma (PDAC) is the third leading cause of cancer-related deaths in the United States of America with an estimated 62,000 new diagnoses and an estimated 50,000 deaths expected in the USA this year [1]. Currently, the only FDA-approved diagnostic for PDAC is serum antigen CA19-9. Patients with cancer of the pancreas, stomach, lung, liver, or colon typically show levels of CA19-9 exceeding 37 U/mL [2]. For PDAC detection, the CA19-9 serum antigen diagnostic test has a sensitivity ranging between 79% and 95% and a specificity ranging between 82% and 91% [3]. However, since non-cancerous conditions such as pancreatitis, gallbladder infection, liver disease, and gallstones may also show increased CA19-9 levels [4,5], there is a need to develop new modalities with increased sensitivity and specificity of detection for PDAC compared to non-cancerous conditions.

Exosomes, and their contents, may offer a more reliable diagnostic alternative to CA19-9. Exosomes are released by all cells in the body, and it is well established that tumor cells release even greater quantities of exosomes [6]. Exosomes are 30–150 nm sized extracellular vesicles that contain proteins, DNA, RNA, and other cellular constituents [7,8]. They are stable in body fluids, allowing for easy collection from patient blood, plasma, serum, saliva, or urine. Recent research has examined both exosomes and their contents for diagnostic feasibility for many diseases [7–11].

The microRNA transcriptome potentially contains diagnostic biomarkers for PDAC that could exceed the sensitivity and specificity of CA19-9 serum markers. MicroRNAs are 19–25-nucleotide-long sequences that have been shown to regulate about a third of human genes, with half being involved in tumor regulation [12]. In non-cancerous cells, miRNAs play roles in a variety of metabolic processes including embryogenesis, growth, repair, cell cycle, proliferation, stress tolerance, and immune response [13,14]. In cancerous tumors, they can play roles in drug resistance, immune evasion, growth, and metastasis. MicroRNAs (miRNAs) can exhibit tumor suppressor or oncogenic roles, with some miRNAs exhibiting both, depending on tissue and tumor type. Functional analyses have examined the roles of miRNA in PDAC progression. For example, miR-196b was implicated in driving PDAC progression by interacting with known PDAC-associated miR-21 and miR-31 [15]. Inhibition of miR-196b resulted in decreased levels of miR-21 and miR-31 as well as a decrease in cell proliferation. Because of these tissue-specific roles, miRNAs associated with abnormalities in cellular metabolic processes characteristic of specific tumors could be used as a more sensitive and specific diagnostic method.

Circulating miRNAs and miRNAs in various body fluids have been extensively researched with promising results. In lung cancer, multiple panels of microRNAs isolated from peripheral blood were used to diagnose early lung cancer compared to the control [16]. Several studies have reported utilizing urinary, plasma, and serum miRNAs to detect bladder cancer [17]. Similarly, there have been several studies that examine circulating miRNAs and miRNAs in body fluids for diagnosis, prognosis, and monitoring response to therapy. Liu et al. reported a serum marker comprised of seven miRNAs that could distinguish PDAC from chronic pancreatitis with 83.6% accuracy [18]. Another study analyzed plasma from PDAC patients and control patients and found miR-21 and miR-483-3p to be significantly increased in PDAC compared to the control [19]. Additional miRs and their roles in PDAC diagnosis have been reported and summarized previously [7,20,21].

Dysregulation of miRNAs and exosomes in cancer have been shown to offer highly sensitive and specific methods for diagnosing respective tumor types, including PDAC [7,8,22]. In our previous study, we developed a workflow strategy to identify a panel of miRNAs for potential pancreatic ductal adenocarcinoma (PDAC) detection [23]. Multiple knowledgebases were accessed to generate a database of 383 PDAC-associated non-coding RNAs (ncRNAs), with the majority belonging to the miRNA subtype. The cBioPortal [24] tool was used to identify 72 miRNAs that exhibited alteration in at least 10% of the University of Texas Southwestern (UTSW) PDAC dataset (N = 109). These 72 miRNAs were enriched for their presence in exosomes, resulting in 50 exosomal miRNA. The cBioPortal batch analysis function was used to test combinations of these exosomal miRNA and identified a final panel of 18 mature miRNAs that exhibited alteration in 90% of the UTSW PDAC dataset. These MIRs provided the basis for the design of a diagnostic panel with the potential for early detection and monitoring of PDAC. These 18 were then analyzed in vitro to provide the basis for testing in plasma derived from PDAC and control patients. In the present study, we used quantitative RT-PCR to measure the levels of these 18 mature miRNA from exosomes harvested from PDAC patient plasma compared to control patient plasma. Four of these 18 candidate exosomal miRNAs exhibited significant expression differences between control and PDAC patient plasma samples. The diagnostic potential of these four miRs was explored and found to be comparable to CA19-9 in sensitivity and specificity, but they exhibited greater potential for detecting early stage PDAC compared to CA19-9.

2. Results

2.1. Patient Information

In our previous study, we analyzed the expression levels of 18 candidate mature miRNAs extracted from exosomes released from in vitro cultured PDAC cell lines versus an immortalized pancreatic cell line [23]. Seven of the 18 candidate mature miRNAs were found to be differentially expressed between the experimental and control groups, suggesting these exosomal miRNAs could potentially be used as a diagnostic panel for PDAC. We sought to design a diagnostic panel for PDAC based on the expression of these same 18 candidate exosomal miRNAs and their differential expression in plasma collected from PDAC patients (N = 15) and control patients (N = 19). Sex, age, ethnicity, CA19-9 levels, and tumor staging are reported in Table 1. Seven of the 15 PDAC patients exhibited CA19-9 levels within normal ranges (<37 U/mL).

Table 1. Patient information obtained from 19 control donors and 15 pancreatic cancer donors.

	Control (N = 19)	Pancreatic Cancer (N = 15)
Sex N (%)		
Male	13 (68.4%)	13 (86.7%)
Female	6 (26.3%)	2 (13.3%)
Mean Age (Range) in years		
Male	42.62 (23–67)	66.54 (40–81)
Female	44.17 (21–67)	70 (62–78)
Ethnicity N (%)		
Caucasian	1 (5%)	3 (20%)
African American	7 (40%)	1 (6.7%)
Hispanic	9 (45%)	11 (66.7%)
Non-white Hispanic	2 (10%)	
CA19-9		Normal: 7 (47%) Elevated: 8 (53%)
Tumor Stage		
I		3 (20%)
II		2 (13.3%)
III		3 (20%)
IV		7 (46.7%)

2.2. RT-qPCR Analysis of Exosomal miRNAs from PDAC Patient Plasma

Exosomal miRNAs were isolated from plasma collected from PDAC and control patients. Expression levels of 18 candidate miRNAs were measured using RT-qPCR. After 40 cycles of PCR, 7 of the 18 candidate miRNAs were not detectable in over 80% of the samples (N = 34) and, thus, were excluded from further analysis as recommended by a previous study [23]. The remaining 11 mature miRNAs (miR-93-5p, miR-93-3p, miR-133a-3p, miR-210-3p, miR-330-5p, miR-330-3p, miR-339-5p, miR-339-3p, miR-425-5p, miR-425-3p, and miR-3620-3p) were further analyzed for differential expression between PDAC and control patients using ΔCq values.

RT-qPCR analysis of exosomal miRNA identified four mature miRNAs that exhibited statistically significant expression differences between the control and PDAC patient plasma samples: miR-93-5p ($p < 0.05$, 99% confidence interval (CI) control ΔCq range: 7.99–9.60 and 99% CI PDAC ΔCq range: 8.92–10.96), miR-339-3p ($p < 0.01$, 99% CI control ΔCq range: 12.75–14.82 and 99% CI PDAC ΔCq: 14.27–16.84), miR-425-5p ($p < 0.001$, 99% CI control ΔCq: 7.67–9.16, 99% CI PDAC ΔCq: 9.59–11.51), and miR-425-3p ($p < 0.01$, 99% CI control ΔCq: 10.93–13.77, 99% CI PDAC ΔCq: 13.15–15.64) (Figure 1). All four miRNAs exhibited significantly increased ΔCq values in PDAC samples compared to the control. These four miRNAs constituted the best candidate miRNAs for an exosomal miRNA PDAC diagnostic and were further analyzed for associated biological functions and pathways.

2.3. Gene Ontology and KEGG Pathway Analysis of Differentially Expressed Plasma Exosome miRNAs

We utilized the DIANA-miRPATH v3 tool [25] to identify biological pathways and functions of the four miRNAs (miR-93-5p, miR-339-3p, miR-425-3p, miR-425-5p) using the DIANA-TarBase 7.0 option. Kyoto Encyclopedia of Genes and Genomes (KEGG) [26] analysis revealed that miR-93-5p was involved in several cancer-specific pathways (glioma, bladder cancer, chronic myeloid leukemia, renal cell carcinoma, colorectal cancer, and pathways in cancer), while miR-93-5p and miR-425-5p were both involved in cancer-regulatory pathways including p53 signaling and HIPPO signaling pathways [27] (Supplementary Figure S1A). Gene ontology analysis revealed that shared functions of three of the four miRs (miR-93-5p, miR-339-3p, and miR-425-5p) included cellular nitrogen compound metabolic processes, gene expression, RNA binding, and protein metabolic processes (Supplementary Figure S1B).

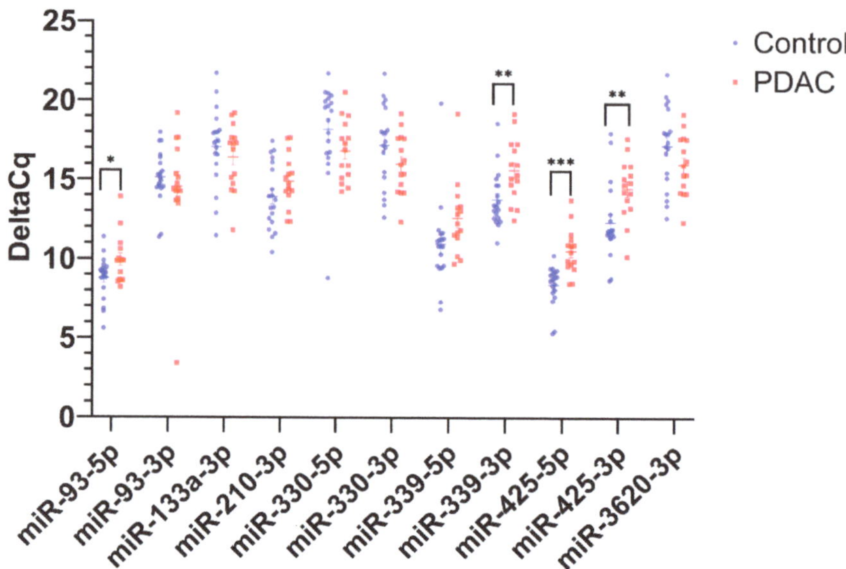

Figure 1. Exosomal miRNA expression levels in PDAC patient plasma compared to control. The scatterplot shows the average ΔCq values for 11 mature miRNAs in PDAC samples (N = 15) and control samples (N = 19). Student's *t*-test was used to establish significance, where *, **, and *** denote $p < 0.05$, $p < 0.01$, and $p < 0.001$, respectively.

2.4. PDAC Stage-Specific Differences in Plasma Exosome miRNA Expression Levels

The RT-qPCR analysis identified four miRNAs (miR-93-5p, miR-339-3p, miR-425-5p, and miR-425-3p) with significantly higher ΔCq in plasma exosomes from PDAC patients compared to the control. PDAC samples were further divided into "Early stage" (stage I and II, N = 5), "Mid stage" (stage III, N= 3), and "Late stage" (stage IV, N = 7) to identify miRNAs with stage-specific differences in expression levels compared to control samples (Figure 2). Analysis of early-stage PDAC samples revealed that miR-425-5p and miR-425-3p had significantly higher ΔCq values compared to control samples ($p < 0.001$, and $p < 0.05$, respectively). Analysis of mid-stage PDAC samples revealed that miR-93-3p had a significantly lower ΔCq value compared to control samples ($p < 0.01$), suggesting greater expression, and miR-425-5p exhibited a significantly higher ΔCq value compared to the control ($p < 0.05$). An analysis of late-stage PDAC samples revealed that miR-93-5p, miR-339-5p, miR-339-3p, miR-425-5p, and miR-425-3p all had significantly higher ΔCq values compared to the control samples ($p < 0.05$, $p < 0.05$, $p < 0.01$, and $p < 0.05$, respectively).

2.5. Diagnostic Value of Plasma Exosomal miRNAs

Receiver operative characteristic (ROC) curve analysis was performed on the four mature miRNAs to assess their combined diagnostic efficacy. Since all four miRNAs (miR-93-5p, miR-339-3p, miR-425-5p, and miR-425-3p) exhibited significantly greater ΔCq values in PDAC samples compared to the control, two threshold values were used to establish a positive "hit" for diagnosing PDAC. If a sample exhibited an miRNA with a ΔCq value that was (1) greater than the upper limit of the 99% confidence interval (CI) of the average control sample and (2) greater than the lower limit of the 99% CI of the average PDAC sample ΔCq value, it was recorded as a positive hit for PDAC. Additionally, if a sample exhibited an miRNA with a ΔCq value greater than both thresholds, it was recorded as two positive hits for PDAC. The upper limit, or the highest average ΔCq value for the control, was determined for each miRNA and used as a threshold for establishing a potential diagnostic. These values (99% CI) were >9.60 for miR-93-5p, >14.82 for miR-339-3p, >9.16 for miR-425-5p, and >13.77 for miR-425-3p. Using these thresholds, a panel consisting of these four miRNAs was assessed via ROC analysis. The AUC was 0.865 ($p < 1 \times 10^{-8}$, Figure 3a) with a sensitivity of 66.7% and a specificity of 94.7%. Similarly, the lower limit of the average PDAC ΔCq value (99% CI) for each miRNA as the threshold was >8.92 for miR-93-5p, >14.27 for miR-339-3p, >9.59 for miR-425-5p, and >13.15 for miR-425-3p. If the values exceeded the minimum average ΔCq for PDAC, it was considered a positive hit for PDAC detection. Using this threshold, the four-miRNA panel yielded an AUC of 0.878 ($p < 1 \times 10^{-9}$, Figure 3b), a sensitivity of 80%, and a specificity of 89.5%. Finally, combining both the upper limits of the control thresholds (99% CI) and the lower limits of the PDAC thresholds (99% CI), (the ROC analysis resulted in an AUC of 0.877 ($p < 1 \times 10^{-9}$), and an overall sensitivity of 80% and specificity of 94.7% (Figure 4). Therefore, this marker allows us to detect PDAC in 12 of the 15 samples, including four out of five early-stage PDAC patients. By comparison, CA19-9 could only detect PDAC in 8 out of the 15 samples and was only able to detect one out of the five early-stage PDAC patients (Figure 5). Interestingly, control sample 8 exhibited the maximum possible of 8 positive hits for PDAC (2 hits for each of the four miRNA), which marks it as a clear outlier sample compared to the other control samples (Figure 5). The provided patient history for the control samples was limited, and, therefore, it could not be verified whether or not control patient 8 may have had a known pancreatic cancer diagnosis.

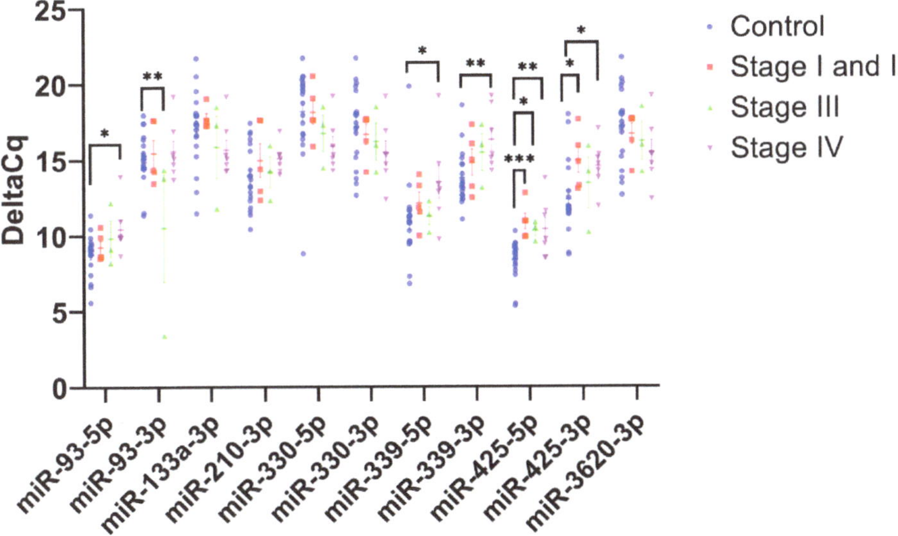

Figure 2. PDAC stage-specific expression levels of plasma exosomal miRNAs. The scatterplot shows the average ΔCq values for 11 mature miRNAs in PDAC samples separated by staging, early stage (stage I and II, N = 5), mid stage (stage III, N = 3) and late stage (stage IV, N = 7), and compared to control samples (N = 19). Student's *t*-test was used to establish significance, where *, **, and *** denote $p < 0.05$, $p < 0.01$, and $p < 0.001$, respectively.

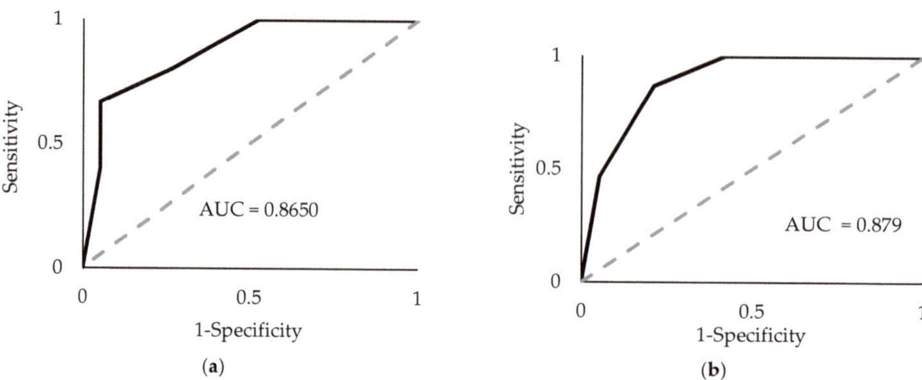

Figure 3. Receiver operator characteristic (ROC) area under the curve (AUC) analyses. The four-miRNA panel (miR-93-5p, miR-339-3p, miR-425-5p, and miR-425-3p) underwent ROC analysis using the upper limits of the average control ΔCq thresholds (**a**) and using the lower limit of the average PDAC ΔCq thresholds (**b**). The values are based off the average ΔCqs for each miRNA in either control or PDAC plasma samples with 99% CI for all such values.

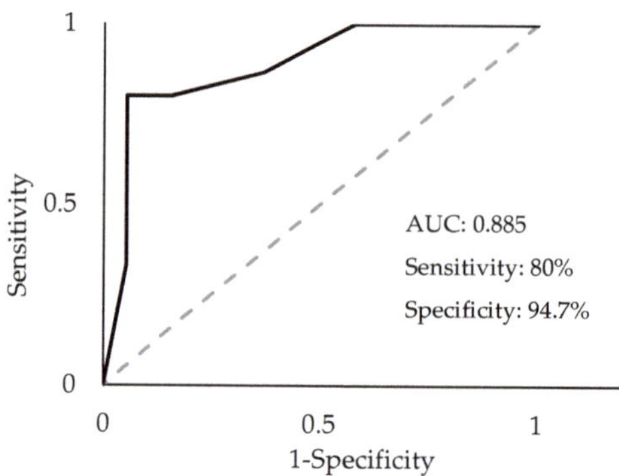

Figure 4. Receiver operator characteristic (ROC) area under the curve (AUC) analysis of the combined four-miRNA panel using two thresholds. Sensitivity = 80%, specificity = 94.7%; AUC = 0.885, CI 99% 0.74–1.00, $p < 1 \times 10^{-10}$.

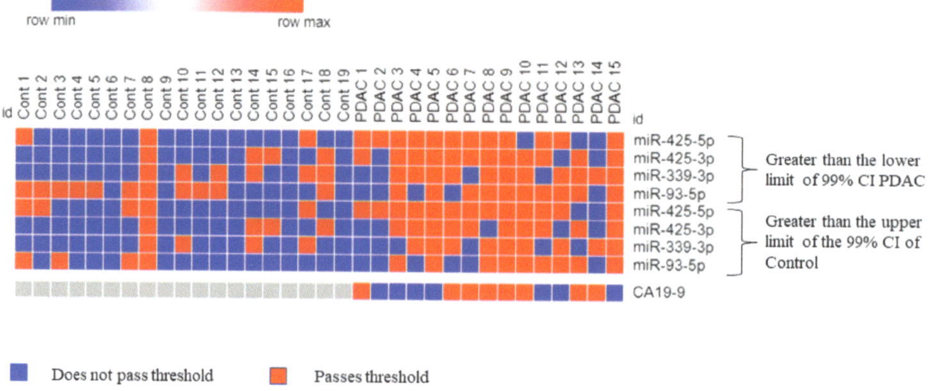

Figure 5. Analyzing PDAC and control samples using the combined four-miRNA biomarker panel. The heatmap shows which samples surpassed the two thresholds (ΔCq is greater than the upper limit

of the control for each miRNA AND ΔCq is greater than the lower limit of PDAC for each miRNA, 99% CI), indicating a positive hit for PDAC. Red indicates a positive hit and blue indicates a negative hit. Columns labeled as "Cont" represent control plasma samples while columns labeled as "PDAC" represent plasma from patients with pancreatic cancer. PDAC 1, 2, 3, 4, and 5 represent early-stage (I and II) pancreatic cancer; PDAC 6, 7, and 12 represent mid-stage (stage III) pancreatic cancer; and PDAC 8, 9, 10, 11, 13, 14, and 15 represent late-stage (stage IV) pancreatic cancer. CA19-9 values are added for comparison where the threshold is the medically established value of >37 U/mL.

3. Discussion

Pancreatic cancer continues to be a difficult disease to diagnose and treat. Mortality rates for PDAC are high, though each year has brought marginal improvements in survival rate, with 2022 reporting a survival rate of 11% [1]. The need to develop more reliable methods of detecting pancreatic cancer at earlier stages remains a top priority. Some avenues of research have turned to freely circulating miRNAs or exosomal miRNAs for an array of diseases, including various cancers [28]. Recently, a study by Zou et al. proposed a panel for the early detection of PDAC comprising six circulating miRNAs (let-7b-5p, miR-192-5p, miR-19a-3p, miR-19b-3p, miR-223-3p, and miR-25-3p) that were identified by machine learning and validated in patient samples [29]. Another study by Wang et al. discovered a single serum exosomal marker, miR-1226-3p, that could diagnose and predict pancreatic cancer invasion and metastases [30]. Despite these important contributions, there remain alternative diagnostic biomarkers to be discovered with the potential for earlier and more reliable detection of pancreatic cancer.

Our previous study identified 18 candidate miRNAs from a bioinformatics analysis of publicly available cBioPortal data collected from pancreatic cancer patients [23]. Seven miRNAs were verified to be differentially expressed in exosomes collected from PDAC cell lines compared to an immortalized pancreatic cell line in vitro. Therefore, we sought to validate the expression levels of these candidate miRNAs in plasma exosomes collected from known PDAC patients. The present study identified four miRNAs, miR-93-5p, miR-339-3p, miR-425-5p, and miR-425-3p, with significantly greater ΔCq values in PDAC plasma samples compared to control plasma samples. A biomarker panel consisting of these four miRNAs and a dual threshold cutoff consisting of the upper limit of the ΔCq values (99% CI) of the miRNAs from control samples and the lower limit of the ΔCq values of the miRNAs (99% CI) from PDAC samples resulted in a diagnostic with an AUC of 0.887, a sensitivity of 80%, and specificity of 94.7%. This is comparable to the current FDA-approved CA19-9 diagnostic, which exhibits both a variable sensitivity (70–90%) and specificity (68–91%) [31]. Interestingly, CA19-9 levels were reported as elevated in only 8 of the 15 (53.33%) PDAC patient samples collected for this study. For comparison, by using the optimal cutoff of the four-miRNA two-threshold biomarker panel proposed in this study, our panel positively identified 12 of the 15 (80.00%) PDAC patient samples. This could indicate a potential improvement in PDAC diagnostic sensitivity using the proposed biomarker panel compared to the established CA19-9 serum diagnostic. However, further studies on a larger sample size are needed to fully evaluate this novel finding.

Our study also identified that both mature forms of miR-425 (miR-425-5p and miR-425-3p) had significantly greater ΔCq values in stage I and II PDAC compared to controls. Consistently, when we applied the two-threshold diagnostic cutoff solely to miR-425-5p, we were able to positively identify five out of five (100% sensitivity) early-stage pancreatic cancer samples. By comparison, CA19-9 was only able to detect one out of five (20%) early-stage pancreatic cancer samples. Therefore, miR-425-5p may offer a more sensitive diagnostic for early-stage PDAC compared to the FDA-approved CA19-9.

Four miRNAs had significantly lower levels of expression in plasma exosomes extracted from PDAC patients compared to control patients. Some of these miRNAs have been previously associated with a variety of cancers and have been implicated in tumor development. For example, miR-93-5p appears to play a role in tumor suppression in ovarian [32] and breast cancers [33,34] by targeting the PD-L1/CCND1 pathway, which is involved in regulating the cell cycle. This is in contrast to the MIR-93 pre-transcript and miR-93-3p, which exhibit oncogenicity. MIR-93 is associated with poor prognosis in PDAC [35], while miR-93-3p predicts poorer outcomes in patients with triple-negative breast cancer [36]. Similarly, overexpression of miR-93-5p appears to promote chemoresistance in PDAC by targeting the PTEN/PI3K/Akt signaling pathway, which is typically involved in tumor suppression [37]. Additional studies confirm the pro-tumorigenic activities of miR-93-5p in endometrial and PDAC tumors [37–40]. Interestingly, our data are consistent with the previous studies for miR-93-3p, which show a decrease in ΔCq values, suggesting an increased expression as PDAC transitions from early to late stages. MicroRNA-425-3p is known to be upregulated in response to cisplatin treatment in non-small cell lung carcinoma (NSCLC) tissue and exosomes [41]. MicroRNA-425-5p has been shown to be expressed at higher levels in PDAC tissue compared to adjacent healthy tissue [42] as well as in NSCLC [43] and in the serum of gastric cancer patients [44]. Exosomal miR-425-5p from MDA-MB-23 breast cancer cells was shown to convert normal fibroblasts to cancer-associated fibroblasts upon uptake by suppressing the expression of TGFβRII, a TGFβ receptor [45]. Additional research has reported that miR-425-5p promotes tumorigenesis in colorectal cancer by inhibiting the PTEN-p53/TGFβ axis [46] and by activating the CTNND1-mediated β-catenin pathway [47]. MicroRNA-339-5p has been shown to suppress colorectal cancer progression by targeting PRL-1 [48], and while miR-339-3p was observed to interfere with CRC progression, its mechanism is unknown [49]. MiR-339-5p has also been implicated in suppressing melanoma by targeting MCL1, which promotes chemoresistance [50]. Little is known about the role miR-339-3p plays in PDAC, though one paper reports that it is downregulated in the PDAC cell line MIA PaCa-2 [51]. Another work implicates miR-339-3p in inhibiting caerulin-induced acute pancreatitis by targeting TRAF3, which promotes inflammation in pancreatitis cells [52]. Although these previous studies indicate potential roles for miR-93-5p, miR-339-3p, miR-425-5p, and miR-425-3p in cancers, to our knowledge, the present study is the first to propose a diagnostic panel for PDAC using all four biomarkers in plasma exosomes.

Although the present findings indicate a potentially promising novel diagnostic panel to detect pancreatic cancer from plasma exosomes, we cannot rule out the risk of a type I error and the chance that the results are a false positive due to a small sample size. Additionally, a history of cancer treatment regimens for each of the PDAC patients was not available. It is possible that some of the observed miRNA expression changes are due to responses to cancer treatment therapies rather than due to cancer progression. Nevertheless, the proposed four-miRNA two-threshold diagnostic biomarker panel was found to perform comparably to the established CA19-9 diagnostic marker. Further studies on a larger sample size would need to be conducted to conclusively evaluate the diagnostic performance of the proposed panel. To rule out the possibility of sex-, age-, and ethnicity-specific differences, chi-squared tests were performed between the control and PDAC patient information. Chi-squared analyses revealed no significant differences for sex ($p > 0.05$) and ethnicity ($p > 0.05$); however, there was a significant difference in the age ($p < 0.05$) of the PDAC group compared to the control group, when all ages are factored in. Our control group comprised 9 age-matched individuals and 11 individuals who were under the age of 40. There is no significant difference in age variation ($p > 0.05$), however, between the nine age-matched control samples (Cont 2 and Cont 9–16, Figure 4) and their PDAC counterparts. In fact, our data show a clear difference in the expression patterns of the miRNAs (miR-93-5p, miR-339-3p, miR-425-5p, and miR-425-3p) between our age-matched control and PDAC samples. However, the sample set remains small, and further studies will require a larger sample size with more closely matched sample demographics.

In summary, the proposed four-miRNA biomarker panel is able to accurately diagnose PDAC (sensitivity = 80%) compared to control (specificity = 94.7%) samples, performs comparably, and is potentially superior to the established CA19-9 diagnostic method. Additionally, miR-425-5p was identified as a potential marker for the early detection of pancreatic cancer at stages I and II. Further investigation is required to fully evaluate this panel for PDAC diagnosis and monitoring.

4. Materials and Methods

4.1. Sample Collection

Plasma samples from 15 patients (male: 13, mean age: 66.54; female: 2, mean age: 70) diagnosed with pancreatic cancer were provided by Baptist Health South Florida. The pancreatic cancer samples comprised 10 previously banked plasma specimens collected between 2018 and 2020 and 5 plasma specimens collected in December 2021 by Baptist Health South Florida Hospital. Sex, age, race, staging, and CA19-9 levels are reported in Table 1. Whole blood samples from 19 control patients (male: 13, mean age: 42.62; female: 6, mean age: 44.17) were provided by Continental Blood Bank, Ft. Lauderdale, FL, between 2020 and 2021. The control samples were tested to ensure they were clean of standard blood-borne pathogens prior to release by the blood bank. Patient age, sex, and race are reported in Table 1. In-depth medical history was not recorded by the provider. Plasma was separated from whole blood by centrifugation for 10 min at 2000 g. All samples were deidentified prior to acquisition and stored at $-80\,^{\circ}\mathrm{C}$ until exosome isolation steps.

4.2. Exosomal miRNA Isolation

Exosomes were isolated from plasma samples using the PEG-based Total Exosome Isolation Kit (from plasma) (Invitrogen, Waltham, MA, USA) as per manufacturer instructions. The Total RNA and Protein Isolation kit (Invitrogen, Waltham, MA, USA) was utilized for the extraction of miRNA from the exosomes isolated from plasma samples. MicroRNA was extracted as per manufacturer instructions, with the addition of an exogenous spike-in control of 1.5 pg of miR-cel-2-3p (Applied Biosystems, Foster City, CA, USA) to monitor RNA extraction efficiency.

4.3. RT-qPCR Analysis

The TaqMan™ Advanced miRNA cDNA Synthesis Kit (Applied Biosystems, Foster City, CA, USA) and TaqMan™ Fast Advanced Master Mix Kit (Applied Biosystems, Foster City, CA, USA) was used to prepare the miRNA for qPCR. The following TaqMan™ Advanced miRNA Assays (Applied Biosystems, Foster City, CA, USA) were used: cel-miR-2-3p (Assay ID: 478291_mir), miR-16-5p (Assay ID: 477860_mir), miR-31-3p (Assay ID: 478012_mir), miR-31-5p (Assay ID: 478015_mir), miR-93-3p (Assay ID: 478209_mir), miR-93-5p (Assay ID: 478210_mir), miR-133a-3p (Assay ID: 478511_mir), miR-133a-5p (Assay ID: 478706_mir), miR-210-3p (Assay ID: 477970_mir), miR-210-5p (Assay ID: 478765_mir), miR-330-3p (Assay ID: 478030_mir), miR-330-5p (Assay ID: 478830_mir), miR-339-3p (Assay ID: 478325_mir), miR-339-5p (Assay ID: 478040_mir), miR-425-3p (Assay ID: 478093_mir) miR-425-5p (Assay ID: 478094_mir), miR-429 (Assay ID: 477849_mir), miR-1208 (Assay ID: 478637_mir), miR-3620-3p (Assay ID: 479690_mir), and miR-3620-5p (Assay ID: 480850_mir).

RT-qPCR was performed using the AriaMX Thermocycler (Agilent, Santa Clara, CA, USA). The PCR settings are described in TaqMan™ Fast Advanced miRNA cDNA Synthesis Kitprotocols. To normalize sample Cq values, the exogenous spiked-in cel-miR-2-3p control and endogenous hsa-miR-16-5p control Cq values were averaged for each sample. All samples were run in duplicate. MicroRNA levels were calculated and expressed as ΔCq between the control miRNA Cq value (average of cel-miR-2-3p and hsa-miR-16-5p) and each of the 11 candidate miRNA Cq values. Due to variability in the initial volume of the plasma samples provided, a Cq adjustment was performed to normalize all samples to 750 uL by using the following formula:

$$Cq_{norm} = Cq_{raw} - \log_2\left(\frac{750\ \mathrm{uL}}{X}\right)$$

where X = Initial sample volume in uL.

To minimize PCR background effects, miRNA with Cq values over 35 or not detected after 40 cycles was adjusted to a Cq value of 36 to test for differential expression between PDAC samples versus control samples. The full protocol has been described elsewhere [53].

4.4. Statistical Validation

Student's two-tailed *t*-test was used to test for statistically significant differences in exosomal miRNA expression levels between control and PDAC samples. The chi-squared test was used to predict the associations between age, sex, or ethnicity in the control vs. PDAC groups.

Supplementary Materials: The supporting information can be downloaded at: https://www.mdpi.com/article/10.3390/ijms24065081/s1.

Author Contributions: Conceptualization, A.M. and W.A.; methodology, A.M. and W.A.; validation, A.M.; formal analysis, A.M.; investigation, A.M. and W.A.; resources, W.A.; data curation, A.M.; writing—original draft preparation, A.M.; writing—review and editing, A.M. and W.A.; visualization, A.M.; supervision, W.A.; project administration, A.M. and W.A.; funding acquisition, W.A. All authors have read and agreed to the published version of the manuscript.

Funding: This research was funded by the Florida Atlantic University (FAU) Institute for Sensing and Embedded Networking Systems Engineering (I-SENSE) Research Initiative Award, and NSF CAREER award 1942487.

Institutional Review Board Statement: Not applicable.

Informed Consent Statement: Not applicable.

Data Availability Statement: The data are within the article and Supplementary Files.

Acknowledgments: We would like to acknowledge Joshua Disatham, for his editorial assistance and guidance on statistical evaluations.

Conflicts of Interest: The authors declare no conflict of interest.

References

1. Siegel, R.L.; Miller, K.D.; Fuchs, H.E.; Jemal, A. Cancer Statistics, 2022. *CA Cancer J. Clin.* **2022**, *72*, 7–33. [CrossRef]
2. Lee, T.; Teng, T.Z.J.; Shelat, V.G. Carbohydrate Antigen 19-9—Tumor Marker: Past, Present, and Future. *World J. Gastrointest. Surg.* **2020**, *12*, 468. [CrossRef]
3. Kim, S.; Park, B.K.; Seo, J.H.; Choi, J.; Choi, J.W.; Lee, C.K.; Chung, J.B.; Park, Y.; Kim, D.W. Carbohydrate Antigen 19-9 Elevation without Evidence of Malignant or Pancreatobiliary Diseases. *Sci. Rep.* **2020**, *10*, 8820. [CrossRef]
4. Bertino, G.; Ardiri, A.M.; Calvagno, G.S.; Malaguarnera, G.; Interlandi, D.; Vacante, M.; Bertino, N.; Lucca, F.; Madeddu, R.; Motta, M. Carbohydrate 19.9 Antigen Serum Levels in Liver Disease. *Biomed. Res. Int.* **2013**, *2013*, 531640. [CrossRef] [PubMed]
5. Teng, D.; Wu, K.; Sun, Y.; Zhang, M.; Wang, D.; Wu, J.; Yin, T.; Gong, W.; Ding, Y.; Xiao, W.; et al. Significant Increased CA199 Levels in Acute Pancreatitis Patients Predicts the Presence of Pancreatic Cancer. *Oncotarget* **2018**, *9*, 12745. [CrossRef] [PubMed]
6. Whiteside, T.L. Tumor-Derived Exosomes and Their Role in Cancer Progression. *Adv. Clin. Chem.* **2016**, *74*, 103. [CrossRef]
7. Makler, A.; Asghar, W. Exosomal Biomarkers for Cancer Diagnosis and Patient Monitoring. *Expert. Rev. Mol. Diagn.* **2020**, *20*, 387–400. [CrossRef] [PubMed]
8. Abhange, K.; Makler, A.; Wen, Y.; Ramnauth, N.; Mao, W.; Asghar, W.; Wan, Y. Small Extracellular Vesicles in Cancer. *Bioact. Mater.* **2021**, *6*, 3705–3743. [CrossRef]
9. Gao, D.; Jiang, L. Exosomes in Cancer Therapy: A Novel Experimental Strategy. *Am. J. Cancer Res.* **2018**, *8*, 2165–2175.
10. Huang, T.; Deng, C.-X.X. Current Progresses of Exosomes as Cancer Diagnostic and Prognostic Biomarkers. *Int. J. Biol. Sci.* **2019**, *15*, 1–11. [CrossRef]
11. Li, B.; Cao, Y.; Sun, M.; Feng, H. Expression, Regulation, and Function of Exosome-Derived MiRNAs in Cancer Progression and Therapy. *FASEB J.* **2021**, *35*, e21916. [CrossRef] [PubMed]
12. Si, W.; Shen, J.; Zheng, H.; Fan, W. The Role and Mechanisms of Action of MicroRNAs in Cancer Drug Resistance. *Clin. Epigenetics* **2019**, *11*, 25. [CrossRef]
13. Saliminejad, K.; Khorram Khorshid, H.R.; Soleymani Fard, S.; Ghaffari, S.H. An Overview of MicroRNAs: Biology, Functions, Therapeutics, and Analysis Methods. *J. Cell Physiol.* **2019**, *234*, 5451–5465. [CrossRef]
14. Hussen, B.M.; Hidayat, H.J.; Salihi, A.; Sabir, D.K.; Taheri, M.; Ghafouri-Fard, S. MicroRNA: A Signature for Cancer Progression. *Biomed. Pharmacother.* **2021**, *138*, 111528. [CrossRef] [PubMed]

15. Kanno, S.; Nosho, K.; Ishigami, K.; Yamamoto, I.; Koide, H.; Kurihara, H.; Mitsuhashi, K.; Shitani, M.; Motoya, M.; Sasaki, S.; et al. MicroRNA-196b Is an Independent Prognostic Biomarker in Patients with Pancreatic Cancer. *Carcinogenesis* **2017**, *38*, 425–431. [CrossRef] [PubMed]
16. Fehlmann, T.; Kahraman, M.; Ludwig, N.; Backes, C.; Galata, V.; Keller, V.; Geffers, L.; Mercaldo, N.; Hornung, D.; Weis, T.; et al. Evaluating the Use of Circulating MicroRNA Profiles for Lung Cancer Detection in Symptomatic Patients. *JAMA Oncol.* **2020**, *6*, 714–723. [CrossRef]
17. Liu, X.; Liu, X.; Wu, Y.; Wu, Q.; Wang, Q.; Yang, Z.; Li, L.; Liu, X.; Liu, X.; Wu, Y.; et al. MicroRNAs in Biofluids Are Novel Tools for Bladder Cancer Screening. *Oncotarget* **2017**, *8*, 32370–32379. [CrossRef]
18. Liu, R.; Chen, X.; Du, Y.; Yao, W.; Shen, L.; Wang, C.; Hu, Z.; Zhuang, R.; Ning, G.; Zhang, C.; et al. Serum MicroRNA Expression Profile as a Biomarker in the Diagnosis and Prognosis of Pancreatic Cancer. *Clin. Chem.* **2012**, *58*, 610–618. [CrossRef]
19. Abue, M.; Yokoyama, M.; Shibuya, R.; Tamai, K.; Yamaguchi, K.; Sato, I.; Tanaka, N.; Hamada, S.; Shimosegawa, T.; Sugamura, K.; et al. Circulating MiR-483-3p and MiR-21 Is Highly Expressed in Plasma of Pancreatic Cancer. *Int. J. Oncol.* **2015**, *46*, 539–547. [CrossRef]
20. Daoud, A.Z.; Mulholland, E.J.; Cole, G.; McCarthy, H.O. MicroRNAs in Pancreatic Cancer: Biomarkers, Prognostic, and Therapeutic Modulators. *BMC Cancer* **2019**, *19*, 1130. [CrossRef]
21. Rachagani, S.; Macha, M.A.; Heimann, N.; Seshacharyulu, P.; Haridas, D.; Chugh, S.; Batra, S.K. Clinical Implications of MiRNAs in the Pathogenesis, Diagnosis and Therapy of Pancreatic Cancer. *Adv. Drug Deliv. Rev.* **2015**, *81*, 16. [CrossRef] [PubMed]
22. Makler, A.; Narayanan, R. Mining Exosomal Genes for Pancreatic Cancer Targets. *Cancer Genom. Proteom.* **2017**, *14*, 161–172. [CrossRef] [PubMed]
23. Makler, A.; Narayanan, R.; Asghar, W. An Exosomal MiRNA Biomarker for the Detection of Pancreatic Ductal Adenocarcinoma. *Biosensors* **2022**, *12*, 831. [CrossRef]
24. Gao, J.; Aksoy, B.A.; Dogrusoz, U.; Dresdner, G.; Gross, B.; Sumer, S.O.; Sun, Y.; Jacobsen, A.; Sinha, R.; Larsson, E.; et al. Integrative Analysis of Complex Cancer Genomics and Clinical Profiles Using the CBioPortal. *Sci. Signal* **2013**, *6*, pl1. [CrossRef]
25. Vlachos, I.S.; Zagganas, K.; Paraskevopoulou, M.D.; Georgakilas, G.; Karagkouni, D.; Vergoulis, T.; Dalamagas, T.; Hatzigeorgiou, A.G. DIANA-MiRPath v3.0: Deciphering MicroRNA Function with Experimental Support. *Nucleic Acids Res.* **2015**, *43*, W460. [CrossRef] [PubMed]
26. Kanehisa, M.; Furumichi, M.; Tanabe, M.; Sato, Y.; Morishima, K. KEGG: New Perspectives on Genomes, Pathways, Diseases and Drugs. *Nucleic Acids Res.* **2017**, *45*, D353–D361. [CrossRef]
27. Med, H.J.T.; Han, Y. Analysis of the Role of the Hippo Pathway in Cancer. *J. Transl. Med.* **2019**, *17*, 116. [CrossRef]
28. Condrat, C.E.; Thompson, D.C.; Barbu, M.G.; Bugnar, O.L.; Boboc, A.; Cretoiu, D.; Suciu, N.; Cretoiu, S.M.; Voinea, S.C. MiRNAs as Biomarkers in Disease: Latest Findings Regarding Their Role in Diagnosis and Prognosis. *Cells* **2020**, *9*, 276. [CrossRef]
29. Zou, X.; Wei, J.; Huang, Z.; Zhou, X.; Lu, Z.; Zhu, W.; Miao, Y. Identification of a Six-miRNA Panel in Serum Benefiting Pancreatic Cancer Diagnosis. *Cancer Med.* **2019**, *8*, 2810. [CrossRef]
30. Wang, C.; Wang, J.; Cui, W.; Liu, Y.; Zhou, H.; Wang, Y.; Chen, X.; Chen, X.; Wang, Z. Serum Exosomal MiRNA-1226 as Potential Biomarker of Pancreatic Ductal Adenocarcinoma. *Onco. Targets Ther.* **2021**, *14*, 1441. [CrossRef] [PubMed]
31. Ermiah, E.; Eddfair, M.; Abdulrahman, O.; Elfagieh, M.; Jebriel, A.; Al-Sharif, M.; Assidi, M.; Buhmeida, A. Prognostic Value of Serum CEA and CA19-9 Levels in Pancreatic Ductal Adenocarcinoma. *Mol. Clin. Oncol.* **2022**, *17*, 2559. [CrossRef] [PubMed]
32. Chen, G.; Yan, Y.; Qiu, X.; Ye, C.; Jiang, X.; Song, S.; Zhang, Y.; Chang, H.; Wang, L.; He, X.; et al. MiR-93-5p Suppresses Ovarian Cancer Malignancy and Negatively Regulate CCND2 by Binding to Its 3′UTR Region. *Discov. Oncol.* **2022**, *13*, 15. [CrossRef]
33. Yang, M.; Xiao, R.; Wang, X.; Xiong, Y.; Duan, Z.; Li, D.; Kan, Q. MiR-93-5p Regulates Tumorigenesis and Tumor Immunity by Targeting PD-L1/CCND1 in Breast Cancer. *Ann. Transl. Med.* **2022**, *10*, 203. [CrossRef]
34. Bao, C.; Chen, J.; Chen, D.; Lu, Y.; Lou, W.; Ding, B.; Xu, L.; Fan, W. MiR-93 Suppresses Tumorigenesis and Enhances Chemosensitivity of Breast Cancer via Dual Targeting E2F1 and CCND1. *Cell Death Dis.* **2020**, *11*, 618. [CrossRef]
35. Vila-Navarro, E.; Fernandez-Castañer, E.; Rovira-Rigau, M.; Raimondi, G.; Vila-Casadesus, M.; Lozano, J.J.; Soubeyran, P.; Iovanna, J.; Castells, A.; Fillat, C.; et al. MiR-93 Is Related to Poor Prognosis in Pancreatic Cancer and Promotes Tumor Progression by Targeting Microtubule Dynamics. *Oncogenesis* **2020**, *9*, 43. [CrossRef]
36. Li, H.Y.; Liang, J.L.; Kuo, Y.L.; Lee, H.H.; Calkins, M.J.; Chang, H.T.; Lin, F.C.; Chen, Y.C.; Hsu, T.I.; Hsiao, M.; et al. MiR-105/93-3p Promotes Chemoresistance and Circulating MiR-105/93-3p Acts as a Diagnostic Biomarker for Triple Negative Breast Cancer. *Breast Cancer Res.* **2017**, *19*, 133. [CrossRef] [PubMed]
37. Wu, Y.; Xu, W.; Yang, Y.; Zhang, Z. MiRNA-93-5p Promotes Gemcitabine Resistance in Pancreatic Cancer Cells by Targeting the PTEN-Mediated PI3K/Akt Signaling Pathway. *Ann. Clin. Lab. Sci.* **2021**, *51*, 310–320. [PubMed]
38. Chen, S.; Chen, X.; Sun, K.X.; Xiu, Y.L.; Liu, B.L.; Feng, M.X.; Sang, X.B.; Zhao, Y. MicroRNA-93 Promotes Epithelial-Mesenchymal Transition of Endometrial Carcinoma Cells. *PLoS ONE* **2016**, *11*, 0165776. [CrossRef]
39. Vila-Navarro, E.; Duran-Sanchon, S.; Vila-Casadesús, M.; Moreira, L.; Gins, A.; Cuatrecasas, M.; Josè Lozano, J.; Bujanda, L.; Castells, A.; Gironella, M. Novel Circulating MiRNA Signatures for Early Detection of Pancreatic Neoplasia. *Clin. Transl. Gastroenterol.* **2019**, *10*, e00029. [CrossRef]
40. Shen, H.; Ye, F.; Xu, D.; Fang, L.; Zhang, X.; Zhu, J. The MYEOV-MYC Association Promotes Oncogenic MiR-17/93-5p Expression in Pancreatic Ductal Adenocarcinoma. *Cell Death Dis.* **2021**, *13*, 15. [CrossRef]

41. Ma, Y.; Yuwen, D.; Chen, J.; Zheng, B.; Gao, J.; Fan, M.; Xue, W.; Wang, Y.; Li, W.; Shu, Y.; et al. Exosomal Transfer Of Cisplatin-Induced MiR-425-3p Confers Cisplatin Resistance In NSCLC Through Activating Autophagy. *Int. J. Nanomed.* **2019**, *14*, 8121. [CrossRef]

42. Lu, Y.; Wu, X.; Wang, J. Correlation of MiR-425-5p and IL-23 with Pancreatic Cancer. *Oncol. Lett.* **2019**, *17*, 4595–4599. [CrossRef] [PubMed]

43. Fu, Y.; Li, Y.; Wang, X.; Li, F.; Lu, Y. Overexpression of MiR-425-5p Is Associated with Poor Prognosis and Tumor Progression in Non-Small Cell Lung Cancer. *Cancer Biomark.* **2020**, *27*, 147–156. [CrossRef] [PubMed]

44. Bie, L.Y.; Li, N.; Deng, W.Y.; Lu, X.Y.; Guo, P.; Luo, S.X. Serum MiR-191 and MiR-425 as Diagnostic and Prognostic Markers of Advanced Gastric Cancer Can Predict the Sensitivity of FOLFOX Chemotherapy Regimen. *Onco. Targets Ther.* **2020**, *13*, 1705. [CrossRef] [PubMed]

45. Zhu, Y.; Dou, H.; Liu, Y.; Yu, P.; Li, F.; Wang, Y.; Xiao, M. Breast Cancer Exosome-Derived MiR-425-5p Induces Cancer-Associated Fibroblast-Like Properties in Human Mammary Fibroblasts by TGF β 1/ROS Signaling Pathway. *Oxid. Med. Cell Longev.* **2022**, *2022*, 5266627. [CrossRef] [PubMed]

46. Hu, X.; Chen, Q.; Guo, H.; Li, K.; Fu, B.; Chen, Y.; Zhao, H.; Wei, M.; Li, Y.; Wu, H. Identification of Target PTEN-Based MiR-425 and MiR-576 as Potential Diagnostic and Immunotherapeutic Biomarkers of Colorectal Cancer With Liver Metastasis. *Front. Oncol.* **2021**, *11*, 3193. [CrossRef]

47. Liu, D.; Zhang, H.; Cui, M.; Chen, C.; Feng, Y. Hsa-MiR-425-5p Promotes Tumor Growth and Metastasis by Activating the CTNND1-Mediated β-Catenin Pathway and EMT in Colorectal Cancer. *Cell Cycle* **2020**, *19*, 1917–1927. [CrossRef]

48. Zhou, C.; Liu, G.; Wang, L.; Lu, Y.; Yuan, L.; Zheng, L.; Chen, F.; Peng, F.; Li, X. MiR-339-5p Regulates the Growth, Colony Formation and Metastasis of Colorectal Cancer Cells by Targeting PRL-1. *PLoS ONE* **2013**, *8*, e63142. [CrossRef] [PubMed]

49. Zhou, C.; Lu, Y.; Li, X. MiR-339-3p Inhibits Proliferation and Metastasis of Colorectal Cancer. *Oncol. Lett.* **2015**, *10*, 2842. [CrossRef]

50. Weber, C.E.M.; Luo, C.; Hotz-Wagenblatt, A.; Gardyan, A.; Kordaß, T.; Holland-Letz, T.; Osen, W.; Eichmüller, S.B. MiR-339-3p Is a Tumor Suppressor in Melanoma. *Cancer Res.* **2016**, *76*, 3562–3571. [CrossRef]

51. Bhutia, Y.D.; Hung, S.W.; Patel, B.; Lovin, D.; Govindarajan, R. CNT1 Expression Influences Proliferation and Chemosensitivity in Drug-Resistant Pancreatic Cancer Cells. *Cancer Res.* **2011**, *71*, 1825–1835. [CrossRef] [PubMed]

52. Wang, Q.; Liu, S.; Han, Z. MiR-339-3p Regulated Acute Pancreatitis Induced by Caerulein through Targeting TNF Receptor-Associated Factor 3 in AR42J Cells. *Open Life Sci.* **2020**, *15*, 912–922. [CrossRef] [PubMed]

53. Gevaert, A.B.; Witvrouwen, I.; Vrints, C.J.; Heidbuchel, H.; van Craenenbroeck, E.M.; van Laere, S.J.; van Craenenbroeck, A.H. MicroRNA Profiling in Plasma Samples Using QPCR Arrays: Recommendations for Correct Analysis and Interpretation. *PLoS ONE* **2018**, *13*, 0193173. [CrossRef] [PubMed]

Disclaimer/Publisher's Note: The statements, opinions and data contained in all publications are solely those of the individual author(s) and contributor(s) and not of MDPI and/or the editor(s). MDPI and/or the editor(s) disclaim responsibility for any injury to people or property resulting from any ideas, methods, instructions or products referred to in the content.

International Journal of
Molecular Sciences

MDPI

Review

Protein Arginine Methyltransferases in Pancreatic Ductal Adenocarcinoma: New Molecular Targets for Therapy

Kritisha Bhandari and Wei-Qun Ding *

Department of Pathology, University of Oklahoma Health Sciences Center, BMSB401A, 940 Stanton L. Young Blvd., Oklahoma City, OK 73104, USA; kritisha-bhandari@ouhsc.edu
* Correspondence: weiqun-ding@ouhsc.edu; Tel.: +405-271-1605

Abstract: Pancreatic ductal adenocarcinoma (PDAC) is a lethal malignant disease with a low 5-year overall survival rate. It is the third-leading cause of cancer-related deaths in the United States. The lack of robust therapeutics, absence of effective biomarkers for early detection, and aggressive nature of the tumor contribute to the high mortality rate of PDAC. Notably, the outcomes of recent immunotherapy and targeted therapy against PDAC remain unsatisfactory, indicating the need for novel therapeutic strategies. One of the newly described molecular features of PDAC is the altered expression of protein arginine methyltransferases (PRMTs). PRMTs are a group of enzymes known to methylate arginine residues in both histone and non-histone proteins, thereby mediating cellular homeostasis in biological systems. Some of the PRMT enzymes are known to be overexpressed in PDAC that promotes tumor progression and chemo-resistance via regulating gene transcription, cellular metabolic processes, RNA metabolism, and epithelial mesenchymal transition (EMT). Small-molecule inhibitors of PRMTs are currently under clinical trials and can potentially become a new generation of anti-cancer drugs. This review aims to provide an overview of the current understanding of PRMTs in PDAC, focusing on their pathological roles and their potential as new therapeutic targets.

Keywords: pancreatic ductal adenocarcinoma (PDAC); arginine methylation; protein arginine methyl-transferases (PRMTs); molecular targets

Citation: Bhandari, K.; Ding, W.-Q.
Protein Arginine Methyltransferases
in Pancreatic Ductal
Adenocarcinoma: New Molecular
Targets for
Therapy. *Int. J. Mol. Sci.* **2024**, *25*, 3958.
https://doi.org/10.3390/ijms25073958

Academic Editors: Donatella
Delle Cave and Claudio Luchini

Received: 29 February 2024
Revised: 28 March 2024
Accepted: 30 March 2024
Published: 2 April 2024

Copyright: © 2024 by the authors.
Licensee MDPI, Basel, Switzerland.
This article is an open access article
distributed under the terms and
conditions of the Creative Commons
Attribution (CC BY) license (https://
creativecommons.org/licenses/by/
4.0/).

1. Introduction: Pancreatic Ductal Adenocarcinoma and the Need for New Therapeutics

Pancreatic ductal adenocarcinoma (PDAC) is a major type of pancreatic neoplasm that originates from ductal or acinar cells, comprising more than 90% of pancreatic cancer cases [1]. It is the third-leading cause of cancer-related death in the United States and is predicted to surpass colorectal cancer by 2040, to become the second-leading cause of cancer-related death [2]. The current 5-year overall survival rate for this disease is 13%, which is lower than that for most solid tumor types [3]. Approximately 80–85% of patients are diagnosed with PDAC when the disease has already metastasized or became locally advanced, making them ineligible for surgical resection [4–7]. For the remaining 15–20% of PDAC patients that are diagnosed early and are eligible for surgical resection, 3 out of 4 patients will develop a relapse within 2 years post-operation [7,8]. In both cases, whether considering patients as surgical candidates or patients with metastatic disease, they usually all undergo intensive chemotherapy. The first line of treatment includes the use of two different regimens: FOLFIRINOX, which is the combination of folinic acid (leucovorin), 5-fluorouracil (5-FU), irinotecan, and oxaliplatin, or gemcitabine combined with nab-paclitaxel [8,9]. The second line of treatment includes liposomal formulation of Irinotecan with 5-Fluorouracil. Patients are eligible to switch to the second line of treatment if their disease progresses during the first line of treatment and they have not received these second-line drugs previously [8].

The major challenge with current PDAC chemotherapy is the development of drug resistance, which has mostly been observed in gemcitabine-treated patients [10]. In addition, the combination of several drugs in FOLFIRINOX is extremely toxic and has a severe impact on the patient's quality of life [11]. Unfortunately, current molecular targeted therapy and immunotherapy, which have shown unprecedented therapeutic benefits for other cancer types, are rarely effective for patients with PDAC [12–15]. The only FDA-approved targeted therapeutic for PDAC is the epidermal growth factor receptor inhibitor erlotinib, which slightly prolongs patient survival [13]. Currently available immunotherapies have shown limited efficacy in improving PDAC patient survival [12]. New strategies in developing effective therapeutics against PDAC are desperately needed.

It should be noted that decades of research in the biology of PDAC have led to the discovery of many promising molecular targets for this disease, such as the KRAS mutation that leads to activation of oncogenic signaling, the desmoplastic tumor microenvironment (TME) that facilitates immune evasion, and the altered tumor metabolism that contributes to chemo-resistance [8]. The potential therapeutic benefit of targeting the mutated KRAS protein in PDAC has been extensively explored, as more than 90% of patients with PDAC harbor this mutation [4,8,16]. However, while the mutated KRAS protein is a "druggable" target when using KRAS specific small-molecule inhibitors (there are no FDA-approved KRAS inhibitors for PDAC as of yet), PDAC cells often find a way to adapt by following an "RAS independent" pathway, compromising the efficacy of the small-molecule inhibitors [17–20].

The altered tumor metabolism for PDAC provides a wide window of opportunities to develop new therapeutic interventions. Targeting autophagy, glutamine metabolism, and glycolysis through lactate dehydrogenase inhibition, in combination with other chemotherapeutics, has been experimentally explored [8,21,22], but clinically applicable therapeutics that target tumor metabolism have not been available for PDAC. Targeting the extracellular matrix (ECM) barrier (the desmoplastic TME) has been expected to improve drug access to the tumor tissues but has not been successful in clinical trials [23,24]. Immunotherapies based on the development of cancer vaccines, checkpoint inhibitors, CAR T-cells, and stroma and myeloid targeting remain ongoing in clinical trials for PDAC [25]; however, it is known that most previous clinical trials on PDAC therapeutics have fallen short of expectations [8]. New molecular targets and therapeutic strategies for PDAC merit further exploration.

Recent advances in our understanding of the biology of PDAC has demonstrated one group of the promising candidate therapeutic targets in PDAC cells: a family of enzymes named protein arginine methyltransferases (PRMTs). PRMTs are enzymes that are responsible for methylating arginine residues in histone as well as non-histone proteins [26]. They play a multitudinous role in the biology of cells and are associated with the progression of diseases, such as cancer. Expression of several PRMTs is upregulated in PDAC cells and tissues, thereby promoting progression of the disease [27]. Drugs targeting PRMTs are known to be efficacious in killing cancer cells in vitro as well as in vivo, either alone or in combination with chemotherapeutic agents. Some of the inhibitors of PRMTs are currently in clinical trials [28]. This review introduces the family of PRMTs in mammalian cells, their involvement in the pathogenesis of PDAC, and the potential of PRMTs as therapeutic targets for PDAC.

2. Arginine Methylation and PRMTs

Protein methylation is the fifth-most abundant post-translational modification (PTM) and is observed in histone as well as non-histone proteins [29]. While several amino acids are known to undergo this modification, the major amino acids that are known to be methylated are lysine and arginine [30]. The lysine methylation in histone and non-histone proteins has been extensively reviewed elsewhere [31–33]. Therefore, this review focuses on arginine methylation of these proteins. PRMTs are the enzymes that catalyze arginine methylation, which is ubiquitously present in both the nuclear as well as cytosolic compartments of the cells [34]. These enzymes can be broadly classified into three different categories: Type I, Type II, and Type III, based on their ability to catalyze different modes of arginine methylation in the proteins. While Type I PRMTs can form mono-methyl arginine (MMA) and asymmetric di-methyl arginine (ADMA), Type II PRMTs produce mono-methyl arginine (MMA) and symmetric di-methyl arginine (SDMA), and Type III PRMTs can only produce mono-methyl arginine (MMA) (Figure 1) [26]. PRMT1, PRMT2, PRMT3, PRMT4/CRAM1, PRMT6, and PRMT8 are Type I PRMTs; PRMT5 and PRMT9 are Type II; and PRMT7 is a solo Type III PRMT. The method by which PRMTs methylate arginine residues has been well described. PRMTs utilize cofactor S-adenosylmethionine (SAM, also known as AdoMet) to catalyze the transfer of methyl groups to the guanidino nitrogen moieties in arginine residues of the substrate protein. This reaction yields the formation of methylarginine, with S-adenosylhomocysteine (SAH) as the side product [35]. All the PRMTs are expressed in the pancreas, with the exception of PRMT8, whose expression is known to be limited to the brain [36]. The expression and localization of different types of PRMTs vary between the endocrine and exocrine regions of the pancreas. Detected by enzyme-specific antibodies, PRMTs are ubiquitously expressed in either the islets of Langerhans or pancreatic acini, with differential expression across cellular compartments, as shown in Table 1.

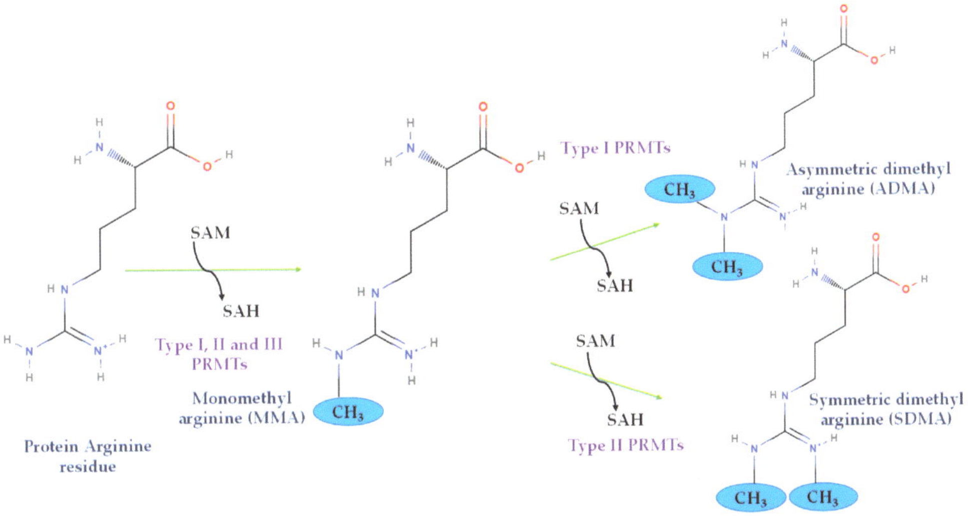

Figure 1. Different types of protein arginine methylation by PRMTs.

Table 1. Expression of PRMTs in the endocrine and exocrine compartments of the pancreas.

PRMTs	Endocrine Region	Exocrine Region
PRMT1	Medium	Low
PRMT2	Low	Medium
PRMT3	Medium	High to medium
PRMT4	Low	Medium to low
PRMT5	Low	Low to medium
PRMT6	Medium	High
PRMT7	Medium	Medium
PRMT9	Negative	Low

2.1. Structural Basis, Localization, and Motif Preference of PRMTs

The canonical structure of PRMTs contains three structural domains: an N-terminal catalytic core (Rossman fold), the α-helical dimerization arm, and the C-terminal ß-barrel domain (Figure 2) [37]. The catalytic core in PRMTs consists of approximately 300 amino acids, containing the SAM-binding site. The ß-barrel domain facilitates the binding of substrates to PRMTs. The dimerization arm is essential in most PRMTs, as PRMTs function as dimers, with the exception of PRMT7 and PRMT9 [38,39]. The structural features of individual PRMTs are explained in detail in Table 2. PRMTs are localized between different cellular compartments and have their own motif of preference for arginine methylation (Table 3). While there is no unanimous recognition motif for different types of PRMTs, scientific findings suggest that the glycine-rich motifs, such as RGG, RxR, and GAR, have a high likelihood of being methylated by PRMTs [34,40]. One recent finding indicates that PRMT5 recognizes a GRG motif to methylate its substrate [41]. Most of the current studies utilize the arginine methylation prediction tools PRmePRed (https://bioinfo.icgeb.res. in/PRmePRed/, accessed on 20 February 2024) [42] or GPS-MSP (http://msp.biocuckoo. org/online.php, accessed on 20 February 2024) [43] to gain a tentative idea of arginine methylation sites in the protein of interest, and then validate these sites using in vitro arginine methylation assays combined with site-directed mutagenesis and proteomics. Interestingly, all of the PRMTs are also highly regulated by different PTMs that may or may not affect their catalytic activity [44–52].

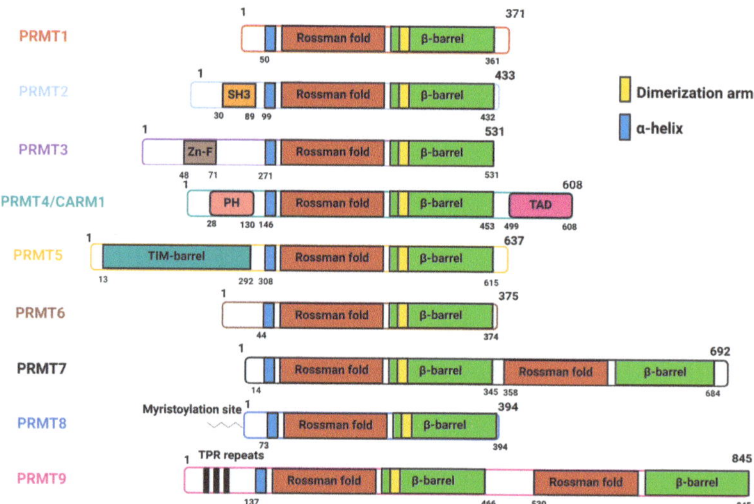

Figure 2. Structure of PRMT proteins (see Table 2 for structure features). Created with Biorender.com.

Table 2. Structural features of PRMTs.

Enzyme	Structural Features
PRMT1	Contain three canonical domains: N-terminal methyltransferase domain (Rossman fold), containing AdoMet binding pocket C-terminal ß-barrel domain, forming cylindrical structure corresponding to the arginine-substrate binding sites α-helical dimerization arm, an N-terminal part of the ß-barrel domain [53]
PRMT2	Three canonical domains, along with a unique Src homology 3 domain (SH3 domain) towards N-terminal extremity [54]
PRMT3	Three canonical domains, along with a unique C_2H_2Zn finger domain at the N-terminus for substrate binding [55,56]
PRMT4/CARM1	Three canonical domains, along with a C-terminal TAD domain and PH-like homology at the N-terminus for substrate recognition [57]

Table 2. *Cont.*

Enzyme	Structural Features
PRMT5	Three canonical domains, with N-terminal TIM barrel, which is essential for formation of complex with MEP50 [58]
PRMT6	Three canonical domains; no unique feature [59]
PRMT7	Three canonical domains, with two tandem methyltransferase domains due to gene duplication [38]
PRMT8	Three canonical domains, with a myristoylation site at the N-terminus that mediates its anchorage to the plasma membrane [36]
PRMT9	Three canonical domains, with two tandem methyltransferase domains and N-terminal TPR repeats [39]

Table 3. Subcellular localization and motif preferences in PRMTs.

PRMT	Cellular Localization *	Enzyme Type	Methylation Product	Motif Preference
PRMT1	**Cytoplasm,** Nucleus	I	MMA/ADMA	RGG or RxR
PRMT2	**Nucleus,** Cytoplasm	I	MMA/ADMA	RGG/RG
PRMT3	**Cytoplasm,** Nucleus	I	MMA/ADMA	RGG/RG
PRMT4/CARM1	**Nucleus,** Cytoplasm	I	MMA/ADMA	PGM
PRMT5	**Cytoplasm,** Nucleus	II	MMA/SDMA	GRG, PGM
PRMT6	**Nucleus,** Cytoplasm	I	MMA/ADMA	RxR
PRMT7	**Cytoplasm,** Nucleus	III	MMA	RxR
PRMT8	**Plasma membrane**	I	MMA/ADMA	RGG or RxR
PRMT9	**Cytoplasm,** Nucleus	II	MMA/SDMA	GAR

* Bold font indicates the dominant localization.

The mechanism of arginine methylation by PRMTs has been extensively studied with SAM as a methyl donor to methylate their substrates [35]. The substrates of PRMTs are present in both the nucleus as well as the cytoplasm. While there has been a debate about the kinetics of these enzymes, most findings suggest a multi-step methylation of the substrate, also called a distributive process, wherein the substrate is first mono-methylated, followed by its dissociation from the SAM–PRMT–substrate complex and a di-methylation of the substrate by re-formation of the donor–enzyme–substrate complex [35,60–62].

2.2. Physiological Role of PRMTs

The arginine methylation of proteins, like any other types of PTMs, increases the diversity of cellular proteome and, therefore, plays a significant role in maintaining cellular homeostasis. The addition of a methyl group in the arginine residues does not seem to alter the charge of the protein; rather, it can facilitate or disrupt the interaction among proteins and nucleic acids, resulting in diverse physiological responses [37]. Given the abundance of protein arginine methylation in eukaryotic cells, there is no doubt that PRMTs are involved in many aspects of cellular function. Studies have revealed that the PRMT substrates are mostly associated with RNAs [63]. Not surprisingly, PRMTs have been shown to mediate gene transcription [64], mRNA splicing [65,66], DNA damage repair (DDR) [67,68], cell stemness [35], etc.

PRMT1 and CARM1 are known to act as transcription coactivators via histone arginine methylation, which facilitates the binding of transcription factors, such as ERα [63,69], p53, YY1 [70], and PPARγ [71], to the promoter of genes. Furthermore, PRMT1 is shown to be able to methylate arginine residues in non-histone proteins to co-activate gene transcription [72]. PRMT1-mediated transcription activation enhances EGFR signaling and promotes colorectal cancer progression [73]. On the other hand, PRMT5 and PRMT6 are shown to have transcription co-repressor activity that suppresses gene transcription via arginine methylation of histone proteins [34]. PRMT5 is considered a general transcription repressor via arginine methylation that interacts with different transcription factors or repressor complexes, such as Snail [74] and BRG1 [75], to repress gene transcription. PRMT5-mediated arginine methylation of histone proteins can repress expression of epithelial junctional genes, thereby promoting cancer cell invasion [76]. However, PRMT5 is also reported to potentiate gene transcription through arginine methylation of histone proteins that enhances the binding of a transcriptional co-activator [77]. Other PRMTs, including PRMT2 and PRMT7, are also reported to mediate gene transcription through arginine methylation of histone proteins [34]. These findings indicate that PRMTs play a critical role in regulating gene transcription through arginine methylation in a context-dependent manner.

The regulation of mRNA splicing is a well-known function of PRMTs, especially CARM1 and PRMT5 [78]. One of the well-explored examples is the arginine methylation of Sm proteins in the GAR motif towards its C-terminal domain by PRMT5, facilitating its interactions with the Tudor domain of the survival of motor neuron (SMN), which plays a key role in maintaining the fidelity of constitutive nuclear splicing events and RNA metabolism. The deletion or inhibition of PRMT5 reduces spliceosome assembly and causes aberrant splicing, such as intron retention [79]. PRMT5 has also been reported to regulate mRNA splicing of the p53 inhibitor MDM4 [66]. Upon deletion of PRMT5, there is a formation of shorter and less stable MDM4 isoform protein due to alternative splicing of the transcript, which is incapable of inhibiting p53. CARM1 has been involved in the regulation of mRNA splicing via methylating RBPs [80]. Direct methylation of splicing factors by CARM1, thereby affecting the RNA spicing process, has been well described [81], and, similar to PRMT5, CARM1 also causes Sm protein methylation in a model system [81]. In contrast to CARM1 and PRMT5, the involvement of PRMT1 in mRNA splicing has been elusive. Specific substrates by which PRMT1 may regulate RNA splicing have not been identified. However, PRMT1 is known to regulate RBP cellular localization [82], and depletion of PRMT1 is associated with aberrant RNA splicing in mouse cardiomyocytes [83].

With regard to DNA damage response (DDR), PRMT5 and PRMT1 have been shown to be critical for the repair of the damaged DNA. PRMT1-null mouse embryonic fibroblasts exhibit genome instability, spontaneous DNA damage, and checkpoint defects [84]. These are mediated by MRE11, an integral component of the MRN complex known to activate the DDR pathways. PRMT1-mediated arginine methylation of MRE11 helps it anchor to the DNA double-strand breaks (DSBs), stimulating nuclease activity, while its demethylated version MRE11^{RK} is defective in DNA end-resection and ATR activation [85]. In the case of PRMT5, it is known that PRMT5 deficiency causes spontaneous DNA damage and defects in homologue recombination-mediated DSB repair [86,87]. PRMT5 stabilizes RPA2, which is one of the three subunits of the RPA complex. The RPA complex binds and protects ssDNA formed during DNA repair. Knockout of PRMT5 results in the depletion of RPA2, causing RPA exhaustion. This leads to impaired homology-directed repair (HDR) of the cells treated with gemcitabine, thereby enhancing the efficacy of gemcitabine treatment in pancreatic cancer cells [68]. PRMT5-mediated arginine methylation has also been shown to regulate several other proteins that are involved in the DDR process, including p53-binding protein 1 [88], RAD9 [89], RUVBL1 [86], and TDP1 [90]. Other PRMTs, such as CARM1 and PRMT6, are also directly or indirectly involved in regulating the DDR process through protein arginine methylation [91,92].

Other biological functions of PRMTs include serving as mediators to promote apoptosis [54], participating in regulating tumor immunity [93,94], and governing stem cell fate and survival during embryogenesis as well as adult homeostasis [35].

3. PRMTs Are Involved in the Pathogenesis and Progression of PDAC

Given the broad implication of PRMTs in the biology of eukaryotic cells, the involvement of PRMTs in the pathogenesis and progression of PDAC is not a surprise. It has been found that some of the PRMTs, especially PRMT1, PRMT3, and PRMT5, are intimately associated with PDAC tumorigenesis, metastasis, and chemo-resistance, as discussed below.

3.1. PRMT1

PRMT1 accounts for about 85% of the total Type I PRMT activities and is responsible for the MMA or ADMA of both histone and non-histone proteins [95]. The consequence of arginine methylation by PRMT1 is observed in the processes of transcription regulation, signal transduction, or DNA damage repair [26]. PRMT1 is known to be involved in the tumorigenesis of different cancer types. For example, it affects several signaling cascades associated with the hormonal receptors for estrogen and progesterone in breast cancer cells [96,97]. Its involvement in other cancer types, such as lung cancer [98] and colorectal cancer [99], is also reported. Notably, expression of PRMT1 is upregulated in PDAC tissues. Tissue microarray by immunohistochemistry involving tissue samples of 90 patients has shown an overexpression of PRMT1 in PDAC tissues compared to the adjacent normal pancreatic tissues. PRMT1 expression level in PDAC is found to be positively correlated with the tumor size and post-operative patient prognosis [100]. In an experimental setting, PRMT1 promotes growth of pancreatic cancer cells in vitro and in vivo via enhancing the ß-catenin level [100]. Individual substrates for PRMT1 that are involved in PDAC progression or chemo-resistance have been described. For example, Gli1, an oncogenic transcription factor, essential for Hedgehog signaling, is a substrate of PRMT1. Gli1 is methylated at the R597 residue, which is critical for its transcriptional activity, and its demethylation sensitizes PDAC cells to gemcitabine [101]. The expression of methylated Gli1 positively correlates with PRMT1 expression, suggesting that PRMT1 methylates Gli1, thereby enhancing its oncogenic activity and promoting PDAC progression [101]. It is also observed that overexpression of PRMT1 in PDAC cells facilitates the arginine methylation of HSP70, which aids in the stabilization of BCL2 mRNA, augmenting the expression of BCL2 protein, which prevents cellular apoptosis and renders PDAC cells chemo-resistant [102]. A recent study demonstrates that PDAC disease maintenance depends on PRMT1-mediated RNA metabolism and cellular processes, further indicating that PRMT1 is a therapeutic target for PDAC [103]. In this study, an RNAi-based screening using patient-derived PDAC cells identifies PRMT1 as a top epigenetic lethality factor. Both knockdown and pharmacological inhibition of PRMT1 in PATC53 cells (PDAC patient-derived cells) results in the reduction of cell proliferation and colony formation in vitro as well as reduced tumor volume in vivo, with a significant decrease in cellular and tissue ADMA levels and an increase in the MMA levels. Knockdown of other Type I PRMTs, including PRMT4 and PRMT6 in the same PDAC model system, has no such effects, indicating the critical dependency of PDAC on PRMT1. Proteomics and transcriptomic analysis upon pharmacological inhibition of PRMT1 reveal that this dependency is most likely due to PRMT1-mediated RNA metabolism, cell cycling, DNA replication, and DNA repair in patient-derived PDAC cells [103]. Specifically, pharmacological inhibition of PRMT1 downregulates expression of genes associated with cell cycle, DNA replication and DNA repair. The binding and methylation of RBPs by PRMT1 regulates RNA splicing, $3'$-end RNA processing and RNA stability, and protein translation efficiency [103].

As stated earlier, immunotherapies are less effective in patients with PDAC [12]. However, the poor therapeutic response of immune checkpoint inhibitors that target programmed death ligand-1 (PD-L1) in PDAC is somewhat rescued by co-administration with the PRMT1 inhibitor PT1001B. PT1001B, in conjunction with an anti-PD-L1 antibody, enhances the inhibition of PD-L1 expression in tumor cells and the infiltration of CD8+ lymphocytes, and reduces PD-1+ leukocytes in vivo, thereby augmenting the efficacy of the immune checkpoint inhibitor in a PDAC model system. These observations support the notion that PRMT1 is a molecular target through which the efficacy of the anti-PD-L1 therapy can be enhanced in PDAC [104].

PRMT1 is also known to regulate the EMT-signaling pathway in cancer cells, suggestive of its involvement in tumor metastasis. The expression of one of the key components of the EMT pathway, ZEB1, is highly associated with the expression of PRMT1 in PDAC cells [105]. Downregulation of PRMT1 in PANC-1 and SW1990 cells reduces cell proliferation and invasion, yet overexpression of PRMT1 did not affect these events, which is explained by the high level of endogenous PRMT1, leading to saturation of the PRMT1 protein in pancreatic cancer cells. The anti-tumor effect of PRMT1 downregulation is reversed by overexpression of ZEB-1, indicating the role of PRMT1–ZEB1-signaling cascade in pancreatic cancer progression [105]. The role of PRMT1 in promoting invasion or metastasis of other type of cancers is also well recognized [106–108].

3.2. PRMT3

PRMT3 is a Type I PRMT. Expression level of PRMT3 is correlated with patient prognosis for PDAC, liver cancer, colorectal cancer, and prostate cancer (https://www.proteinatlas.org accessed on 19 February 2024). Overexpression of PRMT3 is evident in PDAC tissues and is associated with poor survival in PDAC patients [109]. While studies of PRMT3 focus more on the control of apoptosis and tumor progression in breast cancer [110,111], its involvement in PDAC has been mainly associated with chemo-resistance [112], likely due to PRMT3-mediated metabolic reprograming of PDAC cells [109]. Studies have shown that PRMT3 upregulates expression of the multidrug-resistant gene ABCG2 in PDAC cells by enhancing the methylation of hnRNPA1 at the R31 residue that, in turn, increases the binding of hnRNPA1 to ABCG2 mRNA. The binding of hnRNA1 to ABCG2 mRNA facilitates its export to the cytoplasm and enhances its expression level, thereby causing chemo-resistance in PDAC cells [112]. PRMT3 overexpression is also associated with an increased cell proliferation and anchorage-independent cell growth. These observations indicate that inhibition of PRMT3 is a new therapeutic strategy for chemo-resistant PDAC [112]. PRMT3 has been particularly explored for its role in regulating metabolic processes in PDAC cells. This shows that PRMT3 reprograms the metabolic process in PDAC cells via arginine methylation of GAPDH at the R248 residue [109]. This arginine methylation of GAPDH by PRMT3 enhances its catalytic activity, likely due to an enhanced formation of the active tetramer of GAPDH. Thus, overexpression of PRMT3 triggers metabolic reprogramming and enhances glycolysis and mitochondrial respiration in a GAPDH-dependent manner. Consequentially, PRMT3 overexpression sensitizes PDAC cells to the GAPDH inhibitor heptelidic acid [109].

3.3. PRMT5

PRMT5 belongs to the Type II PRMTs and is a major producer of SDMA in histone as well as non-histone proteins. It interacts with methylosome protein 50 (MEP50) to form a heterooctametric complex, eliciting its methyltransferase activity [58]. The expression level of both PRMT5 and MEP50 is often elevated in human cancer [113]. Overexpression of PRMT5 is observed at both mRNA and protein levels in PDAC tissues, which is associated with poor prognosis of PDAC patients [114–116].

The interaction of PRMT5 with cMYC oncogenic signaling in PDAC cells has been well documented. PRMT5 is found to play a critical role in glycolysis and tumorigenesis of PDAC cells via interacting with the F-box/WD repeat-containing protein 7 (FBW7)/cMyc axis [116]. Knockdown of PRMT5 in the pancreatic cancer cell lines MIA-PaCa 2 and SW1990 reduces the viability and colony formation capacity of the cells in vitro and the tumor volume in xenograft nude mouse models. Knockdown of PRMT5 in PDAC cells also inhibits glucose uptake and reduces lactate production. The uptake of 18F-FDG, an indicator of glucose uptake, is higher in subcutaneous tumors, while knockdown of PRMT5 reduces 18F-FDG uptake in these tumors. It turns out that PRMT5 regulates the expression of cMyc at the post-transcriptional level by inhibiting the E3 protein ligase FBW7 [117]. Knockdown of PRMT5 in PDAC cells reduces the protein level of cMYC without affecting its mRNA levels, causing an increased degradation of cMyc via the proteasomal degradation pathway facilitated by FBW7. The expression of FBW7, a tumor suppressor [117], is often reduced in human cancer cells [118]; however, knockdown of PRMT5 in PDAC cells elevated its expression, indicating that PRMT5 regulates FBW7 expression. The suppression of FBW7 expression by PRMT5 is primarily mediated via epigenetic modifications [116]. Thus, PRMT5 stabilizes cMYC protein by suppressing FBW7 expression, thereby promoting PDAC tumorigenesis.

A recent study also shows the connection of cMYC with PRMT5 in PDAC model systems. This study utilized an unbiased pharmacological screening approach in PDAC cells to identify the cMYC-associated epigenetic dependency [119]. PRMT5 inhibitors (PRMT5i) are identified as significant screening hits, where the sensitivity/efficacy of PRMT5i treatment in PDAC cells is directly associated with cMYC overexpression. There is a positive correlation between mRNA expression of PRMT5 and cMYC, and the overexpression of cMYC in patient-derived PDAC tumors is associated with high sensitivity towards PRMT5 inhibition. Evidently, PRMT5i treatment results in lower survival rate for PDAC cells with high cMYC expression (HUPT3, PaTu8988T, PSN1, and DanG) than those with lower cMYC expression (Panc1, PaTu8988S, HPAC, and Panc0504). Apoptosis seemed to be the primary mechanism by which the PRMT5i elicits therapeutic response in cMYC-expressing cells [119].

PRMT5 has been shown to facilitate EMT in PDAC cells via the EGFR/AKT/β-catenin pathway [120], indicative of its involvement in tumor metastasis. In the SW1990 and PaTu8988 cell lines, knockdown of PRMT5 reduces cell proliferation and colony formation, which is rescued upon ectopic re-expression of PRMT5. This is also evident in the xenograft mouse models, in which PRMT5 knockdown reduces the volume of the xenograft tumors. Consistently, the inhibition of the migration and invasion of PDAC cells is observed upon PRMT5 knockdown using the transwell migration and transwell invasion assays, which are rescued by ectopic re-expression of PRMT5. In addition, knockdown of PRMT5 results in an increased expression of the epithelial marker E-cadherin (at both mRNA and protein levels) and decreased expression of the mesenchymal markers Vimentin, Collagen I, and β-catenin, indicating that PRMT5 is involved in tumor metastasis. Mechanistically, knockdown of PRMT5 decreases the phosphorylation level of EGFR at Y1068 and Y1172, indicating that the arginine methylation of EGFR is essential for its phosphorylation in the Y1068 and Y1172 residues, without which the downstream signaling involving phosphorylation and activation of AKT/GSK3β and β-catenin is impaired. Note that arginine methylation of EGFR at the R1175 residue by PRMT5 has been shown to impact downstream signaling [121]. Thus, overexpression of PRMT5 in PDAC cells facilitates tumor EMT, thereby promoting tumor invasion and metastasis.

It is interesting to observe that PRMT5 inhibition affects cell proliferation which is synergized with the loss of Type I PRMTs, particularly PRMT1 [122]. The loss of PRMT1 in MIA PaCa-2 cells makes it highly sensitive towards the treatment with the PRMT5 inhibitor EPZ015666. The synergistic effect is also observed with the use of PRMT1 inhibitor MS023 in combination with the PRMT5 inhibitor EPZ015666. This suggests an overlapped spectrum of substrates of PRMT1 and PRMT5 in the cell proliferation pathway. Furthermore, inhibition of CARM1 or PRMT6 in combination with PRMT5 inhibition also shows similar consequences in these cells, but the effect is less significant compared with the inhibition of PRMT1, suggesting that PRMT1 is a primary Type I PRMT to compensate for the loss of PRMT5 in PDAC cells.

The use of PRMT5 inhibitors along with the chemotherapeutic agent gemcitabine has been shown to have a synergistic effect in tumor growth inhibition in PDAC cells through enhanced DNA damage [68]. Using an in vivo CRISPR gene knockout screening approach to search for the combinatorial targets of gemcitabine in PDAC, PRMT5 is identified as a druggable candidate that may act in synergy with gemcitabine to kill PDAC cells. In both in vitro and in vivo model systems, knockdown of PRMT5 or the use of PRMT5 inhibitors causes excessive DNA damage in PDAC cells when combined with gemcitabine treatment, a synergistic effect likely mediated by RPA exhaustion [68].

However, not all PDAC cells are equally sensitive to PRMT5 inhibition. PRMT5 inhibition appears to be more lethal in PDAC cells with the deletion of the tumor suppressor CDK2NA in the chromosome 9p21 locus. This is because the CDK2NA gene deletion is often associated with co-deletion of its adjacent genes in the genome, and one of the genes is the methylthioadenosine phosphorylase gene (MTAP) [123]. MTAP is essential for the metabolism of its substrate 5′-methylthioadenosine (MTA) to generate methionine and adenosine [124]. In MTAP-deleted cancer cells, the level of MTA is elevated which inhibits the methyltransferase activity of PRMT5 towards all of its substrates, increasing the sensitivity of cancer cells to further PRMT5 inhibition [125]. Therefore, in the MTAP deleted tumors, PRMT5 is a preferred molecular target for therapeutic development, because its inhibition appears to be more lethal than in tumor cells harboring the wild type MTAP gene. This is confirmed using a PDAC patient-derived organoids (PDOs) model with tumors derived from the pancreas that harbored the MTAP gene deletion. Treatment with the PRMT5-specific inhibitor EPZ015556 inhibits growth of the PDOs with the MTAP gene deletion, and this inhibition is significantly reduced in PDOs without the MTAP gene deletion [51].

4. Inhibitors of PRMTs

Since there is no protein arginine demethylase that has been consensually established to this date, arginine methylation is considered a relatively stable PTM that affects several downstream signaling cascades in cancer cells [126]. Given the role of PRMTs in the tumorigenesis and progression of human cancers [96–99,113], including PDAC [114–116], extensive effort has been directed towards the identification, synthesis, and application of PRMT inhibitors as potential cancer therapeutics. Indeed, PRMT inhibitors, especially inhibitors for PRMT1 and PRMT5 due to their established involvement in cancer progression, are currently being explored as therapeutics for hematological malignancies and solid tumors and have entered clinical trials (www.clinicaltrails.gov, accessed on 16 February 2024). The anticancer action of the major inhibitors against PRMT1, PRMT3, and PRMT5 is shown in Figure 3. Note that the cytotoxicity of PRMT inhibitors, especially inhibitors for PRMT5, are shown to be more specific towards cancer cells because of PRMT overexpression in cancer tissues and the reliance of tumor cells on PRMT activity [127]. Some of the recent clinical trials are listed in Table 4.

With the availability of assays to analyze PRMT activity, several groups of small molecules that target PRMTs have been developed, and these inhibitors elicit either specific or non-specific inhibition of different types of PRMTs. Typically, radiometric assays and antibody-based assays are utilized to study PRMT activity upon treatment with prospective small-molecular inhibitors in cellular model systems [128]. The mechanism of action for these small-molecule inhibitors are based on their ability to inhibit the methyltransferase activity of one or multiple PRMTs, mainly by inhibiting the binding of the PRMTs to their substrates or by occupying the SAM binding pockets in the PRMT enzymes, thereby diminishing methyltransferase activity [128]. While the mechanisms of action of most of these inhibitors are clearly described, the downstream effect they exert upon methyltransferase activity inhibition is a challenging task to elucidate because of the broad range of cellular activities mediated by PRMTs [128].

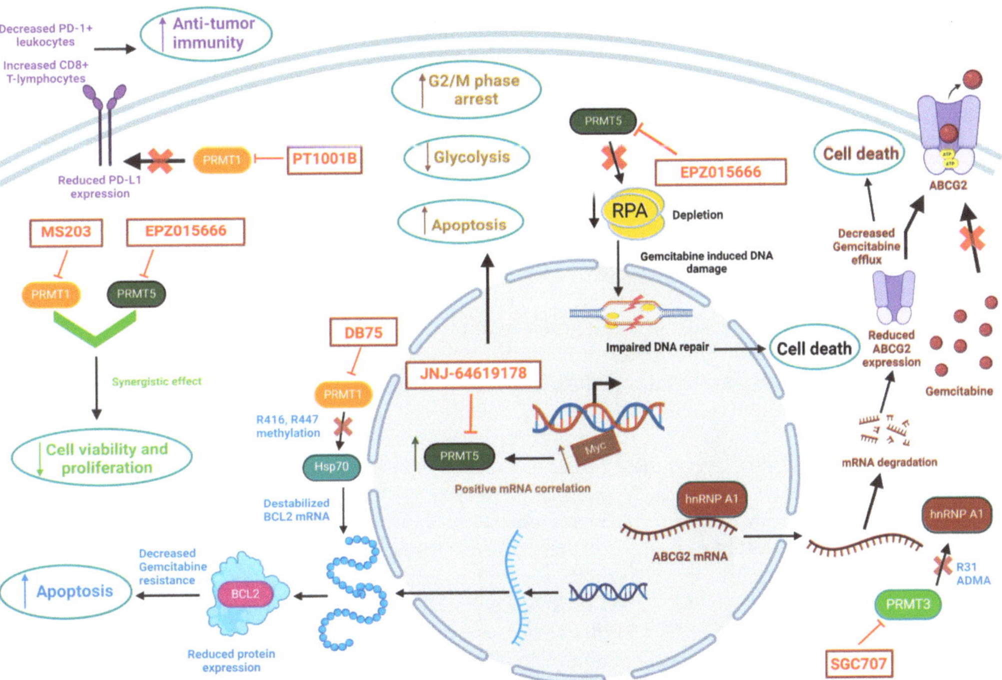

Figure 3. Anticancer action of PRMT inhibitors. Created with Biorender.com.

Table 4. Inhibitors of PRMTs currently in clinical trials.

Inhibitor	Target	Clinical Trial ID	Phase	Tumor
GSK3368715	Type I PRMT	NCT03666988	Phase I	Solid Tumors and Diffuse Large B-cell Lymphoma
AMG 193	PRMT5	NCT05094336	Phase I/II	MTAP-null solid tumors
JNJ-64619178	PRMT5	NCT03573310	Phase I	Advanced solid tumors, B cell non-Hodgkin lymphoma (NHL)
PF-06939999	PRMT5	NCT03854227	Phase I	Advanced solid tumors
PRT543	PRMT5	NCT03886831	Phase I	Relapsed or refractory solid tumors, lymphoma, and leukemia
PRT811	PRMT5	NCT04089449	Phase I	High grade gliomas, anaplastic astrocytoma, and advanced solid tumors
GSK3326595/EPZ015666	PRMT5	NCT04676516 NCT02783300 NCT03614728	Phase I/II Phase I Phase I/II	Early-stage breast cancer Advanced solid tumors, NHL Chronic myelomonocytic leukemia, Adult acute myeloid leukemia

The AMI series of compounds are the first class of PRMT small-molecule inhibitors identified via ELISA-based high-throughput screening that shows selectivity towards Type I PRMTs and specificity in inhibiting arginine, but not lysine methyltransferase activity [129]. Following the identification of AMI compounds, effort in virtual and experimental screening has continued to uncover more PRMT1 inhibitors with high specificity [130]. A series of 1-substituted 1H-tetrazole derivative compounds have been screened for their PRMT1 inhibiting activity and a compound 9a has been deemed most potent, which selectively inhibits the methyltransferase activity of PRMT1 via interfering with the substrate binding site, demonstrated by molecular dynamics simulation. The inhibition of PRMT1 by compound 9a significantly reduces the cellular ADMA level and downregulates the Wnt/b-catenin signaling pathway in MDA-MB-231 cells [131]. MS023, also a Type I PRMT inhibitor, has been identified, displaying high potency to inhibit the activity of PRMT1, -3, -4, -6, and -8, while being completely inactive on the activity of Type II and III PRMTs [132]. A new Type I PRMT inhibitor, GSK3368715, is described as a potent and reversible inhibitor that inhibits all Type I PRMTs except PRMT3 [133]. GSK3368715 has been shown to have anticancer activity and has entered into a Phase I Clinical trial for the treatment of diffuse large B-cell lymphoma and selected solid tumors with MTAP deficiency (NCT03666988). However, this clinical trial has been terminated due to the lack of clinical efficacy [134]. Other inhibitors for Type I PRMTs, such as MS049, a dual inhibitor of CARM1 and PRMT6 [135], and SGC6870, a highly selective inhibitor of PRMT6 [136], have also been developed.

The development of inhibitors for Type II PRMTs, especially for PRMT5, has been quite successful. CMP5 is the first PRMT5-specific inhibitor developed by screening the ChemBridge CNS-Set library of 10,000 small-molecule compounds [137]. Treatment with CMP5 selectively reduces the viability of tumor cells, suggesting that PRMT5 is an ideal therapeutic target against cancer [137]. Another PRMT5 inhibitor EPZ015666 has been identified by using a homogeneous time-resolved fluorescence assay to screen a diverse library containing 370,000 small molecules [138]. This PRMT5 inhibitor acts by disrupting the MEP50:PRMT5 complex that is absolutely crucial for the methyltransferase activity of the enzyme. EPZ015666 is the first orally bioavailable and highly selective inhibitor of PRMT5 with antiproliferative effects in both in vitro and in vivo model systems [138]. Several PRMT5 inhibitors have been further developed, which have shown anticancer activity and have entered into clinical trials. These include GSK3326595 (EPZ015938) for solid tumors and non-Hodgkin's lymphoma (NHL) (NCT02783300), as well as Myelodysplastic Syndrome (MDS) and Acute Myeloid Leukemia (AML) (NCT03614728); JNJ-64619178 for advanced solid tumors, NHL, and low-risk MDS (NCT03573310); PRT543 for advanced solid tumors and hematologic malignancies (NCT03886831); and PRT811 for advanced solid tumors, CNS lymphoma and Gliomas (NCT04089449).

Other PRMT inhibitors, such as EPZ020411 to inhibit PRMT6 by occupying its arginine binding site [139]; Compound II757, a pan-inhibitor for PRMTs [140]; and SGC3027, a potent PRMT7 inhibitor [141], have recently been described. A detailed patent review on PRMT inhibitors, especially inhibitors for PRMT1 and PRMT5, has been recently published [142].

The use of PRMT inhibitors in combination with the existing chemotherapeutics has proven to be beneficial in different cancer types. Combination of the PRMT5 inhibitor EPZ015666 with gemcitabine potentiates the DNA-damaging effect of gemcitabine in PDAC cells and helps to overcome therapeutic resistance via the HDR [68]. Treatment of breast cancer cells (MCF7, T-47D, MDA-MB-231, BT-549, and MDA-MB-468) with the combination of the PRMT5 inhibitor EPZ015666 and chemotherapeutic agents Etoposide/cisplatin demonstrates synergistic effect on the viability of these cells [143]. Notably, the immunotherapeutic efficacy of the anti-PD-L1 mAb in PDAC is enhanced when combined with the PRMT1 inhibitor PT1001B in a mouse model [104]. Some of the PRMT small-molecular inhibitors are orally bioavailable compounds, which do not exert extreme systemic toxicity. They have been reported to work well in combination with standard chemotherapeutics and are implemented in several clinical trials as a monotherapy or in combination (Table 4). Thus, these PRMT inhibitors are attractive therapeutic drug candidates for PDAC, where the major challenges are therapeutic resistance and severe toxicity due to multiple chemotherapeutics utilized during different stages of the treatment course.

5. Perspectives

The arginine methylation of histone and non-histone proteins has not been explored in depth. However, arginine methylation is a vital PTM in eukaryotic cells, where a large number of substrates rely on this PTM to elicit physiological activity. Protein arginine methylation plays critical roles in the initiation and progression of malignant diseases, including PDAC; therefore, targeting PRMTs is a logical strategy for cancer therapeutic development. While PRMT inhibitors have been tested in preclinical models for treatment of PDAC [68,104], clinical trials testing these inhibitors against PDAC have not been initiated. Based on the current status of PRMT inhibitors and the ongoing clinical trials, it is envisioned that new therapeutics targeting PRMTs, used along or in combination with other therapies, are likely to enter into clinical trials and become available in the clinical management of PDAC or other cancer patients in the future. Efforts on developing more enzyme-specific and potent PRMT inhibitors are expected to continue.

While studies using in vitro and in vivo model systems have shown the potential of PRMTs as therapeutic targets, elucidating the mechanism of action for these inhibitors has been a major challenge due to the substrate diversity and the cellular pathways involved. Furthermore, one could imagine that the arginine methylation site in any protein could vary in different cell types, depending on the expression and activity of individual PRMTs in the cell systems. For example, AKT has been recently identified to be methylated by PRMT5 at the R391 residue in MCF7 cells (SDMA formation) [143]. However, in the case of neuroblastoma cells, AKT is observed to be methylated at the R15 residue [144]. These cell type-dependent arginine methylation patterns indicate the intricacy in elucidating the cellular mechanisms related to arginine methylation across different types of cells and tumors. This intricacy has to be considered in our pursuit to better understand PRMT-mediated cellular processes.

Several studies have shown the existence of PRMT5 protein and mRNA in PDAC cell-derived exosomes, a group of small extracellular vesicles (EV) that mediate the transfer and function of biomolecules, including proteins, lipids, and nucleotides [145,146]. Given the involvement of exosomes in intercellular communication and cancer metastasis [147,148], the impact of PRMT5 on cell-to-cell communication via transfer of exosomes needs to be explored. In fact, one of the recent studies has indicated the impact of PRMT5 knockdown on the EV-associated pathways [149]. Expression of the proteins associated with EV biogenesis is upregulated upon PRMT5 knockdown in AML cells. This shows a biological relevance of PRMT5 in EV biogenesis that merits further investigation. In addition, one of the well-known biochemical functions of arginine methylation is in mediating protein phase separation [150–152]. A recent study has demonstrated the role of phase separation in the selective miRNA enrichment in exosomes [153]. This particular area needs to be further explored to establish the link between PRMTs and the biology of EVs. Furthermore, expression of PRMTs is often upregulated in cancer cells and tissues, resulting in different arginine methylation patterns of cellular proteins in normal versus cancerous cells or tissues. The different arginine methylation patterns may serve as biomarkers for early detection or disease monitoring for various malignancies, including PDAC. In particular, since tumor exosomes have been shown to be released into the circulation [154,155], arginine methylation patterns in plasma exosomes are likely indicators of PDAC [156], a lethal malignancy that desperately needs non-invasive biomarkers for early detection.

6. Conclusions

Protein arginine methylation is an important PTM in eukaryotic cells. Expression of PRMTs is often upregulated in PDAC cells and tissues, which facilitates changes in the transcriptional landscape, metabolic processes, EMT, and DNA damage responses, thereby promoting chemo-resistance and tumor progression (Figure 4). Due to the complications of current chemotherapeutics, such as severe systemic toxicity and therapeutic resistance, and the unsatisfactory outcome of immunotherapy and targeted therapy against PDAC, new therapeutic strategies are urgently needed in the clinical management of PDAC to improve patient survival outcomes. Targeting PRMT enzymes using small-molecule inhibitors provides a promising new line of therapy for PDAC. Studies have shown promising anticancer effects when combining PRMT inhibitors with existing chemotherapeutics to reduce tumor burden. Several PRMT inhibitors are now in clinical trials for both solid and hematological malignancies, either as a monotherapy or in combination. These clinical trials may help generate a new generation of therapeutics that can be clinically utilized for PDAC as well as other malignancies.

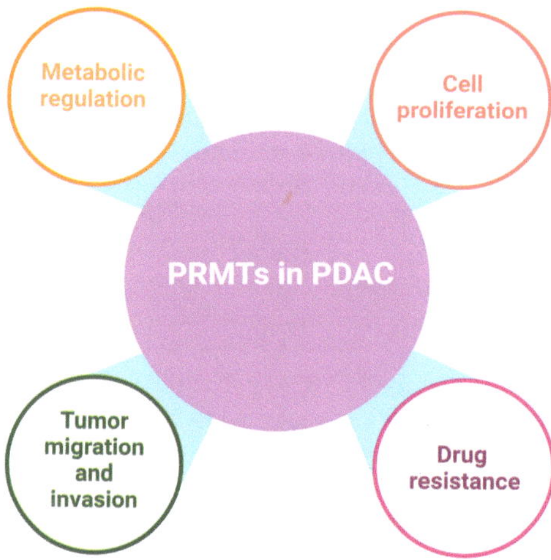

Figure 4. Role of PRMTs in pancreatic ductal adenocarcinoma (PDAC). Created with Biorender.com.

Author Contributions: K.B.: Conceptualization, original draft preparation, W.-Q.D., Conceptualization, review and editing, funding acquisition. All authors have read and agreed to the published version of the manuscript.

Funding: This study was supported in part by grants from Congressionally Directed Medical Research Programs (W81XWH-22-1-0317), Institutional Development Awards (IDeA) from the National

Institute of General Medical Sciences of the National Institutes of Health (P30GM145423), and the Presbyterian Health Foundation.

Conflicts of Interest: The authors declare no conflicts of interest.

References

1. Sarantis, P.; Koustas, E.; Papadimitropoulou, A.; Papavassiliou, A.G.; Karamouzis, M.V. Pancreatic ductal adenocarcinoma: Treatment hurdles, tumor microenvironment and immunotherapy. *World J. Gastrointest. Oncol.* **2020**, *12*, 173–181. [CrossRef]
2. Rahib, L.; Wehner, M.R.; Matrisian, L.M.; Nead, K.T. Estimated Projection of US Cancer Incidence and Death to 2040. *JAMA Netw. Open* **2021**, *4*, e214708. [CrossRef] [PubMed]
3. Siegel, R.L.; Giaquinto, A.N.; Jemal, A. Cancer statistics, 2024. *CA Cancer J. Clin.* **2024**, *74*, 12–49. [CrossRef] [PubMed]
4. Kleeff, J.; Korc, M.; Apte, M.; La Vecchia, C.; Johnson, C.D.; Biankin, A.V.; Neale, R.E.; Tempero, M.; Tuveson, D.A.; Hruban, R.H.; et al. Pancreatic cancer. *Nat. Rev. Dis. Primers* **2016**, *2*, 16022. [CrossRef] [PubMed]
5. Farr, K.P.; Moses, D.; Haghighi, K.S.; Phillips, P.A.; Hillenbrand, C.M.; Chua, B.H. Imaging Modalities for Early Detection of Pancreatic Cancer: Current State and Future Research Opportunities. *Cancers* **2022**, *14*, 2539. [CrossRef] [PubMed]
6. Morani, A.C.; Hanafy, A.K.; Ramani, N.S.; Katabathina, V.S.; Yedururi, S.; Dasyam, A.K.; Prasad, S.R. Hereditary and Sporadic Pancreatic Ductal Adenocarcinoma: Current Update on Genetics and Imaging. *Radiol. Imaging Cancer* **2020**, *2*, e190020. [CrossRef] [PubMed]
7. Singhi, A.D.; Koay, E.J.; Chari, S.T.; Maitra, A. Early Detection of Pancreatic Cancer: Opportunities and Challenges. *Gastroenterology* **2019**, *156*, 2024–2040. [CrossRef] [PubMed]
8. Halbrook, C.J.; Lyssiotis, C.A.; Pasca di Magliano, M.; Maitra, A. Pancreatic cancer: Advances and challenges. *Cell* **2023**, *186*, 1729–1754. [CrossRef] [PubMed]
9. Conroy, T.; Hammel, P.; Hebbar, M.; Ben Abdelghani, M.; Wei, A.C.; Raoul, J.L.; Chone, L.; Francois, E.; Artru, P.; Biagi, J.J.; et al. FOLFIRINOX or Gemcitabine as Adjuvant Therapy for Pancreatic Cancer. *N. Engl. J. Med.* **2018**, *379*, 2395–2406. [CrossRef]
10. Amrutkar, M.; Gladhaug, I.P. Pancreatic Cancer Chemoresistance to Gemcitabine. *Cancers* **2017**, *9*, 157. [CrossRef]
11. de Jesus, V.H.F.; Camandaroba, M.P.G.; Donadio, M.D.S.; Cabral, A.; Muniz, T.P.; de Moura Leite, L.; Sant'Ana, L.F. Retrospective comparison of the efficacy and the toxicity of standard and modified FOLFIRINOX regimens in patients with metastatic pancreatic adenocarcinoma. *J. Gastrointest. Oncol.* **2018**, *9*, 694–707. [CrossRef]
12. Timmer, F.E.F.; Geboers, B.; Nieuwenhuizen, S.; Dijkstra, M.; Schouten, E.A.C.; Puijk, R.S.; de Vries, J.J.J.; van den Tol, M.P.; Bruynzeel, A.M.E.; Streppel, M.M.; et al. Pancreatic Cancer and Immunotherapy: A Clinical Overview. *Cancers* **2021**, *13*, 4138. [CrossRef] [PubMed]
13. Qian, Y.; Gong, Y.; Fan, Z.; Luo, G.; Huang, Q.; Deng, S.; Cheng, H.; Jin, K.; Ni, Q.; Yu, X.; et al. Molecular alterations and targeted therapy in pancreatic ductal adenocarcinoma. *J. Hematol. Oncol.* **2020**, *13*, 130. [CrossRef] [PubMed]
14. Balachandran, V.P.; Beatty, G.L.; Dougan, S.K. Broadening the Impact of Immunotherapy to Pancreatic Cancer: Challenges and Opportunities. *Gastroenterology* **2019**, *156*, 2056–2072. [CrossRef]
15. Fang, Y.T.; Yang, W.W.; Niu, Y.R.; Sun, Y.K. Recent advances in targeted therapy for pancreatic adenocarcinoma. *World J. Gastrointest. Oncol.* **2023**, *15*, 571–595. [CrossRef]
16. Cancer Genome Atlas Research Network. Integrated Genomic Characterization of Pancreatic Ductal Adenocarcinoma. *Cancer Cell* **2017**, *32*, 185–203.e113. [CrossRef] [PubMed]
17. Wang, X.; Allen, S.; Blake, J.F.; Bowcut, V.; Briere, D.M.; Calinisan, A.; Dahlke, J.R.; Fell, J.B.; Fischer, J.P.; Gunn, R.J.; et al. Identification of MRTX1133, a Noncovalent, Potent, and Selective KRAS(G12D) Inhibitor. *J. Med. Chem.* **2022**, *65*, 3123–3133. [CrossRef] [PubMed]
18. Kemp, S.B.; Cheng, N.; Markosyan, N.; Sor, R.; Kim, I.K.; Hallin, J.; Shoush, J.; Quinones, L.; Brown, N.V.; Bassett, J.B.; et al. Efficacy of a Small-Molecule Inhibitor of KrasG12D in Immunocompetent Models of Pancreatic Cancer. *Cancer Discov.* **2023**, *13*, 298–311. [CrossRef]
19. Viale, A.; Pettazzoni, P.; Lyssiotis, C.A.; Ying, H.; Sanchez, N.; Marchesini, M.; Carugo, A.; Green, T.; Seth, S.; Giuliani, V.; et al. Oncogene ablation-resistant pancreatic cancer cells depend on mitochondrial function. *Nature* **2014**, *514*, 628–632. [CrossRef]
20. Kapoor, A.; Yao, W.; Ying, H.; Hua, S.; Liewen, A.; Wang, Q.; Zhong, Y.; Wu, C.J.; Sadanandam, A.; Hu, B.; et al. Yap1 activation enables bypass of oncogenic Kras addiction in pancreatic cancer. *Cell* **2014**, *158*, 185–197. [CrossRef]
21. Bryant, K.L.; Stalnecker, C.A.; Zeitouni, D.; Klomp, J.E.; Peng, S.; Tikunov, A.P.; Gunda, V.; Pierobon, M.; Waters, A.M.; George, S.D.; et al. Combination of ERK and autophagy inhibition as a treatment approach for pancreatic cancer. *Nat. Med.* **2019**, *25*, 628–640. [CrossRef]
22. Daemen, A.; Peterson, D.; Sahu, N.; McCord, R.; Du, X.; Liu, B.; Kowanetz, K.; Hong, R.; Moffat, J.; Gao, M.; et al. Metabolite profiling stratifies pancreatic ductal adenocarcinomas into subtypes with distinct sensitivities to metabolic inhibitors. *Proc. Natl. Acad. Sci. USA* **2015**, *112*, E4410–E4417. [CrossRef] [PubMed]
23. Provenzano, P.P.; Cuevas, C.; Chang, A.E.; Goel, V.K.; Von Hoff, D.D.; Hingorani, S.R. Enzymatic targeting of the stroma ablates physical barriers to treatment of pancreatic ductal adenocarcinoma. *Cancer Cell* **2012**, *21*, 418–429. [CrossRef] [PubMed]

24. Jacobetz, M.A.; Chan, D.S.; Neesse, A.; Bapiro, T.E.; Cook, N.; Frese, K.K.; Feig, C.; Nakagawa, T.; Caldwell, M.E.; Zecchini, H.I.; et al. Hyaluronan impairs vascular function and drug delivery in a mouse model of pancreatic cancer. *Gut* **2013**, *62*, 112–120. [CrossRef] [PubMed]

25. Bear, A.S.; Vonderheide, R.H.; O'Hara, M.H. Challenges and Opportunities for Pancreatic Cancer Immunotherapy. *Cancer Cell* **2020**, *38*, 788–802. [CrossRef] [PubMed]

26. Yang, Y.; Bedford, M.T. Protein arginine methyltransferases and cancer. *Nat. Rev. Cancer* **2013**, *13*, 37–50. [CrossRef] [PubMed]

27. Hwang, J.W.; Cho, Y.; Bae, G.U.; Kim, S.N.; Kim, Y.K. Protein arginine methyltransferases: Promising targets for cancer therapy. *Exp. Mol. Med.* **2021**, *53*, 788–808. [CrossRef]

28. Feustel, K.; Falchook, G.S. Protein Arginine Methyltransferase 5 (PRMT5) Inhibitors in Oncology Clinical Trials: A review. *J. Immunother. Precis. Oncol.* **2022**, *5*, 58–67. [CrossRef] [PubMed]

29. Ramazi, S.; Zahiri, J. Posttranslational modifications in proteins: Resources, tools and prediction methods. *Database* **2021**, *2021*, baab012. [CrossRef]

30. Clarke, S.G. Protein methylation at the surface and buried deep: Thinking outside the histone box. *Trends Biochem. Sci.* **2013**, *38*, 243–252. [CrossRef]

31. Bhat, K.P.; Umit Kaniskan, H.; Jin, J.; Gozani, O. Epigenetics and beyond: Targeting writers of protein lysine methylation to treat disease. *Nat. Rev. Drug Discov.* **2021**, *20*, 265–286. [CrossRef] [PubMed]

32. Lanouette, S.; Mongeon, V.; Figeys, D.; Couture, J.F. The functional diversity of protein lysine methylation. *Mol. Syst. Biol.* **2014**, *10*, 724. [CrossRef] [PubMed]

33. Aziz, N.; Hong, Y.H.; Kim, H.G.; Kim, J.H.; Cho, J.Y. Tumor-suppressive functions of protein lysine methyltransferases. *Exp. Mol. Med.* **2023**, *55*, 2475–2497. [CrossRef] [PubMed]

34. Bedford, M.T.; Clarke, S.G. Protein arginine methylation in mammals: Who, what, and why. *Mol. Cell* **2009**, *33*, 1–13. [CrossRef] [PubMed]

35. Blanc, R.S.; Richard, S. Arginine Methylation: The Coming of Age. *Mol. Cell* **2017**, *65*, 8–24. [CrossRef] [PubMed]

36. Dong, R.; Li, X.; Lai, K.O. Activity and Function of the PRMT8 Protein Arginine Methyltransferase in Neurons. *Life* **2021**, *11*, 1132. [CrossRef]

37. Jarrold, J.; Davies, C.C. PRMTs and Arginine Methylation: Cancer's Best-Kept Secret? *Trends Mol. Med.* **2019**, *25*, 993–1009. [CrossRef] [PubMed]

38. Hasegawa, M.; Toma-Fukai, S.; Kim, J.D.; Fukamizu, A.; Shimizu, T. Protein arginine methyltransferase 7 has a novel homodimer-like structure formed by tandem repeats. *FEBS Lett.* **2014**, *588*, 1942–1948. [CrossRef] [PubMed]

39. Hadjikyriacou, A.; Yang, Y.; Espejo, A.; Bedford, M.T.; Clarke, S.G. Unique Features of Human Protein Arginine Methyltransferase 9 (PRMT9) and Its Substrate RNA Splicing Factor SF3B2. *J. Biol. Chem.* **2015**, *290*, 16723–16743. [CrossRef]

40. Thandapani, P.; O'Connor, T.R.; Bailey, T.L.; Richard, S. Defining the RGG/RG motif. *Mol. Cell* **2013**, *50*, 613–623. [CrossRef]

41. Musiani, D.; Bok, J.; Massignani, E.; Wu, L.; Tabaglio, T.; Ippolito, M.R.; Cuomo, A.; Ozbek, U.; Zorgati, H.; Ghoshdastider, U.; et al. Proteomics profiling of arginine methylation defines PRMT5 substrate specificity. *Sci. Signal.* **2019**, *12*, eaat8388. [CrossRef]

42. Kumar, P.; Joy, J.; Pandey, A.; Gupta, D. PRmePRed: A protein arginine methylation prediction tool. *PLoS ONE* **2017**, *12*, e0183318. [CrossRef]

43. Deng, W.; Wang, Y.; Ma, L.; Zhang, Y.; Ullah, S.; Xue, Y. Computational prediction of methylation types of covalently modified lysine and arginine residues in proteins. *Brief. Bioinform.* **2017**, *18*, 647–658. [CrossRef]

44. Rust, H.L.; Subramanian, V.; West, G.M.; Young, D.D.; Schultz, P.G.; Thompson, P.R. Using unnatural amino acid mutagenesis to probe the regulation of PRMT1. *ACS Chem. Biol.* **2014**, *9*, 649–655. [CrossRef]

45. Wagner, S.; Weber, S.; Kleinschmidt, M.A.; Nagata, K.; Bauer, U.M. SET-mediated promoter hypoacetylation is a prerequisite for coactivation of the estrogen-responsive pS2 gene by PRMT1. *J. Biol. Chem.* **2006**, *281*, 27242–27250. [CrossRef] [PubMed]

46. Fulton, M.D.; Dang, T.; Brown, T.; Zheng, Y.G. Effects of substrate modifications on the arginine dimethylation activities of PRMT1 and PRMT5. *Epigenetics* **2022**, *17*, 1–18. [CrossRef]

47. Hartley, A.V.; Lu, T. Modulating the modulators: Regulation of protein arginine methyltransferases by post-translational modifications. *Drug Discov. Today* **2020**, *25*, 1735–1743. [CrossRef] [PubMed]

48. Feng, Q.; He, B.; Jung, S.Y.; Song, Y.; Qin, J.; Tsai, S.Y.; Tsai, M.J.; O'Malley, B.W. Biochemical control of CARM1 enzymatic activity by phosphorylation. *J. Biol. Chem.* **2009**, *284*, 36167–36174. [CrossRef] [PubMed]

49. Liu, F.; Zhao, X.; Perna, F.; Wang, L.; Koppikar, P.; Abdel-Wahab, O.; Harr, M.W.; Levine, R.L.; Xu, H.; Tefferi, A.; et al. JAK2V617F-mediated phosphorylation of PRMT5 downregulates its methyltransferase activity and promotes myeloproliferation. *Cancer Cell* **2011**, *19*, 283–294. [CrossRef]

50. Halabelian, L.; Barsyte-Lovejoy, D. Structure and Function of Protein Arginine Methyltransferase PRMT7. *Life* **2021**, *11*, 768. [CrossRef]

51. Driehuis, E.; van Hoeck, A.; Moore, K.; Kolders, S.; Francies, H.E.; Gulersonmez, M.C.; Stigter, E.C.A.; Burgering, B.; Geurts, V.; Gracanin, A.; et al. Pancreatic cancer organoids recapitulate disease and allow personalized drug screening. *Proc. Natl. Acad. Sci. USA* **2019**, *116*, 26580–26590. [CrossRef] [PubMed]

52. Singhroy, D.N.; Mesplede, T.; Sabbah, A.; Quashie, P.K.; Falgueyret, J.P.; Wainberg, M.A. Automethylation of protein arginine methyltransferase 6 (PRMT6) regulates its stability and its anti-HIV-1 activity. *Retrovirology* **2013**, *10*, 73. [CrossRef] [PubMed]

53. Thiebaut, C.; Eve, L.; Poulard, C.; Le Romancer, M. Structure, Activity, and Function of PRMT1. *Life* **2021**, *11*, 1147. [CrossRef]

54. Cura, V.; Cavarelli, J. Structure, Activity and Function of the PRMT2 Protein Arginine Methyltransferase. *Life* **2021**, *11*, 1263. [CrossRef] [PubMed]
55. Tang, J.; Gary, J.D.; Clarke, S.; Herschman, H.R. PRMT 3, a type I protein arginine N-methyltransferase that differs from PRMT1 in its oligomerization, subcellular localization, substrate specificity, and regulation. *J. Biol. Chem.* **1998**, *273*, 16935–16945. [CrossRef] [PubMed]
56. Frankel, A.; Clarke, S. PRMT3 is a distinct member of the protein arginine N-methyltransferase family. Conferral of substrate specificity by a zinc-finger domain. *J. Biol. Chem.* **2000**, *275*, 32974–32982. [CrossRef]
57. Suresh, S.; Huard, S.; Dubois, T. CARM1/PRMT4: Making Its Mark beyond Its Function as a Transcriptional Coactivator. *Trends Cell Biol.* **2021**, *31*, 402–417. [CrossRef] [PubMed]
58. Motolani, A.; Martin, M.; Sun, M.; Lu, T. The Structure and Functions of PRMT5 in Human Diseases. *Life* **2021**, *11*, 1074. [CrossRef] [PubMed]
59. Gupta, S.; Kadumuri, R.V.; Singh, A.K.; Chavali, S.; Dhayalan, A. Structure, Activity and Function of the Protein Arginine Methyltransferase 6. *Life* **2021**, *11*, 951. [CrossRef]
60. Kolbel, K.; Ihling, C.; Bellmann-Sickert, K.; Neundorf, I.; Beck-Sickinger, A.G.; Sinz, A.; Kuhn, U.; Wahle, E. Type I Arginine Methyltransferases PRMT1 and PRMT-3 Act Distributively. *J. Biol. Chem.* **2009**, *284*, 8274–8282. [CrossRef]
61. Wang, M.; Fuhrmann, J.; Thompson, P.R. Protein arginine methyltransferase 5 catalyzes substrate dimethylation in a distributive fashion. *Biochemistry* **2014**, *53*, 7884–7892. [CrossRef] [PubMed]
62. Lakowski, T.M.; Frankel, A. A kinetic study of human protein arginine N-methyltransferase 6 reveals a distributive mechanism. *J. Biol. Chem.* **2008**, *283*, 10015–10025. [CrossRef] [PubMed]
63. Pahlich, S.; Zakaryan, R.P.; Gehring, H. Protein arginine methylation: Cellular functions and methods of analysis. *Biochim. Biophys. Acta* **2006**, *1764*, 1890–1903. [CrossRef] [PubMed]
64. Chung, J.; Karkhanis, V.; Baiocchi, R.A.; Sif, S. Protein arginine methyltransferase 5 (PRMT5) promotes survival of lymphoma cells via activation of WNT/beta-catenin and AKT/GSK3beta proliferative signaling. *J. Biol. Chem.* **2019**, *294*, 7692–7710. [CrossRef] [PubMed]
65. Dowhan, D.H.; Harrison, M.J.; Eriksson, N.A.; Bailey, P.; Pearen, M.A.; Fuller, P.J.; Funder, J.W.; Simpson, E.R.; Leedman, P.J.; Tilley, W.D.; et al. Protein arginine methyltransferase 6-dependent gene expression and splicing: Association with breast cancer outcomes. *Endocr. Relat. Cancer* **2012**, *19*, 509–526. [CrossRef] [PubMed]
66. Bezzi, M.; Teo, S.X.; Muller, J.; Mok, W.C.; Sahu, S.K.; Vardy, L.A.; Bonday, Z.Q.; Guccione, E. Regulation of constitutive and alternative splicing by PRMT5 reveals a role for Mdm4 pre-mRNA in sensing defects in the spliceosomal machinery. *Genes Dev.* **2013**, *27*, 1903–1916. [CrossRef] [PubMed]
67. Brobbey, C.; Liu, L.; Yin, S.; Gan, W. The Role of Protein Arginine Methyltransferases in DNA Damage Response. *Int. J. Mol. Sci.* **2022**, *23*, 9780. [CrossRef]
68. Wei, X.; Yang, J.; Adair, S.J.; Ozturk, H.; Kuscu, C.; Lee, K.Y.; Kane, W.J.; O'Hara, P.E.; Liu, D.; Demirlenk, Y.M.; et al. Targeted CRISPR screening identifies PRMT5 as synthetic lethality combinatorial target with gemcitabine in pancreatic cancer cells. *Proc. Natl. Acad. Sci. USA* **2020**, *117*, 28068–28079. [CrossRef]
69. Le Romancer, M.; Treilleux, I.; Leconte, N.; Robin-Lespinasse, Y.; Sentis, S.; Bouchekioua-Bouzaghou, K.; Goddard, S.; Gobert-Gosse, S.; Corbo, L. Regulation of estrogen rapid signaling through arginine methylation by PRMT1. *Mol. Cell* **2008**, *31*, 212–221. [CrossRef]
70. Rezai-Zadeh, N.; Zhang, X.; Namour, F.; Fejer, G.; Wen, Y.D.; Yao, Y.L.; Gyory, I.; Wright, K.; Seto, E. Targeted recruitment of a histone H4-specific methyltransferase by the transcription factor YY1. *Genes Dev.* **2003**, *17*, 1019–1029. [CrossRef]
71. Yadav, N.; Cheng, D.; Richard, S.; Morel, M.; Iyer, V.R.; Aldaz, C.M.; Bedford, M.T. CARM1 promotes adipocyte differentiation by coactivating PPARgamma. *EMBO Rep.* **2008**, *9*, 193–198. [CrossRef]
72. Reintjes, A.; Fuchs, J.E.; Kremser, L.; Lindner, H.H.; Liedl, K.R.; Huber, L.A.; Valovka, T. Asymmetric arginine dimethylation of RelA provides a repressive mark to modulate TNFalpha/NF-kappaB response. *Proc. Natl. Acad. Sci. USA* **2016**, *113*, 4326–4331. [CrossRef]
73. Yao, B.; Gui, T.; Zeng, X.; Deng, Y.; Wang, Z.; Wang, Y.; Yang, D.; Li, Q.; Xu, P.; Hu, R.; et al. PRMT1-mediated H4R3me2a recruits SMARCA4 to promote colorectal cancer progression by enhancing EGFR signaling. *Genome Med.* **2021**, *13*, 58. [CrossRef] [PubMed]
74. Hou, Z.; Peng, H.; Ayyanathan, K.; Yan, K.P.; Langer, E.M.; Longmore, G.D.; Rauscher, F.J., 3rd. The LIM protein AJUBA recruits protein arginine methyltransferase 5 to mediate SNAIL-dependent transcriptional repression. *Mol. Cell. Biol.* **2008**, *28*, 3198–3207. [CrossRef]
75. Pal, S.; Yun, R.; Datta, A.; Lacomis, L.; Erdjument-Bromage, H.; Kumar, J.; Tempst, P.; Sif, S. mSin3A/histone deacetylase 2- and PRMT5-containing Brg1 complex is involved in transcriptional repression of the Myc target gene cad. *Mol. Cell. Biol.* **2003**, *23*, 7475–7487. [CrossRef]
76. Chen, H.; Lorton, B.; Gupta, V.; Shechter, D. A TGFbeta-PRMT5-MEP50 axis regulates cancer cell invasion through histone H3 and H4 arginine methylation coupled transcriptional activation and repression. *Oncogene* **2017**, *36*, 373–386. [CrossRef] [PubMed]
77. Migliori, V.; Muller, J.; Phalke, S.; Low, D.; Bezzi, M.; Mok, W.C.; Sahu, S.K.; Gunaratne, J.; Capasso, P.; Bassi, C.; et al. Symmetric dimethylation of H3R2 is a newly identified histone mark that supports euchromatin maintenance. *Nat. Struct. Mol. Biol.* **2012**, *19*, 136–144. [CrossRef]

78. Guccione, E.; Richard, S. The regulation, functions and clinical relevance of arginine methylation. *Nat. Rev. Mol. Cell. Biol.* **2019**, *20*, 642–657. [CrossRef] [PubMed]

79. Meister, G.; Fischer, U. Assisted RNP assembly: SMN and PRMT5complexes cooperate in the formation ofspliceosomal UsnRNPs. *EMBO J.* **2002**, *21*, 5853–5863. [CrossRef]

80. Larsen, S.C.; Sylvestersen, K.B.; Mund, A.; Lyon, D.; Mullari, M.; Madsen, M.V.; Daniel, J.A.; Jensen, L.J.; Nielsen, M.L. Proteome-wide analysis of arginine monomethylation reveals widespread occurrence in human cells. *Sci. Signal.* **2016**, *9*, rs9. [CrossRef]

81. Cheng, D.; Cote, J.; Shaaban, S.; Bedford, M.T. The arginine methyltransferase CARM1 regulates the coupling of transcription and mRNA processing. *Mol. Cell* **2007**, *25*, 71–83. [CrossRef] [PubMed]

82. Tradewell, M.L.; Yu, Z.; Tibshirani, M.; Boulanger, M.C.; Durham, H.D.; Richard, S. Arginine methylation by PRMT1 regulates nuclear-cytoplasmic localization and toxicity of FUS/TLS harbouring ALS-linked mutations. *Hum. Mol. Genet.* **2012**, *21*, 136–149. [CrossRef] [PubMed]

83. Murata, K.; Lu, W.; Hashimoto, M.; Ono, N.; Muratani, M.; Nishikata, K.; Kim, J.D.; Ebihara, S.; Ishida, J.; Fukamizu, A. PRMT1 Deficiency in Mouse Juvenile Heart Induces Dilated Cardiomyopathy and Reveals Cryptic Alternative Splicing Products. *iScience* **2018**, *8*, 200–213. [CrossRef]

84. Yu, Z.; Chen, T.; Hebert, J.; Li, E.; Richard, S. A mouse PRMT1 null allele defines an essential role for arginine methylation in genome maintenance and cell proliferation. *Mol. Cell. Biol.* **2009**, *29*, 2982–2996. [CrossRef] [PubMed]

85. Yu, Z.; Vogel, G.; Coulombe, Y.; Dubeau, D.; Spehalski, E.; Hebert, J.; Ferguson, D.O.; Masson, J.Y.; Richard, S. The MRE11 GAR motif regulates DNA double-strand break processing and ATR activation. *Cell Res.* **2012**, *22*, 305–320. [CrossRef] [PubMed]

86. Clarke, T.L.; Sanchez-Bailon, M.P.; Chiang, K.; Reynolds, J.J.; Herrero-Ruiz, J.; Bandeiras, T.M.; Matias, P.M.; Maslen, S.L.; Skehel, J.M.; Stewart, G.S.; et al. PRMT5-Dependent Methylation of the TIP60 Coactivator RUVBL1 Is a Key Regulator of Homologous Recombination. *Mol. Cell* **2017**, *65*, 900–916 e907. [CrossRef]

87. Hamard, P.J.; Santiago, G.E.; Liu, F.; Karl, D.L.; Martinez, C.; Man, N.; Mookhtiar, A.K.; Duffort, S.; Greenblatt, S.; Verdun, R.E.; et al. PRMT5 Regulates DNA Repair by Controlling the Alternative Splicing of Histone-Modifying Enzymes. *Cell Rep.* **2018**, *24*, 2643–2657. [CrossRef] [PubMed]

88. Hwang, J.W.; Kim, S.N.; Myung, N.; Song, D.; Han, G.; Bae, G.U.; Bedford, M.T.; Kim, Y.K. PRMT5 promotes DNA repair through methylation of 53BP1 and is regulated by Src-mediated phosphorylation. *Commun. Biol.* **2020**, *3*, 428. [CrossRef]

89. He, W.; Ma, X.; Yang, X.; Zhao, Y.; Qiu, J.; Hang, H. A role for the arginine methylation of Rad9 in checkpoint control and cellular sensitivity to DNA damage. *Nucleic Acids Res.* **2011**, *39*, 4719–4727. [CrossRef]

90. Rehman, I.; Basu, S.M.; Das, S.K.; Bhattacharjee, S.; Ghosh, A.; Pommier, Y.; Das, B.B. PRMT5-mediated arginine methylation of TDP1 for the repair of topoisomerase I covalent complexes. *Nucleic Acids Res.* **2018**, *46*, 5601–5617. [CrossRef]

91. El-Andaloussi, N.; Valovka, T.; Toueille, M.; Steinacher, R.; Focke, F.; Gehrig, P.; Covic, M.; Hassa, P.O.; Schar, P.; Hubscher, U.; et al. Arginine methylation regulates DNA polymerase beta. *Mol. Cell* **2006**, *22*, 51–62. [CrossRef] [PubMed]

92. Lee, Y.H.; Bedford, M.T.; Stallcup, M.R. Regulated recruitment of tumor suppressor BRCA1 to the p21 gene by coactivator methylation. *Genes Dev.* **2011**, *25*, 176–188. [CrossRef] [PubMed]

93. Sengupta, S.; Kennemer, A.; Patrick, K.; Tsichlis, P.; Guerau-de-Arellano, M. Protein Arginine Methyltransferase 5 in T Lymphocyte Biology. *Trends Immunol.* **2020**, *41*, 918–931. [CrossRef] [PubMed]

94. Kim, H.; Kim, H.; Feng, Y.; Li, Y.; Tamiya, H.; Tocci, S.; Ronai, Z.A. PRMT5 control of cGAS/STING and NLRC5 pathways defines melanoma response to antitumor immunity. *Sci. Transl. Med.* **2020**, *12*, eaaz5683. [CrossRef] [PubMed]

95. Tang, J.; Frankel, A.; Cook, R.J.; Kim, S.; Paik, W.K.; Williams, K.R.; Clarke, S.; Herschman, H.R. PRMT1 is the predominant type I protein arginine methyltransferase in mammalian cells. *J. Biol. Chem.* **2000**, *275*, 7723–7730. [CrossRef] [PubMed]

96. Choucair, A.; Pham, T.H.; Omarjee, S.; Jacquemetton, J.; Kassem, L.; Tredan, O.; Rambaud, J.; Marangoni, E.; Corbo, L.; Treilleux, I.; et al. The arginine methyltransferase PRMT1 regulates IGF-1 signaling in breast cancer. *Oncogene* **2019**, *38*, 4015–4027. [CrossRef] [PubMed]

97. Malbeteau, L.; Jacquemetton, J.; Languilaire, C.; Corbo, L.; Le Romancer, M.; Poulard, C. PRMT1, a Key Modulator of Unliganded Progesterone Receptor Signaling in Breast Cancer. *Int. J. Mol. Sci.* **2022**, *23*, 9509. [CrossRef] [PubMed]

98. Avasarala, S.; Van Scoyk, M.; Karuppusamy Rathinam, M.K.; Zerayesus, S.; Zhao, X.; Zhang, W.; Pergande, M.R.; Borgia, J.A.; DeGregori, J.; Port, J.D.; et al. PRMT1 Is a Novel Regulator of Epithelial-Mesenchymal-Transition in Non-small Cell Lung Cancer. *J. Biol. Chem.* **2015**, *290*, 13479–13489. [CrossRef]

99. Liao, H.W.; Hsu, J.M.; Xia, W.; Wang, H.L.; Wang, Y.N.; Chang, W.C.; Arold, S.T.; Chou, C.K.; Tsou, P.H.; Yamaguchi, H.; et al. PRMT1-mediated methylation of the EGF receptor regulates signaling and cetuximab response. *J. Clin. Investig.* **2015**, *125*, 4529–4543. [CrossRef]

100. Song, C.; Chen, T.; He, L.; Ma, N.; Li, J.A.; Rong, Y.F.; Fang, Y.; Liu, M.; Xie, D.; Lou, W. PRMT1 promotes pancreatic cancer growth and predicts poor prognosis. *Cell. Oncol.* **2020**, *43*, 51–62. [CrossRef]

101. Wang, Y.; Hsu, J.M.; Kang, Y.; Wei, Y.; Lee, P.C.; Chang, S.J.; Hsu, Y.H.; Hsu, J.L.; Wang, H.L.; Chang, W.C.; et al. Oncogenic Functions of Gli1 in Pancreatic Adenocarcinoma Are Supported by Its PRMT1-Mediated Methylation. *Cancer Res.* **2016**, *76*, 7049–7058. [CrossRef] [PubMed]

102. Wang, L.; Jia, Z.; Xie, D.; Zhao, T.; Tan, Z.; Zhang, S.; Kong, F.; Wei, D.; Xie, K. Methylation of HSP70 Orchestrates Its Binding to and Stabilization of BCL2 mRNA and Renders Pancreatic Cancer Cells Resistant to Therapeutics. *Cancer Res.* **2020**, *80*, 4500–4513. [CrossRef] [PubMed]

103. Giuliani, V.; Miller, M.A.; Liu, C.Y.; Hartono, S.R.; Class, C.A.; Bristow, C.A.; Suzuki, E.; Sanz, L.A.; Gao, G.; Gay, J.P.; et al. PRMT1-dependent regulation of RNA metabolism and DNA damage response sustains pancreatic ductal adenocarcinoma. *Nat. Commun.* **2021**, *12*, 4626. [CrossRef] [PubMed]
104. Zheng, N.N.; Zhou, M.; Sun, F.; Huai, M.X.; Zhang, Y.; Qu, C.Y.; Shen, F.; Xu, L.M. Combining protein arginine methyltransferase inhibitor and anti-programmed death-ligand-1 inhibits pancreatic cancer progression. *World J. Gastroenterol.* **2020**, *26*, 3737–3749. [CrossRef] [PubMed]
105. Lin, Z.; Chen, Y.; Lin, Z.; Chen, C.; Dong, Y. Overexpressing PRMT1 Inhibits Proliferation and Invasion in Pancreatic Cancer by Inverse Correlation of ZEB1. *IUBMB Life* **2018**, *70*, 1032–1039. [CrossRef] [PubMed]
106. Liu, Y.; Liu, H.; Ye, M.; Jiang, M.; Chen, X.; Song, G.; Ji, H.; Wang, Z.W.; Zhu, X. Methylation of BRD4 by PRMT1 regulates BRD4 phosphorylation and promotes ovarian cancer invasion. *Cell Death Dis.* **2023**, *14*, 624. [CrossRef] [PubMed]
107. Li, Z.; Wang, D.; Lu, J.; Huang, B.; Wang, Y.; Dong, M.; Fan, D.; Li, H.; Gao, Y.; Hou, P.; et al. Methylation of EZH2 by PRMT1 regulates its stability and promotes breast cancer metastasis. *Cell Death Differ.* **2020**, *27*, 3226–3242. [CrossRef] [PubMed]
108. Zheng, D.; Chen, D.; Lin, F.; Wang, X.; Lu, L.; Luo, S.; Chen, J.; Xu, X. LncRNA NNT-AS1 promote glioma cell proliferation and metastases through miR-494-3p/PRMT1 axis. *Cell Cycle* **2020**, *19*, 1621–1631. [CrossRef]
109. Hsu, M.C.; Tsai, Y.L.; Lin, C.H.; Pan, M.R.; Shan, Y.S.; Cheng, T.Y.; Cheng, S.H.; Chen, L.T.; Hung, W.C. Protein arginine methyltransferase 3-induced metabolic reprogramming is a vulnerable target of pancreatic cancer. *J. Hematol. Oncol.* **2019**, *12*, 79. [CrossRef]
110. Jiang, W.; Newsham, I.F. The tumor suppressor DAL-1/4.1B and protein methylation cooperate in inducing apoptosis in MCF-7 breast cancer cells. *Mol. Cancer* **2006**, *5*, 4. [CrossRef]
111. Zhi, R.; Wu, K.; Zhang, J.; Liu, H.; Niu, C.; Li, S.; Fu, L. PRMT3 regulates the progression of invasive micropapillary carcinoma of the breast. *Cancer Sci.* **2023**, *114*, 1912–1928. [CrossRef] [PubMed]
112. Hsu, M.C.; Pan, M.R.; Chu, P.Y.; Tsai, Y.L.; Tsai, C.H.; Shan, Y.S.; Chen, L.T.; Hung, W.C. Protein Arginine Methyltransferase 3 Enhances Chemoresistance in Pancreatic Cancer by Methylating hnRNPA1 to Increase ABCG2 Expression. *Cancers* **2018**, *11*, 8. [CrossRef] [PubMed]
113. Stopa, N.; Krebs, J.E.; Shechter, D. The PRMT5 arginine methyltransferase: Many roles in development, cancer and beyond. *Cell. Mol. Life Sci.* **2015**, *72*, 2041–2059. [CrossRef] [PubMed]
114. Berglund, L.; Bjorling, E.; Oksvold, P.; Fagerberg, L.; Asplund, A.; Szigyarto, C.A.; Persson, A.; Ottosson, J.; Wernerus, H.; Nilsson, P.; et al. A genecentric Human Protein Atlas for expression profiles based on antibodies. *Mol. Cell Proteom.* **2008**, *7*, 2019–2027. [CrossRef] [PubMed]
115. Lee, M.K.C.; Grimmond, S.M.; McArthur, G.A.; Sheppard, K.E. PRMT5: An Emerging Target for Pancreatic Adenocarcinoma. *Cancers* **2021**, *13*, 5136. [CrossRef] [PubMed]
116. Qin, Y.; Hu, Q.; Xu, J.; Ji, S.; Dai, W.; Liu, W.; Xu, W.; Sun, Q.; Zhang, Z.; Ni, Q.; et al. PRMT5 enhances tumorigenicity and glycolysis in pancreatic cancer via the FBW7/cMyc axis. *Cell Commun. Signal* **2019**, *17*, 30. [CrossRef] [PubMed]
117. Welcker, M.; Orian, A.; Jin, J.; Grim, J.E.; Harper, J.W.; Eisenman, R.N.; Clurman, B.E. The Fbw7 tumor suppressor regulates glycogen synthase kinase 3 phosphorylation-dependent c-Myc protein degradation. *Proc. Natl. Acad. Sci. USA* **2004**, *101*, 9085–9090. [CrossRef] [PubMed]
118. Fan, J.; Bellon, M.; Ju, M.; Zhao, L.; Wei, M.; Fu, L.; Nicot, C. Clinical significance of FBXW7 loss of function in human cancers. *Mol. Cancer* **2022**, *21*, 87. [CrossRef] [PubMed]
119. Orben, F.; Lankes, K.; Schneeweis, C.; Hassan, Z.; Jakubowsky, H.; Krauss, L.; Boniolo, F.; Schneider, C.; Schafer, A.; Murr, J.; et al. Epigenetic drug screening defines a PRMT5 inhibitor-sensitive pancreatic cancer subtype. *JCI Insight* **2022**, *7*, e151353. [CrossRef]
120. Ge, L.; Wang, H.; Xu, X.; Zhou, Z.; He, J.; Peng, W.; Du, F.; Zhang, Y.; Gong, A.; Xu, M. PRMT5 promotes epithelial-mesenchymal transition via EGFR-beta-catenin axis in pancreatic cancer cells. *J. Cell. Mol. Med.* **2020**, *24*, 1969–1979. [CrossRef]
121. Hsu, J.M.; Chen, C.T.; Chou, C.K.; Kuo, H.P.; Li, L.Y.; Lin, C.Y.; Lee, H.J.; Wang, Y.N.; Liu, M.; Liao, H.W.; et al. Crosstalk between Arg 1175 methylation and Tyr 1173 phosphorylation negatively modulates EGFR-mediated ERK activation. *Nat. Cell Biol.* **2011**, *13*, 174–181. [CrossRef] [PubMed]
122. Gao, G.; Zhang, L.; Villarreal, O.D.; He, W.; Su, D.; Bedford, E.; Moh, P.; Shen, J.; Shi, X.; Bedford, M.T.; et al. PRMT1 loss sensitizes cells to PRMT5 inhibition. *Nucleic Acids Res.* **2019**, *47*, 5038–5048. [CrossRef] [PubMed]
123. Zhang, H.; Chen, Z.H.; Savarese, T.M. Codeletion of the genes for p16INK4, methylthioadenosine phosphorylase, interferon-alpha1, interferon-beta1, and other 9p21 markers in human malignant cell lines. *Cancer Genet. Cytogenet.* **1996**, *86*, 22–28. [CrossRef]
124. Zappia, V.; Della Ragione, F.; Pontoni, G.; Gragnaniello, V.; Carteni-Farina, M. Human 5'-deoxy-5'-methylthioadenosine phosphorylase: Kinetic studies and catalytic mechanism. *Adv. Exp. Med. Biol.* **1988**, *250*, 165–177. [CrossRef] [PubMed]
125. Kryukov, G.V.; Wilson, F.H.; Ruth, J.R.; Paulk, J.; Tsherniak, A.; Marlow, S.E.; Vazquez, F.; Weir, B.A.; Fitzgerald, M.E.; Tanaka, M.; et al. MTAP deletion confers enhanced dependency on the PRMT5 arginine methyltransferase in cancer cells. *Science* **2016**, *351*, 1214–1218. [CrossRef] [PubMed]
126. Xu, J.; Richard, S. Cellular pathways influenced by protein arginine methylation: Implications for cancer. *Mol. Cell* **2021**, *81*, 4357–4368. [CrossRef] [PubMed]
127. Guccione, E.; Schwarz, M.; Di Tullio, F.; Mzoughi, S. Cancer synthetic vulnerabilities to protein arginine methyltransferase inhibitors. *Curr. Opin. Pharmacol.* **2021**, *59*, 33–42. [CrossRef] [PubMed]

128. Hu, H.; Qian, K.; Ho, M.C.; Zheng, Y.G. Small Molecule Inhibitors of Protein Arginine Methyltransferases. *Expert. Opin. Investig. Drugs* **2016**, *25*, 335–358. [CrossRef] [PubMed]

129. Cheng, D.; Yadav, N.; King, R.W.; Swanson, M.S.; Weinstein, E.J.; Bedford, M.T. Small molecule regulators of protein arginine methyltransferases. *J. Biol. Chem.* **2004**, *279*, 23892–23899. [CrossRef]

130. Spannhoff, A.; Heinke, R.; Bauer, I.; Trojer, P.; Metzger, E.; Gust, R.; Schule, R.; Brosch, G.; Sippl, W.; Jung, M. Target-based approach to inhibitors of histone arginine methyltransferases. *J. Med. Chem.* **2007**, *50*, 2319–2325. [CrossRef]

131. Sun, Y.; Wang, Z.; Yang, H.; Zhu, X.; Wu, H.; Ma, L.; Xu, F.; Hong, W.; Wang, H. The Development of Tetrazole Derivatives as Protein Arginine Methyltransferase I (PRMT I) Inhibitors. *Int. J. Mol. Sci.* **2019**, *20*, 3840. [CrossRef]

132. Eram, M.S.; Shen, Y.; Szewczyk, M.; Wu, H.; Senisterra, G.; Li, F.; Butler, K.V.; Kaniskan, H.U.; Speed, B.A.; Dela Sena, C.; et al. A Potent, Selective, and Cell-Active Inhibitor of Human Type I Protein Arginine Methyltransferases. *ACS Chem. Biol.* **2016**, *11*, 772–781. [CrossRef] [PubMed]

133. Fedoriw, A.; Rajapurkar, S.R.; O'Brien, S.; Gerhart, S.V.; Mitchell, L.H.; Adams, N.D.; Rioux, N.; Lingaraj, T.; Ribich, S.A.; Pappalardi, M.B.; et al. Anti-tumor Activity of the Type I PRMT Inhibitor, GSK3368715, Synergizes with PRMT5 Inhibition through MTAP Loss. *Cancer Cell* **2019**, *36*, 100–114.e125. [CrossRef]

134. El-Khoueiry, A.B.; Clarke, J.; Neff, T.; Crossman, T.; Ratia, N.; Rathi, C.; Noto, P.; Tarkar, A.; Garrido-Laguna, I.; Calvo, E.; et al. Phase 1 study of GSK3368715, a type I PRMT inhibitor, in patients with advanced solid tumors. *Br. J. Cancer* **2023**, *129*, 309–317. [CrossRef] [PubMed]

135. Shen, Y.; Szewczyk, M.M.; Eram, M.S.; Smil, D.; Kaniskan, H.U.; de Freitas, R.F.; Senisterra, G.; Li, F.; Schapira, M.; Brown, P.J.; et al. Discovery of a Potent, Selective, and Cell-Active Dual Inhibitor of Protein Arginine Methyltransferase 4 and Protein Arginine Methyltransferase 6. *J. Med. Chem.* **2016**, *59*, 9124–9139. [CrossRef]

136. Shen, Y.; Li, F.; Szewczyk, M.M.; Halabelian, L.; Park, K.S.; Chau, I.; Dong, A.; Zeng, H.; Chen, H.; Meng, F.; et al. Discovery of a First-in-Class Protein Arginine Methyltransferase 6 (PRMT6) Covalent Inhibitor. *J. Med. Chem.* **2020**, *63*, 5477–5487. [CrossRef]

137. Alinari, L.; Mahasenan, K.V.; Yan, F.; Karkhanis, V.; Chung, J.H.; Smith, E.M.; Quinion, C.; Smith, P.L.; Kim, L.; Patton, J.T.; et al. Selective inhibition of protein arginine methyltransferase 5 blocks initiation and maintenance of B-cell transformation. *Blood* **2015**, *125*, 2530–2543. [CrossRef] [PubMed]

138. Chan-Penebre, E.; Kuplast, K.G.; Majer, C.R.; Boriack-Sjodin, P.A.; Wigle, T.J.; Johnston, L.D.; Rioux, N.; Munchhof, M.J.; Jin, L.; Jacques, S.L.; et al. A selective inhibitor of PRMT5 with in vivo and in vitro potency in MCL models. *Nat. Chem. Biol.* **2015**, *11*, 432–437. [CrossRef]

139. Mitchell, L.H.; Drew, A.E.; Ribich, S.A.; Rioux, N.; Swinger, K.K.; Jacques, S.L.; Lingaraj, T.; Boriack-Sjodin, P.A.; Waters, N.J.; Wigle, T.J.; et al. Aryl Pyrazoles as Potent Inhibitors of Arginine Methyltransferases: Identification of the First PRMT6 Tool Compound. *ACS Med. Chem. Lett.* **2015**, *6*, 655–659. [CrossRef]

140. Iyamu, I.D.; Al-Hamashi, A.A.; Huang, R. A Pan-Inhibitor for Protein Arginine Methyltransferase Family Enzymes. *Biomolecules* **2021**, *11*, 854. [CrossRef]

141. Szewczyk, M.M.; Ishikawa, Y.; Organ, S.; Sakai, N.; Li, F.; Halabelian, L.; Ackloo, S.; Couzens, A.L.; Eram, M.; Dilworth, D.; et al. Pharmacological inhibition of PRMT7 links arginine monomethylation to the cellular stress response. *Nat. Commun.* **2020**, *11*, 2396. [CrossRef]

142. Dong, J.; Duan, J.; Hui, Z.; Garrido, C.; Deng, Z.; Xie, T.; Ye, X.Y. An updated patent review of protein arginine N-methyltransferase inhibitors (2019–2022). *Expert. Opin. Ther. Pat.* **2022**, *32*, 1185–1205. [CrossRef] [PubMed]

143. Yin, S.; Liu, L.; Brobbey, C.; Palanisamy, V.; Ball, L.E.; Olsen, S.K.; Ostrowski, M.C.; Gan, W. PRMT5-mediated arginine methylation activates AKT kinase to govern tumorigenesis. *Nat. Commun.* **2021**, *12*, 3444. [CrossRef] [PubMed]

144. Huang, L.; Zhang, X.O.; Rozen, E.J.; Sun, X.; Sallis, B.; Verdejo-Torres, O.; Wigglesworth, K.; Moon, D.; Huang, T.; Cavaretta, J.P.; et al. PRMT5 activates AKT via methylation to promote tumor metastasis. *Nat. Commun.* **2022**, *13*, 3955. [CrossRef]

145. Kugeratski, F.G.; Hodge, K.; Lilla, S.; McAndrews, K.M.; Zhou, X.; Hwang, R.F.; Zanivan, S.; Kalluri, R. Quantitative proteomics identifies the core proteome of exosomes with syntenin-1 as the highest abundant protein and a putative universal biomarker. *Nat. Cell Biol.* **2021**, *23*, 631–641. [CrossRef] [PubMed]

146. Hinzman, C.P.; Singh, B.; Bansal, S.; Li, Y.; Iliuk, A.; Girgis, M.; Herremans, K.M.; Trevino, J.G.; Singh, V.K.; Banerjee, P.P.; et al. A multi-omics approach identifies pancreatic cancer cell extracellular vesicles as mediators of the unfolded protein response in normal pancreatic epithelial cells. *J. Extracell. Vesicles* **2022**, *11*, e12232. [CrossRef] [PubMed]

147. Hannafon, B.N.; Ding, W.Q. Intercellular Communication by Exosome-Derived microRNAs in Cancer. *Int. J. Mol. Sci.* **2013**, *14*, 14240–14269. [CrossRef]

148. Hoshino, A.; Costa-Silva, B.; Shen, T.L.; Rodrigues, G.; Hashimoto, A.; Tesic Mark, M.; Molina, H.; Kohsaka, S.; Di Giannatale, A.; Ceder, S.; et al. Tumour exosome integrins determine organotropic metastasis. *Nature* **2015**, *527*, 329–335. [CrossRef] [PubMed]

149. Radzisheuskaya, A.; Shliaha, P.V.; Grinev, V.; Lorenzini, E.; Kovalchuk, S.; Shlyueva, D.; Gorshkov, V.; Hendrickson, R.C.; Jensen, O.N.; Helin, K. PRMT5 methylome profiling uncovers a direct link to splicing regulation in acute myeloid leukemia. *Nat. Struct. Mol. Biol.* **2019**, *26*, 999–1012. [CrossRef]

150. Wang, Q.; Li, Z.; Zhang, S.; Li, Y.; Wang, Y.; Fang, Z.; Ma, Y.; Liu, Z.; Zhang, W.; Li, D.; et al. Global profiling of arginine dimethylation in regulating protein phase separation by a steric effect-based chemical-enrichment method. *Proc. Natl. Acad. Sci. USA* **2022**, *119*, e2205255119. [CrossRef]

151. Tsai, W.C.; Gayatri, S.; Reineke, L.C.; Sbardella, G.; Bedford, M.T.; Lloyd, R.E. Arginine Demethylation of G3BP1 Promotes Stress Granule Assembly. *J. Biol. Chem.* **2016**, *291*, 22671–22685. [CrossRef] [PubMed]

152. Hofweber, M.; Hutten, S.; Bourgeois, B.; Spreitzer, E.; Niedner-Boblenz, A.; Schifferer, M.; Ruepp, M.D.; Simons, M.; Niessing, D.; Madl, T.; et al. Phase Separation of FUS Is Suppressed by Its Nuclear Import Receptor and Arginine Methylation. *Cell* **2018**, *173*, 706–719.e713. [CrossRef]

153. Liu, X.M.; Ma, L.; Schekman, R. Selective sorting of microRNAs into exosomes by phase-separated YBX1 condensates. *eLife* **2021**, *10*, e71982. [CrossRef] [PubMed]

154. Suetsugu, A.; Honma, K.; Saji, S.; Moriwaki, H.; Ochiya, T.; Hoffman, R.M. Imaging exosome transfer from breast cancer cells to stroma at metastatic sites in orthotopic nude-mouse models. *Adv. Drug Deliv. Rev.* **2013**, *65*, 383–390. [CrossRef] [PubMed]

155. Taylor, D.D.; Gercel-Taylor, C. MicroRNA signatures of tumor-derived exosomes as diagnostic biomarkers of ovarian cancer. *Gynecol. Oncol.* **2008**, *110*, 13–21. [CrossRef]

156. Bhandari, K.; Kong, J.S.; Morris, K.; Xu, C.; Ding, W.Q. Protein Arginine Methylation Patterns in Plasma Small Extracellular Vesicles Are Altered in Patients with Early-Stage Pancreatic Ductal Adenocarcinoma. *Cancers* **2024**, *16*, 654. [CrossRef]

Disclaimer/Publisher's Note: The statements, opinions and data contained in all publications are solely those of the individual author(s) and contributor(s) and not of MDPI and/or the editor(s). MDPI and/or the editor(s) disclaim responsibility for any injury to people or property resulting from any ideas, methods, instructions or products referred to in the content.

International Journal of
Molecular Sciences

MDPI

Review

Copy Number Variations in Pancreatic Cancer: From Biological Significance to Clinical Utility

Daisy J. A. Oketch, Matteo Giulietti * and Francesco Piva *

Department of Specialistic Clinical and Odontostomatological Sciences, Polytechnic University of Marche, 60131 Ancona, Italy
* Correspondence: m.giulietti@univpm.it (M.G.); f.piva@univpm.it (F.P.)

Abstract: Pancreatic ductal adenocarcinoma (PDAC) is the most common type of pancreatic cancer, characterized by high tumor heterogeneity and a poor prognosis. Inter- and intra-tumoral heterogeneity in PDAC is a major obstacle to effective PDAC treatment; therefore, it is highly desirable to explore the tumor heterogeneity and underlying mechanisms for the improvement of PDAC prognosis. Gene copy number variations (CNVs) are increasingly recognized as a common and heritable source of inter-individual variation in genomic sequence. In this review, we outline the origin, main characteristics, and pathological aspects of CNVs. We then describe the occurrence of CNVs in PDAC, including those that have been clearly shown to have a pathogenic role, and further highlight some key examples of their involvement in tumor development and progression. The ability to efficiently identify and analyze CNVs in tumor samples is important to support translational research and foster precision oncology, as copy number variants can be utilized to guide clinical decisions. We provide insights into understanding the CNV landscapes and the role of both somatic and germline CNVs in PDAC, which could lead to significant advances in diagnosis, prognosis, and treatment. Although there has been significant progress in this field, understanding the full contribution of CNVs to the genetic basis of PDAC will require further research, with more accurate CNV assays such as single-cell techniques and larger cohorts than have been performed to date.

Keywords: pancreatic ductal adenocarcinoma (PDAC); copy number variations (CNVs); non-allelic homologous recombination (NAHR); patient stratification

Citation: Oketch, D.J.A.; Giulietti, M.; Piva, F. Copy Number Variations in Pancreatic Cancer: From Biological Significance to Clinical Utility. *Int. J. Mol. Sci.* **2024**, *25*, 391. https://doi.org/10.3390/ijms25010391

Academic Editors: Claudio Luchini and Donatella Delle Cave

Received: 24 November 2023
Revised: 20 December 2023
Accepted: 24 December 2023
Published: 27 December 2023

Copyright: © 2023 by the authors. Licensee MDPI, Basel, Switzerland. This article is an open access article distributed under the terms and conditions of the Creative Commons Attribution (CC BY) license (https://creativecommons.org/licenses/by/4.0/).

1. Introduction

Pancreatic ductal adenocarcinoma (PDAC) is the most prevalent type of pancreatic cancer. Due to the absence of early symptoms and the lack of effective and reliable meth-ods for early diagnosis and screening, the majority of the patients (80–85%) present distant metastatic or locally advanced disease that is not resectable [1], with an overall 5-year survival rate of 12% [2]. PDAC thus remains one of the most challenging and aggressive malignancies facing oncologists today and has been projected to become the second leading cause of cancer death by 2030 [3]. A comprehensive understanding of the biology of the disease is therefore urgently needed as part of an effort to develop more effective therapy and improve survival.

Genetic variations have been appreciated since the emergence of molecular genetics. In the human genome, they are present in various forms, such as mutations, variable number of tandem repeats (VNTRs), transposable elements, structural alterations, insertion and deletion variations (indels), and single nucleotide polymorphisms/variations (SNPs/SNVs). SNPs were previously believed to be the predominant type of genomic variation responsible for most of the phenotypic variability. However, the Human Genome Project identified DNA sequence variations other than SNPs and collectively named them copy number variations (CNVs). They include translocations of various segments of a chromosome and deletions and insertions of nucleotides [4].

Among the cancer-associated genetic variations, mutations have been the best characterized. More recently, however, thanks to new sequencing techniques, the roles of ge-nomic recombinations, such as CNVs, in tumor onset, heterogeneity, and prognosis have also emerged [5]. For this reason, we report the involvement of CNVs in PDAC develop-ment and progression.

1.1. Classification of CNVs

Copy number variations refer to a phenomenon in which segments of the genome are repeated or deleted, with varying numbers of these repeats among different individuals' genomes. Observations made in 2006, when the first comprehensive human haplotype map (HapMap) project Phase II of the human genome was constructed by Redon et al. [6], revealed that CNVs cover 12% of the human genome (about 360 Mb pairs), most of which are small-size rearrangements (<20 kb). The CNVs lay in both coding and non-coding regions, encompassing hundreds of genes and other functional elements. When the frequency of a CNV is less than 1%, it is a rare CNV, as opposed to common or polymorphic CNVs, which have a frequency >1% [7].

Researchers generally distinguish CNVs into two categories, depending on the length of the sequence affected [8]. The first category consists of copy number polymorphisms (CNPs), which are prevalent in the general population, with the majority being less than 10 kb in length and frequently enriched for genes encoding proteins that are important in immunity and drug detoxification. Therefore, these CNVs have well documented roles in evolutionary adaptation to new environmental niches [4,9].

The second category consists of relatively rare variants that are longer than CNPs, having up to over a million base pairs. These variants, also referred to as microduplications (smaller than 5 Mb) and microdeletions [8], can arise within a family during the development of the oocyte or spermatozoa that give rise to a specific individual and be passed down to offspring.

Copy number variants have also been divided into three groups depending on their origin: (i) de novo CNVs, newly acquired but not present in a parent; (ii) germline CNVs, inherited and present in a parent; and (iii) somatic CNVs, meaning that they occurred after the single-cell stage of an embryo [10]. For example, although monozygotic (MZ) twins are expected to be genetically identical, one study on 19 pairs of MZ twins revealed many different CNVs among them and suggested that these variations may have occurred during somatic development [11]. Somatic mutations were also observed in 10–20% of the nucleated blood cells of the MZ twins [11]. CNVs have also been observed between different tissues of the same individual, further supporting the idea that CNVs can occur in either somatic or meiotic tissues [12]. Further studies on age-stratified MZ twins and single-born subjects [13] as well as on DNA samples (mainly from peripheral blood) of more than 50,000 individuals genotyped for the Gene-Environment Association Studies (GENEVA) consortium [14,15] have revealed the accumulation of CNVs with age in the nuclear genome of blood cells [13,15]. Data from population genetics analysis of CNVs and SNPs, collected in the HapMap project, showed that over 99% of the observed copy number variations of individuals are due to inheritance rather than new mutations, and nearly 80% of the former are due to common CNVs [10].

1.2. Mechanisms of CNV Formation

To date, several different mechanisms have been shown to be involved in the development of CNVs, including germline genomic rearrangements that result in losses or gains of DNA segments [16].

1.2.1. Genomic Factors and Molecular Mechanisms of CNV Formation

CNVs are produced through a variety of mutational mechanisms, including those connected to DNA replication, repair, and recombination. Although the mechanisms underlying the formation of CNVs are not completely understood, the fact that they preferentially occur within or near duplicated sequences such as long interspersed nuclear elements (LINEs) and short interspersed nuclear elements (SINEs) has provided some clues to their origin [8]. During meiosis, the presence of different repetitive DNA sequences (low copy repeats, LCRs) in male and female homologous chromosomes at non-corresponding positions (i.e., that are not alleles but share significant sequence homology) can "mislead" the recombination machinery and result in an unequal crossing-over event. This aberrant recombination, known as non-allelic homologous recombination (NAHR) [17], leads to the loss or gain of copies of genomic segments [18] (Figure 1).

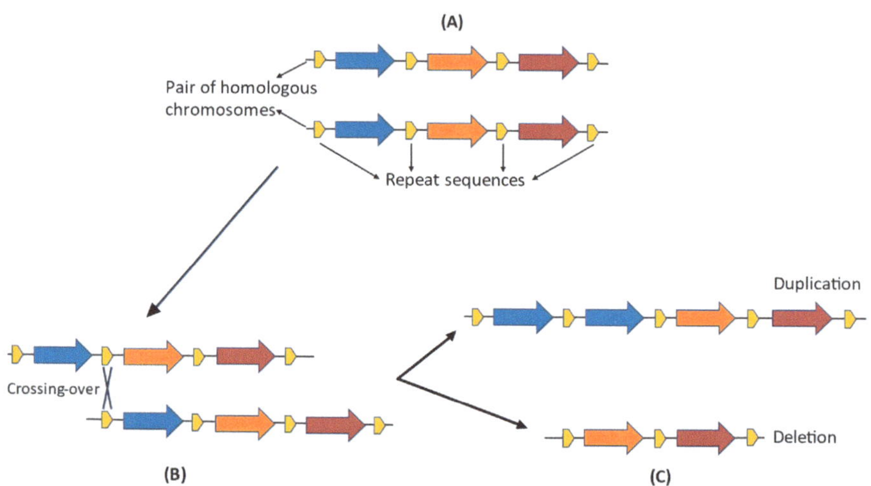

Figure 1. Non-allelic homologous recombination (NAHR). (**A**) Normal alignment of homologous chromosomes. (**B**) Misalignment of homologous chromosomes before crossing over, for example, due to abnormal pairing between repetitive sequences with high sequence identity. (**C**) Duplication and deletion that result from the unequal crossing over.

Other molecular mechanisms proposed to be responsible for the formation of CNVs include the (i) replication Fork Stalling and Template Switching (FoSTeS) model [19,20], which suggests that the stalling of a replication fork can cause the lagging strand to disengage from its original template and, owing to microhomology, invade and switch to another active replication fork's template, where it restarts DNA synthesis. The occurrence of a deletion or duplication is determined by the location of the ectopic association, and the nascent lagging strand has the potential for further disengagement and invasion of other replication forks. FoSTeS happens during DNA replication and can therefore occur either in mitosis or meiosis. (ii) Microhomology-mediated end joining (MMEJ) and non-homologous end joining (NHEJ) mechanisms [19,21], which can lead to some chromosomal rearrangements by joining nonhomologous sequences during the repair of DNA double-strand breaks (DSBs). In particular, these damages prompt NHEJ- and MMEJ-associated proteins to repair and ligate DNA sequences together. Sequence deletions, or duplications, can occur when fragments from different chromosomes are joined together. NHEJ and MMEJ occur throughout the cell cycle. Not all DSBs result in chromosomal rearrangements since they can be repaired through homologous recombination (HR) [22,23].

1.2.2. Environmental Factors in CNV Formation

It is unclear how environmental factors contribute to the emergence of CNVs. However, various studies have demonstrated that chemical and physical mutagens can induce the formation of CNVs and that chemical mutagens generate copy number losses more frequently than gains, while ionizing radiation induces deletions and duplications equally across the human genome [16].

Replication stress caused by chemical mutagens, such as hydroxyurea (HU), which is a ribonucleotide-reductase inhibitor as well as an important drug for the treatment of various diseases, including sickle-cell disease, has been demonstrated to induce the formation of de novo CNVs in the human genome [24].

Physical mutagens such as ionizing radiation (including X-ray, gamma ray, and ultraviolet light) can also induce de novo CNVs through a replication-dependent mechanism because the DNA strand breaks due to radiation, which may cause the replication fork to collapse [16,25,26]. A study by Costa et al. [27] demonstrated that the offspring of parents who had exposure to low doses of ionizing radiation had a 1.5-fold higher germline CNV mutation load. In a retrospective analysis of human populations exposed to low doses of ionizing radiation, the load of de novo CNVs has been demonstrated to be a helpful biomarker of parental exposure. Another study on CNVs in papillary thyroid carcinoma (PTC) among victims of the Chernobyl accident [28] reported that following their exposure to radiation from this disaster, PTC significantly increased in the irradiated individuals. Further studies of these radiation-induced PTCs revealed more multiple aberrations of the chromosome structure than in spontaneous thyroid tumors [29,30].

1.3. Distribution of CNVs

Research has revealed that CNVs are more common in genes that play a role in brain development and activity and in the immune system [6], functions that have evolved rapidly in humans. In contrast, CNVs tend to be rare in genes involved in early development and basic cellular activities since the alteration of essential cellular functions can have adverse effects, suggesting that their genes could have been subjected to powerful purification selection associated with copy number variation [31].

Some scientists propose that CNVs are not random in the human genome but rather tend to cluster in areas of complex genomic architecture. These proposed hotspot regions where CNVs are enriched [20] comprise complex patterns of inverted and direct low-copy repeats (LCRs) as well as high-copy repeats (e.g., SINEs, LINEs). LCRs provide the homology required for recombination that causes NAHR-mediated modifications. LINEs and SINEs are retrotransposons that contribute to the CNVs by NAHR either because of persistent single-strandedness (e.g., due to replication pausing, secondary structures, or extensive transcription) or frequent DNA breaks in these regions (e.g., due to live transposon activity), which make them potential sites for annealing by single-stranded DNA ends [11,20].

The question of whether the localization of CNVs in the human genome is random or not is still a highly debated topic [20,32–35], but more recent studies highlight a random distribution [9,36,37].

1.4. Identification and Detection of CNVs

Over the years, "targeted" approaches (single gene or single panel testing) or "whole" approaches (whole genome or whole exome) have been used to detect CNVs.

1.4.1. "Whole" Approaches

The process of microarray technology involves the immobilization of specific probes on a solid support, which then hybridize with target DNA segments. The two most widely used microarray technologies are array-CGH and SNP-array. In aCGH, a test sample and a reference sample are compared by labeling their genomic DNA (gDNA) with two different fluorescent dyes and applying them to an array of probes to detect differences in fluorescence intensity. On the contrary, SNP arrays consist of oligonucleotide DNA probes that correspond to regions in the genome exhibiting SNPs among individuals and do not require the use of reference sample DNA. CNV location and organization of structural variants (SVs) are not determined by microarray methods, making it necessary to subsequently perform FISH [38,39].

Next Generation Sequencing (NGS) technology involves the sequencing of highly fragmented DNA molecules to produce "reads", which are then mapped to a human reference genome using bioinformatics software. After alignment, any differences between the newly sequenced reads and the reference genome can be identified, and the "dosage" of that specific DNA fragment and the presence of CNVs may be calculated using the number of reads generated [40]. Currently, there are four distinct methods used for the detection of CNVs from NGS data [41,42]. These methods include read-depth-based detection (RD), paired-end mapping-based detection (PE), de novo assembly-based detection (DA), and split read-based detection (SR), of which RD is the most used. Details regarding the operation of these methods have been described in other studies [43–45].

1.4.2. "Targeted" Approaches

These involve the analysis of only one single gene or a group of genes, whether they are scattered across the genome or adjacent, and identifying a long chromosomal trait. Targeted approaches include Southern blot, fluorescent in situ hybridization, quantitative polymerase chain reaction, and multiplex ligation-dependent probe amplification.

Born in the seventies, Southern blot is still useful in the detection of some CNVs with high or extremely high numbers of repeats, particularly in the diagnosis of repeat expansion diseases. Restrictions endonucleases are used to fragment the target DNA, followed by electrophoresis to separate the resulting fragments [46]. These fragments are then incubated with DNA probes labeled either by incorporating radioactivity or by tagging the molecules with a chromogenic or fluorescent dye. CNVs are detected by comparing the hybridization intensities between a normal control and unknown samples [46] and/or by observing changes in fragment sizes (differentiated by length) and mobility following the hybridization and electrophoresis steps [39].

The fluorescent in situ hybridization (FISH) technique utilizes fluorochrome-labeled probes to match with chromosomes on a plate in order to detect any CNVs or translocations affecting a specific chromosomal region. FISH has high levels of sensitivity and specificity and is capable of detecting deletions, duplications, and translocations. However, FISH is limited in its ability to detect small imbalances and cannot be used to scan the entire karyotype without prior knowledge of the target region and appropriate probe selection. FISH can determine the location of CNVs identified by microarrays, NGS, and WGS [38,39].

Quantitative polymerase chain reaction (qPCR), also known as real-time PCR (rt-PCR), measures the accumulation of PCR amplicons in real time [39] by use of fluorescent probes. For the quantification of CNVs, a test locus with an unknown copy number and a reference locus with a known copy number are amplified in qPCR. Fluorescence intensity increases in direct proportion to the quantity of amplicon generated in each PCR cycle, and by determining the number of cycles needed to reach a specific threshold level of fluorescence, the quantity of the initial template can be determined [47].

The multiplex ligation-dependent probe amplification (MLPA) technique is based on the hybridization and ligation of specific DNA regions with two adjacently located complementary probes, followed by multiple PCR using a single pair of fluorescent primers. In particular, primer pairs containing identical 5′ sequences are used to amplify the target DNA sequences, followed by pooling into a probe mix [48] since all probes possess the same 5′ sequences. The PCR products are then separated by a capillary sequencer based on their size, and the resulting fluorescence intensities are exported for further analysis [48]. This method's capability to analyze sequences of high identity is greatly attributed to the sensitivity of the ligation step, which allows for the design of probes containing mismatches at the ligation site. MLPA can detect CNVs at multiple loci (>40) from relatively low amounts of genomic DNA [49] and is gaining popularity due to its simplicity, fast execution, cost efficiency, and robustness.

1.5. Implications of CNVs

Significant human inter-population variations in gene copy number, as reported by Redon et al. [6] and Jakobsson et al. [50], suggest that CNVs may be involved in adaptation to various environments, evolution, and susceptibility to common diseases.

In several organisms, a large number of CNVs have been reported in genes with tissue-specific expression rather than in genes that are widely expressed and may have housekeeping activities. The evolution of myoglobin, hemoglobin, trichromatic vision, and olfactory genes are a few of the most often mentioned instances of evolutionarily significant CNVs in humans that conform to this concept. The amylase (AMY) gene family, an enzyme that digests starch, can be used as a multifaceted example in understanding evolutionary processes mediated by CNVs. The number of copies of the AMY genes in modern humans differs from those in other primates and even other species of early humans. The current human population has up to 20 copies of the alpha-amylase 1 gene (AMY1) [51], unlike Neanderthals, who had only two copies. Given that gene expression is affected by the number of its copies and that the copy number variation of the AMY1 gene has been linked to diet, it is an example of recent human evolutionary adaptation. This suggests that our lineage evolved specific adaptations to digest foods rich in starch, foods of increasing importance in our diet [52].

CNVs and other variations of the human genome play an important role in human health and disease. Considering that CNVs occur throughout the genome and can cover a large number of genes and regulatory regions, pathogenic CNVs have been associated with genomic disorders and syndromes as well as complex multifactorial diseases including neurodevelopmental, neurodegenerative, autoimmune, and cardiovascular diseases [16].

There is a common basis and high similarity in the mechanisms through which CNVs can cause disease and yet contribute to evolution. Given that copies of redundant genes can acquire new roles, duplications (or multiplications in any number) are the most commonly mentioned mechanisms considered as key sources of evolutionary variation. If fitness is not compromised because the duplicated gene is not dosage-sensitive, one copy of a gene may retain its original function while the other copy escapes selective pressure, continuously undergoes mutation, and can even develop a new and different function [16].

CNVs also act on evolution and disease through other processes [35], including:

(i)   direct influence on the expression of a gene product, giving rise to changing levels of a protein. For example, Miller et al. [53] demonstrated an almost perfect correlation between the α-synuclein (SNCA) gene dosage and its mRNA and protein levels in Parkinson disease. SNCA triplication resulted in a doubling in the effective load of the normal gene and increased deposition of aggregated forms of the protein level in the brain into insoluble fractions.

(ii)   alteration of regulatory regions due to CNVs on non-coding sequences. This directly influences the levels and timing of expression and the cellular localization of the related protein. For example, the regulation of SOX9 gene expression in the testis is governed by a set of regulatory elements (RevSex and XYSR) located upstream of its promoter [54,55]. Loss of one or both of these regions in an XY individual results in a loss of SOX9 expression and male-to-female sex reversal [55], while duplication of the RevSex region in an XX individual could increase SOX9 expression and lead to female-to-male sex reversal [56–59].

(iii)   recombination of functional domains of different genes, leading to the formation of modified or new products with newly acquired functions, as seen in the example of glucocorticoid-remediable aldosteronism (GRA). Some researchers have shown that it is caused by a chimeric 11 β-hydroxylase (CYP11B1)/aldosterone synthase (CYP11B2) gene formed when a gene duplication resulting from unequal crossing over fuses the 5′ regulatory region of 11/β-hydroxylase to the coding sequences of aldosterone synthase [60]. The ectopic expression of CYP11B2 in the adrenal zona fasciculata may be responsible for these abnormalities because the gene is normally only expressed in the adrenal zona glomerulosa [35,60].

1.6. CNVs and Cancer

Cancer refers to a group of diseases characterized by the uncontrolled proliferation of certain cells in the body with the possibility of invasion or spreading to other parts of the body. The uncontrolled proliferation is due to dysregulation in the activity and expression of genes that control this function [61]. Somatic or germline mutations in tumor suppressor genes and oncogenes are the most well-known causes of cancer. With the increasing use of whole-genome techniques, somatic and germline CNVs have also been recognized as genomic alterations that lead to cancer development [5].

Germline CNVs are present in egg or sperm cells and can be passed down from parent to offspring. If they involve particular genes, an individual can be significantly predisposed to inherited cancers [35] as a result of alterations in DNA repair processes [20,62] or variations in the gene dosage of oncogenes and tumor suppressor genes [63]. Using a hereditary cancer panel to detect cancer susceptibility, Genekor's Medical S.A. laboratory evaluated a total of 2163 patients [62]. Of these, 1785 had breast cancer, 267 had ovarian cancer, and 111 had colon cancer. NGS and MLPA techniques revealed 464 samples (21.5%) to have pathogenic/likely pathogenic variants (P/LP), referring to alterations in DNA that are predicted to result in a known genetic condition, of which 10.8% (50/464) were attributed to CNVs. Notably, CNVs accounted for 10.2% (37/362) and 6.8% (5/74) of pathogenic variants in breast and ovarian cancer patients, respectively. Meanwhile, in colorectal cancer patients, CNVs were responsible for 28.6% (8/28) of P/LP variants. Out of the 50 CNVs found, 8% were in a low-risk cancer gene (8% FANCA), 20% in moderate-risk genes (4% ATM, 16% CHEK2), and 72% in high-risk genes (2% BRCA2, 8% MSH2, 8% PMS2, and 54% BRCA1) [62].

Somatic CNVs are those present only in particular cells and are primarily nonhereditary. They are acquired during an individual's lifespan, mostly as a result of environmental factors or errors in cell division. Somatic CNVs are classified as either large-scale variants or focal variants based on their size. Both types are important in the context of disease, but focal variants are considered more suitable for identifying candidate driver genes due to their relatively small size and low gene content [16]. Genome-wide analysis using high-resolution SNP arrays is currently being used to define the extent of somatic CNVs in cancer genomes. This has enabled the observation of a more immediate and direct role of these CNVs in the cancer cells themselves, whereby the cancer cells often display differential gene expression, especially of oncogenes and tumor suppressor genes [64].

Common cancer CNVs. In addition to phenotypic influence, CNVs that are common in the healthy population are also likely to play a role in carcinogenesis. In one correlation study between common CNVs and malignancy [5], all known CNVs in the normal human genome whose loci coincide with those of cancer-related genes such as ERBB2 and TP53 (as cataloged by [65]) were mapped and named common cancer CNVs. Although all gene regions are usually thought to be little affected by CNVs [6], it was surprising that 49 cancer-related genes were found to be directly overlapped or encompassed by a CNV in many individuals from the large reference population of 770 healthy genomes [5]. Each of the common cancer CNVs only slightly increases the risk of disease, but collectively, they can induce a significantly elevated risk [5].

Rare cancers are CNVs. These are the rare CNVs (with a population frequency of <1%) observed in cancer-related genes. Most are associated with hereditary cancer syndromes and involve genes such as FANCA in Fanconi anemia A, CHEK2 in familial breast cancer, RB1 in familial retinoblastoma, and MSH6 in hereditary non-polyposis colorectal cancer [5,66]. There are more than 200 cancer syndromes, most of which arise infrequently, and they account for approximately 5–10% of all cancers [67]. Rare cancer CNVs are often highly penetrant on their own, exhibit autosomal dominant inheritance, and will most often show co-segregation with the disease in families in contrast to low-penetrance alleles [67].

2. CNVs and Pancreatic Ductal Adenocarcinoma

Although mutations are recognized as the most commonly known genetic altera-tions able to cause cancer, genomic alterations such as CNVs are also playing an emergent role. Here we focus on the implications of CNVs in PDAC.

2.1. Mechanisms of Pancreatic Cancer Pathogenesis

Whole-genome sequencing has revealed that somatic mutations in oncogenic genes such as KRAS and loss-of-function mutations in tumor suppressor genes such as TP53, CDNK2A, and SMAD4 are the main drivers of PDAC [68]. Other causes of PDAC include (i) epigenetic modifications, which in turn lead to altered transcriptional reprogramming; and (ii) chromosomal alterations [69].

While most PDACs arise sporadically, up to 10% occur in patients with familial and hereditary predispositions. For instance, patients are more likely to develop PDAC if they have germline mutations in the BRCA1, BRCA2, PRSS1, or mismatch repair genes [70]. Even though initial correlation studies showed no significant association between CNVs and PDAC tumorigenesis and progression [71], current research has revealed associations between sporadic and familial pancreatic cancer (FPC) with CNVs [72,73] (Figure 2).

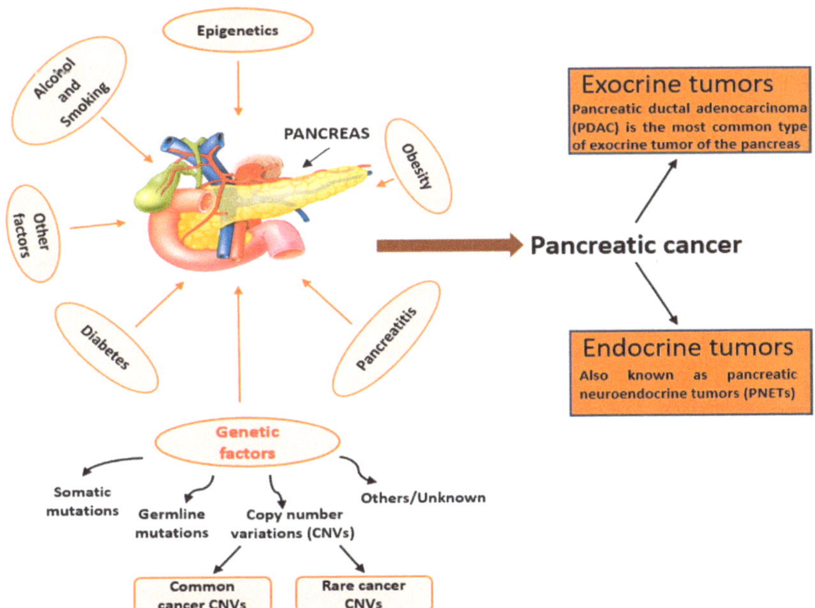

Figure 2. Associations in the development of pancreatic cancer, including the relationship between PDAC and copy number variations (CNVs).

CNVs in sporadic pancreatic cancer. CNV analysis in PDAC has revealed common cancer CNVs, including amplifications of KRAS (12p12.1), GATA6 (18q11.2), MYC (8q24.2), ERBB2 (17q12), PAK4 (19q13), NCOA3/AIB1 (20q13.12), SKAP2/SCAP2 (7p15.2), and AKT2 (19q13), as well as deletions of SMAD4 (18q21.2), CDKN2A (9p21.3), CDKN2B (9p21.3), PTEN (10q23.31), MAP2K4 (17p12), RUNX3 (1p36.11), TP53 (17p13.1), DCC (18q21.1), and ARID1A (1p36.11) [74–76]. Other CNVs are reported in Supplementary Table S1. However, it has not been established whether these CNVs are the cause or effect of cancer.

CNVs in familial pancreatic cancer. Hereditary CNVs in the genome may also contribute to a genetic susceptibility to PDAC. One study used representational oligonucleotide microarray analysis (ROMA) to characterize germline CNVs in 60 cancer patients from 57 FPC families, i.e., those in which at least two first-degree relatives have been diagnosed with pancreatic cancer. A total of 56 distinct genomic areas, including 25 deletions and 31 amplifications, were found to have CNVs that were not present in the healthy controls [77]. Among these CNVs, functionally interesting candidate genes were selected whose germline amplification (e.g., JunD, MAFK, RND1, WNT10B, WNT1, MAP2K2, and BIRC6) or deletion (e.g., ANKRD3, PDZRN3, and FHIT) may contribute to tumor development. These CNVs may define potential candidate loci for familial PDAC.

2.2. Identification and Analysis of CNVs in PDAC

Overexpressed proteins in PDAC due to CNVs could be therapeutic targets as well as diagnostic and prognostic markers. For example, in a genome-wide analysis of 27 microdissected PDAC samples using high-density microarrays representing $\sim$116,000 single nucleotide polymorphism (SNP) loci, frequent gains of 1q, 2, 3, 5, 7p, 8q, 11, 14q, and 17q ($\geq$78% of cases) and losses of 1p, 3p, 6, 9p, 13q, 14q, 17p, and 18q ($\geq$44%) were detected [76]. Quantitative real-time PCR revealed that the SKAP2 gene (7p15.2), a member of the src family kinases, was the most frequently amplified ($\geq$3 copies found in 59–63% of cases), and reverse transcription PCR was used to confirm its recurrent overexpression in eight out of 12 PDAC cases (67%). Moreover, in situ RNA hybridization (ISH) and FISH analyses revealed a significant correlation between SKAP2 DNA copy number and its mRNA expression level, suggesting that SKAP2 upregulation is due to CNVs [76,78]. The overexpression of SKAP2 was observed consistently from early-stage (I–II) to late-stage (III–IVb) tumors, suggesting a potential involvement of this gene in the development of PDAC, including control of the growth and differentiation of PDAC cells via α-Synuclein [79], as well as modulation of their motility and spread by interacting with the focal adhesion kinase RAFTK [80]. Based on these findings, scientists proposed that the SKAP2 gene could be used as a potential target for therapeutic intervention as well as a potential marker gene for early diagnosis in PDAC [76].

GATA binding protein 6 (GATA6), a zinc-finger transcription factor that plays an important role in the normal development of endodermal and mesodermal tissues, including the pancreas, is amplified in PDAC due to CNVs [81]. Since the progression of normal pancreatic ductal epithelium to infiltrating cancer is believed to occur through a series of morphologically defined precursors known as pancreatic intraepithelial neoplasia (PanIN-1, 2, and 3) [82], the GATA6 copy number was assessed in microdissected samples of normal duct epithelium, PanIN, and human PDAC to investigate its role in PDAC. Quantitative PCR revealed no gain of GATA6 in normal duct epithelium (0 of 4), PanIN-1 (0 of 13), or PanIN-2 (0 of 10) lesions when compared to the haploid genome [83]. However, an increased GATA6 copy number ($\geq$2.3 copies) was identified in 6/17 samples (35%) of PanIN-3 and in 18/55 samples (33%) of PDAC, and confirmed through FISH in paraffin-embedded sections of 10 PDAC samples and one PanIN-3. This GATA6 amplification and consequent transcriptional upregulation observed late in PDAC carcinogenesis suggest that detectable GATA6 copy number gain may have value as a diagnostic marker [83]. Early findings from Comprehensive Molecular Characterization of Advanced Pancreatic Ductal Adenocarcinoma for Better Treatment Selection (COMPASS; a prospective study: NCT02750657) further demonstrated that molecular profiling can predict how different patients with locally advanced or metastatic PDAC and with different genomic and transcriptome subtypes will respond to chemotherapy. Patients with the transcriptomic "basal-like subtype", a highly chemoresistant phenotype, have a shorter median overall survival than those with the "classical" subtype. The latter are easily identified by positive GATA6 staining by an RNAscope in situ hybridization (ISH) assay and high GATA6 expression. GATA6 could therefore be a useful marker for the classical subtype [84,85].

Another target of gene amplification in PDAC is MYC, a member of a family of transcription factors that work together to control cell proliferation, metabolism, and the expression of genes necessary for these processes [86]. Pre-clinical experimental evidence has shown that MYC is an essential and non-redundant node of oncogenic signaling and therefore should be a therapeutic target [87–90]. It is usually upregulated by gene amplifications, and consequently, it can enhance the progression of cancer by promoting cell competition, survival signals in hypoxic settings, and altered metabolic pathways. This amplification is inversely correlated to that of GATA6, and a high MYC expression level is typical in the basal-like PDAC subtype [91]. MYC also has an emerging role in remodeling the tumor microenvironment (TME). TME is a distinctive feature of PDAC that makes up about 90% of the tumor mass and is characterized by a prominent desmoplastic reaction [92]. MYC amplification in PDAC induces the depletion of CD3 T cells while increasing the recruitment of immune cells such as neutrophils, macrophages, B cells, and granulocytic myeloid suppressor cells [93], collectively enhancing an immunosuppressive phenotype [93,94]. Genomic and transcriptomic analyses have further linked MYC to a high number of metastases in patients (>10 metastases in a patient) in PDAC [95]. In terms of drug resistance, MYC overexpression has been linked to the resistance to inhibitors of the serine/threonine protein kinase mammalian target of rapamycin (mTOR) [96–101].

Alteration of regulatory regions due to CNVs on non-coding sequences can also influence the level and timing of expression of the related protein [102]. In a case–control cohort consisting of 1031 controls and 1027 pancreatic cancer cases, researchers demonstrated that CNVR2966.1, a CNV located in a gene desert region on 6q13, is significantly associated with the risk of developing disease and functions as a potential trans-acting regulator of the CDKN2B (p15 or INK4B) gene located on 9p21.3. CNVR2966.1 is an insertion/deletion and chromosome conformation capture-on-chip (4C), and other functional experiments have shown that it may contain a transcriptional activation element and regulate CDKN2B transcription through interchromosomal long-range interaction. CDKN2B is a tumor suppressor that encodes a cyclin-dependent kinase inhibitor that regulates cell growth and the cell cycle G1 progression by preventing the activation of cyclin-D-dependent kinases [103]. It has been found to be frequently co-deleted with the neighboring tumor suppressor gene CDKN2A (which codes p16-INK4a and p14ARF) in various tumors, and its deletion has been reported in a significantly high proportion in pancreatic cancer [76,104]. Therefore, CNVR2966.1 may be important for risk assessment, early detection, and a better understanding of PDAC [72].

In another study aimed at exploring potential biomarkers of PDAC, analysis of transcriptomic and clinical data from The Cancer Genome Atlas Program (TCGA) revealed high expressions of the COL17A1 and ECT2 genes and associated this expression with CNVs [105]. The highly expressed genes of these patients were also related to the cell cycle and proteasome pathways. COL17A1 is a transmembrane protein that can affect the proliferation and differentiation of epithelial cells and therefore acts as an important factor in the formation and maintenance of multilayered epithelial structures in PDAC [106], while ECT2 is an oncogene that plays an important role in cell proliferation and metastasis. Clinical correlations further showed that the expression of these two genes was significantly associated with tumor grade and that the overall survival (OS) rate decreased with an increase in their expressions. Since several research studies have demonstrated the success of combining anti-PD-1 antibody immunotherapy with chemotherapy in treating PDAC [107,108], this study further demonstrated that the high-ECT2 group exhibited greater sensitivity towards anti-PD-1 therapy and 20 chemotherapeutic agents (e.g., bortezomib and rapamycin). These discoveries suggest that ECT2 and COL17A1 are potential diagnostic and prognostic markers for PDAC that can also facilitate innovative approaches for personalized treatment [105].

2.3. CNV-Based Classifications of PDAC

Researchers proposed that reclassifying PDAC into subtypes based on genetic and molecular characteristics may guide novel treatment choices with prognostic and biological significance [109]. According to this hypothesis, PDAC has further been classified into structural [75] and molecular [110] subtypes, for example, based on CNVs as highlighted below.

2.3.1. Structural Variation Profiles

Some researchers performed a deep WGS and CNV analysis using SNP arrays in 100 normal and tumor-derived samples obtained from patients with PDAC. After retrieval, validation of the presence of carcinoma in the samples to be sequenced, and estimation of the ratio of malignant epithelial nuclei to stromal nuclei, the samples were removed, followed by processing in formalin or full-face sectioning using optical coherence tomography (OCT). Macrodissection was carried out, when necessary, to excise non-malignant tissue areas, followed by the extraction of nucleic acids. CNVs analyses led to a classification of the disease into four subtypes based on the number, frequency, and distribution of structural rearrangement events across the genome in each patient [75]. The majority of these structural rearrangements were due to a copy number change (events classified as deletion, duplication, tandem duplication, amplified inversion, and foldback inversion).

Stable (subtype 1). Tumors contain a few structural rearrangements (<50) located randomly throughout the genome. They exhibit aneuploidy, suggesting cell cycle/mitosis defects, given that although aneuploidy was classically defined as whole chromosome numerical aberrations, this definition has recently been expanded in the cancer genome literature to include losses or gains of chromosome arms [111–113].

Locally rearranged (Subtype 2). Tumors exhibit non-random intra-chromosomal rearrangements on one or a few chromosomes. These are further classified as either: (i) focal amplifications where most of the events are gains in known oncogenes including GATA6, SOX9, and KRAS, as well as therapeutic targets like CDK6, MET, ERBB2, PIK3R3, and PIK3CA; or (ii) complex rearrangements involving complex genomic events like breakage–fusion–bridge (BFB) or chromothripsis (i.e., the simultaneous occurrence of multiple structural alterations in a single mitotic event) [114].

Although in this subtype the most known oncogene copy-number increases in tumors were observed in a few patients, most of these oncogenes are well-recognized therapeutic targets (MET, FGFR1, ERBB2) with readily available inhibitors. The other oncogene amplifications identified include GATA6, which is known to be amplified in PDAC and correlates with a poor prognosis [75].

Scattered (Subtype 3). Tumors contain 50–200 structural rearrangements scattered throughout the genome.

Unstable (Subtype 4). Tumors contain many structural rearrangements (>200) scattered throughout the genome. Such a large scale of genomic instability suggests defects in DNA maintenance, in addition to potentially highlighting sensitivity to DNA-damaging agents.

Notably, these authors did not perform clinical correlation analyses.

2.3.2. Molecular Subtypes

Some researchers profiled genomic alterations in a Chinese cohort of 608 PDAC patients from a database containing somatic mutations, CNVs, and pathogenic germline variants [110]. Targeted-region capture and sequencing were performed using two gene panels specifically designed for cancer gene detection, comprising 566 and 764 genes, respectively. Germline and somatic CNVs were identified, and this information was used to perform unsupervised consensus clustering of the patients as well as differential CNV analysis. Functional/pathway enrichment analysis was then conducted for genes with significantly higher CNV values in each cluster or group. More specifically, consensus clustering revealed two groups, namely CNV-G1 and CNV-G2. Based on the CNV of genes involved in DNA repair and receptor tyrosine kinase (RTK)-related signaling, patients from CNV-G1 were further subdivided into two subtypes: the proliferation-active subtype and the repair-deficient subtype. Patients from CNV-G2 were also subdivided into two subtypes: the repair-enhanced and the repair-proficient subtypes [110].

CNV-G1 is characterized by deletions predominantly in DNA repair genes, higher copy number instability (CNI), and defects in DNA-DSB (double-strand break) repair by homologous recombination (HR). It consists of the (i) proliferation-active group with a high

CNV score and amplification of genes in the RTK-related signaling pathway, and the (ii) repair-deficiency group with a low CNV score.

CNV-G2 is characterized by amplifications predominantly in DNA repair genes, a higher tumor mutational burden (TMB), and defects in polymerase POLE. It consists of the (i) repair-enhanced group with a low CNV score and amplification of genes in the HRR pathway, and the (ii) repair-proficient group with a high CNV score.

The prognosis of the repair-deficient subtype was better (median survival time of 410 days) than that of the other three subtypes, suggesting that deletion of genes in the DNA repair pathway (specifically the HRR pathway) causes greater genomic instability and is detrimental to the survival of cancer cells. On the contrary, patients in the proliferation-active and repair-enhanced subtypes showed worse prognoses, with median survival times of 197 and 239 days, respectively. Furthermore, the prognosis of the proliferation-active subgroup was worse than that of the repair-deficient subgroup, suggesting that genetic amplification in RTK-related signaling would promote cancer cell proliferation and thereby confer a worse prognosis [110].

Together with the evidence from genomic footprint analysis, the study proposes that repair-proficient and repair-enhanced subtypes are better suited for immunotherapy, while DNA-damage therapies (such as platinum-based chemotherapy and PARPi) are highly recommended for repair-deficient and proliferation-active subtypes [110].

3. CNV Studies in PDAC

3.1. Literature Review

We carried out a literature search for several published papers on copy number variations and pancreatic cancer and highlighted the CNV landscape in the disease (Supplementary Table S1). We analyzed a total of 41 published articles from PubMed and SCOPUS in which researchers examined the expression levels of the genes that were discovered to have amplifications or deletions (in pooled public datasets or samples) in normal pancreatic tissues in comparison to malignant tissues (Figure 3). We further analyzed the biological and clinical importance of these studies, particularly whether these genes displayed dysregulated expression linked to survival outcomes.

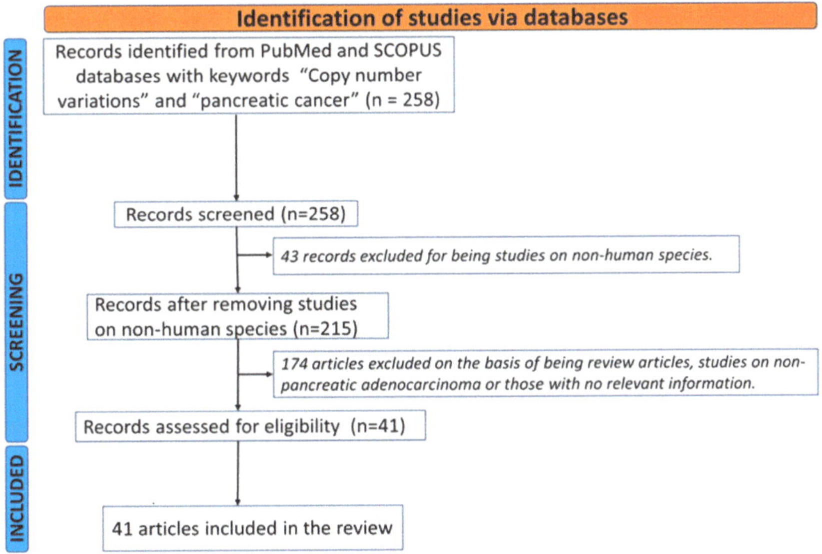

Figure 3. A flow chart of the identification process of the included articles: 258 papers were initially identified by title, of which 41 fulfilled the inclusion criteria after full-text evaluation. A total of 41 papers were included.

Samples were obtained from diverse sources, with most studies being carried out on samples from primary tumors only (21/41). Other studies, however, included both primary and metastatic samples (5/41). Other sample sources included were from public databases such as TCGA and NCBI GEO (13/41), tissue microarrays (TMAs) samples (1/41) and peripheral leukocytes from patients and controls (1/41). Some studies included cell lines (8/41) in the verification of identified CNVs, while others (33/41) did not.

Most of the studies were performed using "whole" (whole genome or whole exome) approaches (63.5%), including SNP-arrays (10/41), aCGH (8/41), tissue microarrays (TMA) (2/41), NGS (4/41) and both SNP/aCGH (2/41) techniques. Targeted approaches were used in the rest of the studies we considered (36.5%). Use of the "whole" approach resulted in the identification of numerous CNVs in the whole PDAC genome in comparison to normal controls; however, subsequent studies such as the roles of the identified CNVs in the development and progression of disease as well as their effects on currently available therapy were focused only on a few selected genes.

We noted that the use of diverse techniques for analyzing CNVs in different PDAC samples, as well as confirming their effect on levels of mRNA expression, did not significantly affect the consistency of the results. Moreover, most of the CNVs detected could be verified in various publicly available datasets, such as NCBI GEO and The Cancer Genome Atlas Program (TCGA). Recurrent gains on chromosomes 1q, 2p, 3q, 5p, 6p, 7q, 8q, 11q, 12p, 15q, 17q, 18q, 19q, and 20q included several known or suspected oncogenes, and recurrent losses on chromosomes 1p, 3p, 6, 8p, 9p, 10q, 12q, 13q, 15q, 17, 18, 19p, 20p, 21 and 22, which included several known or suspected tumor suppressor genes. In general, tumors with more copy number alterations (an indicator of chromosomal instability) trended toward a poor prognosis.

As expected, CNVs were almost always observed in the classical mutation genes in PDAC, including amplification of the oncogene KRAS (12p12.1) and deletions of the tumor suppressor genes TP53 (17p13.1), CDNK2A (9p21.3), and SMAD4 (18q21.2) to further confirm their role in the disease [75,76,115–129]. Interestingly, the frequencies of CNVs are consistent throughout various ethnicities, even though disparities have been observed in the frequency of driver mutations in PDAC, such as a lower frequency of KRAS mutations in Korea [130,131] and Japan [125]. For instance, one study performed microarray and CNV analyses of 93 pancreatic cancer data derived from the Japanese version of the Cancer Genome Atlas (JCGA) and revealed frequent CNVs as gains in 3q, 7q, and 2q and losses in 7q, 12q, 19q, and 19p [125], which are consistent with CNVs in other ethnicities [75,76,115–129].

The most frequent CNVs reported were amplifications of MYC (8q24) (15/41) and GATA6 (18q11.2) (7/41), and deletions of CDKN2A (9p21.3) (16/41), CDKN2B (9p21.3) (7/41), and SMAD4 (18q21.2) (14/41). MYC overexpression is typical in the basal-like PDAC subtype, which exhibits poor prognosis and chemoresistance. GATA6 overexpression is typical of the classical PDAC subtype, and its expression is observed late in PDAC carcinogenesis, suggesting that detectable GATA6 copy number gain may have value as a diagnostic marker. GATA6 overexpression has been associated with poor prognosis, but interestingly, it has been shown to correlate to a better prognosis after resection and adjuvant therapy, where it was believed to act as a suppressor of mutant KRAS^{G12V}-driven PDAC [75,117,121,123,126,132,133].

In chromosome 1, the amplifications of 1p12 (NOTCH2) [115,118,123] and 1p13.1-p12 (REG4) [75,116,117] and the deletion of 1p36.11 (ARID1A) [75,117,126] were the more recurrent CNVs. REG4 overexpression was associated with poor prognosis and resistance to gemcitabine treatment in one study, suggesting that adjuvant therapies that target reg4 could enhance the usual gemcitabine-based treatment of pancreatic cancer [75,116,117].

ASAP2 (2p25.1) amplification has been associated with lower overall survival (OS) as well as lower relapse-free survival (RFS) [134,135]. FHIT (3p14.2) and ATR (3q23) deletions were mostly reported in both sporadic familial cases, indicating their possible role in PDAC susceptibility as well as progression [77,110,118,133,136]. FHIT (3p14.2) deletions and ECT (3q26.31) amplifications have also been correlated with poor prognosis in PDAC [73,77,105,118,133,136].

The most recurrent amplification on chromosome 8 was 8p11.21 (FGFR1, IDO1, ZNF703) [110,123,126,129], observed in both sporadic and familial PDAC. One study demonstrated that the loss of 8p was exclusively observed in patients with shorter survival and associated this with specific CNV acquisitions due to potential positive selection and genetic drift. The genes that have been associated with this location are 8p, 8p23.2 (CSMD1) in sporadic PDAC, 8p23.1 (MCPH1 and ANGPT2), and 8p22 (NAT1) in FPC [116,117,136].

Some researchers examined a patient's complicated evolutionary history and very long postsurgical survival period (43 months) and proposed that this could be due to the amplification of a segment 9p.22 covering the FREM1 gene, which has recently been linked to increased immune cell (IC) infiltration. They suggested that an active immune response could improve the outcome. FREM1 in this particular context could be further explored both as a molecular target and/or immune checkpoint-blocking therapeutic strategy and as a biomarker of an active local immunological response [116].

No deletions were reported on chromosome 11, but there were amplifications on 11q13.3 (CCND1, TMEM16H), 11q13.5 (EMSY), and 11p14.1 (LGR4), of which the CCND1 gene was the most recurrent amplification [118,120,132,134,137]. One study showed that the presence of elevated EMSY copy numbers in relatively large, clustered cells surrounded by tumor cells expressing normal copy numbers suggests that the mutation occurred later in the carcinogenesis process rather than at an early stage. This could clarify a previous investigation that discovered a negative correlation between this mutation and the course of the disease [137].

In one study, TMEM132E (17q12) amplification was prevalent in a relapse (within 1 year after resection) subgroup ($n = 15$) compared with a non-relapse subgroup ($n = 15$) of 47% vs. 7% [138].

Some researchers demonstrated that the loss of a specific cytoband, 18q22.3, which encompasses only five genes, including the carboxypeptidase of glutamate-like (CPGL) gene, is linked to a poorer prognosis in both a testing cohort and an independent validation cohort of surgically resected pancreatic cancers. Further experiments involving reintroducing the CPGL gene, or its splicing variant CPGL-B, into CPGL-deficient pancreatic cancer cells showed a reduction in anchorage-independent cell growth and migration while promoting G1 accumulation. These findings imply that CPGL is a novel growth suppressor for pancreatic cancer cells and that risk classification in pancreatic cancer patients who have had their tumors removed could be based on the CPGL gene [132].

In one study, the perineural invasion in PDAC was linked to gains of 4q13.3, 4q35.2, 7p12.2, 10q26.3, 11q13.3, 17q23.1, 22q13.32, and loss of 6p21.32, whereas the amplification of 8q24.13 was strongly correlated with the T, N, and M stages simultaneously [134].

Some CNVs have also been associated with the increased glycolysis observed in PDAC. One study compared the SNP microarray data of glycolysis-high samples to glycolysis-low samples and found substantial amplifications of MYC (8q24.2), GATA6 (18q11.2), FGFR1 (8p11.21), and IDO1 (8p11.21), as well as deletions of SMAD4 (18q21.2) that were associated with the aerobic glycolysis phenotype characteristic of PDAC [123].

Another study demonstrated that the assessment of overall CNV burden through genome-wide methylation profiling could be a valuable prognostic tool in patients with surgically treated PDAC [129]. By analyzing DNA extracted from 108 chemotherapy-naïve, surgical PDAC specimens, the researchers were able to gather data on the DNA methylation status of more than 850,000 CpG sites located in various regions such as the promoter, enhancer, and gene body. Morphological subtyping, as per Kalimuthu et al. [139], classified PDAC into Group A tumors, which showed a dominant conventional and/or tubulopapillary growth pattern, and Group B tumors, which showed a dominant composite and/or squamous growth pattern. CNV profiles were then generated from the accumulated CpG methylation signal distributed throughout the genome (except for 13p, 14p, 15p, 21p, and 22p, X, and Y), and all the PDACs were classified into three distinct groups based on the number of chromosomal arm-level alterations: high ($\geq$17), moderate (5–16), or low (0–4). The most prevalent chromosomal arm-level aberrations included gains of 1q (19%) and 8q (29%), as well as losses of 8p (25%), 19p (26%), 6p (26%), 9p (36%), 6q (37%), 18q (43%), and 17p (55%). In particular, the CNVs involved deletions of CDKN2A/B, KDM6A, and SMAD4 and focal amplifications of MYC, FGFR1, or CDK6. Overall, low CNV burden was observed in Group A tumors, while high CNV burden was observed in Group B tumors, and this higher CNV burden in Group B was further associated with a poor prognosis and shorter overall survival [129]. Notably, this study was performed on PDAC-enriched FFPE tissues, and further studies are necessary to establish the possibility of performing CNV burden analysis on endoscopic ultrasound-guided fine-needle biopsies from non-resectable PDAC patients, as well as whether this has any prognostic value [140,141].

In another study, by analyzing 21 FFPE tumor tissues of PDAC patients, the authors analyzed the mutational spectrum of the disease and assessed the therapeutic relevance of OncoPan, a previously developed and validated NGS panel of 37 genes [142]. This panel includes the evaluation of indels, SNVs, and CNVs of various actionable genes for the identification of therapeutic targets as well as inherited cancer syndromes. Oncopan led to the discovery of biomarkers for personalized therapy in five PDAC patients. Among these patients, two exhibited HER2 amplification, making them potentially eligible for immunotherapy [142]. Numerous ongoing clinical studies are utilizing trastuzumab for pancreatic cancer treatment, and in a recent study, Hirokawa et al. reported that patients with HER2-positive heterotopic pancreatic cancer responded well to trastuzumab treatment [143]. These kinds of studies are a practical example of the clinical relevance of CNVs, and therefore it is expected that further/new panels will be developed to evaluate CNVs of a greater number of genes.

In the future, these assessments will also be less invasive thanks to the possibility of carrying out liquid biopsies. In fact, the use of circulating tumor DNA (ctDNA) is gaining significant popularity in molecular diagnosis, observation of clonal evolution, evaluation of treatment response, identification of cancer recurrence, and evaluation of drug resistance [144–146]. One study in 48 late-stage non-small cell lung cancer (NSCLC) patients analyzed matched tumor tissues and blood samples and determined gene-level CNVs from ctDNA [147]. Although the identification of somatic CNVs from ctDNA samples using targeted sequencing is challenging, amplifications of the EGFR, ERBB2, and MET genes were observed. Further comparison of these amplifications between tissue WES and ctDNA showed significantly high concordance and sensitivity, with 100% specificity observed for all three genes. Although the study was performed on NSCLC, the pipeline can be extended to other cancers, including PDAC [147], where liquid biopsies sequencing provides an alternative to obtaining the patient's genomic information in cases where tissue biopsies are not available [146].

3.2. CNVs in PDAC Stages and Grades

Research is being carried out to identify CNVs that may be useful clinical markers, and a recent study on ovarian cancer has provided results that encourage continued in-vestigation of these relationships. In particular, CNV-profiling analyses have been successfully used to distinguish between malignant and nonmalignant, as well as early and late stages in ovarian tumors [148]. The possible roles of CNVs in the early or late stages of pancreatic cancer have also been studied to assess their usefulness as potential markers of the various stages as well as the grades of the disease. For example, a gain in copy number at the 7p15.2 locus that causes the overexpression of the SKAP2 gene characterizes both PanIN lesions and early- (I–II) and late-stage (III–IVb) PDAC tumors. Therefore, it could be a potential marker gene for early diagnosis as well as a possible target for therapeutic intervention [76].

Some researchers also studied the relationships between 19q13 amplification and clinicopathological characteristics in PDAC and observed that the frequency of 19q13 gains increased from G1-G2 (low/moderate) to G3 (high grade) tumors and from pT1-pT2 (early) to pT3-pT4 (late) stage tumors. Moreover, none of the G1 tumors exhibited 19q13 copy number changes, while 11% of G2 tumors and 16.8% of G3 tumors displayed an in-crease in 19q13 copy number [149].

Among the CNVs in PDAC, amplification and overexpression of the PSCA and HMGA2 genes have further been associated with lymph node metastasis (N0) and in-vasive depth of the disease, respectively [124].

In another study, SNP arrays on 20 PDAC tumors identified two different CNV groups with different genetic profiles: group 1 (*n* = 9) showed losses at Xp22.33, 17p13.3, 9p24.3, 9p22.1, 6q25.2, and 1p36.11 chromosomal regions and gains at 1q21.1, while group 2 (*n* = 11) showed gains at 22q13.32, 22q13.31, 22q13.1, 16q24.3, 16q24.1, 11q13.4, 11q13.3, 11q13.1, 10q26.3, 10q26.13, 5q32, 3q22.1, and 2q14.2 chromosomal regions. From a clinical and histological perspective, grade I/II PDAC tumors that were smaller and well- or moderately-differentiated were linked to group 1 cases, while grade III carcinomas that were primarily poorly-differentiated made up group 2 PDAC cases, which were bigger in size [150]. Further analyses of these CNV regions showed that they harbor various cancer-associated genes, including those that have been specifically associated with PDAC, such as the TNFRSF6B gene, whose amplification has been observed in many tumors [151–154] and whose overexpression is known to block growth inhibition signals in PDAC [155], and the MAPRE2 gene, whose deletion has been observed in leukemic cells [156] as well as pancreatic cancer [157]. In this study, deletions of other genes, such as MYOCD [158] and PTAFR [159], were found to be recurrent in PDAC, although the association of these genes with PDAC pathogenesis should be further investigated [150].

3.3. CNVs in PDAC Chemoresistance

Using aCGH and qPCR in 14 PDAC samples, some researchers detected and confirmed gains in the copy number of the REG4 gene (1p13.1-p12) in all the analyzed samples [160]. CNV analysis in six pancreatic precancerous lesions (PanINs) also revealed an increase in REG4 copy number (in 6/7, 1/7, and 0/6 of PanIN3, PanIN2, and PanIN1 lesions, respec-tively), suggesting that this amplification is an early event in PDAC development [160]. REG4, a member of the multigenic family named reg, plays a role in the resistance of cells to anticancer drugs like 5-fluorouracil and methotrexate [161,162], and it promotes over-expression of the antiapoptotic proteins Bcl-xL, Bcl-2, and survivin, as well as the phosphorylation of AKT [162,163]. Its overexpression is observed in cancerous tissues of the stomach [164], colon [161,165], and pancreas [166]. In this study, PDAC-derived cells with REG4 protein overexpression grew more rapidly and were more resistant to gemcitabine treatment, and this enhanced growth was also confirmed in PDAC cell lines. Circulating REG4 protein is therefore a potential target to make PDAC sensitive to gemcitabine [160].

4. Challenges and Limitations in Clinical Application

Interpretation of any detected CNVs is important because they could have clinical im-plications [167,168], but this is faced with various challenges. Determining the pathogenicity of CNVs is difficult, and accurate interpretation often depends on the amount of infor-mation available in databases [7]. However, there are several important considerations when utilizing public databases. Firstly, there may be variations in the reported sizes of identical CNVs due to the usage of various array platforms [169]. For example, a large number of the previously reported benign CNVs may be overestimated in size because they are based on the bacterial artificial chromosome (BAC) microarray technique [170]. Secondly, it is not always possible to obtain sex information about the individuals included in these databases. This is particularly significant when studying X-linked CNVs in males, as many of the reported benign variants found in the databases are observed in females. However, the same alteration may already be pathogenic in males who possess only one X chromosome. Thirdly, the majority of CNVs reported in large population studies have not undergone validation. Lastly, factors such as incomplete penetrance, variable expressivity, age of onset, and parent of origin imprinting effects were not recorded [7,171].

It is also important to note that the interpretation of CNVs is heavily reliant on the specific clinical indications, and therefore clinicians must provide detailed clinical phenotypes to enable accurate interpretation of the results [172]. To facilitate this process, several groups have devised graphical workflows for CNV interpretation, which prove invaluable in routine diagnostic work. However, interlaboratory comparisons and external quality control schemes (such as the European Molecular Genetics Quality Network (EMQN) and the USA quality assessment scheme CAP (College of American Pathologists)) on the use of some technologies, such as arrays, in diagnostic laboratories show that there are differences in the interpretation, quality, and reporting among laboratories [172]. Therefore, the minimum detection resolution, reporting, and interpretation of CNVs should be standardized among laboratories.

The different types of CNV analysis software used are also unique, frequently employing varying default settings and/or statistical methodologies [173–178]. Additionally, each laboratory implements its own experimental and analysis protocols. These variations in protocols and software directly affect the sensitivity and resolution of a test, giving results that are very different from each other or only partially in agreement [172].

5. Conclusions

Copy Number Variations (CNVs) are the most frequent genetic structural alterations, making up approximately 12% of the human genome [6].

Currently, many lines of evidence have also shown that CNVs play important pathogenic roles in a variety of human disorders, from causative high-penetrance CNVs in rare genomic disorders to intermediate or low-penetrance CNVs in complex multifactorial diseases such as cancer [16–18]. The identification of these amplification and deletion events is therefore one of the main goals of medical genetics research.

Indeed, CNVs have been observed in patients with pancreatic ductal adenocarcinoma (PDAC) [74–77,179–187]. However, the detection of CNVs and their subsequent association with functional and clinical phenotypes remains very challenging. With the increasing use of whole-genome technologies to detect CNVs, germline and somatic CNVs are now recognized as frequent contributors to the spectrum of mutations leading to PDAC development, progression, and drug resistance.

Recent advances in technology have provided powerful tools for the detection and analysis of CNVs at the level of the genome as well as for targeted loci. For example, single-cell RNA-seq (scRNA-seq) studies in human tumors have revealed new insights into tumor heterogeneity and distinct subpopulations, which are pivotal for comprehensively dissecting tumor-related mechanisms [188]. In PDAC, scRNA-seq has been used to acquire the transcriptomic atlas of individual pancreatic cells from primary and metastatic tumors, as well as control pancreases and identify diverse stromal and malignant cell types. This has facilitated the comprehensive delineation of PDAC intratumoral heterogeneity and the underlying mechanisms for PDAC progression [188].

The correlation between CNV and gene expression suggests that the analysis of cancer genome CNVs may be useful in informing therapeutic decisions on the management of individual patients with particular patterns of mutations [109].

Although it is evident that CNVs have a significant impact on inter-individual variation in gene expression, the full extent to which they contribute to the molecular basis of PDAC remains to be established. This is due to persistent technical challenges in the accurate measurement of CNVs [189]. Further studies, using accurate genotyping assays in large population cohorts, will help to define the overall role of CNVs in PDAC pathogenesis more precisely [63].

Supplementary Materials: The following supporting information can be downloaded at: https://www.mdpi.com/article/10.3390/ijms25010391/s1.

Author Contributions: Conceptualization, F.P.; investigation, D.J.A.O. and M.G.; writing—original draft preparation, D.J.A.O.; writing—review and editing, M.G. and F.P.; supervision, F.P. All authors have read and agreed to the published version of the manuscript.

Funding: This research received no external funding.

Data Availability Statement: Data is contained within the article or Supplementary Material.

Conflicts of Interest: The authors declare no conflicts of interest.

References

1. Hruban, R.H.; Gaida, M.M.; Thompson, E.; Hong, S.M.; Noe, M.; Brosens, L.A.; Jongepier, M.; Offerhaus, G.J.A.; Wood, L.D. Why is pancreatic cancer so deadly? The pathologist's view. *J. Pathol.* **2019**, *248*, 131–141. [CrossRef] [PubMed]
2. Siegel, R.L.; Miller, K.D.; Wagle, N.S.; Jemal, A. Cancer statistics, 2023. *CA Cancer J. Clin.* **2023**, *73*, 17–48. [CrossRef]
3. Rahib, L.; Smith, B.D.; Aizenberg, R.; Rosenzweig, A.B.; Fleshman, J.M.; Matrisian, L.M. Projecting cancer incidence and deaths to 2030: The unexpected burden of thyroid, liver, and pancreas cancers in the United States. *Cancer Res.* **2014**, *74*, 2913–2921. [CrossRef] [PubMed]
4. Feuk, L.; Carson, A.R.; Scherer, S.W. Structural variation in the human genome. *Nat. Rev. Genet.* **2006**, *7*, 85–97. [CrossRef] [PubMed]
5. Shlien, A.; Malkin, D. Copy number variations and cancer. *Genome Med.* **2009**, *1*, 62. [CrossRef] [PubMed]
6. Redon, R.; Ishikawa, S.; Fitch, K.R.; Feuk, L.; Perry, G.H.; Andrews, T.D.; Fiegler, H.; Shapero, M.H.; Carson, A.R.; Chen, W.; et al. Global variation in copy number in the human genome. *Nature* **2006**, *444*, 444–454. [CrossRef] [PubMed]
7. Nowakowska, B. Clinical interpretation of copy number variants in the human genome. *J. Appl. Genet.* **2017**, *58*, 449–457. [CrossRef]
8. Zhang, F.; Gu, W.; Hurles, M.E.; Lupski, J.R. Copy number variation in human health, disease, and evolution. *Annu. Rev. Genom. Hum. Genet.* **2009**, *10*, 451–481. [CrossRef]
9. Nguyen, D.Q.; Webber, C.; Ponting, C.P. Bias of selection on human copy-number variants. *PLoS Genet.* **2006**, *2*, e20. [CrossRef]
10. Choy, K.W.; Setlur, S.R.; Lee, C.; Lau, T.K. The impact of human copy number variation on a new era of genetic testing. *BJOG* **2010**, *117*, 391–398. [CrossRef]
11. Bruder, C.E.; Piotrowski, A.; Gijsbers, A.A.; Andersson, R.; Erickson, S.; Diaz de Stahl, T.; Menzel, U.; Sandgren, J.; von Tell, D.; Poplawski, A.; et al. Phenotypically concordant and discordant monozygotic twins display different DNA copy-number-variation profiles. *Am. J. Hum. Genet.* **2008**, *82*, 763–771. [CrossRef] [PubMed]
12. Piotrowski, A.; Bruder, C.E.; Andersson, R.; Diaz de Stahl, T.; Menzel, U.; Sandgren, J.; Poplawski, A.; von Tell, D.; Crasto, C.; Bogdan, A.; et al. Somatic mosaicism for copy number variation in differentiated human tissues. *Hum. Mutat.* **2008**, *29*, 1118–1124. [CrossRef] [PubMed]
13. Forsberg, L.A.; Rasi, C.; Razzaghian, H.R.; Pakalapati, G.; Waite, L.; Thilbeault, K.S.; Ronowicz, A.; Wineinger, N.E.; Tiwari, H.K.; Boomsma, D.; et al. Age-related somatic structural changes in the nuclear genome of human blood cells. *Am. J. Hum. Genet.* **2012**, *90*, 217–228. [CrossRef] [PubMed]
14. Cornelis, M.C.; Agrawal, A.; Cole, J.W.; Hansel, N.N.; Barnes, K.C.; Beaty, T.H.; Bennett, S.N.; Bierut, L.J.; Boerwinkle, E.; Doheny, K.F.; et al. The Gene, Environment Association Studies consortium (GENEVA): Maximizing the knowledge obtained from GWAS by collaboration across studies of multiple conditions. *Genet. Epidemiol.* **2010**, *34*, 364–372. [CrossRef] [PubMed]
15. Laurie, C.C.; Laurie, C.A.; Rice, K.; Doheny, K.F.; Zelnick, L.R.; McHugh, C.P.; Ling, H.; Hetrick, K.N.; Pugh, E.W.; Amos, C.; et al. Detectable clonal mosaicism from birth to old age and its relationship to cancer. *Nat. Genet.* **2012**, *44*, 642–650. [CrossRef] [PubMed]
16. Pos, O.; Radvanszky, J.; Buglyo, G.; Pos, Z.; Rusnakova, D.; Nagy, B.; Szemes, T. DNA copy number variation: Main characteristics, evolutionary significance, and pathological aspects. *Biomed. J.* **2021**, *44*, 548–559. [CrossRef] [PubMed]
17. Gu, W.; Zhang, F.; Lupski, J.R. Mechanisms for human genomic rearrangements. *Pathogenetics* **2008**, *1*, 4. [CrossRef]
18. Lupski, J.R. Genomic disorders: Structural features of the genome can lead to DNA rearrangements and human disease traits. *Trends Genet.* **1998**, *14*, 417–422. [CrossRef]
19. Shaw, C.J.; Lupski, J.R. Implications of human genome architecture for rearrangement-based disorders: The genomic basis of disease. *Hum. Mol. Genet.* **2004**, *13* (Suppl. S1), R57–R64. [CrossRef]
20. Hastings, P.J.; Lupski, J.R.; Rosenberg, S.M.; Ira, G. Mechanisms of change in gene copy number. *Nat. Rev. Genet.* **2009**, *10*, 551–564. [CrossRef]
21. Hastings, P.J.; Ira, G.; Lupski, J.R. A microhomology-mediated break-induced replication model for the origin of human copy number variation. *PLoS Genet.* **2009**, *5*, e1000327. [CrossRef] [PubMed]
22. Thompson, L.H.; Schild, D. Homologous recombinational repair of DNA ensures mammalian chromosome stability. *Mutat. Res.* **2001**, *477*, 131–153. [CrossRef] [PubMed]

23. Li, X.; Heyer, W.D. Homologous recombination in DNA repair and DNA damage tolerance. *Cell Res.* **2008**, *18*, 99–113. [CrossRef] [PubMed]
24. Arlt, M.F.; Ozdemir, A.C.; Birkeland, S.R.; Wilson, T.E.; Glover, T.W. Hydroxyurea induces de novo copy number variants in human cells. *Proc. Natl. Acad. Sci. USA* **2011**, *108*, 17360–17365. [CrossRef] [PubMed]
25. Mullenders, L.H.F. Solar UV damage to cellular DNA: From mechanisms to biological effects. *Photochem. Photobiol. Sci.* **2018**, *17*, 1842–1852. [CrossRef] [PubMed]
26. Arlt, M.F.; Wilson, T.E.; Glover, T.W. Replication stress and mechanisms of CNV formation. *Curr. Opin. Genet. Dev.* **2012**, *22*, 204–210. [CrossRef] [PubMed]
27. Costa, E.O.A.; Pinto, I.P.; Goncalves, M.W.; da Silva, J.F.; Oliveira, L.G.; da Cruz, A.S.; Silva, D.M.E.; da Silva, C.C.; Pereira, R.W.; da Cruz, A.D. Small de novo CNVs as biomarkers of parental exposure to low doses of ionizing radiation of caesium-137. *Sci. Rep.* **2018**, *8*, 5914. [CrossRef] [PubMed]
28. Zitzelsberger, H.; Unger, K. DNA copy number alterations in radiation-induced thyroid cancer. *Clin. Oncol.* **2011**, *23*, 289–296. [CrossRef]
29. Zitzelsberger, H.; Lehmann, L.; Hieber, L.; Weier, H.U.; Janish, C.; Fung, J.; Negele, T.; Spelsberg, F.; Lengfelder, E.; Demidchik, E.P.; et al. Cytogenetic changes in radiation-induced tumors of the thyroid. *Cancer Res.* **1999**, *59*, 135–140.
30. Hovhannisyan, G.; Harutyunyan, T.; Aroutiounian, R.; Liehr, T. DNA Copy Number Variations as Markers of Mutagenic Impact. *Int. J. Mol. Sci.* **2019**, *20*, 4723. [CrossRef]
31. Kim, P.M.; Korbel, J.O.; Gerstein, M.B. Positive selection at the protein network periphery: Evaluation in terms of structural constraints and cellular context. *Proc. Natl. Acad. Sci. USA* **2007**, *104*, 20274–20279. [CrossRef] [PubMed]
32. Yatsenko, S.A.; Brundage, E.K.; Roney, E.K.; Cheung, S.W.; Chinault, A.C.; Lupski, J.R. Molecular mechanisms for subtelomeric rearrangements associated with the 9q34.3 microdeletion syndrome. *Hum. Mol. Genet.* **2009**, *18*, 1924–1936. [CrossRef] [PubMed]
33. Shao, L.; Shaw, C.A.; Lu, X.Y.; Sahoo, T.; Bacino, C.A.; Lalani, S.R.; Stankiewicz, P.; Yatsenko, S.A.; Li, Y.; Neill, S.; et al. Identification of chromosome abnormalities in subtelomeric regions by microarray analysis: A study of 5380 cases. *Am. J. Med. Genet. A* **2008**, *146A*, 2242–2251. [CrossRef] [PubMed]
34. She, X.; Horvath, J.E.; Jiang, Z.; Liu, G.; Furey, T.S.; Christ, L.; Clark, R.; Graves, T.; Gulden, C.L.; Alkan, C.; et al. The structure and evolution of centromeric transition regions within the human genome. *Nature* **2004**, *430*, 857–864. [CrossRef] [PubMed]
35. Agiannitopoulos, K.; Pepe, G.; Tsaousis, G.N.; Potska, K.; Bouzarelou, D.; Katseli, A.; Ntogka, C.; Meintani, A.; Tsoulos, N.; Giassas, S.; et al. Copy Number Variations (CNVs) Account for 10.8% of Pathogenic Variants in Patients Referred for Hereditary Cancer Testing. *Cancer Genom. Proteom.* **2023**, *20*, 448–455. [CrossRef] [PubMed]
36. Sharp, A.J.; Locke, D.P.; McGrath, S.D.; Cheng, Z.; Bailey, J.A.; Vallente, R.U.; Pertz, L.M.; Clark, R.A.; Schwartz, S.; Segraves, R.; et al. Segmental duplications and copy-number variation in the human genome. *Am. J. Hum. Genet.* **2005**, *77*, 78–88. [CrossRef]
37. McCarroll, S.A.; Altshuler, D.M. Copy-number variation and association studies of human disease. *Nat. Genet.* **2007**, *39* (Suppl. S7), S37–S42. [CrossRef] [PubMed]
38. Weckselblatt, B.; Rudd, M.K. Human Structural Variation: Mechanisms of Chromosome Rearrangements. *Trends Genet.* **2015**, *31*, 587–599. [CrossRef]
39. Cantsilieris, S.; Baird, P.N.; White, S.J. Molecular methods for genotyping complex copy number polymorphisms. *Genomics* **2013**, *101*, 86–93. [CrossRef]
40. Singh, A.K.; Olsen, M.F.; Lavik, L.A.S.; Vold, T.; Drablos, F.; Sjursen, W. Detecting copy number variation in next generation sequencing data from diagnostic gene panels. *BMC Med. Genom.* **2021**, *14*, 214. [CrossRef]
41. Medvedev, P.; Stanciu, M.; Brudno, M. Computational methods for discovering structural variation with next-generation sequencing. *Nat. Methods* **2009**, *6* (Suppl. S11), S13–S20. [CrossRef] [PubMed]
42. Mills, R.E.; Walter, K.; Stewart, C.; Handsaker, R.E.; Chen, K.; Alkan, C.; Abyzov, A.; Yoon, S.C.; Ye, K.; Cheetham, R.K.; et al. Mapping copy number variation by population-scale genome sequencing. *Nature* **2011**, *470*, 59–65. [CrossRef] [PubMed]
43. Alkan, C.; Kidd, J.M.; Marques-Bonet, T.; Aksay, G.; Antonacci, F.; Hormozdiari, F.; Kitzman, J.O.; Baker, C.; Malig, M.; Mutlu, O.; et al. Personalized copy number and segmental duplication maps using next-generation sequencing. *Nat. Genet.* **2009**, *41*, 1061–1067. [CrossRef] [PubMed]
44. Yoon, S.; Xuan, Z.; Makarov, V.; Ye, K.; Sebat, J. Sensitive and accurate detection of copy number variants using read depth of coverage. *Genome Res.* **2009**, *19*, 1586–1592. [CrossRef] [PubMed]
45. Pirooznia, M.; Goes, F.S.; Zandi, P.P. Whole-genome CNV analysis: Advances in computational approaches. *Front. Genet.* **2015**, *6*, 138. [CrossRef] [PubMed]
46. Mellars, G.; Gomez, K. Mutation detection by Southern blotting. *Methods Mol. Biol.* **2011**, *688*, 281–291. [PubMed]
47. Higuchi, R.; Fockler, C.; Dollinger, G.; Watson, R. Kinetic PCR analysis: Real-time monitoring of DNA amplification reactions. *Biotechnology* **1993**, *11*, 1026–1030. [CrossRef]
48. Sellner, L.N.; Taylor, G.R. MLPA and MAPH: New techniques for detection of gene deletions. *Hum. Mutat.* **2004**, *23*, 413–419. [CrossRef]
49. Schouten, J.P.; McElgunn, C.J.; Waaijer, R.; Zwijnenburg, D.; Diepvens, F.; Pals, G. Relative quantification of 40 nucleic acid sequences by multiplex ligation-dependent probe amplification. *Nucleic Acids Res.* **2002**, *30*, e57. [CrossRef]
50. Jakobsson, M.; Scholz, S.W.; Scheet, P.; Gibbs, J.R.; VanLiere, J.M.; Fung, H.C.; Szpiech, Z.A.; Degnan, J.H.; Wang, K.; Guerreiro, R.; et al. Genotype, haplotype and copy-number variation in worldwide human populations. *Nature* **2008**, *451*, 998–1003. [CrossRef]

51.   Janiak, M.C. Of starch and spit. *elife* **2019**, *8*, e47523. [CrossRef] [PubMed]
52.   Hardy, K.; Brand-Miller, J.; Brown, K.D.; Thomas, M.G.; Copeland, L. The Importance of Dietary Carbohydrate in Human Evolution. *Q. Rev. Biol.* **2015**, *90*, 251–268. [CrossRef] [PubMed]
53.   Miller, D.W.; Hague, S.M.; Clarimon, J.; Baptista, M.; Gwinn-Hardy, K.; Cookson, M.R.; Singleton, A.B. Alpha-synuclein in blood and brain from familial Parkinson disease with SNCA locus triplication. *Neurology* **2004**, *62*, 1835–1838. [CrossRef] [PubMed]
54.   Xiao, B.; Ji, X.; Xing, Y.; Chen, Y.W.; Tao, J. A rare case of 46, XX SRY-negative male with approximately 74-kb duplication in a region upstream of SOX9. *Eur. J. Med. Genet.* **2013**, *56*, 695–698. [CrossRef] [PubMed]
55.   Kim, G.J.; Sock, E.; Buchberger, A.; Just, W.; Denzer, F.; Hoepffner, W.; German, J.; Cole, T.; Mann, J.; Seguin, J.H.; et al. Copy number variation of two separate regulatory regions upstream of SOX9 causes isolated 46,XY or 46,XX disorder of sex development. *J. Med. Genet.* **2015**, *52*, 240–247. [CrossRef] [PubMed]
56.   Benko, S.; Gordon, C.T.; Mallet, D.; Sreenivasan, R.; Thauvin-Robinet, C.; Brendehaug, A.; Thomas, S.; Bruland, O.; David, M.; Nicolino, M.; et al. Disruption of a long distance regulatory region upstream of SOX9 in isolated disorders of sex development. *J. Med. Genet.* **2011**, *48*, 825–830. [CrossRef] [PubMed]
57.   Cox, J.J.; Willatt, L.; Homfray, T.; Woods, C.G. A SOX9 duplication and familial 46,XX developmental testicular disorder. *N. Engl. J. Med.* **2011**, *364*, 91–93. [CrossRef]
58.   Hyon, C.; Chantot-Bastaraud, S.; Harbuz, R.; Bhouri, R.; Perrot, N.; Peycelon, M.; Sibony, M.; Rojo, S.; Piguel, X.; Bilan, F.; et al. Refining the regulatory region upstream of SOX9 associated with 46,XX testicular disorders of Sex Development (DSD). *Am. J. Med. Genet. A* **2015**, *167A*, 1851–1858. [CrossRef]
59.   Vetro, A.; Dehghani, M.R.; Kraoua, L.; Giorda, R.; Beri, S.; Cardarelli, L.; Merico, M.; Manolakos, E.; Parada-Bustamante, A.; Castro, A.; et al. Testis development in the absence of SRY: Chromosomal rearrangements at SOX9 and SOX3. *Eur. J. Hum. Genet.* **2015**, *23*, 1025–1032. [CrossRef]
60.   Deng, N.; Goh, L.K.; Wang, H.; Das, K.; Tao, J.; Tan, I.B.; Zhang, S.; Lee, M.; Wu, J.; Lim, K.H.; et al. A comprehensive survey of genomic alterations in gastric cancer reveals systematic patterns of molecular exclusivity and co-occurrence among distinct therapeutic targets. *Gut* **2012**, *61*, 673–684. [CrossRef]
61.   Brown, J.S.; Amend, S.R.; Austin, R.H.; Gatenby, R.A.; Hammarlund, E.U.; Pienta, K.J. Updating the Definition of Cancer. *Mol. Cancer Res.* **2023**, *21*, 1142–1147. [CrossRef] [PubMed]
62.   Liu, W.; Sun, J.; Li, G.; Zhu, Y.; Zhang, S.; Kim, S.T.; Sun, J.; Wiklund, F.; Wiley, K.; Isaacs, S.D.; et al. Association of a germ-line copy number variation at 2p24.3 and risk for aggressive prostate cancer. *Cancer Res.* **2009**, *69*, 2176–2179. [CrossRef] [PubMed]
63.   Fanciulli, M.; Petretto, E.; Aitman, T.J. Gene copy number variation and common human disease. *Clin. Genet.* **2010**, *77*, 201–213. [CrossRef]
64.   Shao, X.; Lv, N.; Liao, J.; Long, J.; Xue, R.; Ai, N.; Xu, D.; Fan, X. Copy number variation is highly correlated with differential gene expression: A pan-cancer study. *BMC Med. Genet.* **2019**, *20*, 175. [CrossRef] [PubMed]
65.   Higgins, M.E.; Claremont, M.; Major, J.E.; Sander, C.; Lash, A.E. CancerGenes: A gene selection resource for cancer genome projects. *Nucleic Acids Res.* **2007**, *35*, D721–D726. [CrossRef] [PubMed]
66.   Park, R.W.; Kim, T.M.; Kasif, S.; Park, P.J. Identification of rare germline copy number variations over-represented in five human cancer types. *Mol. Cancer* **2015**, *14*, 25. [CrossRef] [PubMed]
67.   Nagy, R.; Sweet, K.; Eng, C. Highly penetrant hereditary cancer syndromes. *Oncogene* **2004**, *23*, 6445–6470. [CrossRef]
68.   Thompson, E.D.; Roberts, N.J.; Wood, L.D.; Eshleman, J.R.; Goggins, M.G.; Kern, S.E.; Klein, A.P.; Hruban, R.H. The genetics of ductal adenocarcinoma of the pancreas in the year 2020: Dramatic progress, but far to go. *Mod. Pathol.* **2020**, *33*, 2544–2563. [CrossRef]
69.   Wood, L.D.; Canto, M.I.; Jaffee, E.M.; Simeone, D.M. Pancreatic Cancer: Pathogenesis, Screening, Diagnosis, and Treatment. *Gastroenterology* **2022**, *163*, 386–402.e1. [CrossRef]
70.   Morani, A.C.; Hanafy, A.K.; Ramani, N.S.; Katabathina, V.S.; Yeduluri, S.; Dasyam, A.K.; Prasad, S.R. Hereditary and Sporadic Pancreatic Ductal Adenocarcinoma: Current Update on Genetics and Imaging. *Radiol. Imaging Cancer* **2020**, *2*, e190020. [CrossRef]
71.   Willis, J.A.; Mukherjee, S.; Orlow, I.; Viale, A.; Offit, K.; Kurtz, R.C.; Olson, S.H.; Klein, R.J. Genome-wide analysis of the role of copy-number variation in pancreatic cancer risk. *Front. Genet.* **2014**, *5*, 29. [CrossRef] [PubMed]
72.   Huang, L.; Yu, D.; Wu, C.; Zhai, K.; Jiang, G.; Cao, G.; Wang, C.; Liu, Y.; Sun, M.; Li, Z.; et al. Copy number variation at 6q13 functions as a long-range regulator and is associated with pancreatic cancer risk. *Carcinogenesis* **2012**, *33*, 94–100. [CrossRef] [PubMed]
73.   Al-Sukhni, W.; Joe, S.; Lionel, A.C.; Zwingerman, N.; Zogopoulos, G.; Marshall, C.R.; Borgida, A.; Holter, S.; Gropper, A.; Moore, S.; et al. Identification of germline genomic copy number variation in familial pancreatic cancer. *Hum. Genet.* **2012**, *131*, 1481–1494. [CrossRef] [PubMed]
74.   Cancer Genome Atlas Research Network. Integrated Genomic Characterization of Pancreatic Ductal Adenocarcinoma. *Cancer Cell* **2017**, *32*, 185–203.e13. [CrossRef] [PubMed]
75.   Waddell, N.; Pajic, M.; Patch, A.M.; Chang, D.K.; Kassahn, K.S.; Bailey, P.; Johns, A.L.; Miller, D.; Nones, K.; Quek, K.; et al. Whole genomes redefine the mutational landscape of pancreatic cancer. *Nature* **2015**, *518*, 495–501. [CrossRef] [PubMed]
76.   Harada, T.; Chelala, C.; Bhakta, V.; Chaplin, T.; Caulee, K.; Baril, P.; Young, B.D.; Lemoine, N.R. Genome-wide DNA copy number analysis in pancreatic cancer using high-density single nucleotide polymorphism arrays. *Oncogene* **2008**, *27*, 1951–1960. [CrossRef]

77. Lucito, R.; Suresh, S.; Walter, K.; Pandey, A.; Lakshmi, B.; Krasnitz, A.; Sebat, J.; Wigler, M.; Klein, A.P.; Brune, K.; et al. Copy-number variants in patients with a strong family history of pancreatic cancer. *Cancer Biol. Ther.* **2007**, *6*, 1592–1599. [CrossRef]
78. Harada, T.; Baril, P.; Gangeswaran, R.; Kelly, G.; Chelala, C.; Bhakta, V.; Caulee, K.; Mahon, P.C.; Lemoine, N.R. Identification of genetic alterations in pancreatic cancer by the combined use of tissue microdissection and array-based comparative genomic hybridisation. *Br. J. Cancer* **2007**, *96*, 373–382. [CrossRef]
79. Fujita, M.; Sugama, S.; Nakai, M.; Takenouchi, T.; Wei, J.; Urano, T.; Inoue, S.; Hashimoto, M. alpha-Synuclein stimulates differentiation of osteosarcoma cells: Relevance to down-regulation of proteasome activity. *J. Biol. Chem.* **2007**, *282*, 5736–5748. [CrossRef]
80. McLean, G.W.; Carragher, N.O.; Avizienyte, E.; Evans, J.; Brunton, V.G.; Frame, M.C. The role of focal-adhesion kinase in cancer—A new therapeutic opportunity. *Nat. Rev. Cancer* **2005**, *5*, 505–515. [CrossRef]
81. Fu, B.; Luo, M.; Lakkur, S.; Lucito, R.; Iacobuzio-Donahue, C.A. Frequent genomic copy number gain and overexpression of GATA-6 in pancreatic carcinoma. *Cancer Biol. Ther.* **2008**, *7*, 1593–1601. [CrossRef] [PubMed]
82. Maitra, A.; Hruban, R.H. Pancreatic cancer. *Annu. Rev. Pathol.* **2008**, *3*, 157–188. [CrossRef]
83. Zhong, Y.; Wang, Z.; Fu, B.; Pan, F.; Yachida, S.; Dhara, M.; Albesiano, E.; Li, L.; Naito, Y.; Vilardell, F.; et al. GATA6 activates Wnt signaling in pancreatic cancer by negatively regulating the Wnt antagonist Dickkopf-1. *PLoS ONE* **2011**, *6*, e22129. [CrossRef]
84. O'Kane, G.M.; Grunwald, B.T.; Jang, G.H.; Masoomian, M.; Picardo, S.; Grant, R.C.; Denroche, R.E.; Zhang, A.; Wang, Y.; Lam, B.; et al. GATA6 Expression Distinguishes Classical and Basal-like Subtypes in Advanced Pancreatic Cancer. *Clin. Cancer Res.* **2020**, *26*, 4901–4910. [CrossRef] [PubMed]
85. Aung, K.L.; Fischer, S.E.; Denroche, R.E.; Jang, G.H.; Dodd, A.; Creighton, S.; Southwood, B.; Liang, S.B.; Chadwick, D.; Zhang, A.; et al. Genomics-Driven Precision Medicine for Advanced Pancreatic Cancer: Early Results from the COMPASS Trial. *Clin. Cancer Res.* **2018**, *24*, 1344–1354. [CrossRef] [PubMed]
86. Dang, C.V. MYC on the path to cancer. *Cell* **2012**, *149*, 22–35. [CrossRef] [PubMed]
87. Schneider, G.; Wirth, M.; Keller, U.; Saur, D. Rationale for MYC imaging and targeting in pancreatic cancer. *EJNMMI Res.* **2021**, *11*, 104. [CrossRef]
88. Soucek, L.; Whitfield, J.; Martins, C.P.; Finch, A.J.; Murphy, D.J.; Sodir, N.M.; Karnezis, A.N.; Swigart, L.B.; Nasi, S.; Evan, G.I. Modelling Myc inhibition as a cancer therapy. *Nature* **2008**, *455*, 679–683. [CrossRef]
89. Dang, C.V.; Reddy, E.P.; Shokat, K.M.; Soucek, L. Drugging the 'undruggable' cancer targets. *Nat. Rev. Cancer* **2017**, *17*, 502–508. [CrossRef]
90. Beaulieu, M.E.; Jauset, T.; Masso-Valles, D.; Martinez-Martin, S.; Rahl, P.; Maltais, L.; Zacarias-Fluck, M.F.; Casacuberta-Serra, S.; Serrano Del Pozo, E.; Fiore, C.; et al. Intrinsic cell-penetrating activity propels Omomyc from proof of concept to viable anti-MYC therapy. *Sci. Transl. Med.* **2019**, *11*, eaar5012. [CrossRef]
91. Hayashi, A.; Fan, J.; Chen, R.; Ho, Y.J.; Makohon-Moore, A.P.; Lecomte, N.; Zhong, Y.; Hong, J.; Huang, J.; Sakamoto, H.; et al. A unifying paradigm for transcriptional heterogeneity and squamous features in pancreatic ductal adenocarcinoma. *Nat. Cancer* **2020**, *1*, 59–74. [CrossRef] [PubMed]
92. Hessmann, E.; Buchholz, S.M.; Demir, I.E.; Singh, S.K.; Gress, T.M.; Ellenrieder, V.; Neesse, A. Microenvironmental Determinants of Pancreatic Cancer. *Physiol. Rev.* **2020**, *100*, 1707–1751. [CrossRef] [PubMed]
93. Sodir, N.M.; Kortlever, R.M.; Barthet, V.J.A.; Campos, T.; Pellegrinet, L.; Kupczak, S.; Anastasiou, P.; Swigart, L.B.; Soucek, L.; Arends, M.J.; et al. MYC Instructs and Maintains Pancreatic Adenocarcinoma Phenotype. *Cancer Discov.* **2020**, *10*, 588–607. [CrossRef] [PubMed]
94. Ischenko, I.; D'Amico, S.; Rao, M.; Li, J.; Hayman, M.J.; Powers, S.; Petrenko, O.; Reich, N.C. KRAS drives immune evasion in a genetic model of pancreatic cancer. *Nat. Commun.* **2021**, *12*, 1482. [CrossRef] [PubMed]
95. Maddipati, R.; Norgard, R.J.; Baslan, T.; Rathi, K.S.; Zhang, A.; Saeid, A.; Higashihara, T.; Wu, F.; Kumar, A.; Annamalai, V.; et al. MYC Levels Regulate Metastatic Heterogeneity in Pancreatic Adenocarcinoma. *Cancer Discov.* **2022**, *12*, 542–561. [CrossRef]
96. Hassan, Z.; Schneeweis, C.; Wirth, M.; Veltkamp, C.; Dantes, Z.; Feuerecker, B.; Ceyhan, G.O.; Knauer, S.K.; Weichert, W.; Schmid, R.M.; et al. MTOR inhibitor-based combination therapies for pancreatic cancer. *Br. J. Cancer* **2018**, *118*, 366–377. [CrossRef] [PubMed]
97. Conway, J.R.; Herrmann, D.; Evans, T.J.; Morton, J.P.; Timpson, P. Combating pancreatic cancer with PI3K pathway inhibitors in the era of personalised medicine. *Gut* **2019**, *68*, 742–758. [CrossRef]
98. Driscoll, D.R.; Karim, S.A.; Sano, M.; Gay, D.M.; Jacob, W.; Yu, J.; Mizukami, Y.; Gopinathan, A.; Jodrell, D.I.; Evans, T.R.; et al. mTORC2 Signaling Drives the Development and Progression of Pancreatic Cancer. *Cancer Res.* **2016**, *76*, 6911–6923. [CrossRef]
99. Morran, D.C.; Wu, J.; Jamieson, N.B.; Mrowinska, A.; Kalna, G.; Karim, S.A.; Au, A.Y.; Scarlett, C.J.; Chang, D.K.; Pajak, M.Z.; et al. Targeting mTOR dependency in pancreatic cancer. *Gut* **2014**, *63*, 1481–1489. [CrossRef]
100. Knudsen, E.S.; Kumarasamy, V.; Ruiz, A.; Sivinski, J.; Chung, S.; Grant, A.; Vail, P.; Chauhan, S.S.; Jie, T.; Riall, T.S.; et al. Cell cycle plasticity driven by MTOR signaling: Integral resistance to CDK4/6 inhibition in patient-derived models of pancreatic cancer. *Oncogene* **2019**, *38*, 3355–3370. [CrossRef]
101. Allen-Petersen, B.L.; Risom, T.; Feng, Z.; Wang, Z.; Jenny, Z.P.; Thoma, M.C.; Pelz, K.R.; Morton, J.P.; Sansom, O.J.; Lopez, C.D.; et al. Activation of PP2A and Inhibition of mTOR Synergistically Reduce MYC Signaling and Decrease Tumor Growth in Pancreatic Ductal Adenocarcinoma. *Cancer Res.* **2019**, *79*, 209–219. [CrossRef] [PubMed]

102. Stranger, B.E.; Forrest, M.S.; Dunning, M.; Ingle, C.E.; Beazley, C.; Thorne, N.; Redon, R.; Bird, C.P.; de Grassi, A.; Lee, C.; et al. Relative impact of nucleotide and copy number variation on gene expression phenotypes. *Science* **2007**, *315*, 848–853. [CrossRef] [PubMed]

103. Krimpenfort, P.; Ijpenberg, A.; Song, J.Y.; van der Valk, M.; Nawijn, M.; Zevenhoven, J.; Berns, A. p15Ink4b is a critical tumour suppressor in the absence of p16Ink4a. *Nature* **2007**, *448*, 943–946. [CrossRef] [PubMed]

104. Roussel, M.F. The INK4 family of cell cycle inhibitors in cancer. *Oncogene* **1999**, *18*, 5311–5317. [CrossRef] [PubMed]

105. Huang, W.L.; Wu, S.F.; Huang, X.; Zhou, S. Integrated Analysis of ECT2 and COL17A1 as Potential Biomarkers for Pancreatic Cancer. *Dis. Markers* **2022**, *2022*, 9453549. [CrossRef] [PubMed]

106. Kozawa, K.; Sekai, M.; Ohba, K.; Ito, S.; Sako, H.; Maruyama, T.; Kakeno, M.; Shirai, T.; Kuromiya, K.; Kamasaki, T.; et al. The CD44/COL17A1 pathway promotes the formation of multilayered, transformed epithelia. *Curr. Biol.* **2021**, *31*, 3086–3097.e7. [CrossRef]

107. Ho, T.T.B.; Nasti, A.; Seki, A.; Komura, T.; Inui, H.; Kozaka, T.; Kitamura, Y.; Shiba, K.; Yamashita, T.; Yamashita, T.; et al. Combination of gemcitabine and anti-PD-1 antibody enhances the anticancer effect of M1 macrophages and the Th1 response in a murine model of pancreatic cancer liver metastasis. *J. Immunother. Cancer* **2020**, *8*, e001367. [CrossRef]

108. Pu, N.; Gao, S.; Yin, H.; Li, J.A.; Wu, W.; Fang, Y.; Zhang, L.; Rong, Y.; Xu, X.; Wang, D.; et al. Cell-intrinsic PD-1 promotes proliferation in pancreatic cancer by targeting CYR61/CTGF via the hippo pathway. *Cancer Lett.* **2019**, *460*, 42–53. [CrossRef]

109. Hayashi, H.; Higashi, T.; Miyata, T.; Yamashita, Y.I.; Baba, H. Recent advances in precision medicine for pancreatic ductal adenocarcinoma. *Ann. Gastroenterol. Surg.* **2021**, *5*, 457–466. [CrossRef]

110. Zhan, Q.; Wen, C.; Zhao, Y.; Fang, L.; Jin, Y.; Zhang, Z.; Zou, S.; Li, F.; Yang, Y.; Wu, L.; et al. Identification of copy number variation-driven molecular subtypes informative for prognosis and treatment in pancreatic adenocarcinoma of a Chinese cohort. *eBioMedicine* **2021**, *74*, 103716. [CrossRef]

111. Ben-David, U.; Amon, A. Context is everything: Aneuploidy in cancer. *Nat. Rev. Genet.* **2020**, *21*, 44–62. [CrossRef]

112. Taylor, A.M.; Shih, J.; Ha, G.; Gao, G.F.; Zhang, X.; Berger, A.C.; Schumacher, S.E.; Wang, C.; Hu, H.; Liu, J.; et al. Genomic and Functional Approaches to Understanding Cancer Aneuploidy. *Cancer Cell* **2018**, *33*, 676–689.e3. [CrossRef] [PubMed]

113. Zack, T.I.; Schumacher, S.E.; Carter, S.L.; Cherniack, A.D.; Saksena, G.; Tabak, B.; Lawrence, M.S.; Zhsng, C.Z.; Wala, J.; Mermel, C.H.; et al. Pan-cancer patterns of somatic copy number alteration. *Nat. Genet.* **2013**, *45*, 1134–1140. [CrossRef]

114. Notta, F.; Chan-Seng-Yue, M.; Lemire, M.; Li, Y.; Wilson, G.W.; Connor, A.A.; Denroche, R.E.; Liang, S.B.; Brown, A.M.; Kim, J.C.; et al. A renewed model of pancreatic cancer evolution based on genomic rearrangement patterns. *Nature* **2016**, *538*, 378–382. [CrossRef] [PubMed]

115. Balli, D.; Rech, A.J.; Stanger, B.Z.; Vonderheide, R.H. Immune Cytolytic Activity Stratifies Molecular Subsets of Human Pancreatic Cancer. *Clin. Cancer Res.* **2017**, *23*, 3129–3138. [CrossRef] [PubMed]

116. Petersson, A.; Andersson, N.; Hau, S.O.; Eberhard, J.; Karlsson, J.; Chattopadhyay, S.; Valind, A.; Elebro, J.; Nodin, B.; Leandersson, K.; et al. Branching Copy-Number Evolution and Parallel Immune Profiles across the Regional Tumor Space of Resected Pancreatic Cancer. *Mol. Cancer Res.* **2022**, *20*, 749–761. [CrossRef] [PubMed]

117. Xu, D.; Wang, Y.; Liu, X.; Zhou, K.; Wu, J.; Chen, J.; Chen, C.; Chen, L.; Zheng, J. Development and clinical validation of a novel 9-gene prognostic model based on multi-omics in pancreatic adenocarcinoma. *Pharmacol. Res.* **2021**, *164*, 105370. [CrossRef]

118. Chen, S.; Auletta, T.; Dovirak, O.; Hutter, C.; Kuntz, K.; El-ftesi, S.; Kendall, J.; Han, H.; Von Hoff, D.D.; Ashfaq, R.; et al. Copy number alterations in pancreatic cancer identify recurrent PAK4 amplification. *Cancer Biol. Ther.* **2008**, *7*, 1793–1802. [CrossRef]

119. Zheng, Q.; Yu, X.; Zhang, Q.; He, Y.; Guo, W. Genetic characteristics and prognostic implications of m1A regulators in pancreatic cancer. *Biosci. Rep.* **2021**, *41*, BSR20210337. [CrossRef]

120. Witkiewicz, A.K.; McMillan, E.A.; Balaji, U.; Baek, G.; Lin, W.C.; Mansour, J.; Mollaee, M.; Wagner, K.U.; Koduru, P.; Yopp, A.; et al. Whole-exome sequencing of pancreatic cancer defines genetic diversity and therapeutic targets. *Nat. Commun.* **2015**, *6*, 6744. [CrossRef]

121. Campbell, P.J.; Yachida, S.; Mudie, L.J.; Stephens, P.J.; Pleasance, E.D.; Stebbings, L.A.; Morsberger, L.A.; Latimer, C.; McLaren, S.; Lin, M.L.; et al. The patterns and dynamics of genomic instability in metastatic pancreatic cancer. *Nature* **2010**, *467*, 1109–1113. [CrossRef] [PubMed]

122. Zhou, Z.; Zhang, J.; Xu, C.; Yang, J.; Zhang, Y.; Liu, M.; Shi, X.; Li, X.; Zhan, H.; Chen, W.; et al. An integrated model of N6-methyladenosine regulators to predict tumor aggressiveness and immune evasion in pancreatic cancer. *eBioMedicine* **2021**, *65*, 103271. [CrossRef] [PubMed]

123. Zhu, L.L.; Wu, Z.; Li, R.K.; Xing, X.; Jiang, Y.S.; Li, J.; Wang, Y.H.; Hu, L.P.; Wang, X.; Qin, W.T.; et al. Deciphering the genomic and lncRNA landscapes of aerobic glycolysis identifies potential therapeutic targets in pancreatic cancer. *Int. J. Biol. Sci.* **2021**, *17*, 107–118. [CrossRef] [PubMed]

124. Liang, J.W.; Shi, Z.Z.; Shen, T.Y.; Che, X.; Wang, Z.; Shi, S.S.; Xu, X.; Cai, Y.; Zhao, P.; Wang, C.F.; et al. Identification of genomic alterations in pancreatic cancer using array-based comparative genomic hybridization. *PLoS ONE* **2014**, *9*, e114616. [CrossRef] [PubMed]

125. Imamura, T.; Ashida, R.; Ohshima, K.; Uesaka, K.; Sugiura, T.; Okamura, Y.; Ohgi, K.; Ohnami, S.; Nagashima, T.; Yamaguchi, K. Genomic landscape of pancreatic cancer in the Japanese version of the Cancer Genome Atlas. *Ann. Gastroenterol. Surg.* **2023**, *7*, 491–502. [CrossRef] [PubMed]

126. Cao, L.; Huang, C.; Cui Zhou, D.; Hu, Y.; Lih, T.M.; Savage, S.R.; Krug, K.; Clark, D.J.; Schnaubelt, M.; Chen, L.; et al. Proteogenomic characterization of pancreatic ductal adenocarcinoma. *Cell* **2021**, *184*, 5031–5052.e26. [CrossRef]

127. Zhang, X.; Mao, T.; Zhang, B.; Xu, H.; Cui, J.; Jiao, F.; Chen, D.; Wang, Y.; Hu, J.; Xia, Q.; et al. Characterization of the genomic landscape in large-scale Chinese patients with pancreatic cancer. *eBioMedicine* **2022**, *77*, 103897. [CrossRef] [PubMed]

128. Yang, Y.; Ding, Y.; Gong, Y.; Zhao, S.; Li, M.; Li, X.; Song, G.; Zhai, B.; Liu, J.; Shao, Y.; et al. The genetic landscape of pancreatic head ductal adenocarcinoma in China and prognosis stratification. *BMC Cancer* **2022**, *22*, 186. [CrossRef]

129. Detlefsen, S.; Boldt, H.B.; Burton, M.; Thomsen, M.M.; Rasmussen, L.G.; Orbeck, S.V.; Pfeiffer, P.; Mortensen, M.B.; de Stricker, K. High overall copy number variation burden by genome-wide methylation profiling holds negative prognostic value in surgically treated pancreatic ductal adenocarcinoma. *Hum. Pathol.* **2023**, *142*, 68–80. [CrossRef]

130. Kwon, M.J.; Jeon, J.Y.; Park, H.R.; Nam, E.S.; Cho, S.J.; Shin, H.S.; Kwon, J.H.; Kim, J.S.; Han, B.; Kim, D.H.; et al. Low frequency of KRAS mutation in pancreatic ductal adenocarcinomas in Korean patients and its prognostic value. *Pancreas* **2015**, *44*, 484–492. [CrossRef]

131. Shin, S.H.; Kim, S.C.; Hong, S.M.; Kim, Y.H.; Song, K.B.; Park, K.M.; Lee, Y.J. Genetic alterations of K-ras, p53, c-erbB-2, and DPC4 in pancreatic ductal adenocarcinoma and their correlation with patient survival. *Pancreas* **2013**, *42*, 216–222. [CrossRef] [PubMed]

132. Lee, J.H.; Giovannetti, E.; Hwang, J.H.; Petrini, I.; Wang, Q.; Voortman, J.; Wang, Y.; Steinberg, S.M.; Funel, N.; Meltzer, P.S.; et al. Loss of 18q22.3 involving the carboxypeptidase of glutamate-like gene is associated with poor prognosis in resected pancreatic cancer. *Clin. Cancer Res.* **2012**, *18*, 524–533. [CrossRef] [PubMed]

133. Barrett, M.T.; Deiotte, R.; Lenkiewicz, E.; Malasi, S.; Holley, T.; Evers, L.; Posner, R.G.; Jones, T.; Han, H.; Sausen, M.; et al. Clinical study of genomic drivers in pancreatic ductal adenocarcinoma. *Br. J. Cancer* **2017**, *117*, 572–582. [CrossRef] [PubMed]

134. Guo, S.; Shi, X.; Gao, S.; Hou, Q.; Jiang, L.; Li, B.; Shen, J.; Wang, H.; Shen, S.; Zhang, G.; et al. The Landscape of Genetic Alterations Stratified Prognosis in Oriental Pancreatic Cancer Patients. *Front. Oncol.* **2021**, *11*, 717989. [CrossRef] [PubMed]

135. Fujii, A.; Masuda, T.; Iwata, M.; Tobo, T.; Wakiyama, H.; Koike, K.; Kosai, K.; Nakano, T.; Kuramitsu, S.; Kitagawa, A.; et al. The novel driver gene ASAP2 is a potential druggable target in pancreatic cancer. *Cancer Sci.* **2021**, *112*, 1655–1668. [CrossRef]

136. Birnbaum, D.J.; Bertucci, F.; Finetti, P.; Adelaide, J.; Giovannini, M.; Turrini, O.; Delpero, J.R.; Raoul, J.L.; Chaffanet, M.; Moutardier, V.; et al. Expression of Genes with Copy Number Alterations and Survival of Patients with Pancreatic Adenocarcinoma. *Cancer Genom. Proteom.* **2016**, *13*, 191–200.

137. van Hattem, W.A.; Carvalho, R.; Li, A.; Offerhaus, G.J.; Goggins, M. Amplification of EMSY gene in a subset of sporadic pancreatic adenocarcinomas. *Int. J. Clin. Exp. Pathol.* **2008**, *1*, 343–351.

138. Huang, L.; Yuan, X.; Zhao, L.; Han, Q.; Yan, H.; Yuan, J.; Guan, S.; Xu, X.; Dai, G.; Wang, J.; et al. Gene signature developed for predicting early relapse and survival in early-stage pancreatic cancer. *BJS Open* **2023**, *7*, zrad031. [CrossRef]

139. Kalimuthu, S.N.; Wilson, G.W.; Grant, R.C.; Seto, M.; O'Kane, G.; Vajpeyi, R.; Notta, F.; Gallinger, S.; Chetty, R. Morphological classification of pancreatic ductal adenocarcinoma that predicts molecular subtypes and correlates with clinical outcome. *Gut* **2020**, *69*, 317–328. [CrossRef]

140. Thomsen, M.M.; Larsen, M.H.; Di Caterino, T.; Hedegaard Jensen, G.; Mortensen, M.B.; Detlefsen, S. Accuracy and clinical outcomes of pancreatic EUS-guided fine-needle biopsy in a consecutive series of 852 specimens. *Endosc. Ultrasound* **2022**, *11*, 306–318.

141. Fitzpatrick, M.J.; Hernandez-Barco, Y.G.; Krishnan, K.; Casey, B.; Pitman, M.B. Evaluating triage protocols for endoscopic ultrasound-guided fine needle biopsies of the pancreas. *J. Am. Soc. Cytopathol.* **2020**, *9*, 396–404. [CrossRef]

142. Tibiletti, M.G.; Carnevali, I.; Pensotti, V.; Chiaravalli, A.M.; Facchi, S.; Volorio, S.; Mariette, F.; Mariani, P.; Fortuzzi, S.; Pierotti, M.A.; et al. OncoPan((R)): An NGS-Based Screening Methodology to Identify Molecular Markers for Therapy and Risk Assessment in Pancreatic Ductal Adenocarcinoma. *Biomedicines* **2022**, *10*, 1208. [CrossRef] [PubMed]

143. Hirokawa, Y.S.; Iwata, T.; Okugawa, Y.; Tanaka, K.; Sakurai, H.; Watanabe, M. HER2-positive adenocarcinoma arising from heterotopic pancreas tissue in the duodenum: A case report. *World J. Gastroenterol.* **2021**, *27*, 4738–4745. [CrossRef] [PubMed]

144. Abbosh, C.; Birkbak, N.J.; Wilson, G.A.; Jamal-Hanjani, M.; Constantin, T.; Salari, R.; Le Quesne, J.; Moore, D.A.; Veeriah, S.; Rosenthal, R.; et al. Phylogenetic ctDNA analysis depicts early-stage lung cancer evolution. *Nature* **2017**, *545*, 446–451. [CrossRef]

145. Goldberg, S.B.; Narayan, A.; Kole, A.J.; Decker, R.H.; Teysir, J.; Carriero, N.J.; Lee, A.; Nemati, R.; Nath, S.K.; Mane, S.M.; et al. Early Assessment of Lung Cancer Immunotherapy Response via Circulating Tumor DNA. *Clin. Cancer Res.* **2018**, *24*, 1872–1880. [CrossRef] [PubMed]

146. Siravegna, G.; Mussolin, B.; Buscarino, M.; Corti, G.; Cassingena, A.; Crisafulli, G.; Ponzetti, A.; Cremolini, C.; Amatu, A.; Lauricella, C.; et al. Clonal evolution and resistance to EGFR blockade in the blood of colorectal cancer patients. *Nat. Med.* **2015**, *21*, 795–801. [CrossRef] [PubMed]

147. Peng, H.; Lu, L.; Zhou, Z.; Liu, J.; Zhang, D.; Nan, K.; Zhao, X.; Li, F.; Tian, L.; Dong, H.; et al. CNV Detection from Circulating Tumor DNA in Late Stage Non-Small Cell Lung Cancer Patients. *Genes* **2019**, *10*, 926. [CrossRef] [PubMed]

148. Chen, L.; Ma, R.; Luo, C.; Xie, Q.; Ning, X.; Sun, K.; Meng, F.; Zhou, M.; Sun, J. Noninvasive early differential diagnosis and progression monitoring of ovarian cancer using the copy number alterations of plasma cell-free DNA. *Transl. Res.* **2023**, *262*, 12–24. [CrossRef]

149. Kuuselo, R.; Simon, R.; Karhu, R.; Tennstedt, P.; Marx, A.H.; Izbicki, J.R.; Yekebas, E.; Sauter, G.; Kallioniemi, A. 19q13 amplification is associated with high grade and stage in pancreatic cancer. *Genes Chromosomes Cancer* **2010**, *49*, 569–575. [CrossRef]

150. Gutierrez, M.L.; Munoz-Bellvis, L.; Abad Mdel, M.; Bengoechea, O.; Gonzalez-Gonzalez, M.; Orfao, A.; Sayagues, J.M. Association between genetic subgroups of pancreatic ductal adenocarcinoma defined by high density 500 K SNP-arrays and tumor histopathology. *PLoS ONE* **2011**, *6*, e22315. [CrossRef]

151. Sung, H.Y.; Wu, H.G.; Ahn, J.H.; Park, W.Y. Dcr3 inhibit p53-dependent apoptosis in gamma-irradiated lung cancer cells. *Int. J. Radiat. Biol.* **2010**, *86*, 780–790. [CrossRef] [PubMed]

152. Chen, C.; Zhang, C.; Zhuang, G.; Luo, H.; Su, J.; Yin, P.; Wang, J. Decoy receptor 3 overexpression and immunologic tolerance in hepatocellular carcinoma (HCC) development. *Cancer Investig.* **2008**, *26*, 965–974. [CrossRef]

153. Ho, C.H.; Chen, C.L.; Li, W.Y.; Chen, C.J. Decoy receptor 3, upregulated by Epstein-Barr virus latent membrane protein 1, enhances nasopharyngeal carcinoma cell migration and invasion. *Carcinogenesis* **2009**, *30*, 1443–1451. [CrossRef] [PubMed]

154. Chen, G.; Luo, D. Over-expression of decoy receptor 3 in gastric precancerous lesions and carcinoma. *Ups. J. Med. Sci.* **2008**, *113*, 297–304. [CrossRef] [PubMed]

155. Tsuji, S.; Hosotani, R.; Yonehara, S.; Masui, T.; Tulachan, S.S.; Nakajima, S.; Kobayashi, H.; Koizumi, M.; Toyoda, E.; Ito, D.; et al. Endogenous decoy receptor 3 blocks the growth inhibition signals mediated by Fas ligand in human pancreatic adenocarcinoma. *Int. J. Cancer* **2003**, *106*, 17–25. [CrossRef] [PubMed]

156. Casagrande, G.; te Kronnie, G.; Basso, G. The effects of siRNA-mediated inhibition of E2A-PBX1 on EB-1 and Wnt16b expression in the 697 pre-B leukemia cell line. *Haematologica* **2006**, *91*, 765–771. [PubMed]

157. Abiatari, I.; Gillen, S.; DeOliveira, T.; Klose, T.; Bo, K.; Giese, N.A.; Friess, H.; Kleeff, J. The microtubule-associated protein MAPRE2 is involved in perineural invasion of pancreatic cancer cells. *Int. J. Oncol.* **2009**, *35*, 1111–1116. [PubMed]

158. Kimura, Y.; Morita, T.; Hayashi, K.; Miki, T.; Sobue, K. Myocardin functions as an effective inducer of growth arrest and differentiation in human uterine leiomyosarcoma cells. *Cancer Res.* **2010**, *70*, 501–511. [CrossRef]

159. de Oliveira, S.I.; Andrade, L.N.; Onuchic, A.C.; Nonogaki, S.; Fernandes, P.D.; Pinheiro, M.C.; Rohde, C.B.; Chammas, R.; Jancar, S. Platelet-activating factor receptor (PAF-R)-dependent pathways control tumour growth and tumour response to chemotherapy. *BMC Cancer* **2010**, *10*, 200. [CrossRef]

160. Legoffic, A.; Calvo, E.; Cano, C.; Folch-Puy, E.; Barthet, M.; Delpero, J.R.; Ferres-Maso, M.; Dagorn, J.C.; Closa, D.; Iovanna, J. The reg4 gene, amplified in the early stages of pancreatic cancer development, is a promising therapeutic target. *PLoS ONE* **2009**, *4*, e7495. [CrossRef]

161. Violette, S.; Festor, E.; Pandrea-Vasile, I.; Mitchell, V.; Adida, C.; Dussaulx, E.; Lacorte, J.M.; Chambaz, J.; Lacasa, M.; Lesuffleur, T. Reg IV, a new member of the regenerating gene family, is overexpressed in colorectal carcinomas. *Int. J. Cancer* **2003**, *103*, 185–193. [CrossRef] [PubMed]

162. Bishnupuri, K.S.; Luo, Q.; Murmu, N.; Houchen, C.W.; Anant, S.; Dieckgraefe, B.K. Reg IV activates the epidermal growth factor receptor/Akt/AP-1 signaling pathway in colon adenocarcinomas. *Gastroenterology* **2006**, *130*, 137–149. [CrossRef] [PubMed]

163. Kuniyasu, H.; Oue, N.; Sasahira, T.; Yi, L.; Moriwaka, Y.; Shimomoto, T.; Fujii, K.; Ohmori, H.; Yasui, W. Reg IV enhances peritoneal metastasis in gastric carcinomas. *Cell Prolif.* **2009**, *42*, 110–121. [CrossRef] [PubMed]

164. Oue, N.; Hamai, Y.; Mitani, Y.; Matsumura, S.; Oshimo, Y.; Aung, P.P.; Kuraoka, K.; Nakayama, H.; Yasui, W. Gene expression profile of gastric carcinoma: Identification of genes and tags potentially involved in invasion, metastasis, and carcinogenesis by serial analysis of gene expression. *Cancer Res.* **2004**, *64*, 2397–2405. [CrossRef] [PubMed]

165. Zhang, Y.; Lai, M.; Lv, B.; Gu, X.; Wang, H.; Zhu, Y.; Zhu, Y.; Shao, L.; Wang, G. Overexpression of Reg IV in colorectal adenoma. *Cancer Lett.* **2003**, *200*, 69–76. [CrossRef] [PubMed]

166. Takehara, A.; Eguchi, H.; Ohigashi, H.; Ishikawa, O.; Kasugai, T.; Hosokawa, M.; Katagiri, T.; Nakamura, Y.; Nakagawa, H. Novel tumor marker REG4 detected in serum of patients with resectable pancreatic cancer and feasibility for antibody therapy targeting REG4. *Cancer Sci.* **2006**, *97*, 1191–1197. [CrossRef] [PubMed]

167. South, S.T.; Brothman, A.R. Clinical laboratory implementation of cytogenomic microarrays. *Cytogenet. Genome Res.* **2011**, *135*, 203–211. [CrossRef] [PubMed]

168. De Leeuw, N.; Dijkhuizen, T.; Hehir-Kwa, J.Y.; Carter, N.P.; Feuk, L.; Firth, H.V.; Kuhn, R.M.; Ledbetter, D.H.; Martin, C.L.; van Ravenswaaij-Arts, C.M.; et al. Diagnostic interpretation of array data using public databases and internet sources. *Hum. Mutat.* **2012**, *33*, 930–940. [CrossRef]

169. Haraksingh, R.R.; Abyzov, A.; Gerstein, M.; Urban, A.E.; Snyder, M. Genome-wide mapping of copy number variation in humans: Comparative analysis of high resolution array platforms. *PLoS ONE* **2011**, *6*, e27859. [CrossRef]

170. Perry, G.H.; Ben-Dor, A.; Tsalenko, A.; Sampas, N.; Rodriguez-Revenga, L.; Tran, C.W.; Scheffer, A.; Steinfeld, I.; Tsang, P.; Yamada, N.A.; et al. The fine-scale and complex architecture of human copy-number variation. *Am. J. Hum. Genet.* **2008**, *82*, 685–695. [CrossRef]

171. Kearney, H.M.; Thorland, E.C.; Brown, K.K.; Quintero-Rivera, F.; South, S.T.; Working Group of the American College of Medical Genetics Laboratory Quality Assurance Committee. American College of Medical Genetics standards and guidelines for interpretation and reporting of postnatal constitutional copy number variants. *Genet. Med.* **2011**, *13*, 680–685. [CrossRef] [PubMed]

172. Vermeesch, J.R.; Brady, P.D.; Sanlaville, D.; Kok, K.; Hastings, R.J. Genome-wide arrays: Quality criteria and platforms to be used in routine diagnostics. *Hum. Mutat.* **2012**, *33*, 906–915. [CrossRef] [PubMed]

173. Xu, Q.; Chen, S.; Hu, Y.; Huang, W. Single-cell RNA transcriptome reveals the intra-tumoral heterogeneity and regulators underlying tumor progression in metastatic pancreatic ductal adenocarcinoma. *Cell Death Discov.* **2021**, *7*, 331. [CrossRef]

174. Mahdipour-Shirayeh, A.; Erdmann, N.; Leung-Hagesteijn, C.; Tiedemann, R.E. sciCNV: High-throughput paired profiling of transcriptomes and DNA copy number variations at single-cell resolution. *Brief. Bioinform.* **2022**, *23*, bbab413. [CrossRef] [PubMed]
175. Gao, R.; Bai, S.; Henderson, Y.C.; Lin, Y.; Schalck, A.; Yan, Y.; Kumar, T.; Hu, M.; Sei, E.; Davis, A.; et al. Delineating copy number and clonal substructure in human tumors from single-cell transcriptomes. *Nat. Biotechnol.* **2021**, *39*, 599–608. [CrossRef]
176. Fu, X.; Patel, H.P.; Coppola, S.; Xu, L.; Cao, Z.; Lenstra, T.L.; Grima, R. Quantifying how post-transcriptional noise and gene copy number variation bias transcriptional parameter inference from mRNA distributions. *eLife* **2022**, *11*, e82493. [CrossRef]
177. Erickson, A.; He, M.; Berglund, E.; Marklund, M.; Mirzazadeh, R.; Schultz, N.; Kvastad, L.; Andersson, A.; Bergenstrahle, L.; Bergenstrahle, J.; et al. Spatially resolved clonal copy number alterations in benign and malignant tissue. *Nature* **2022**, *608*, 360–367. [CrossRef]
178. Mu, Q.; Wang, J. CNAPE: A Machine Learning Method for Copy Number Alteration Prediction from Gene Expression. *IEEE/ACM Trans. Comput. Biol. Bioinform.* **2021**, *18*, 306–311. [CrossRef]
179. Canto, M.I.; Goggins, M.; Yeo, C.J.; Griffin, C.; Axilbund, J.E.; Brune, K.; Ali, S.Z.; Jagannath, S.; Petersen, G.M.; Fishman, E.K.; et al. Screening for pancreatic neoplasia in high-risk individuals: An EUS-based approach. *Clin. Gastroenterol. Hepatol.* **2004**, *2*, 606–621. [CrossRef]
180. Fanale, D.; Iovanna, J.L.; Calvo, E.L.; Berthezene, P.; Belleau, P.; Dagorn, J.C.; Bronte, G.; Cicero, G.; Bazan, V.; Rolfo, C.; et al. Germline copy number variation in the YTHDC2 gene: Does it have a role in finding a novel potential molecular target involved in pancreatic adenocarcinoma susceptibility? *Expert Opin. Ther. Targets* **2014**, *18*, 841–850. [CrossRef]
181. Lin, B.; Pan, Y.; Yu, D.; Dai, S.; Sun, H.; Chen, S.; Zhang, J.; Xiang, Y.; Huang, C. Screening and Identifying m6A Regulators as an Independent Prognostic Biomarker in Pancreatic Cancer Based on The Cancer Genome Atlas Database. *Biomed. Res. Int.* **2021**, *2021*, 5573628. [CrossRef] [PubMed]
182. Laurila, E.; Savinainen, K.; Kuuselo, R.; Karhu, R.; Kallioniemi, A. Characterization of the 7q21-q22 amplicon identifies ARPC1A, a subunit of the Arp2/3 complex, as a regulator of cell migration and invasion in pancreatic cancer. *Genes Chromosomes Cancer.* **2009**, *48*, 330–339. [CrossRef]
183. Lin, Z.; Liu, J.; Xiao, X.; Meng, J. Expression and prognostic significance of epithelial cell transforming sequence 2 in invasive breast cancer. *Chin. J. Phys. Train.* **2023**, *46*, 780–785.
184. Rausch, V.; Krieg, A.; Camps, J.; Behrens, B.; Beier, M.; Wangsa, D.; Heselmeyer-Haddad, K.; Baldus, S.E.; Knoefel, W.T.; Ried, T.; et al. Array comparative genomic hybridization of 18 pancreatic ductal adenocarcinomas and their autologous metastases. *BMC Res. Notes.* **2017**, *10*, 560. [CrossRef] [PubMed]
185. Qi, B.; Liu, H.; Dong, Y.; Shi, X.; Zhou, Q.; Zeng, F.; Bao, N.; Li, Q.; Yuan, Y.; Yao, L.; et al. The nine ADAMs family members serve as potential biomarkers for immune infiltration in pancreatic adenocarcinoma. *PeerJ.* **2020**, *8*, e9736. [CrossRef] [PubMed]
186. Xu, X.; Yu, Y.; Zong, K.; Lv, P.; Gu, Y. Up-regulation of IGF2BP2 by multiple mechanisms in pancreatic cancer promotes cancer proliferation by activating the PI3K/Akt signaling pathway. *J. Exp. Clin. Cancer. Res.* **2019**, *38*, 497. [CrossRef]
187. Zhong, H.; Shi, Q.; Wen, Q.; Chen, J.; Li, X.; Ruan, R.; Zeng, S.; Dai, X.; Xiong, J.; Li, L.; et al. Pan-cancer analysis reveals potential of FAM110A as a prognostic and immunological biomarker in human cancer. *Front. Immunol.* **2023**, *14*, 1058627. [CrossRef]
188. Peng, J.; Sun, B.F.; Chen, C.Y.; Zhou, J.Y.; Chen, Y.S.; Chen, H.; Liu, L.; Huang, D.; Jiang, J.; Cui, G.S.; et al. Single-cell RNA-seq highlights intra-tumoral heterogeneity and malignant progression in pancreatic ductal adenocarcinoma. *Cell Res.* **2019**, *29*, 725–738. [CrossRef]
189. McCarroll, S.A. Extending genome-wide association studies to copy-number variation. *Hum. Mol. Genet.* **2008**, *17*, R135–R142. [CrossRef]

Disclaimer/Publisher's Note: The statements, opinions and data contained in all publications are solely those of the individual author(s) and contributor(s) and not of MDPI and/or the editor(s). MDPI and/or the editor(s) disclaim responsibility for any injury to people or property resulting from any ideas, methods, instructions or products referred to in the content.

International Journal of
Molecular Sciences

Review

Locoregional Therapies and Remodeling of Tumor Microenvironment in Pancreatic Cancer

Maria Caterina De Grandis [1,†], Velio Ascenti [2,†], Carolina Lanza [2], Giacomo Di Paolo [1], Barbara Galassi [3], Anna Maria Ierardi [4], Gianpaolo Carrafiello [4,5], Antonio Facciorusso [6,*] and Michele Ghidini [3]

[1] Oncology Unit 1, Veneto Institute of Oncology IOV-IRCCS, 35128 Padua, Italy; mariacaterina.degrandis@iov.veneto.it (M.C.D.G.); giacomo.dipaolo@iov.veneto.it (G.D.P.)
[2] Postgraduate School of Diagnostic and Interventional Radiology, University of Milan, 20122 Milan, Italy; velio.ascenti@gmail.com (V.A.); carolinalanza92@gmail.com (C.L.)
[3] Oncology Unit, Fondazione IRCCS Ca' Granda Ospedale Maggiore Policlinico, 20122 Milan, Italy; barbara.galassi@policlinico.mi.it (B.G.); michele.ghidini@policlinico.mi.it (M.G.)
[4] Radiology Unit, Fondazione IRCCS Ca' Granda Ospedale Maggiore Policlinico, 20122 Milan, Italy; annamaria.ierardi@policlinico.mi.it (A.M.I.); gianpaolo.carrafiello@policlinico.mi.it (G.C.)
[5] Department of Oncology and Haemato-Oncology, University of Milan, 20122 Milan, Italy
[6] Section of Gastroenterology, Department of Medical and Surgical Sciences, University of Foggia, 71122 Foggia, Italy
* Correspondence: antonio.facciorusso@unifg.it
† These authors contributed equally to this work.

Citation: De Grandis, M.C.; Ascenti, V.; Lanza, C.; Di Paolo, G.; Galassi, B.; Ierardi, A.M.; Carrafiello, G.; Facciorusso, A.; Ghidini, M. Locoregional Therapies and Remodeling of Tumor Microenvironment in Pancreatic Cancer. *Int. J. Mol. Sci.* **2023**, *24*, 12681. https://doi.org/10.3390/ijms241612681

Academic Editors: Claudio Luchini and Donatella Delle Cave

Received: 29 June 2023
Revised: 5 August 2023
Accepted: 8 August 2023
Published: 11 August 2023

Copyright: © 2023 by the authors. Licensee MDPI, Basel, Switzerland. This article is an open access article distributed under the terms and conditions of the Creative Commons Attribution (CC BY) license (https://creativecommons.org/licenses/by/4.0/).

Abstract: Despite the advances made in treatment, the prognosis of pancreatic ductal adenocarcinoma (PDAC) remains dismal, even in the locoregional and locally advanced stages, with high relapse rates after surgery. PDAC exhibits a chemoresistant and immunosuppressive phenotype, and the tumor microenvironment (TME) surrounding cancer cells actively participates in creating a stromal barrier to chemotherapy and an immunosuppressive environment. Recently, there has been an increasing use of interventional radiology techniques for the treatment of PDAC, although they do not represent a standard of care and are not included in clinical guidelines. Local approaches such as radiation therapy, hyperthermia, microwave or radiofrequency ablation, irreversible electroporation and high-intensity focused ultrasound exert their action on the tumor tissue, altering the composition and structure of TME and potentially enhancing the action of chemotherapy. Moreover, their action can increase antigen release and presentation with T-cell activation and reduction tumor-induced immune suppression. This review summarizes the current evidence on locoregional therapies in PDAC and their effect on remodeling TME to make it more susceptible to the action of antitumor agents.

Keywords: pancreatic cancer; locoregional treatments; tumor microenvironment; ablation therapies; radiotherapy

1. Introduction

According to the Global Cancer Observatory, in 2020, pancreatic cancer ranked as the 12th most common cancer for incidence and the 7th in terms of annual number of deaths worldwide [1]. Incidence and mortality have shown an increasing trend in recent years [2], and pancreatic cancer will become the second cause of cancer-related death by 2030 in the United States [3]. At diagnosis, no more than 15–20% of patients are eligible for upfront surgery, while approximately 30–40% present with borderline resectable or locally advanced disease, and the rest of the patients are diagnosed with metastases [4].

The treatment of resectable pancreatic ductal adenocarcinoma (PDAC) involves surgery followed by adjuvant chemotherapy. Preferred regimens include Gemcitabine-Capecitabine and mFOLFIRINOX for patients with good performance status [5,6]. For patients with borderline resectable PDAC, various treatment regimens have been tested, but currently, there is still no consensus on the optimal therapeutic approach [7–10]. For advanced disease, first-line chemotherapy options for patients in good overall condition include gemcitabine plus nab-paclitaxel or FOLFIRINOX, as an alternative to gemcitabine alone [5,11,12].

Despite the advances made in treatment, the prognosis of PDAC remains poor, with a median overall survival (OS) of approximately 12 months for metastatic disease at diagnosis [5,12]. PDAC is an aggressive tumor, and the fact that it is often diagnosed at advanced stages limits the prospects for treatment. Additionally, tumor cells frequently exhibit mechanisms of resistance to available treatments, reducing the effectiveness of drugs [13]. Therefore, the development of novel therapeutic strategies is a major challenge. The efficacy and feasibility of locoregional treatments are currently under investigation. Local approaches such as radiation therapy, hyperthermia, microwave or radiofrequency ablation, irreversible electroporation, and high-intensity focused ultrasound exert their action on the tumor tissue, limiting toxicity to healthy tissues and altering the composition and structure of the tumor microenvironment (TME), potentially enhancing the action of other anticancer agents.

This review aims to summarize the current evidence on locoregional therapies in PDAC and their effect on remodeling the tumor environment to make it more susceptible to the action of antitumor agents.

2. Tumor Microenvironment in PDAC

The biological behavior of PDAC is strongly dependent on its interaction with the adjacent tissues. TME refers to all the normal cells, molecules and blood vessels that surround cancer cells. TME may be considered as a dynamic network of cells and stroma constituents; thus, its composition and functions are extremely various and it shows both an intratumor and intertumor heterogeneity [14,15] (Figure 1).

Figure 1. Schematic cellular composition of tumor microenvironment (TME) in PDAC. Tumor fibrosis is sustained by the action of pancreatic stellate cells (PSCs), which are activated from quiescent PSCs, as well as myofibroblast cancer-associated fibroblasts (CAFs). CAFs also play immunomodulatory roles (inflammatory and antigen-presenting CAFs). The immune-cell infiltrate is heterogeneous and includes B, T cells and tumor-associated macrophages. ECM: extracellular matrix.

The cross talk between cancer cells and TME is responsible for tumor growth, metastatic potential and therapeutic resistance [16–19]. PDAC TME is made of various cell types: pancreatic stellate cells (PSCs), cancer-associated fibroblasts (CAFs), myeloid cells, regulatory

T cells and B cells, endothelial cells and neuronal cells, which contribute to the formation of a tumor microenvironment characterized, among other things, by fibrosis, hypoxia and immunosuppression. Therefore, resistance to antitumor agents is sustained not only by intracellular mechanisms of the tumor cell (such as the lack of intracellular transporters) but also by the presence of a physical barrier that limits drug delivery [20]. Both PSCs and myofibroblast CAF contribute to the abundant production of extracellular matrix (ECM) molecules such as collagen, fibronectin, proteoglycan, that lead to fibrosis and high-grade tissue stiffness [21,22] (Figure 1).

PSCs are star-like-shaped cells that normally exist in a quiescent state but can be activated by various stimuli such as reactive oxygen species (ROS), cytokines, hypoxia and growth factors. When activated, PSCs can release components of extracellular matrix (ECM), metalloproteinases and maintain their activation through a loop sustained by the secretion of autocrine cytokines [23,24].

The Sonic Hedgehog (Shh) signaling pathway is often hyperactivated in many solid tumors, including PDAC, and appears to promote the activation of PSCs and desmoplasia [25,26]. In a murine model of PDAC, Shh inhibition through cyclopamine has been associated with improved survival [27]. Some trials have, therefore, assessed the role of Shh inhibitors in combination with chemotherapy in patients with advanced PDAC, yielding unsatisfactory results [28,29].

Furthermore, TGF-β, a pleiotropic cytokine that performs various functions, both tumor-promoting and antitumoral, is also involved in the deposition of ECM components, including fibronectin and collagens. TGF-β promotes the activation and proliferation of PSCs, which, in turn, release additional TGF-β through a positive feedback mechanism [30,31]. The use of TGF-β inhibitors has been evaluated in preclinical models and in some clinical studies, including a phase II study that compared gemcitabine plus galunisertib (an inhibitor of type I TGF-β receptor) versus gemcitabine plus placebo. The addition of galunisertib has been shown to increase survival, with a favorable tolerability profile [32]. CAFs are derived from activated PSCs, quiescent resident fibroblast and mesenchymal stem cells. They display distinct functions according to which we can distinguish myofibroblast CAFs that produce stroma, inflammatory CAFs that release cytokines involved in immune response, and antigen-presenting CAFs that are able to process and present antigens through the MHC-II complex [33,34].

Dense desmoplasia is a hallmark of the PDAC microenvironment and, together with the increased fluid pressure partly due to proteoglycan and hyaluronan, alters the organ architecture and represents an important obstacle to the delivery of therapeutic agents [35–38]. Moreover, the stroma of PDAC shows paucity of vascularity which similarly limits the diffusion of drugs [39,40]. Hypoxia and desmoplastic reaction support each other. In hypoxic conditions, hypoxia-inducible factor 1 alfa subunit (HIF-1α), unstable in normal oxygenation conditions, translocates in the nucleus, promoting the transcription of a variety of genes (including proangiogenic factors such as VEGF), thus creating a proinflammatory microenvironment. In response to such environment, PSCs are activated into myofibroblast-like cells. PSC under hypoxia increase the secretion of type I collagen, periostin and fibronectin [41,42], and hypoxia helps the release of Sonic Hedgehog ligand whose pathway promotes the secretion of type I collagen and fibronectin as well [43]. Furthermore, prolonged hypoxia is demonstrated to promote autophagy by HIF-1α and AMPK pathways in PSCs, thus decreasing production of inhibiting factors such as lumican (an extracellular matrix protein both secreted and present in PSCs cytoplasm), known for slowing PDAC cells proliferation [44–46]. Considering the role of hyaluronic acid in causing an increase in interstitial pressure, leading to vascular collapse and reduced perfusion, some trials have evaluated the addition of recombinant human hyaluronidase (PEGPH20) to chemotherapy in patients with metastatic PDAC. However, these trials did not show a better outcome despite increased toxicity [47,48].

PDAC exhibits an immunosuppressive phenotype, and the TME actively participates in creating an immunosuppressive environment by triggering mechanisms that promote immune evasion and restrict the activation of an effective antitumor immune response. CAFs may inhibit cytotoxic T lymphocyte (CTL) function with the release of immunosuppressive cytokines such as IL-10 and TGFB [49,50]. CAFs also restrict the movement of CTLs to the peri-tumoral stromal compartments through the activation of focal adhesion kinase (FAK) and the overproduction of C-X-C Motif Chemokine Ligand 12 (CXCL12), which binds the C-X-C Motif Chemokine Receptor 4 (CXCR4) [51]. Additionally, it has been demonstrated that, in the early stages of carcinogenesis, an infiltration of cells that facilitate mechanisms of immune evasion can be observed, such as tumor-associated macrophages (TAMs), myeloid-derived suppressor (MDSCs), and regulatory T cells (Tregs) [52,53]. Inhibition of the Hedgehog pathway leads to a reduction in the proportion of myCAF and an increase in inflammatory CAFs, resulting in a modification of the inflammatory infiltrate (reduction of CD8 T cells and an increase in regulatory T cells) [54] (Figure 2).

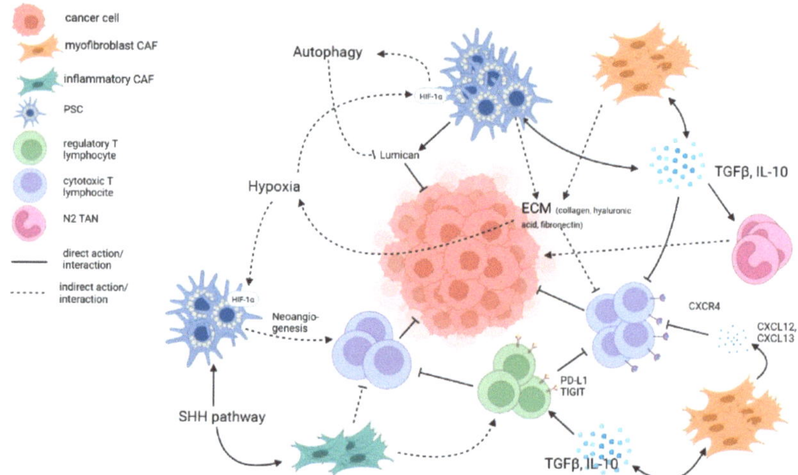

Figure 2. Schematization of key molecular interactions in PDAC microenvironment. Pancreatic stellate cells (PSCs) and cancer-associated fibroblasts (CAFs) secrete immunosuppressive cytokines (IL-10, TGFβ), furthering their proliferation through positive feedback, polarizing tumor-associated neutrophiles (TANs) towards their pro-tumorigenic N2 phenotype and inhibiting cytotoxic T lymphocytes (CTLs) function. Additionally, the deposition of extracellular matrix (ECM) and the overproduction of adhesion molecules ligands (CXCL12, CXCL13) impede normal lymphocytes motility, promoting immune evasion. Abundancy of ECM, moreover, facilitates hypoxia, which activates the HIF-1α pathway. This, on one hand, induces autophagy in cancer-inhibiting cell populations, but, on the other hand, promotes neoangiogenesis and CTLs infiltration.

3. Role of Locoregional Therapies in PC and Effects on TME

Given the clinical need to improve treatment efficacy in PDAC, the development of new therapeutic strategies should take into consideration the important role of TME in supporting tumor growth and promoting treatment resistance. In this regard, locoregional treatments could reshape the TME, sensitizing tumor cells to systemic treatments by enhancing the delivery of cytotoxic agents or altering the composition of the immune infiltrate.

Microwave ablation (MWA), radiofrequency ablation (RFA), radiation therapy (RT), irreversible electroporation (IRE), high-intensity focused ultrasound (HIFU), and intra-arterial infusion chemotherapy are becoming increasingly popular due to their possibility to specifically target the tumor while limiting adverse events [55].

3.1. Ablation Therapies

Nonmetastatic locally advanced pancreatic cancer (LAPC) includes 30% of all newly diagnosed PDAC [56]. Standard-of-care treatment for LAPC is chemotherapy with or without radiation therapy. Neoadjuvant chemotherapy can downstage 20% of LAPC, leading to potential resectability with a significantly improved survival (35.3 months (mo)) compared to patients who do not become surgical candidates after treatment (16.2 mo) [57]. Patients with LAPC who do not become surgical candidates may benefit from local ablation therapies.

Locoregional ablative therapies are minimally invasive techniques that use a generator and a needle-like electrode to transmit energy directly to the target location, with the aim of producing tissue necrosis. These interventions are often performed percutaneously with the insertion of an electrode under imaging guidance; however, they can also be performed during laparoscopic/open surgery or endoscopy. The variety of minimally invasive ablative treatments may be classified according to their use of thermal energy, such as radiofrequency (RF), microwave (MO), and cryoablation (CA), or non-thermal energy, such as irreversible electroporation (IRE) [58]. High-Intensity Focused Ultrasound (HIFU) technology is another minimally invasive ablative method that, unlike the others described above, does not entail needle-like electrodes at the target tissue [59].

3.1.1. Radiofrequency Ablation

In RFA, one or more electrodes create alternating currents at high frequencies, which cause very high local temperatures, resulting in thermal coagulation, protein denaturation and, consequently, thermo-coagulative necrosis of the target tissue. An ablated spherical area, generally 2 to 5 cm in diameter, is generated in about 10 to 30 min [60]. With RFA, the tissue heating zone is limited to a few millimeters surrounding the active electrode due to the high flow of electrical current, while the rest of the ablation zone is heated by thermal conduction [61]. Therefore, when the size of the target area increases, treatment efficacy is reduced, with a maximum result for volumes less than 3.5 cm [62]. This method of heat generation is dependent on conductivity with a close correlation to the water content of the tissue [63,64].

So far, there have been no randomized controlled trials (RCT) regarding RFA effectiveness. However, there is one ongoing RCT (PELICAN RCT) that aims to evaluate if the combination of chemotherapy and RFA prolongs OS compared to chemotherapy alone in patients with LAPC with absence of progression following two months of systemic treatment [65]. In a nonrandomized study, Giardino et al. enrolled 107 patients and divided them equally into two groups: group 1 underwent RFA as a first treatment, while group 2 received neoadjuvant therapy followed by laparotomic RFA. Twenty-nine patients also received intra-arterial chemotherapy (epirubicin and cisplatin into the celiac trunk). The median OS was 14.7 mo in group 1 and 25.6 mo in group 2. Patients treated with a combination of RFA, neoadjuvant therapy and intra-arterial chemotherapy had a prolonged median OS of 34 mo. Adverse events were reported in 25% of the cases [66,67]. In a recent systematic review on RFAs performed with endoscopic ultrasound (EUS) guidance, Spadaccini et al. evaluated a total of 120 patients from 14 studies with primary endpoints of adverse events and mortality. The pooled analysis showed a success rate of 99%, with an adverse events rate of 8% without mortality related to the procedure [68]. D'onofrio et al. reported their experience on percutaneous RFA of LAPC. They evaluated 30 patients with nonresectable PDAC nonresponsive to first-line chemotherapy treated with RFA percutaneously. At 30-days follow-up, no adverse events were reported. The mean survival was 310 days (65–718) [69].

Studies on PDAC treated with RFA are mostly focused on treatment via laparotomy or endoscopy, with only a small proportion of treatments performed percutaneously. When percutaneous RFA is feasible, it may avoid laparotomy, thus reducing the risk of infection, bleeding and other surgical complications [70]. In addition, while surgery involves impaired immune response, it is known that, in patients treated percutaneously, there is an enhanced immune system activation [71,72].

RFA can cause stroma denaturation and tissue permeability modifications in order to implement drug delivery. Ware et al. showed that RF affects molecular transport in a 3D model of PDAC with higher diffusion of DAPI fluorescence in spheroids in comparison to no RF. This could influence the response to medications that are passively diffusing in the TME, since the drug molecules would be able to extravasate in large quantities from the vasculature and disseminate deeper and more uniformly throughout the tissue [73].

RFA has been shown to induce a modification in the composition of the inflammatory infiltrate both in the treated tumor areas and in nontarget lesions distant from the locally treated zone, known as the "abscopal effect". The "abscopal effect" refers to the response in a nontarget lesion distant from the treated site, which occurs due to immune activation and increased immune-responsiveness [74].

Faraoni et al. [75] used murine models with PDAC tumors, distinguishing between lesions treated with RFA and those without RFA treatment. Their results showed that RFA reduced PDAC tumor progression in vivo and favored a strong remodeling of the tumor microenvironment (TME), with an associated abscopal effect observed in 87.5% of non-RFA-treated tumors. A key role in the immune response in non-treated sites seems to be played by tumor-associated neutrophils (TANs), whose intralesional infiltration increases, and they are polarized into an antitumor phenotype. RFA also increases the expression of chemokines such as CXCL12, CXCL12 and CXCL13, recruiting B and T cells. Furthermore, they found an increase in PD-L1 levels compared to the control, with higher levels in RFA-treated tumors compared to non-RFA-treated ones, where the trend was increasing, although not significant. Similarly, Fei et al. evaluated immunological changes in RFA-treated and non-RFA-treated tumors using murine models. In tumors subjected to RFA, an increase in activated CD4 and CD8 cells, macrophages expressing NOS2 (associated with antitumor properties), and dendritic cells promoting proliferation and differentiation of T cells was observed. On the non-RFA side, the infiltration of CD8+ T cells, measured at 3, 5 and 8 days after the procedure, initially showed an increase compared to the control, but then reduced from the 3rd day onwards, suggesting that an immune response occurs, but it is likely transient and not effective [76]. Lawrence et al. conducted serial biopsies of LAPC in human patients before and after treatment with RFA on three occasions. They observed an upregulation of the CD1E gene, which is involved in antigen presentation, in both patients compared to the baseline. In one patient, an increase in genes related to T cells and their cytolytic function was detected. In both patients, there were alterations in the expression of genes related to the activity of CAFs [77]. It is not clear whether these changes can be effectively exploited to enhance the antitumor action of treatments, but they serve as evidence that RFA induces diverse modifications not only at the treatment site but also involving various aspects of tumor immune regulation and its interaction with the microenvironment.

3.1.2. Microwave Ablation

Similar to RFA, microwave ablation (MWA) causes target tissue necrosis with a thermo-coagulative process. The majority of the heat produced by MWA is caused by the excitation of polar water molecules, while ionic polarization has a significantly lower impact. MWA produces a greater zone of active heating in a shorter time in comparison to RFA, allowing a more uniform necrosis in the target lesion. The two primary frequency bands utilized are 915 and 2450 MHz, with the latter being the more frequently employed [78]. The theoretical advantages of MWA over RFA are various: first of all, the ability to treat a larger lesion because the area of necrosis created is larger; moreover, the procedure has a shorter procedural time, less influence of vaporization and carbonization mechanisms and reduced heat dissipation in the presence of surrounding blood vessels (reduced heat sink effect) [67]. Although microwave ablation is a well-established intervention in liver, kidney, lung and bone malignancies [79–84], studies regarding this technology in PDAC are very limited. Carrafiello et al. evaluated the effectiveness of MWA (45 W power, 915 MHz frequency) in 10 unresectable PDAC (5 with percutaneous and 5 with laparotomic approach). The follow-up was on average 9.2 months (3 to 16). One late major complication was observed in one patient (gastroduodenal artery pseudoaneurysm, successfully treated with endovascular embolization); two patients had pancreatitis resolved during the hospital stay. The one-year survival rate was 80% with improvements in quality of life (QoL) for all patients in the first 6 weeks after treatment [85]. Vogl et al. treated with percutaneous MWA (5–100 W power, 2450 MHz frequency) 22 unresectable PDAC. Tumors were in the pancreatic head in 17 (77.3%) patients and in the pancreatic tail in 5 (22.7%). MWA's technical success rate was 100%. There were no significant adverse events reported. Only patients who did not receive further neoadjuvant treatments 10/22 were evaluated for local tumor progression (LTP). Out of the patients evaluable, LTP was detected in one case (10%) at 3-months follow-up [86]. Ierardi et al. examined the viability and safety of percutaneous MWA (100 W power, 2450 MHz frequency) in 5 LAPC situated in the pancreatic head. At follow-up CT performed at 1, 3 and 12 months, no major adverse events were reported. An improvement in QoL was observed in all patients despite a tendency to come back to preoperative conditions in the months following the procedure [87]. In the context of MWA, percutaneous approach was the most frequently employed for the treatment of PDAC, which likely contributed to the decreased complication rates. However, there were some differences in the rates of MWA complications across the evaluated studies. This result, in addition to the difference in patient selection, can also be partly explained by the different technology used in the aforementioned studies.

3.1.3. High-Intensity Focused Ultrasound

High-Intensity Focused Ultrasound (HIFU) technology is a noninvasive technique that uses high-energy ultrasound waves to ablate a limited target volume with US or MRI guide. Focused ultrasound devices are made of a generator that produces ultrasound energy and a transducer that focuses the waves into a beam aimed at a well-defined target region. HIFU has a dual effect on tissues, inducing both thermal and mechanical damage. The thermal effect generated by the absorption of sound waves is different whether the dose of deposited energy is low or high. At low energies (<55 °C), the hyperthermia induced does not generate cell death but increases cell membrane permeability. At high temperatures (>55 °C), cell death is induced by coagulation necrosis [88]. The mechanical damage, on the other hand, includes radiation force, increased pressure and acoustic cavitation [89]. In contrast to other heat-based ablation technologies, HIFU does not require the usage of needle-like electrodes and does not require routine employment of anesthesia even if the use of antispasmodic and anxiolytic drugs can be used to minimize involuntary movements of the patient [90,91]. Due to its heating effects, the most commonly explored use of HIFU is thermal ablation.

Numerous large-volume studies have shown that HIFU has a considerable positive impact on patients' QoL, with improved tumor responsiveness and low rate of adverse events. A recent meta-analysis conducted by Fergadi et al. evaluated 939 patients with PDAC. They assessed that HIFU combined with neoadjuvant chemotherapy is a safe strategy that increases OS and causes less discomfort when compared to chemotherapy alone [92]. Another study by Ning et al. evaluated 523 unresectable PDAC treated with gemcitabine + HIFU or gemcitabine only. The median OS of patients receiving HIFU combined with gemcitbine vs. gemcitabine alone was 7.4 vs. 6.0 mo ($p = 0.002$), without any severe complication related to HIFU reported [93].

Besides thermal ablation, another application of HIFU that is recently gaining attention is its use as a means to provide targeted drug delivery. This is possible since ultrasound administered at high intensity promotes the creation of transient openings in the cellular membrane, increasing cellular permeability. This process is known as sonoporation [94]. There are two main motivations for using nanoparticles in combination with ultrasound. Firstly, the mechanical effects of HIFU are amplified by nanoparticles, reducing the cavitation threshold during the generation of microbubbles, leading to more effective therapeutic applications [95]. Secondly, carrier particles themselves can be loaded with drug molecules in order to be ablated (i.e., "activated") at the proper delivery location by administering selective HIFU to that area [96]. In vitro experiments using PDAC spheroids composed of DT66066 cancer cells and normal fibroblasts showed that the same dose of gemcitabine is less cytotoxic in the presence of fibroblasts, supporting the hypothesis that the microenvironment can negatively influence the action of chemotherapeutic drugs. Moreover, they reported that cavitation generated by HIFU increased gemcitabine delivery and its therapeutic efficiency attributable to an increased cell membrane permeability, the damage of the cell membrane, an enhanced drug intake through sonoporation or ultrasound thermal effects and the synthesis of reactive hydroxyl species (ROS) [97]. In a mouse model of PDAC, Li et al. employed pulsed HIFU (pHIFU) to increase doxorubicin penetration via ultrasound-induced cavitation. They discovered that, in comparison to controls, the concentration of doxorubicin increased up to 4.5-fold. Additionally, normalized doxorubicin concentration was linked to the cavitation metrics (P 0.01), demonstrating that persistent high cavitation increases the penetration of treatment [98]. In a phase I clinical trial by Dimcevski et al., 10 patients with LAPC were treated with pHIFU at low intensity in conjunction with exogenously administered microbubbles to facilitate cavitation. This treatment, in combination with gemcitabine, doubled the median OS of gemcitabine single agent (17.6 months versus 8.9 months) [99].

Regarding temperature-dependent drug delivery, in vitro and in vivo animal models showed that doxorubicin was released more quickly and concentratedly after HIFU treatment in conjunction with injection of temperature-sensitive liposomes. Another work by Liang et al. showed that temperature-sensitive cerasomes released drug molecules in their target area when the temperature has increased by 5 °C [89,100]. The use of HIFU for facilitating drug delivery has certain limitations that should be considered, such as its short duration of action and variable drug uptake associated with treatment. Moreover, the efficacy of successful delivery of nanoparticles using HIFU can vary significantly within heterogeneous tumors and from patient to patient, leading to dramatic differences in the penetration and uptake of drugs [101]. In addition to the aforementioned properties of HIFU, an immunomodulating role for this therapy has recently been proposed. Wang et al. evaluated blood samples of 15 patients before and after HIFU therapy revealing larger percentages of circulating CD3+ and CD4 T cells (in 66% of patients), an increased CD4+/CD8+ T-cell ratio, and increased NK cell activity [102]. These results were supported by a recent meta-analysis of 3022 clinical cases of PDAC that had been thermally ablated using HIFU. Furthermore, hyperthermia induces the upregulation of heat shock proteins (HSP), which, in turn, stimulate the host's immune system, and pancreatic necrosis in areas subjected to HIFU ablation results in the accumulation of IL-1 and IL-2, which are implicated in immune regulation [103].

3.1.4. Cryoablation

Cryoablation (CA) is a thermoablative technology that is based on multiple cycles of freezing and thawing that results in the development of intra- and extracellular ice crystals, osmotic pressure fluctuations, disruption of cells membrane, and eventually cellular death [56]. One of the advantages of cryoablation over different thermal ablation techniques is the possibility to monitor the ablation zone during the procedure. In fact, during freezing, water in the tissue undergoes transition from liquid to solid, forming an "ice-ball" that is visible on ultrasound, computed tomography and magnetic resonance imaging [67]. In addition, because the cooling of tissue and nerves provides an anesthetic effect, CA tends to be less painful than heat-based thermal ablation technologies and, therefore, could theoretically be performed safely with moderate sedation only [104]. Experiments in vitro have shown that the temperature in which irreversible cell death is present in PANC-1 cell line after a single exposure is −25 °C; if repeated freezing cycles are performed, cellular death is present even at higher temperatures [105,106]. Interestingly, a complete cell death was found even after a single freeze at −15 °C if performed in combination with chemotherapy (100 nM of gemcitabine or 8.8 µM of oxaliplatin) before the ablation [106]. As for MWA, the available data on PDAC are very poor in literature. Xu et al. explored the possibility of treatment with CA (36 percutaneous and 13 intraoperative) in conjunction with 125I seed implantation in 49 patients with LAPC (12 of whom had liver metastases). Simultaneous CA was carried out for liver metastases positioning additional cryoprobes with an intercostal approach. During a median follow-up of 18 mo (range of 5–40 mo), the median OS was 16.2 mo. Complete response was recorded in 20.4% of patients, partial response in 38.8% of patients, stable disease in 30.6%, and progressive disease in 10.2% (5/49) of patients [107]. Niu et al. evaluated 67 patients with stage IV (metastatic) PDAC divided into four treatment groups: 22 of them had chemotherapy, 36 underwent CA alone, 17 had immunotherapy alone, and 31 received both CA and immunotherapy. Compared to the cryotherapy (7 mo), immunotherapy (5 mo), and chemotherapy (3.5 mo) groups, the CA-immunotherapy group's median OS was considerably longer (13 mo). They performed tests for the immunologic index before treatment, both in patients treated with cryoimmunotherapy and in the immunotherapy group. There were no differences in any immunologic index between the groups, and they found that patients with normal immune function had a higher survival rate compared to patients with reduced immune activity. Unfortunately, the evaluation of the immunologic index was not repeated after the treatment, so it is not possible to assess any alterations of the tumor microenvironment induced by cryoimmunotherapy [108].

3.1.5. Irreversible Electroporation

Irreversible electroporation (IRE) is an innovative ablative method employed in the clinical treatment of LAPC that provides intratumorally high-voltage electric pulses, causing the death of tumor cells by destroying the cell membrane integrity, creating nanopores. Compared to conventional ablative modalities, cell death in IRE is based on electrical energy rather than thermal energy and has different advantages: it is not influenced by the "heat sink effect", avoiding incomplete ablation due to the energy reduction caused by blood flow; it preserves the extracellular matrix of vasculature and shows better safety profiles next to vital structures, especially for vital nerves, vessels, and cavity structures [109]. The preservation of vessels, which aids in the passage of immune molecules or cells, may result in a higher immune response [110].

The procedure can be performed percutaneously under guidance (typically CT-guided), laparoscopically, or through an open approach following a midline laparotomy [111] (Figure 3). The percutaneous approach can be performed also with a transgastric approach [77,112].

Following the treatment, a contrast-enhanced CT scan should be done to confirm the correct ablation zone and to check for any early complications [77]. The mechanism of action is based on repeated cycles of brief, extremely high-voltage electrical pulses that change the transmembrane potential of tumor cells, causing the lipid bilayer of the cell membrane to develop nanoscale holes that increase membrane permeability. The membrane permeability becomes permanent under the right electrical conditions (90 pulses of 70 μs; electric field strength of 1500 V cm^{-1}; delivered current of 20–50 A), and the cell dies due to loss of homeostasis [113]. The pulsatile application of electrical pulses at very high voltages poses particular difficulties for anesthesiologists, including the potential for inducing cardiac arrhythmias due to the tissue's enhanced cell membrane permeability, which creates a pathway for ion transportation. Additionally, the activation of muscular or neurological tissue may result in strong muscle contractions and epileptic seizures. As a result, all IRE operations require general anesthesia and the use of neuromuscular blocking medications since total muscle paralysis is required to stop muscle contractions [113]. For these reasons, it is crucial to consider that several cardiac-related illnesses are absolute contraindications to this procedure [114]. The intensity of the electric pulses, which is inversely proportional to the distance between the electrodes and the tumor cells, determines the cytotoxicity of IRE. Additionally, the intratumoral heterogeneity may create low-pulse-strength areas where tumor cells can persist. In fact, insufficient ablation is frequently a cause responsible for local tumor recurrence after IRE [109].

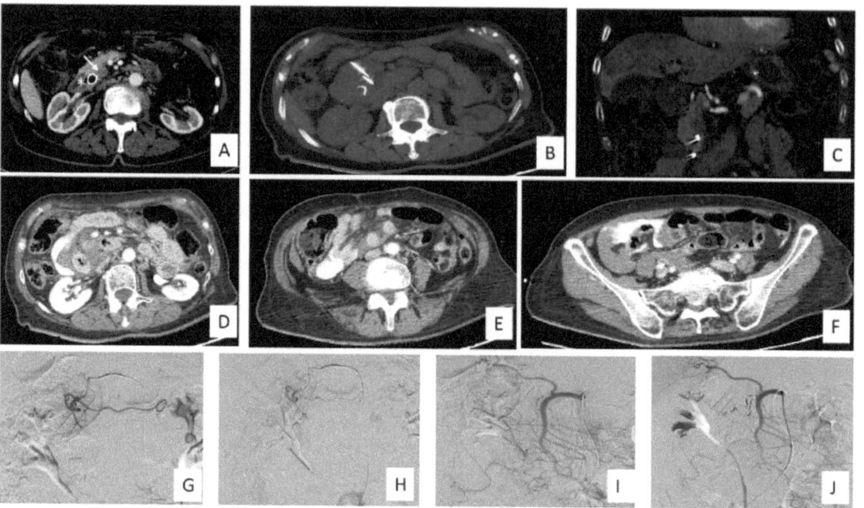

Figure 3. IRE for LAPC treatment timeline and complication management in a 76-year-old female patient. (**A**) CECT in arterial phase demonstrates the presence of the LAPC in the head of the pancreas (white arrows) and biliary stent (asterisk) prior to IRE treatment. (**B**) Axial view of noncontrast scan shows two needle electrodes in situ. (**C**) Coronal view in arterial phase shows the two electrodes in situ. (**D–F**) CECT scan immediately post procedure shows the presence of intraabdominal hematic fluid and extravasation of contrast media without any visible source of bleeding. (**G–J**) Angiography of celiac trunk, SMA, GDA, right renal artery and phlebography of the inferior vena cava and right renal vein did not demonstrate any source of bleeding. A preventive endovascular embolization of GDA and PDA were made using 3, 4 and 5 mm micro coils. Abbreviations: contrast-enhanced CT (CECT), irreversible electroporation (IRE), gastroduodenal artery (GDA), locally advanced pancreatic cancer (LAPC), pancreatic duodenal artery (PDA), superior mesenteric artery (SMA).

It has been demonstrated that IRE not only destroys the tumor itself directly, but also has been shown to boost antitumor immunity and temporarily lessen the stroma-induced immunosuppression [109]. PDAC is known to have a microenvironment that is extremely immunosuppressive and has a low mutational burden, which results in a small number of neoantigens and permits the tumor to spread unhindered [115]. Compared to other ablation modalities, IRE might exhibit larger immune enhancing abilities in terms of protein release and T-cell activation compared to cryo- or heat ablation with enhanced antigen presentation, cause inflammation, and reduce tumor-induced immune suppression [115,116]. IRE induces a systemic immune response that results in the release of antigens and damage-associated molecular pattern molecules (DAMPs). These DAMPs are absorbed by dendritic cells (DCs) residing in tumor tissues; after that, the DCs go to draining lymph nodes where they become mature by the binding of toll-like receptor 9 on the plasmocytoid DCs. The DC maturation leads to the release of IFN-g and to the activation of cytotoxic and helper T cells that are specific for the tumor antigen, helping the development of a systemic immune response and therefore an active in vivo antitumor vaccination [115,117]. This systemic T-cell response could then lead to regression in distant metastases [74,114]. However, PDAC are well known for having a microenvironment that is extremely immunosuppressive, which makes it challenging for the proinflammatory T cells that have been activated to contribute to tumor elimination. A combination with immunotherapy in the form of checkpoint inhibitors or other active immune-enhancing therapies may have a synergistic impact to leverage the patients' own activated immune system through the IRE technique [114]. O'Neill et al. tested the hypothesis that that IFN-g would cause expression of PD-L1 PDAC, and they performed an in vitro study where human pancreatic cell lines were cultured with interferon-g, and murine models of PDAC were treated with IRE, and PD-L1 expression was measured. They revealed that IRE induces expression of PD-L1 in vitro, and the combination therapy with concurrent nivolumab was well tolerated [118]. Zhao et al. [119], using murine models of PDAC treated with anti-PD1 or IRE or IRE + anti-PD1, demonstrated increased survival in those treated with the combination treatment. The authors highlighted an increase in CD8+ T cells in the group treated with IRE + anti-PD1 compared to the others, while no significant differences were observed in the frequency of CD4+ T cells, NK cells, B cells, DCs, or MDSCs among the groups. The same authors investigated the mechanisms of IRE-induced damage in an in vitro model and found that, at high voltages, there was an increase in the concentration of adenosine triphosphate (ATP) and high-mobility group protein B1 (HMGB1), known as damage-associated molecular patterns (DAMPs). Furthermore, IRE modulates the stroma by inducing necrosis in the tumor center, with an increase in microvascular density and reduced expression of HIF-1α, at the fourth day after the procedure, leading to increased blood vessel permeability, likely promoting the infiltration of cytotoxic lymphocytes. Additionally, several components of the tumor microenvironment were found to be downregulated, including FAPα, hyaluronic acid and LOX. FAPα+ CAFs produce CXCL12, which limits the intratumoral entry of lymphocytes. Hyaluronic acid, as previously discussed, increases interstitial fluid pressure, restricting the extravasation of immune cells, and LOX is implicated in the formation of a fibrotic network that restricts the infiltration of T cells. He et al. evaluated the role of IRE plus anti-PD-1 antibody versus IRE alone for patients with LAPC and demonstrated that IRE plus toripalimab had acceptable toxic effects and might improve survival in LAPC compared with IRE alone [110]. The tumor-associated neutrophils (TANs) can be divided into two phenotype groups: antitumor phenotype N and pro-tumor phenotype N2. Immunosuppressive molecules in the tumor microenvironment can polarize TANs into N2 phenotype, inducing the tumor activity. Recently, a tunable glutathione (GSH)-responsive mesoporous silica nanoformulation (dMSN-SB) was developed. It was shown to inhibit intratumoral TGF-β signaling, promoting TAN polarization toward the antitumor N1 phenotypes [109]. Peng et al. conducted a study demonstrating that dMSN-SB inhibits the TGFB signaling pathway and prevents the polarization of neutrophils into the pro-tumoral N2 phenotype in cell cultures. Subsequently, they investigated the use of dMSN-SB in combination with IRE and anti-PD1 in murine models and found that, after treatment, the infiltrate of CD8+ T cells was much more abundant compared to models treated with IRE + anti-PD1 agents or with IRE + dMSN-SB or the control. The group treated with the triple therapy showed lower expression of CXCL12, IL-1β, C5A, IL-1α, compared to the group treated with IRE + anti-PD1 [109].

3.2. Radiotherapy

Radiation therapy (RT) has been utilized in managing patients with PDAC both in the resectable and in the unresectable PDAC. Theoretically, preoperative radiation therapy has advantages over postoperative therapy, including better tissue oxygenation (which could increase RT effectiveness), sterilization of the operating field (which could reduce iatrogenic tumor seeding), and improved patient tolerance and compliance with treatment setting of resectable disease and unresectable disease [120]. The role of radiotherapy in unresectable LAPC is still debatable. RT has the potential to reduce the evolution of disease and potentially alleviate or avoid symptoms such as pain, biliary blockage, hemorrhage, and bowel obstruction. Nevertheless, the possibility of micrometastatic disease is significant, treatment is unlikely to be curative, and radiation can be toxic [121]. Recent advancements in RT techniques have resulted in an increased use of specific techniques including stereotactic body RT (SBRT), intensity-modulated RT (IMRT), and intraoperative RT (IORT). SBRT delivers 1 to 5 high-dose radiation fractions, while conventional RT delivers 25 to 28 fractions of 1.8/2 Gy each. The rationale behind SBRT is that tissues within the radiation field receive extremely high doses and are expected to suffer significant radiation-related damage. Some studies showed that SBRT in PDAC treatment is related to good local control rate but high rates of gastrointestinal toxicities [122–124]. In a recent series of 13 LAPC, a COMBO strategy was used, with a final SBRT boost given after initial chemotherapy first and subsequent concomitant chemoradiotherapy. The treatment strategy was well tolerated with good survival outcomes (median OS 21.5 mo, median PFS 17.5 mo) [125]. Intensity-modulated RT is delivered with conventional fractionation, but unlike conventional RT, the intensity of the radiation is nonuniform. Dose distribution is designed to minimize the radiation dose to normal tissues. Up to now, IMRT has not been proven superior to conventional three-dimensional RT, but IMRT techniques showed a reduction in dose to normal adjacent tissue during treatment of pancreas tumor [126,127]. Intraoperative RT (IORT) has been reached with the aim to increase radiation dose in the target tumor and, at the same time, to limit the radiation dose to adjacent normal structures such as bowel [121]. Although IORT is expected to improve local control, it has not been shown to increase survival. Consequently, despite the fact that IORT shows promise, the technique has not been widely used [128].

Different studies have investigated the role of RT in the immunomodulation in the human PDAC TME following therapy. Mills et al. evaluated the role of stereotactic body radiotherapy (SBRT) as a treatment modality for PDAC to prime cytotoxic T cells by inducing immunogenic tumor cell death in preclinical models [129]. They revealed that SBRT reduced PDAC cell density and induced immunogenic cell death (ICD) without damaging vasculature. However, the barrier to SBRT-induced antitumor immune responses in PDAC is an abundance of immunosuppressive myeloid populations. In conclusion, although SBRT may induce anticancer immune responses against human PDAC, the survival benefits are likely to be neutralized by long-term immune suppression mechanisms in the tumor microenvironment and future immunotherapy strategies are likely to be crucial to improve the treatment strategy.

3.3. Pancreatic Intra-Arterial Infusion Chemotherapy

Advanced PDAC has a very poor prognosis due to its chemoresistant nature. Chemoresistance of the tumor depends on the presence of a very dense and poorly vascularized fibrotic tumor stroma that involves tumor environment, the poor vascularization of PDAC and the high expression of the membrane-bound P-170 glycoprotein, part of an ATP-dependent drug efflux enzyme system [130].

Small case series have examined locoregional chemotherapy as an option for PDAC, demonstrating dose-dependent tumor sensitivity. In fact, when compared to systemic chemotherapy, pancreatic artery infusion (PAI) chemotherapy, using different chemotherapeutic agents such as gemcitabine and 5-fluorouracil, gives higher local concentrations of chemotherapeutic drugs while preserving healthy tissues and having a lower rate of side effects. Wang et al. evaluated the efficacy and safety of PAI with nab-paclitaxel in patients with advanced PDAC and demonstrated that PAI is safe, well tolerated, and effective for the relief of clinical symptoms [131]. Qiu et al. evaluated the efficacy and safety of PAI for the treatment of PDAC in patients unfit for chemotherapy or refractory, demonstrating that PAI is an effective and safe choice for this population, and, in addition, patients with a better performance status had better treatment outcomes [132]. Table 1 summarizes the available ablative treatments in PDAC.

Table 1. Characteristics of ablative therapies in pancreatic cancer.

Technology	Mechanism of Action	Modalities of Intervention	Advantages	Disadvantages
RFA	Heat-based technology that produces coagulative necrosis through the creation of high-frequency alternating current	Laparotomy–Percutaneous–EUS	Economic Several long-term studies in literature	The use of ground pads is required. The ablation zone is limited to the immediate vicinity of the antenna.
MWA	Heat-based technology that generates coagulative necrosis through dielectric hysteresis of polar water molecules	Laparotomy–Percutaneous	It reaches remarkable ablation zones in a short time. A single antenna is usually sufficient.	There is little literature in the pancreatic field. Risk of damaging vascular structures
HIFU	High-frequency ultrasound-based technology that generates cellular damage either by temperature increase or through a mechanical damage	No needle placement is needed. US- or MRI-guided procedure	Ablation performed with very high degree of precision No placement of antennas required	Still not very widespread Limited treatment volume
CA	Cold-based technology that uses repeated freezing and thawing cycles to cause cell death	Laparotomy–Percutaneous	Real-time monitoring of the volume of the "ice ball" during treatment Less painful than heat-dependent treatment due to tissue and nerve cooling	Simultaneous use of multiple antennas often necessary Treatment is quite time-consuming.
IRE	Technology based on the use of high-voltage electrical pulses that allow the creation of nanopores in the cell membrane and subsequent cell death	Laparotomy–Percutaneous	Preserves surrounding structures like great vessels, bile ducts, and intestinal loops Systemic immune response against target tumor tissue potentially greater than other techniques	Expensive Need at least two needles parallel to each other to generate the circuit

Abbreviations: RFA: Radiofrequency Ablation; MWA: Microwave Ablation; HIFU: High-Intensity Focused Ultrasound; CA: Cryoablation; IRE: Irreversible Electroporation; EUS: Endoscopy Ultrasound.

4. Discussion and Future Perspectives

Apart from RT, locoregional treatments are not included in guidelines for the treatment of localized and advanced PDAC. Indeed, NCCN guidelines consider RT only in the case of PDAC amenable to receive preoperative treatment together with induction chemotherapy or alone (stereotactic body RT) before surgery. Moreover, in the case of LAPC, RT may be an option as a single treatment or together with chemotherapy [133]. However, in recent years, there has been an increasing use of interventional radiology techniques for the treatment of PDAC. In LAPC, interventional therapies must not be considered as an alternative to surgery. However, they may be used in patients unfit for or unwilling to undergo surgery or in the case of inadequate disease shrinkage after neoadjuvant treatment with subsequent impossibility to perform a radical resection. As far as advanced disease is concerned, locoregional treatments may be used as palliative strategies in patients with obstructive jaundice, liver metastases, in case of severe lower back pain and those who cannot tolerate systemic chemotherapy [134].

There is increasing evidence that locoregional treatments may remodel tumor TME and contribute to different mechanisms, as demonstrated through high-quality samples [135–137]. TME creates a stromal barrier to chemotherapy and actively participates in creating an immunosuppressive environment.

Locoregional treatments can modulate the composition of TME in multiple ways: on one hand, they promote structural alterations that facilitate the delivery of a higher amount of drug and, consequently, potentially enhance therapeutic efficacy. In this regard, several studies are evaluating the role of postlocoregional treatment chemotherapy for patients with nonmetastatic PDAC. In a phase II study, all patients undergoing IRE for the treatment of LAPC will receive either FOLFIRINOX or gemcitabine as peri-ablation treatment (NCT03484299). Another phase II study is determining the feasibility, tolerability, and treatment effect of EUS-RFA plus standard-of-care neoadjuvant chemotherapy with FOLFIRINOX or gemcitabine/nab-paclitaxel in LAPC (NCT04990609). A phase II/III trial has been completed, and patients were randomized to receive either IRE and synchronous chemotherapy with gemcitabine or IRE and subsequent gemcitabine starting on day 7 following ablative procedure in LAPC (NCT03673137). Differently, a phase I study is testing the tolerability of chemotherapy and EUS-RFA using the RF Electrode in patients receiving palliative second- or third-line therapy for unresectable nonmetastatic PDAC (NCT05723107). In addition, a phase II study has been completed and evaluated the safety and feasibility of the NanoKnife Low-Energy Direct Current (LEDC) System when used to treat unresectable PDAC (NCT01369420). On the other hand, locoregional therapies often modify the composition of immune response effectors, not only in the treated area, but sometimes also at a distance, as described in the abscopal effect promoted by ablative therapies. This leads to a polarization towards an antitumoral immune response with several observed modifications, including an increase in tumor-associated neutrophils (TANs) and the activation of B and T cells. Furthermore, some studies have documented an increase in PD-L1 expression following local treatments. These findings provide the rationale for the use of immune checkpoint inhibitors (ICI) in combination with locoregional therapies. In this direction, a phase II study is enrolling patients who will receive IRE for LAPC and will be treated with nivolumab postoperatively (NCT03080974). However, it should be noted that the duration and the actual antitumoral effectiveness of immune microenvironment modifications have not yet been established, and it remains to be seen whether they can be exploited therapeutically.

5. Conclusions

PDAC is a chemo- and immune-resistant disease with poor prognosis. Locoregional treatments, even if not considered standard of care, may have a role in interfering with TME with subsequent improvement of chemotherapy outcomes and increased immune sensitivity. Early phase and phase III studies are warranted in order to understand the impact of ablative therapies alone or in combination with chemotherapy/immunotherapy on disease response and survival outcomes.

Author Contributions: Conceptualization, M.C.D.G. and M.G.; methodology, M.C.D.G.; software, M.G.; validation, A.F., B.G. and G.C.; formal analysis, V.A.; investigation, C.L. and G.D.P.; resources, A.F.; data curation, A.M.I., V.A. and C.L.; writing—original draft preparation, M.C.D.G., V.A., C.L., G.D.P. and M.G.; writing—review and editing, M.G.; visualization, M.G.; supervision, G.C.; project administration, M.C.D.G.; funding acquisition, A.F. All authors have read and agreed to the published version of the manuscript.

Funding: This research received no external funding.

Institutional Review Board Statement: Not applicable.

Informed Consent Statement: Not applicable.

Data Availability Statement: Data sharing not applicable.

Conflicts of Interest: The authors declare no conflict of interest.

References

1. Sung, H.; Ferlay, J.; Siegel, R.L.; Laversanne, M.; Soerjomataram, I.; Jemal, A.; Bray, F. Global Cancer Statistics 2020: GLOBOCAN Estimates of Incidence and Mortality Worldwide for 36 Cancers in 185 Countries. *CA Cancer J. Clin.* **2021**, *71*, 209–249. [CrossRef]
2. Cai, J.; Chen, H.; Lu, M.; Zhang, Y.; Lu, B.; You, L.; Zhang, T.; Dai, M.; Zhao, Y. Advances in the epidemiology of pancreatic cancer: Trends, risk factors, screening, and prognosis. *Cancer Lett.* **2021**, *520*, 1–11. [CrossRef]
3. Rahib, L.; Smith, B.D.; Aizenberg, R.; Rosenzweig, A.B.; Fleshman, J.M.; Matrisian, L.M. Projecting cancer incidence and deaths to 2030: The unexpected burden of thyroid, liver, and pancreas cancers in the United States. *Cancer Res.* **2014**, *74*, 2913–2921. [CrossRef] [PubMed]

4. Shinde, R.S.; Bhandare, M.; Chaudhari, V.; Shrikhande, S.V. Cutting-edge strategies for borderline resectable pancreatic cancer. *Ann. Gastroenterol. Surg.* **2019**, *3*, 368–372. [CrossRef]
5. Conroy, T.; Desseigne, F.; Ychou, M.; Bouché, O.; Guimbaud, R.; Bécouarn, Y.; Adenis, A.; Raoul, J.-L.; Gourgou-Bourgade, S.; De La Fouchardière, C.; et al. FOLFIRINOX versus Gemcitabine for Metastatic Pancreatic Cancer. *N. Engl. J. Med.* **2011**, *364*, 1817–1825. [CrossRef]
6. Neoptolemos, J.P.; Palmer, D.H.; Ghaneh, P.; Psarelli, E.E.; Valle, J.W.; Halloran, C.M.; Faluyi, O.; O'Reilly, D.A.; Cunningham, D.; Wadsley, J.; et al. Comparison of adjuvant gemcitabine and capecitabine with gemcitabine monotherapy in patients with resected pancreatic cancer (ESPAC-4): A multicentre, open-label, randomised, phase 3 trial. *Lancet* **2017**, *389*, 1011–1024. [CrossRef] [PubMed]
7. Janssen, Q.P.; O'Reilly, E.M.; van Eijck, C.H.J.; Koerkamp, B.G. Neoadjuvant Treatment in Patients With Resectable and Borderline Resectable Pancreatic Cancer. *Front. Oncol.* **2020**, *10*, 41. [CrossRef] [PubMed]
8. Reni, M.; Cereda, S.; Rognone, A.; Belli, C.; Ghidini, M.; Longoni, S.; Fugazza, C.; Rezzonico, S.; Passoni, P.; Slim, N.; et al. A randomized phase II trial of two different 4-drug combinations in advanced pancreatic adenocarcinoma: Cisplatin, capecitabine, gemcitabine plus either epirubicin or docetaxel (PEXG or PDXG regimen). *Cancer Chemother. Pharmacol.* **2011**, *69*, 115–123. [CrossRef]
9. Sohal, D.P.S.; Duong, M.; Ahmad, S.A.; Gandhi, N.S.; Beg, M.S.; Wang-Gillam, A.; Wade, J.L., 3rd; Chiorean, E.G.; Guthrie, K.A.; Lowy, A.M.; et al. Efficacy of Perioperative Chemotherapy for Resectable Pancreatic Adenocarcinoma: A Phase 2 Randomized Clinical Trial. *JAMA Oncol.* **2021**, *7*, 421–427. [CrossRef] [PubMed]
10. van Dam, J.L.; Janssen, Q.P.; Besselink, M.G.; Homs, M.Y.; van Santvoort, H.C.; van Tienhoven, G.; de Wilde, R.F.; Wilmink, J.W.; van Eijck, C.H.; Koerkamp, B.G.; et al. Neoadjuvant therapy or upfront surgery for resectable and borderline resectable pancreatic cancer: A meta-analysis of randomised controlled trials. *Eur. J. Cancer* **2021**, *160*, 140–149. [CrossRef] [PubMed]
11. Burris, H.A., 3rd; Moore, M.J.; Andersen, J.; Green, M.R.; Rothenberg, M.L.; Modiano, M.R.; Cripps, M.C.; Portenoy, R.K.; Storniolo, A.M.; Tarassoff, P.; et al. Improvements in survival and clinical benefit with gemcitabine as first-line therapy for patients with advanced pancreas cancer: A randomized trial. *J. Clin. Oncol.* **1997**, *15*, 2403–2413. [CrossRef] [PubMed]
12. Von Hoff, D.D.; Ervin, T.; Arena, F.P.; Chiorean, E.G.; Infante, J.; Moore, M.; Seay, T.; Tjulandin, S.A.; Ma, W.W.; Saleh, M.N.; et al. Increased Survival in Pancreatic Cancer with nab-Paclitaxel plus Gemcitabine. *N. Engl. J. Med.* **2013**, *369*, 1691–1703. [CrossRef] [PubMed]
13. Quiñonero, F.; Mesas, C.; Doello, K.; Cabeza, L.; Perazzoli, G.; Jimenez-Luna, C.; Rama, A.R.; Melguizo, C.; Prados, J. The challenge of drug resistance in pancreatic ductal adenocarcinoma: A current overview. *Cancer Biol. Med.* **2019**, *16*, 688–699. [CrossRef]
14. Anderson, N.M.; Simon, M.C. The tumor microenvironment. *Curr. Biol.* **2020**, *30*, R921–R925. [CrossRef]
15. Baghban, R.; Roshangar, L.; Jahanban-Esfahlan, R.; Seidi, K.; Ebrahimi-Kalan, A.; Jaymand, M.; Kolahian, S.; Javaheri, T.; Zare, P. Tumor microenvironment complexity and therapeutic implications at a glance. *Cell Commun. Signal.* **2020**, *18*, 59. [CrossRef]
16. Ho, W.J.; Jaffee, E.M.; Zheng, L. The tumour microenvironment in pancreatic cancer—Clinical challenges and opportunities. *Nat. Rev. Clin. Oncol.* **2020**, *17*, 527–540. [CrossRef] [PubMed]
17. Murakami, T.; Hiroshima, Y.; Matsuyama, R.; Homma, Y.; Hoffman, R.M.; Endo, I. Role of the tumor microenvironment in pancreatic cancer. *Ann. Gastroenterol. Surg.* **2019**, *3*, 130–137. [CrossRef] [PubMed]
18. Ren, B.; Cui, M.; Yang, G.; Wang, H.; Feng, M.; You, L.; Zhao, Y. Tumor microenvironment participates in metastasis of pancreatic cancer. *Mol. Cancer* **2018**, *17*, 108. [CrossRef]
19. Sherman, M.H.; Beatty, G.L. Tumor Microenvironment in Pancreatic Cancer Pathogenesis and Therapeutic Resistance. *Annu. Rev. Pathol. Mech. Dis.* **2023**, *18*, 123–148. [CrossRef] [PubMed]
20. Spratlin, J.; Sangha, R.; Glubrecht, D.; Dabbagh, L.; Young, J.D.; Dumontet, C.; Cass, C.; Lai, R.; Mackey, J.R. The Absence of Human Equilibrative Nucleoside Transporter 1 Is Associated with Reduced Survival in Patients With Gemcitabine-Treated Pancreas Adenocarcinoma. *Clin. Cancer Res.* **2004**, *10*, 6956–6961. [CrossRef] [PubMed]
21. Erkan, M.; Adler, G.; Apte, M.V.; Bachem, M.G.; Buchholz, M.; Detlefsen, S.; Esposito, I.; Friess, H.; Gress, T.M.; Habisch, H.J.; et al. StellaTUM: Current consensus and discussion on pancreatic stellate cell research. *Gut* **2012**, *61*, 172–178. [CrossRef]
22. Vaish, U.; Jain, T.; Are, A.C.; Dudeja, V. Cancer-Associated Fibroblasts in Pancreatic Ductal Adenocarcinoma: An Update on Heterogeneity and Therapeutic Targeting. *Int. J. Mol. Sci.* **2021**, *22*, 13408. [CrossRef] [PubMed]
23. Deng, D.; Patel, R.; Chiang, C.-Y.; Hou, P. Role of the Tumor Microenvironment in Regulating Pancreatic Cancer Therapy Resistance. *Cells* **2022**, *11*, 2952. [CrossRef]
24. Wu, Y.; Zhang, C.; Jiang, K.; Werner, J.; Bazhin, A.V.; D'haese, J.G. The Role of Stellate Cells in Pancreatic Ductal Adenocarcinoma: Targeting Perspectives. *Front. Oncol.* **2021**, *10*, 621937. [CrossRef]
25. Hidalgo, M.; Maitra, A. The Hedgehog Pathway and Pancreatic Cancer. *N. Engl. J. Med.* **2009**, *361*, 2094–2096. [CrossRef]
26. Bailey, J.M.; Swanson, B.J.; Hamada, T.; Eggers, J.P.; Singh, P.K.; Caffery, T.; Ouellette, M.M.; Hollingsworth, M.A. Sonic Hedgehog Promotes Desmoplasia in Pancreatic Cancer. *Clin. Cancer Res.* **2008**, *14*, 5995–6004. [CrossRef] [PubMed]
27. Feldmann, G.; Habbe, N.; Dhara, S.; Bisht, S.; Alvarez, H.; Fendrich, V.; Beaty, R.; Mullendore, M.; Karikari, C.; Bardeesy, N.; et al. Hedgehog inhibition prolongs survival in a genetically engineered mouse model of pancreatic cancer. *Gut* **2008**, *57*, 1420–1430. [CrossRef]

28. Catenacci, D.V.T.; Junttila, M.R.; Karrison, T.; Bahary, N.; Horiba, M.N.; Nattam, S.R.; Marsh, R.; Wallace, J.; Kozloff, M.; Rajdev, L.; et al. Randomized Phase Ib/II Study of Gemcitabine Plus Placebo or Vismodegib, a Hedgehog Pathway Inhibitor, in Patients With Metastatic Pancreatic Cancer. *J. Clin. Oncol.* **2015**, *33*, 4284–4292. [CrossRef] [PubMed]
29. De Jesus-Acosta, A.; Sugar, E.A.; O'dwyer, P.J.; Ramanathan, R.K.; Von Hoff, D.D.; Rasheed, Z.; Zheng, L.; Begum, A.; Anders, R.; Maitra, A.; et al. Phase 2 study of vismodegib, a hedgehog inhibitor, combined with gemcitabine and nab-paclitaxel in patients with untreated metastatic pancreatic adenocarcinoma. *Br. J. Cancer* **2019**, *122*, 498–505. [CrossRef]
30. Marzoq, A.J.; Mustafa, S.A.; Heidrich, L.; Hoheisel, J.D.; Alhamdani, M.S.S. Impact of the secretome of activated pancreatic stellate cells on growth and differentiation of pancreatic tumour cells. *Sci. Rep.* **2019**, *9*, 5303. [CrossRef] [PubMed]
31. Principe, D.R.; Timbers, K.E.; Atia, L.G.; Koch, R.M.; Rana, A. TGFβ Signaling in the Pancreatic Tumor Microenvironment. *Cancers* **2021**, *13*, 5086. [CrossRef]
32. Melisi, D.; Garcia-Carbonero, R.; Macarulla, T.; Pezet, D.; Deplanque, G.; Fuchs, M.; Trojan, J.; Oettle, H.; Kozloff, M.; Cleverly, A.; et al. Galunisertib plus gemcitabine vs. gemcitabine for first-line treatment of patients with unresectable pancreatic cancer. *Br. J. Cancer* **2018**, *119*, 1208–1214. [CrossRef] [PubMed]
33. Geng, X.; Chen, H.; Zhao, L.; Hu, J.; Yang, W.; Li, G.; Cheng, C.; Zhao, Z.; Zhang, T.; Li, L.; et al. Cancer-Associated Fibroblast (CAF) Heterogeneity and Targeting Therapy of CAFs in Pancreatic Cancer. *Front. Cell Dev. Biol.* **2021**, *9*, 655152. [CrossRef] [PubMed]
34. Sunami, Y.; Häußler, J.; Kleeff, J. Cellular Heterogeneity of Pancreatic Stellate Cells, Mesenchymal Stem Cells, and Cancer-Associated Fibroblasts in Pancreatic Cancer. *Cancers* **2020**, *12*, 3770. [CrossRef] [PubMed]
35. Cannon, A.; Thompson, C.; Hall, B.R.; Jain, M.; Kumar, S.; Batra, S.K. Desmoplasia in pancreatic ductal adenocarcinoma: Insight into pathological function and therapeutic potential. *Genes Cancer* **2018**, *9*, 78–86. [CrossRef]
36. Heldin, C.-H.; Rubin, K.; Pietras, K.; Östman, A. High interstitial fluid pressure—An obstacle in cancer therapy. *Nat. Rev. Cancer* **2004**, *4*, 806–813. [CrossRef] [PubMed]
37. Kanat, O.; Ertas, H. Shattering the castle walls: Anti-stromal therapy for pancreatic cancer. *World J. Gastrointest. Oncol.* **2018**, *10*, 202–210. [CrossRef] [PubMed]
38. Thomas, D.; Radhakrishnan, P. Tumor-stromal crosstalk in pancreatic cancer and tissue fibrosis. *Mol. Cancer* **2019**, *18*, 14. [CrossRef] [PubMed]
39. Erler, J.T.; Cawthorne, C.J.; Williams, K.J.; Koritzinsky, M.; Wouters, B.G.; Wilson, C.; Miller, C.; Demonacos, C.; Stratford, I.J.; Dive, C. Hypoxia-Mediated Down-Regulation of Bid and Bax in Tumors Occurs via Hypoxia-Inducible Factor 1-Dependent and -Independent Mechanisms and Contributes to Drug Resistance. *Mol. Cell. Biol.* **2004**, *24*, 2875–2889. [CrossRef]
40. Katsuta, E.; Qi, Q.; Peng, X.; Hochwald, S.N.; Yan, L.; Takabe, K. Pancreatic adenocarcinomas with mature blood vessels have better overall survival. *Sci. Rep.* **2019**, *9*, 1310. [CrossRef] [PubMed]
41. Erkan, M.; Reiser-Erkan, C.; Michalski, C.W.; Deucker, S.; Sauliunaite, D.; Streit, S.; Esposito, I.; Friess, H.; Kleeff, J. Cancer-Stellate Cell Interactions Perpetuate the Hypoxia-Fibrosis Cycle in Pancreatic Ductal Adenocarcinoma. *Neoplasia* **2009**, *11*, 497–508. [CrossRef]
42. Masamune, A.; Kikuta, K.; Watanabe, T.; Satoh, K.; Hirota, M.; Shimosegawa, T. Hypoxia stimulates pancreatic stellate cells to induce fibrosis and angiogenesis in pancreatic cancer. *Am. J. Physiol. Gastrointest. Liver Physiol.* **2008**, *295*, G709–G717. [CrossRef]
43. Spivak-Kroizman, T.R.; Hostetter, G.; Posner, R.; Aziz, M.; Hu, C.; Demeure, M.J.; Von Hoff, D.; Hingorani, S.R.; Palculict, T.B.; Izzo, J.; et al. Hypoxia Triggers Hedgehog-Mediated Tumor–Stromal Interactions in Pancreatic Cancer. *Cancer Res.* **2013**, *73*, 3235–3247. [CrossRef]
44. Joshi, S.; Kumar, S.; Ponnusamy, M.P.; Batra, S.K. Hypoxia-induced oxidative stress promotes MUC4 degradation via autophagy to enhance pancreatic cancer cells survival. *Oncogene* **2016**, *35*, 5882–5892. [CrossRef]
45. Li, X.; Lee, Y.; Kang, Y.; Dai, B.; Perez, M.R.; Pratt, M.; Koay, E.J.; Kim, M.; Brekken, R.A.; Fleming, J.B. Hypoxia-induced autophagy of stellate cells inhibits expression and secretion of lumican into microenvironment of pancreatic ductal adenocarcinoma. *Cell Death Differ.* **2018**, *26*, 382–393. [CrossRef] [PubMed]
46. Piffoux, M.; Eriau, E.; Cassier, P.A. Autophagy as a therapeutic target in pancreatic cancer. *Br. J. Cancer* **2020**, *124*, 333–344. [CrossRef]
47. Ramanathan, R.K.; McDonough, S.L.; Philip, P.A.; Hingorani, S.R.; Lacy, J.; Kortmansky, J.S.; Thumar, J.; Chiorean, E.G.; Shields, A.F.; Behl, D.; et al. Phase IB/II Randomized Study of FOLFIRINOX Plus Pegylated Recombinant Human Hyaluronidase Versus FOLFIRINOX Alone in Patients With Metastatic Pancreatic Adenocarcinoma: SWOG S1313. *J. Clin. Oncol.* **2019**, *37*, 1062–1069. [CrossRef]
48. Van Cutsem, E.; Tempero, M.A.; Sigal, D.; Oh, D.-Y.; Fazio, N.; Macarulla, T.; Hitre, E.; Hammel, P.; Hendifar, A.E.; Bates, S.E.; et al. Randomized Phase III Trial of Pegvorhyaluronidase Alfa With Nab-Paclitaxel Plus Gemcitabine for Patients With Hyaluronan-High Metastatic Pancreatic Adenocarcinoma. *J. Clin. Oncol.* **2020**, *38*, 3185–3194. [CrossRef]
49. Moo-Young, T.A.; Larson, J.W.; Belt, B.A.; Tan, M.C.; Hawkins, W.G.; Eberlein, T.J.; Goedegebuure, P.S.; Linehan, D.C. Tumor-derived TGF-β Mediates Conversion of CD4+Foxp3+ Regulatory T Cells in a Murine Model of Pancreas Cancer. *J. Immunother.* **2009**, *32*, 12–21. [CrossRef]
50. Principe, D.R.; DeCant, B.; Mascariñas, E.; Wayne, E.A.; Diaz, A.M.; Akagi, N.; Hwang, R.; Pasche, B.; Dawson, D.W.; Fang, D.; et al. TGFβ Signaling in the Pancreatic Tumor Microenvironment Promotes Fibrosis and Immune Evasion to Facilitate Tumorigenesis. *Cancer Res.* **2016**, *76*, 2525–2539. [CrossRef]

51. Ene–Obong, A.; Clear, A.J.; Watt, J.; Wang, J.; Fatah, R.; Riches, J.C.; Marshall, J.F.; Chin-Aleong, J.; Chelala, C.; Gribben, J.G.; et al. Activated Pancreatic Stellate Cells Sequester CD8+ T Cells to Reduce Their Infiltration of the Juxtatumoral Compartment of Pancreatic Ductal Adenocarcinoma. *Gastroenterology* **2013**, *145*, 1121–1132. [CrossRef]

52. Clark, C.E.; Hingorani, S.R.; Mick, R.; Combs, C.; Tuveson, D.A.; Vonderheide, R.H. Dynamics of the Immune Reaction to Pancreatic Cancer from Inception to Invasion. *Cancer Res.* **2007**, *67*, 9518–9527. [CrossRef]

53. Zhao, F.; Obermann, S.; von Wasielewski, R.; Haile, L.; Manns, M.P.; Korangy, F.; Greten, T.F. Increase in frequency of myeloid-derived suppressor cells in mice with spontaneous pancreatic carcinoma. *Immunology* **2009**, *128*, 141–149. [CrossRef]

54. Steele, N.G.; Biffi, G.; Kemp, S.B.; Zhang, Y.; Drouillard, D.; Syu, L.; Hao, Y.; Oni, T.E.; Brosnan, E.; Elyada, E.; et al. Inhibition of Hedgehog Signaling Alters Fibroblast Composition in Pancreatic Cancer. *Clin. Cancer Res.* **2021**, *27*, 2023–2037. [CrossRef]

55. Lambin, T.; Lafon, C.; Drainville, R.A.; Pioche, M.; Prat, F. Locoregional therapies and their effects on the tumoral microenvironment of pancreatic ductal adenocarcinoma. *World J. Gastroenterol.* **2022**, *28*, 1288–1303. [CrossRef]

56. Bibok, A.; Kim, D.W.; Malafa, M.; Kis, B. Minimally invasive image-guided therapy of primary and metastatic pancreatic cancer. *World J. Gastroenterol.* **2021**, *27*, 4322–4341. [CrossRef]

57. Gemenetzis, G.; Groot, V.P.; Blair, A.B.; Laheru, D.A.; Zheng, L.; Narang, A.K.; Fishman, E.K.; Hruban, R.H.; Yu, J.; Burkhart, R.A.; et al. Survival in Locally Advanced Pancreatic Cancer After Neoadjuvant Therapy and Surgical Resection. *Ann. Surg.* **2019**, *270*, 340–347. [CrossRef]

58. Ahmed, M.; Brace, C.L.; Lee, F.T., Jr.; Goldberg, S.N. Principles of and advances in percutaneous ablation. *Radiology* **2011**, *258*, 351–369. [CrossRef]

59. Rossi, M.; Orgera, G.; Hatzidakis, A.; Krokidis, M. Minimally Invasive Ablation Treatment for Locally Advanced Pancreatic Adenocarcinoma. *Cardiovasc. Interv. Radiol.* **2013**, *37*, 586–591. [CrossRef]

60. Crocetti, L.; De Baere, T.; Lencioni, R. Quality Improvement Guidelines for Radiofrequency Ablation of Liver Tumours. *Cardiovasc. Intervent. Radiol.* **2010**, *33*, 11–17. [CrossRef]

61. Yousaf, M.N.; Ehsan, H.; Muneeb, A.; Wahab, A.; Sana, M.K.; Neupane, K.; Chaudhary, F.S. Role of Radiofrequency Ablation in the Management of Unresectable Pancreatic Cancer. *Front. Med.* **2021**, *7*, 624997. [CrossRef] [PubMed]

62. Ierardi, A.M.; Lucchina, N.; Petrillo, M.; Floridi, C.; Piacentino, F.; Bacuzzi, A.; Fonio, P.; Fontana, F.; Fugazzola, C.; Brunese, L.; et al. Systematic review of minimally invasive ablation treatment for locally advanced pancreatic cancer. *Radiol. Med.* **2014**, *119*, 483–498. [CrossRef] [PubMed]

63. Hong, K.; Georgiades, C.; Fereydooni, A.; Letzen, B.; Ghani, M.A.; Miszczuk, M.A.; Huber, S.; Chapiro, J. Others Radiofrequency Ablation: Mechanism of Action and Devices. *J. Vasc. Interv. Radiol.* **2010**, *21*, S179–S186. [CrossRef] [PubMed]

64. Solazzo, S.A.; Liu, Z.; Lobo, S.M.; Ahmed, M.; Hines-Peralta, A.U.; Lenkinski, R.E.; Goldberg, S.N. Radiofrequency Ablation: Importance of Background Tissue Electrical Conductivity—An Agar Phantom and Computer Modeling Study. *Radiology* **2005**, *236*, 495–502. [CrossRef]

65. Walma, M.S.; Rombouts, S.J.; Brada, L.J.H.; Borel Rinkes, I.H.; Bosscha, K.; Bruijnen, R.C.; Busch, O.R.; Creemers, G.J.; Daams, F.; van Dam, R.M.; et al. Radiofrequency ablation and chemotherapy versus chemotherapy alone for locally advanced pancreatic cancer (PELICAN): Study protocol for a randomized controlled trial. *Trials* **2021**, *22*, 313. [CrossRef]

66. Giardino, A.; Girelli, R.; Frigerio, I.; Regi, P.; Cantore, M.; Alessandra, A.; Lusenti, A.; Salvia, R.; Bassi, C.; Pederzoli, P. Triple approach strategy for patients with locally advanced pancreatic carcinoma. *HPB* **2013**, *15*, 623–627. [CrossRef]

67. Punzi, E.; Carrubba, C.; Contegiacomo, A.; Posa, A.; Barbieri, P.; De Leoni, D.; Mazza, G.; Tanzilli, A.; Cina, A.; Natale, L.; et al. Interventional Radiology in the Treatment of Pancreatic Adenocarcinoma: Present and Future Perspectives. *Life* **2023**, *13*, 835. [CrossRef]

68. Spadaccini, M.; Di Leo, M.; Iannone, A.; von den Hoff, D.; Fugazza, A.; Galtieri, P.A.; Pellegatta, G.; Maselli, R.; Anderloni, A.; Colombo, M.; et al. Endoscopic ultrasound-guided ablation of solid pancreatic lesions: A systematic review of early outcomes with pooled analysis. *World J. Gastrointest. Oncol.* **2022**, *14*, 533–542. [CrossRef]

69. D'onofrio, M.; Beleù, A.; Sarno, A.; De Robertis, R.; Paiella, S.; Viviani, E.; Frigerio, I.; Girelli, R.; Salvia, R.; Bassi, C. US-Guided Percutaneous Radiofrequency Ablation of Locally Advanced Pancreatic Adenocarcinoma: A 5-Year High-Volume Center Experience. *Ultraschall. Med.* **2020**, *43*, 380–386. [CrossRef]

70. Granata, V.; Grassi, R.; Fusco, R.; Belli, A.; Palaia, R.; Carrafiello, G.; Miele, V.; Grassi, R.; Petrillo, A.; Izzo, F. Local ablation of pancreatic tumors: State of the art and future perspectives. *World J. Gastroenterol.* **2021**, *27*, 3413–3428. [CrossRef]

71. Gao, S.; Pu, N.; Yin, H.; Li, J.; Chen, Q.; Yang, M.; Lou, W.; Chen, Y.; Zhou, G.; Li, C.; et al. Radiofrequency ablation in combination with an mTOR inhibitor restrains pancreatic cancer growth induced by intrinsic HSP70. *Ther. Adv. Med. Oncol.* **2020**, *12*, 1758835920953728. [CrossRef] [PubMed]

72. Waitz, R.; Solomon, S.B. Can Local Radiofrequency Ablation of Tumors Generate Systemic Immunity against Metastatic Disease? *Radiology* **2009**, *251*, 1–2. [CrossRef] [PubMed]

73. Ware, M.J.; Curtis, L.T.; Wu, M.; Ho, J.C.; Corr, S.J.; Curley, S.A.; Godin, B.; Frieboes, H.B. Pancreatic adenocarcinoma response to chemotherapy enhanced with non-invasive radio frequency evaluated via an integrated experimental/computational approach. *Sci. Rep.* **2017**, *7*, 3437. [CrossRef]

74. Imran, K.M.; Nagai-Singer, M.A.; Brock, R.M.; Alinezhadbalalami, N.; Davalos, R.V.; Allen, I.C. Exploration of Novel Pathways Underlying Irreversible Electroporation Induced Anti-Tumor Immunity in Pancreatic Cancer. *Front. Oncol.* **2022**, *12*, 853779. [CrossRef]

75. Faraoni, E.Y.; O'Brien, B.J.; Strickland, L.N.; Osborn, B.K.; Mota, V.; Chaney, J.; Atkins, C.L.; Cen, P.; Rowe, J.; Cardenas, J.; et al. Radiofrequency Ablation Remodels the Tumor Microenvironment and Promotes Neutrophil-Mediated Abscopal Immunomodulation in Pancreatic Cancer. *Cancer Immunol. Res.* **2022**, *11*, 4–12. [CrossRef]

76. Fei, Q.; Pan, Y.; Lin, W.; Zhou, Y.; Yu, X.; Hou, Z.; Yu, X.; Lin, X.; Lin, R.; Lu, F.; et al. High-dimensional single-cell analysis delineates radiofrequency ablation induced immune microenvironmental remodeling in pancreatic cancer. *Cell Death Dis.* **2020**, *11*, 589. [CrossRef]

77. Lawrence, P.V.; Desai, K.; Wadsworth, C.; Mangal, N.; Kocher, H.M.; Habib, N.; Sadanandam, A.; Sodergren, M.H. A Case Report on Longitudinal Collection of Tumour Biopsies for Gene Expression-Based Tumour Microenvironment Analysis from Pancreatic Cancer Patients Treated with Endoscopic Ultrasound Guided Radiofrequency Ablation. *Curr. Oncol.* **2022**, *29*, 6754–6763. [CrossRef]

78. Carrafiello, G.; Lagana, D.; Mangini, M.; Fontana, F.; Dionigi, G.; Boni, L.; Rovera, F.; Cuffari, S.; Fugazzola, C. Microwave tumors ablation: Principles, clinical applications and review of preliminary experiences. *Int. J. Surg.* **2008**, *6* (Suppl. 1), S65–S69. [CrossRef]

79. Izzo, F.; Granata, V.; Grassi, R.; Fusco, R.; Palaia, R.; Delrio, P.; Carrafiello, G.; Azoulay, D.; Petrillo, A.; Curley, S.A. Radiofrequency Ablation and Microwave Ablation in Liver Tumors: An Update. *Oncologist* **2019**, *24*, e990–e1005. [CrossRef]

80. Carriero, S.; Lanza, C.; Pellegrino, G.; Ascenti, V.; Sattin, C.; Pizzi, C.; Angileri, S.A.; Biondetti, P.; Ianniello, A.A.; Piacentino, F.; et al. Ablative Therapies for Breast Cancer: State of Art. *Technol. Cancer Res. Treat.* **2023**, *22*, 15330338231157193. [CrossRef]

81. Cazzato, R.L.; de Rubeis, G.; de Marini, P.; Dalili, D.; Koch, G.; Auloge, P.; Garnon, J.; Gangi, A. Percutaneous microwave ablation of bone tumors: A systematic review. *Eur. Radiol.* **2020**, *31*, 3530–3541. [CrossRef]

82. Floridi, C.; De Bernardi, I.; Fontana, F.; Muollo, A.; Ierardi, A.M.; Agostini, A.; Fonio, P.; Squillaci, E.; Brunese, L.; Fugazzola, C.; et al. Microwave ablation of renal tumors: State of the art and development trends. *Radiol. Med.* **2014**, *119*, 533–540. [CrossRef] [PubMed]

83. Ierardi, A.M.; Floridi, C.; Fontana, F.; Chini, C.; Giorlando, F.; Piacentino, F.; Brunese, L.; Pinotti, G.; Bacuzzi, A.; Carrafiello, G. Microwave ablation of liver metastases to overcome the limitations of radiofrequency ablation. *Radiol. Med.* **2013**, *118*, 949–961. [CrossRef]

84. Lassandro, G.; Picchi, S.; Corvino, A.; Gurgitano, M.; Carrafiello, G.; Lassandro, F. Ablation of pulmonary neoplasms: Review of literature and future perspectives. *Pol. J. Radiol.* **2023**, *88*, 216–224. [CrossRef] [PubMed]

85. Carrafiello, G.; Ierardi, A.M.; Fontana, F.; Petrillo, M.; Floridi, C.; Lucchina, N.; Cuffari, S.; Dionigi, G.; Rotondo, A.; Fugazzola, C. Microwave Ablation of Pancreatic Head Cancer: Safety and Efficacy. *J. Vasc. Interv. Radiol.* **2013**, *24*, 1513–1520. [CrossRef]

86. Vogl, T.J.; Panahi, B.; Albrecht, M.H.; Naguib, N.N.N.; Nour-Eldin, N.-E.A.; Gruber-Rouh, T.; Thompson, Z.M.; Basten, L.M. Microwave ablation of pancreatic tumors. *Minim. Invasive Ther. Allied Technol.* **2017**, *27*, 33–40. [CrossRef] [PubMed]

87. Ierardi, A.M.; Biondetti, P.; Coppola, A.; Fumarola, E.M.; Biasina, A.M.; Angileri, S.A.; Carrafiello, G. Percutaneous microwave thermosphere ablation of pancreatic tumours. *Gland. Surg.* **2018**, *7*, 59–66. [CrossRef]

88. ter Haar, G.; Coussios, C. High intensity focused ultrasound: Physical principles and devices. *Int. J. Hyperth.* **2007**, *23*, 89–104. [CrossRef]

89. Bachu, V.S.; Kedda, J.; Suk, I.; Green, J.J.; Tyler, B. High-Intensity Focused Ultrasound: A Review of Mechanisms and Clinical Applications. *Ann. Biomed. Eng.* **2021**, *49*, 1975–1991. [CrossRef]

90. Sofuni, A.; Asai, Y.; Mukai, S.; Yamamoto, K.; Itoi, T. High-intensity focused ultrasound therapy for pancreatic cancer. *J. Med. Ultrason* **2022**, *Online Ahead of Print*. [CrossRef]

91. Zhang, L.; Wang, Z.-B. High-intensity focused ultrasound tumor ablation: Review of ten years of clinical experience. *Front. Med. China* **2010**, *4*, 294–302. [CrossRef]

92. Fergadi, M.P.; Magouliotis, D.E.; Rountas, C.; Vlychou, M.; Athanasiou, T.; Symeonidis, D.; Pappa, P.A.; Zacharoulis, D. A meta-analysis evaluating the role of high-intensity focused ultrasound (HIFU) as a fourth treatment modality for patients with locally advanced pancreatic cancer. *Abdom. Radiol.* **2021**, *47*, 254–264. [CrossRef]

93. Ning, Z.; Xie, J.; Chen, Q.; Zhang, C.; Xu, L.; Song, L.; Meng, Z. HIFU is safe, effective, and feasible in pancreatic cancer patients: A monocentric retrospective study among 523 patients. *OncoTargets Ther.* **2019**, *12*, 1021–1029. [CrossRef]

94. McClure, A. Using High-Intensity Focused Ultrasound as a Means to Provide Targeted Drug Delivery. *J. Diagn. Med. Sonogr.* **2016**, *32*, 343–350. [CrossRef]

95. Khirallah, J.; Schmieley, R.; Demirel, E.; Rehman, T.U.; Howell, J.; Durmaz, Y.Y.; Vlaisavljevich, E. Nanoparticle-mediated histotripsy (NMH) using perfluorohexane 'nanocones'. *Phys. Med. Biol.* **2019**, *64*, 125018. [CrossRef]

96. Airan, R.D.; Meyer, R.A.; Ellens, N.P.K.; Rhodes, K.R.; Farahani, K.; Pomper, M.G.; Kadam, S.D.; Green, J.J. Noninvasive Targeted Transcranial Neuromodulation via Focused Ultrasound Gated Drug Release from Nanoemulsions. *Nano Lett.* **2017**, *17*, 652–659. [CrossRef]

97. Leenhardt, R.; Camus, M.; Mestas, J.L.; Jeljeli, M.; Ali, E.A.; Chouzenoux, S.; Bordacahar, B.; Nicco, C.; Batteux, F.; Lafon, C.; et al. Ultrasound-induced Cavitation enhances the efficacy of Chemotherapy in a 3D Model of Pancreatic Ductal Adenocarcinoma with its microenvironment. *Sci. Rep.* **2019**, *9*, 18916. [CrossRef]

98. Li, T.; Wang, Y.-N.; Khokhlova, T.D.; D'Andrea, S.; Starr, F.; Chen, H.; McCune, J.S.; Risler, L.J.; Mashadi-Hossein, A.; Hingorani, S.R.; et al. Pulsed High-Intensity Focused Ultrasound Enhances Delivery of Doxorubicin in a Preclinical Model of Pancreatic Cancer. *Cancer Res.* **2015**, *75*, 3738–3746. [CrossRef]

99. Dimcevski, G.; Kotopoulis, S.; Bjånes, T.; Hoem, D.; Schjøtt, J.; Gjertsen, B.T.; Biermann, M.; Molven, A.; Sorbye, H.; McCormack, E.; et al. A human clinical trial using ultrasound and microbubbles to enhance gemcitabine treatment of inoperable pancreatic cancer. *J. Control. Release* **2016**, *243*, 172–181. [CrossRef]

100. Liang, X.; Gao, J.; Jiang, L.; Luo, J.; Jing, L.; Li, X.; Jin, Y.; Dai, Z. Nanohybrid Liposomal Cerasomes with Good Physiological Stability and Rapid Temperature Responsiveness for High Intensity Focused Ultrasound Triggered Local Chemotherapy of Cancer. *ACS Nano* **2015**, *9*, 1280–1293. [CrossRef]

101. Payne, A.; Vyas, U.; Blankespoor, A.; Christensen, D.; Roemer, R. Minimisation of HIFU pulse heating and interpulse cooling times. *Int. J. Hyperth.* **2010**, *26*, 198–208. [CrossRef]

102. Wang, X.; Sun, J. High-intensity focused ultrasound in patients with late-stage pancreatic carcinoma. *Chin. Med. J.* **2002**, *115*, 1332–1335.

103. Zhou, Y. High-Intensity Focused Ultrasound Treatment for Advanced Pancreatic Cancer. *Gastroenterol. Res. Pract.* **2014**, *2014*, 205325. [CrossRef]

104. Luo, X.-M.; Niu, L.-Z.; Chen, J.-B.; Xu, K.-C. Advances in cryoablation for pancreatic cancer. *World J. Gastroenterol.* **2016**, *22*, 790–800. [CrossRef]

105. Baumann, K.W.; Baust, J.M.; Snyder, K.K.; Baust, J.G.; Van Buskirk, R.G. Characterization of Pancreatic Cancer Cell Thermal Response to Heat Ablation or Cryoablation. *Technol. Cancer Res. Treat.* **2016**, *16*, 393–405. [CrossRef]

106. Baust, J.M.; Santucci, K.L.; Van Buskirk, R.G.; Raijman, I.; Fisher, W.E.; Baust, J.G.; Snyder, K.K. An In Vitro Investigation into Cryoablation and Adjunctive Cryoablation/Chemotherapy Combination Therapy for the Treatment of Pancreatic Cancer Using the PANC-1 Cell Line. *Biomedicines* **2022**, *10*, 450. [CrossRef]

107. Xu, K.-C.; Niu, L.-Z.; Hu, Y.-Z.; He, W.-B.; He, Y.-S.; Li, Y.-F.; Zuo, J.-S. A pilot study on combination of cryosurgery and 125iodine seed implantation for treatment of locally advanced pancreatic cancer. *World J. Gastroenterol.* **2008**, *14*, 1603–1611. [CrossRef]

108. Niu, L.; Chen, J.; He, L.; Liao, M.; Yuan, Y.; Zeng, J.; Li, J.; Zuo, J.; Xu, K. Combination Treatment With Comprehensive Cryoablation and Immunotherapy in Metastatic Pancreatic Cancer. *Pancreas* **2013**, *42*, 1143–1149. [CrossRef]

109. Peng, H.; Shen, J.; Long, X.; Zhou, X.; Zhang, J.; Xu, X.; Huang, T.; Xu, H.; Sun, S.; Li, C.; et al. Local Release of TGF-β Inhibitor Modulates Tumor-Associated Neutrophils and Enhances Pancreatic Cancer Response to Combined Irreversible Electroporation and Immunotherapy. *Adv. Sci.* **2022**, *9*, e2105240. [CrossRef]

110. He, C.; Sun, S.; Zhang, Y.; Li, S. Irreversible Electroporation Plus Anti-PD-1 Antibody versus Irreversible Electroporation Alone for Patients with Locally Advanced Pancreatic Cancer. *J. Inflamm. Res.* **2021**, *14*, 4795–4807. [CrossRef]

111. Rai, Z.L.; Feakins, R.; Pallett, L.J.; Manas, D.; Davidson, B.R. Irreversible Electroporation (IRE) in Locally Advanced Pancreatic Cancer: A Review of Current Clinical Outcomes, Mechanism of Action and Opportunities for Synergistic Therapy. *J. Clin. Med.* **2021**, *10*, 1609. [CrossRef]

112. Ruarus, A.; Vroomen, L.; Puijk, R.; Scheffer, H.; Meijerink, M. Locally Advanced Pancreatic Cancer: A Review of Local Ablative Therapies. *Cancers* **2018**, *10*, 16. [CrossRef]

113. Nielsen, K.; Scheffer, H.J.; Vieveen, J.M.; van Tilborg, A.A.J.M.; Meijer, S.; van Kuijk, C.; van den Tol, M.P.; Meijerink, M.R.; Bouwman, R.A. Anaesthetic management during open and percutaneous irreversible electroporation. *Br. J. Anaesth.* **2014**, *113*, 985–992. [CrossRef]

114. Timmer, F.E.; Geboers, B.; Ruarus, A.H.; Schouten, E.A.; Nieuwenhuizen, S.; Puijk, R.S.; de Vries, J.J.; Meijerink, M.R.; Scheffer, H.J. Irreversible Electroporation for Locally Advanced Pancreatic Cancer. *Tech. Vasc. Interv. Radiol.* **2020**, *23*, 100675. [CrossRef]

115. Geboers, B.; Timmer, F.E.F.; Ruarus, A.H.; Pouw, J.E.E.; Schouten, E.A.C.; Bakker, J.; Puijk, R.S.; Nieuwenhuizen, S.; Dijkstra, M.; van den Tol, M.P.; et al. Irreversible Electroporation and Nivolumab Combined with Intratumoral Administration of a Toll-Like Receptor Ligand, as a Means of In Vivo Vaccination for Metastatic Pancreatic Ductal Adenocarcinoma (PANFIRE-III). A Phase-I Study Protocol. *Cancers* **2021**, *13*, 3902. [CrossRef]

116. White, S.B.; Zhang, Z.; Chen, J.; Gogineni, V.R.; Larson, A.C. Early Immunologic Response of Irreversible Electroporation versus Cryoablation in a Rodent Model of Pancreatic Cancer. *J. Vasc. Interv. Radiol.* **2018**, *29*, 1764–1769. [CrossRef] [PubMed]

117. Zhang, N.; Li, Z.; Han, X.; Zhu, Z.; Li, Z.; Zhao, Y.; Liu, Z.; Lv, Y. Irreversible Electroporation: An Emerging Immunomodulatory Therapy on Solid Tumors. *Front. Immunol.* **2022**, *12*, 811726. [CrossRef]

118. O'neill, C.; Hayat, T.; Hamm, J.; Healey, M.; Zheng, Q.; Li, Y.; Martin, R.C.G., 2nd. A phase 1b trial of concurrent immunotherapy and irreversible electroporation in the treatment of locally advanced pancreatic adenocarcinoma. *Surgery* **2020**, *168*, 610–616. [CrossRef] [PubMed]

119. Zhao, J.; Wen, X.; Tian, L.; Li, T.; Xu, C.; Wen, X.; Melancon, M.P.; Gupta, S.; Shen, B.; Peng, W.; et al. Irreversible electroporation reverses resistance to immune checkpoint blockade in pancreatic cancer. *Nat. Commun.* **2019**, *10*, 899. [CrossRef] [PubMed]

120. Malla, M.; Fekrmandi, F.; Malik, N.; Hatoum, H.; George, S.; Goldberg, R.M.; Mukherjee, S. The evolving role of radiation in pancreatic cancer. *Front. Oncol.* **2023**, *12*, 1060885. [CrossRef] [PubMed]

121. Hazard, L. The role of radiation therapy in pancreas cancer. *Gastrointest. Cancer Res.* **2009**, *3*, 20–28.
122. Hoyer, M.; Roed, H.; Traberg Hansen, A.; Ohlhuis, L.; Petersen, J.; Nellemann, H.; Kill Berthelsen, A.; Grau, C.; Aage Engelholm, S.; Von Der Maase, H. Phase II study on stereotactic body radiotherapy of colorectal metastases. *Acta Oncol.* **2006**, *45*, 823–830. [CrossRef]
123. Koong, A.C.; Christofferson, E.; Le, Q.-T.; Goodman, K.A.; Ho, A.; Kuo, T.; Ford, J.M.; Fisher, G.A.; Greco, R.; Norton, J.; et al. Phase II study to assess the efficacy of conventionally fractionated radiotherapy followed by a stereotactic radiosurgery boost in patients with locally advanced pancreatic cancer. *Int. J. Radiat. Oncol. Biol. Phys.* **2005**, *63*, 320–323. [CrossRef]
124. Koong, A.C.; Le, Q.T.; Ho, A.; Fong, B.; Fisher, G.; Cho, C.; Ford, J.; Poen, J.; Gibbs, I.C.; Mehta, V.K.; et al. Phase I study of stereotactic radiosurgery in patients with locally advanced pancreatic cancer. *Int. J. Radiat. Oncol. Biol. Phys.* **2004**, *58*, 1017–1021. [CrossRef]
125. Parisi, S.; Ferini, G.; Cacciola, A.; Lillo, S.; Tamburella, C.; Santacaterina, A.; Bottari, A.; Brogna, A.; Ferrantelli, G.; Pontoriero, A.; et al. A non-surgical COMBO-therapy approach for locally advanced unresectable pancreatic adenocarcinoma: Preliminary results of a prospective study. *Radiol. Med.* **2022**, *127*, 214–219. [CrossRef]
126. Brown, M.W.; Ning, H.; Arora, B.; Albert, P.S.; Poggi, M.; Camphausen, K.; Citrin, D. A dosimetric analysis of dose escalation using two intensity-modulated radiation therapy techniques in locally advanced pancreatic carcinoma. *Int. J. Radiat. Oncol. Biol. Phys.* **2006**, *65*, 274–283. [CrossRef]
127. Milano, M.T.; Chmura, S.J.; Garofalo, M.C.; Rash, C.; Roeske, J.C.; Connell, P.P.; Kwon, O.-H.; Jani, A.B.; Heimann, R. Intensity-modulated radiotherapy in treatment of pancreatic and bile duct malignancies: Toxicity and clinical outcome. *Int. J. Radiat. Oncol.* **2004**, *59*, 445–453. [CrossRef]
128. Tepper, J.E.; Noyes, D.; Krall, J.M.; Sause, W.T.; Wolkov, H.B.; Dobelbower, R.R.; Thomson, J.; Owens, J.; Hanks, G.E. Intraoperative radiation therapy of pancreatic carcinoma: A report of RTOG-8505. *Int. J. Radiat. Oncol.* **1991**, *21*, 1145–1149. [CrossRef]
129. Mills, B.N.; Qiu, H.; Drage, M.G.; Chen, C.; Mathew, J.S.; Garrett-Larsen, J.; Ye, J.; Uccello, T.P.; Murphy, J.D.; Belt, B.A.; et al. Modulation of the Human Pancreatic Ductal Adenocarcinoma Immune Microenvironment by Stereotactic Body Radiotherapy. *Clin. Cancer Res.* **2022**, *28*, 150–162. [CrossRef]
130. Laface, C.; Laforgia, M.; Molinari, P.; Foti, C.; Ambrogio, F.; Gadaleta, C.D.; Ranieri, G. Intra-Arterial Infusion Chemotherapy in Advanced Pancreatic Cancer: A Comprehensive Review. *Cancers* **2022**, *14*, 450. [CrossRef]
131. Ye, X.; Wang, N.; Xu, J.; Wang, G.; Cao, P. Pancreatic intra-arterial infusion chemotherapy for the treatment of patients with advanced pancreatic carcinoma: A pilot study. *J. Cancer Res. Ther.* **2022**, *18*, 1945–1951. [CrossRef] [PubMed]
132. Qiu, B.; Zhang, X.; Tsauo, J.; Zhao, H.; Gong, T.; Li, J.; Li, X. Transcatheter arterial infusion for pancreatic cancer: A 10-year National Cancer Center experience in 115 patients and literature review. *Abdom. Radiol.* **2019**, *44*, 2801–2808. [CrossRef] [PubMed]
133. National Comprehensive Cancer Network. *NCCN Clinical Practice Guidelines in Oncology—Pancreatic Adenocarcinoma*; National Comprehensive Cancer Network: Plymouth Meeting, PA, USA, 2023.
134. Li, M. Clinical practice guidelines for the interventional treatment of advanced pancreatic cancer (5th edition). *J. Interv. Med.* **2021**, *4*, 159–171. [CrossRef] [PubMed]
135. Gkolfakis, P.; Crinò, S.F.; Tziatzios, G.; Ramai, D.; Papaefthymiou, A.; Papanikolaou, I.S.; Triantafyllou, K.; Arvanitakis, M.; Lisotti, A.; Fusaroli, P.; et al. Comparative diagnostic performance of end-cutting fine-needle biopsy needles for EUS tissue sampling of solid pancreatic masses: A network meta-analysis. *Gastrointest. Endosc.* **2022**, *95*, 1067–1077.e15. [CrossRef]
136. Facciorusso, A.; Bajwa, H.; Menon, K.; Buccino, V.; Muscatiello, N. Comparison between 22G aspiration and 22G biopsy needles for EUS-guided sampling of pancreatic lesions: A meta-analysis. *Endosc. Ultrasound* **2020**, *9*, 167–174. [CrossRef]
137. Facciorusso, A.; Mohan, B.P.; Crinò, S.F.; Ofosu, A.; Ramai, D.; Lisotti, A.; Chandan, S.; Fusaroli, P. Contrast-enhanced harmonic endoscopic ultrasound-guided fine-needle aspiration versus standard fine-needle aspiration in pancreatic masses: A meta-analysis. *Expert Rev. Gastroenterol. Hepatol.* **2021**, *15*, 821–828. [CrossRef]

Disclaimer/Publisher's Note: The statements, opinions and data contained in all publications are solely those of the individual author(s) and contributor(s) and not of MDPI and/or the editor(s). MDPI and/or the editor(s) disclaim responsibility for any injury to people or property resulting from any ideas, methods, instructions or products referred to in the content.

International Journal of
Molecular Sciences

Review

The Tango between Cancer-Associated Fibroblasts (CAFs) and Immune Cells in Affecting Immunotherapy Efficacy in Pancreatic Cancer

Imke Stouten, Nadine van Montfoort *,† and Lukas J. A. C. Hawinkels *,†

Department of Gastroenterology and Hepatology, Leiden University Medical Center, 2333 ZA Leiden, The Netherlands; i.stouten@lumc.nl
* Correspondence: nvanmontfoort@lumc.nl (N.v.M.); l.j.a.c.hawinkels@lumc.nl (L.J.A.C.H.)
† These authors contributed equally to this work.

Abstract: The lack of response to therapy in pancreatic ductal adenocarcinoma (PDAC) patients has contributed to PDAC having one of the lowest survival rates of all cancer types. The poor survival of PDAC patients urges the exploration of novel treatment strategies. Immunotherapy has shown promising results in several other cancer types, but it is still ineffective in PDAC. What sets PDAC apart from other cancer types is its tumour microenvironment (TME) with desmoplasia and low immune infiltration and activity. The most abundant cell type in the TME, cancer-associated fibroblasts (CAFs), could be instrumental in why low immunotherapy responses are observed. CAF heterogeneity and interactions with components of the TME is an emerging field of research, where many paths are to be explored. Understanding CAF–immune cell interactions in the TME might pave the way to optimize immunotherapy efficacy for PDAC and related cancers with stromal abundance. In this review, we discuss recent discoveries on the functions and interactions of CAFs and how targeting CAFs might improve immunotherapy.

Keywords: tumour microenvironment; desmoplasia; cancer-associated fibroblasts; immune cells; T cells; pancreatic ductal adenocarcinoma; immunotherapy; checkpoint inhibitor

Citation: Stouten, I.; van Montfoort, N.; Hawinkels, L.J.A.C. The Tango between Cancer-Associated Fibroblasts (CAFs) and Immune Cells in Affecting Immunotherapy Efficacy in Pancreatic Cancer. *Int. J. Mol. Sci.* **2023**, *24*, 8707. https://doi.org/10.3390/ijms24108707

Academic Editors: Claudio Luchini and Donatella Delle Cave

Received: 4 April 2023
Revised: 9 May 2023
Accepted: 11 May 2023
Published: 13 May 2023

Copyright: © 2023 by the authors. Licensee MDPI, Basel, Switzerland. This article is an open access article distributed under the terms and conditions of the Creative Commons Attribution (CC BY) license (https://creativecommons.org/licenses/by/4.0/).

1. Introduction

Pancreatic ductal adenocarcinoma (PDAC) is the most common type of pancreatic cancer, accounting for more than 90% of pancreatic cancer cases [1]. Of all cancer types, pancreatic cancer has the lowest 5-year survival of only 8% [2]. Treatment for pancreatic cancer consists of a combination of surgery and systemic therapies. Chemotherapy is the most common therapy for PDAC since 80–90% of the patients are precluded from surgical resection due to the advanced state of their disease [3,4]. Nonetheless, chemoresistance is a recurrent problem [5]. Recently, immunotherapy as an alternative approach to target the tumour has become of interest. The goal of immunotherapy is (re-)activating the immune system directed against cancer. Multiple novel immunotherapies have been studied in PDAC, which often aim to activate T cells either by blocking their inhibitory signals or by enhancing their anti-tumour activity. Among these therapies are immune checkpoint inhibitors, cancer vaccines, CD40 agonists, chimeric antigen receptor (CAR) T cells and oncolytic viruses [6,7].

Immunotherapies have yet to reach a clinical effect in PDAC. The low mutational burden in PDAC is one of the potential explanations underlying the low response rates [8]. In addition, the nature of the tumour microenvironment (TME) of PDAC is believed to play a decisive role in response to various therapies. This TME is characterized by desmoplasia (enhanced extracellular matrix deposition), poor vascularization, low oxygen levels and low immune cell infiltration [9]. The TME is composed of a variety of cells, including cancer-associated fibroblasts (CAFs), endothelial cells and immune cells, accompanied by a few other cell types with lower abundance [10]. The most abundant cell type, CAFs, can be subdivided into different subsets [11]. However, the classification of different CAF subtypes in PDAC is still in its infancy. An increasing number of studies suggest that CAFs could play a part in why immunotherapies have not reached their full potential [12]. The presence of CAFs could for example contribute to exclusion of T cells from the tumour, often referred to as an excluded or cold tumour. Since immunotherapy relies on the presence of its effector cells, the T cells, T cell exclusion in the TME impairs efficacy [13]. Hence, studying CAF heterogeneity, their functions and how to therapeutically target them could provide insight into immunotherapy resistance and provide directions to improve immunotherapy efficacy. In this review, we summarize the current knowledge of how the interactions of CAFs in the TME affect immunotherapy responses in PDAC.

2. CAFs

A consensus statement defined CAFs as cells with an elongated morphology that lack cancer-specific mutations and are negative for epithelial, leukocyte or endothelial markers [14]. CAFs are thought to predominantly originate from local resident fibroblasts with a contribution from several other cell types [15]. In a cancerous environment, fibroblasts can become stimulated by high local concentrations of cytokines and adopt a different, activated phenotype [16]. Solid tumours can consist, for a large part, of CAFs, representing up to 80% of the PDAC and breast cancer tumour mass [17]. Studies have shown that CAFs in different cancer types can have different molecular and functional characteristics. For instance, CAFs in PDAC have been found to express different markers and secrete other factors than CAFs in breast cancer, colorectal cancer and lung cancer [18]. The pro-tumourigenic properties of CAFs can be exerted through extracellular matrix (ECM) remodelling, direct cell–cell interactions, as well as secretion of soluble factors [19]. In PDAC, effects of CAFs on tumour cells include increased proliferation, migration and invasion, for instance via indirect interleukin (IL)-6 signalling [20,21]. These interactions between CAFs and tumour cells contribute to the formation of distant metastasis. Besides tumour cells, CAFs interact with endothelial cells through the upregulation of vascular endothelial growth factor (VEGF) [22]. Hypoxia stimulates CAFs to induce the proliferation and migration of endothelial cells, resulting in angiogenesis in vitro and in vivo [23]. CAFs importantly could impact the immune status of the tumour as well [13,24]. Thus far, the role of CAFs in immune deserted (cold) tumours is quite unknown. As for inflamed (hot) tumours, the secretory profile of CAFs could influence immune activity via inhibition of cytotoxic functions or recruitment of immunosuppressive cells [13]. CAFs have been correlated with T cell exclusion in solid tumours, which suggests CAFs being linked to an immune excluded phenotype [24]. Activation of CAFs by transforming growth factor β (TGF-β) promotes mediation of T cell exclusion and has been correlated to checkpoint inhibitor resistance [13,24]. Furthermore, CAFs increase levels of fibronectin and collagen that alter the composition of the ECM, which promotes the migratory properties of tumour cells [25]. Conceivably, the relationship between CAFs and other cells in the TME plays a key role in tumour progression.

2.1. The Good or the Bad Guys?

In PDAC, as in many other cancer types, data on the role of CAFs in tumour progression pointed towards the pro-tumourigenic effects of CAFs [26–28]. In vitro, pancreatic cancer cells displayed enhanced cell proliferation and migration in the presence of CAFs [26,28]. Moreover, in vivo depletion of fibroblast activation protein (FAP)+ stromal cells, a marker abundantly expressed by CAFs, resulted in the inhibition of tumour growth [27]. Recently, the presence of a CAF-based gene expression signature created with PDAC expression data from the Cancer Genome Atlas (TCGA), Gene Expression Omnibus (GEO) and ArrayExpress was related to worse patient survival rates [29]. Even though the vast majority of published studies exclusively reported pro-tumourigenic properties of CAFs in PDAC, more recent work showed opposite findings. Transgenic mice with α-smooth muscle actin

(α-SMA)+ driven CAF depletion developed undifferentiated, aggressive tumours. The overall survival (OS) of the mice was decreased, and the number of regulatory T cells in the tumour was significantly enhanced [30]. In another study, inhibition of sonic hedgehog (shh) signalling in mice with PDAC resulted in a decrease in α-SMA+ CAFs in the TME, accompanied by the presence of aggressive and undifferentiated tumours [31]. These data suggest that α-SMA+ CAFs might pose anti-tumourigenic effects in PDAC and other cancer models.

The discovery that the presence/depletion of different CAF subsets can lead to opposite outcomes reveals the heterogeneity of the cell population. CAFs in PDAC are most often divided into three subtypes: myofibroblast-type CAFs (myCAFs), inflammatory CAFs (iCAFs) and antigen-presenting CAFs (apCAFs) [32,33]. Defining these subsets could improve our understanding of their specific effects on the TME and their role in affecting immunotherapy resistance.

2.2. MyCAFs, iCAFs and apCAFs

iCAFs and myCAFs were the first subsets of CAFs recognized as two distinct types in PDAC. ECM-producing α-SMA+ CAFs located close to epithelial tumour cells in human PDAC tissue and mouse models for pancreatic cancer were defined as myCAFs [32]. The most used model for PDAC is the KPC mouse. These mice spontaneously develop pancreatic cancer lesions due to pancreas-specific point mutations in protein 53 (P53) and an activating mutation in Kirsten rat sarcoma virus (KRAS) to recapitulate human pancreatic cancer [34]. The research was continued with a co-culture of pancreatic stellate cells, a possible precursor of CAFs, and PDAC organoids [32]. Pancreatic stellate cells adjacent to the organoid became myCAFs, and a noticeable increase in inflammatory cytokines such as interleukin 6 (IL-6) by another CAF subset was observed. The cytokine-secreting CAFs did not express α-SMA. This led to the revelation of two distinct subtypes: myCAFs with high α-SMA expression and iCAFs with low α-SMA expression and elevated cytokine expression. By RNA sequencing, the division of the iCAFs and myCAFs was confirmed since the two subclasses had unique clusters of upregulated genes [32]. Later on, the signalling pathways of the two different phenotypes were studied. Induction of iCAFs is promoted by IL-1 and involves JAK/STAT signalling, whereas TGF-β counteracts this cascade by downregulating IL-1 receptor expression, which leads to differentiation into myCAFs. The two subclasses exhibit a degree of plasticity since it has been shown that iCAFs switch to a myCAF phenotype once IL-1 signalling is inhibited. Along with these results, a small CAF population with both α-SMA and activated STAT signalling was identified, which also supports the plasticity between the two subcategories [35]. More recently, a third subset has been added. A subpopulation of CAFs that express major histocompatibility complex class II (MHC II) was able to activate CD4 T cells, which gave rise to their name, antigen-presenting CAFs (apCAFs) [33]. To conclude, to date, three well-described subtypes of CAFs can be distinguished that are associated with different functions in the TME (Figure 1). However, these phenotypes are plastic, and their functions intertwine with each other.

2.3. The Different Colours beyond Black and White

As research evolved, it became more apparent that the division into three subtypes of CAFs is probably somewhat too simplistic. In other cancer types, such as melanoma, lung cancer and breast cancer, more CAF subsets have been reported [36,37]. In PDAC, recent work, for example, identified CAF heterogeneity by the presence or absence of endoglin (CD105), a glycoprotein that is part of the TGF-β receptor complex [30]. Notably, CD105− CAFs included myCAFs, iCAFs and apCAFs, while CD105+ CAFs included only myCAFs and iCAFs. CD105+ CAFs are the largest in number and exhibit tumour-permissive properties in a pancreatic cancer mouse model. CD105− CAFs appear more tumour suppressive through activation of helper T cells and cytotoxic T cells. In contrast, evidence from multiple studies supports that apCAFs, which are observed in the CD105−

group, cause immunosuppression by inhibiting T cells and are therefore considered protumourigenic [33,38–40]. However, the functional role of CD105+ and CD105− subsets is still not firmly established, as data from our group show that therapeutic inhibition or fibroblast-specific genetic deletion of CD105 did not affect mouse pancreatic tumour growth [41,42].

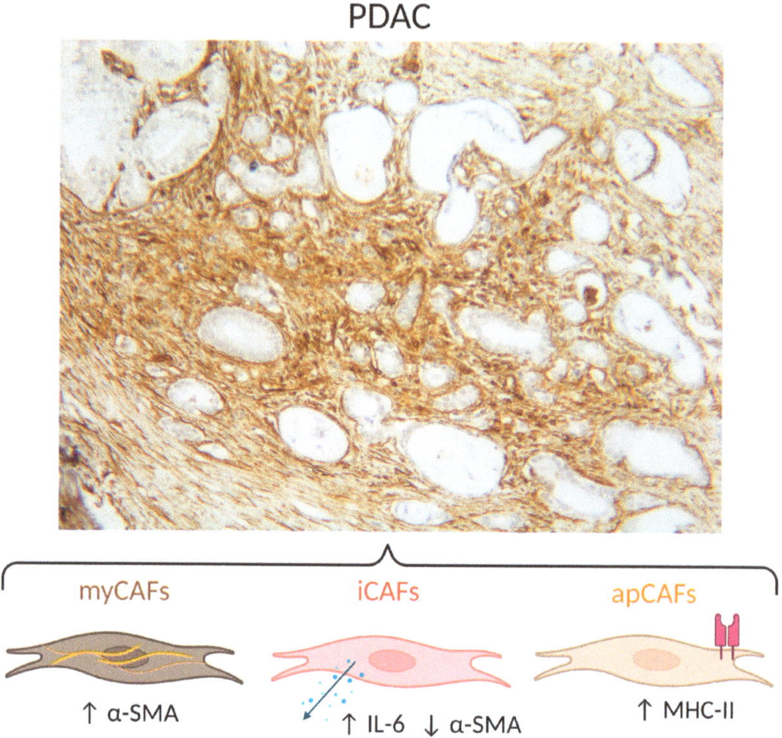

Figure 1. The most common subclassification of CAFs in PDAC consists of myCAFs (α-SMA+), iCAFs (IL-6 secretion, α-SMA−) and apCAFs (MHC-II). PDAC: pancreatic ductal adenocarcinoma; CAF: cancer-associated fibroblast; myCAF: myofibroblast-type CAF; iCAF: inflammatory CAF; apCAF: antigen-presenting CAF; α-SMA: α-smooth muscle actin; IL-6: interleukin 6; MHC-II: major histocompatibility complex class II. Histology picture: human PDAC sample stained for vimentin. Created with BioRender.com (www.biorender.com, accessed on 3 April 2023).

Another study has endeavoured on subcategorizing CAFs based on their metabolic state. In PDAC patients with low desmoplasia, CAFs with a highly active metabolic state (meCAFs) were increased in comparison to patients with highly desmoplastic tumours [43]. The risk of metastasis in patients with an abundant presence of meCAFs was increased, yet their response to immunotherapy was profoundly better. Recently, a follow-up study from the same group identified phospholipase A2 group 2A (PLA2G2A) as a prominent marker for meCAFs [44]. Besides research on CAFs in the primary tumour, CAFs in metastases have been studied and are referred to as a separate subclass, metastases-associated fibroblasts (MAFs), due to observed differences in function. The interactions between MAFs and tumour cells mainly support angiogenesis and tumour progression [45]. Overall, the deviations from the status quo of CAF subclassifications show that the definition of existing subclasses is not definite. These subclasses are probably intertwined, and specific subsets could be more causally involved in the low response to immunotherapies than others.

3. CAF-Immune Cell Interactions

Immunotherapy is aimed at inducing or exploiting immune cell reactivity towards malignant cells. In PDAC, the immunosuppressive TME limits the efficacy of immunotherapeutic approaches. Both immune cells of the myeloid and lymphoid lineage take part in creating this immunosuppressive environment, and accumulating evidence suggests that interactions with CAFs can play a determining role [46,47]. Myeloid cells are a component of the innate immune system that rapidly infiltrate local tissue sites upon their recruitment. Myeloid cells that are associated with PDAC progression are myeloid-derived suppressor cells (MDSCs), tumour-associated macrophages (TAMs) and tumour-associated neutrophils (TANs) [46]. Within the lymphoid lineage, T cells have been studied extensively in PDAC. Cytotoxic T cells are direct anti-cancer effector cells, whereas helper T cells act indirectly by activating other immune cells, including cytotoxic T cells. Regulatory T cells, in contrast, suppress T cell activity. The balance between these T cell subsets plays a crucial role in anti-tumour immunity and therefore tumour progression, and their activity heavily depends on the other cells present in the stroma [48,49]. Reported mechanisms on how CAF–immune cell interactions can shape the immunosuppressive TME are depicted in Figure 2 and are described below.

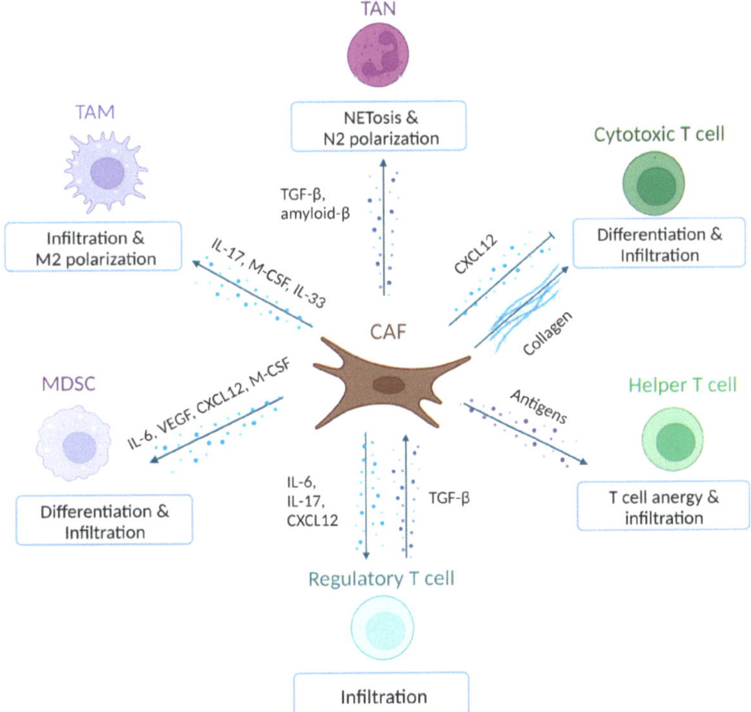

Figure 2. Schematic representation of the observed CAF–immune cell interactions in PDAC. CAFs alter the function, differentiation and infiltration of various myeloid cells, such as TANs, TAMs and MDSCs, as well as T cells. CAF: cancer-associated fibroblast; PDAC: pancreatic ductal adenocarcinoma; TAN: tumour-associated neutrophil; TAM: tumour-associated macrophage; MDSC: myeloid-derived suppressor cell; TME: tumour microenvironment; TGF-β: transforming growth factor β; IL-17: interleukin 17; M-CSF: macrophage colony-stimulating factor; VEGF: vascular endothelial growth factor; CXCL12: C-X-C motif chemokine ligand 12. Created with BioRender.com (www.biorender.com, accessed on 3 April 2023).

3.1. CAF-MDSC Interactions

In PDAC patients, the number of MDSCs in the peripheral blood is related to disease progression [50]. In several studies, the involvement of CAFs in regulating MDSC infiltration and activation has been investigated. iCAF secretion of C-X-C motif chemokine ligand 12 (CXCL12) increases MDSC infiltration, while signalling activation and differentiation are promoted via additional soluble factors IL-6, VEGF, C-C chemokine (CC-motif) ligand 2 (CCL2) and macrophage colony-stimulating factor (M-CSF) [51,52]. Subsequently, MDSCs can inhibit T cells via the depletion of the amino acid L-arginine in the TME [53,54]. The immunosuppression by MDSCs in PDAC patients is dependent on the activation of STAT3, which can be activated through CAF-derived IL-6 [55]. Taken together, CAFs induce MDSC activity and in doing so suppress immune activity in the TME.

3.2. CAF-Macrophage Interactions

Circulating monocytes in addition to tissue-resident macrophages give rise to TAMs [56]. TAMs are commonly subclassified into M1 and M2 macrophages due to their opposite polarization states. In cancer, M1 macrophages have been identified as pro-inflammatory with anti-neoplastic effects and M2 macrophages as anti-inflammatory with pro-neoplastic effects [57]. In PDAC specifically, a high-density infiltration of M2 macrophages has been associated with a shorter OS [58]. The presence of CAFs in the TME increases signalling molecules such as IL-17, which can mediate monocyte recruitment [59,60]. In addition, convincing evidence reported in multiple studies suggests that CAFs can polarize macrophages to an M2 state. For instance, monocytes added to a 3D co-culture of pancreatic cancer cells and fibroblasts differentiated into TAMs with an M2 phenotype [61]. These TAMs could inhibit helper T cell and cytotoxic T cell activity and their proliferative capability in vitro. In line with these results, a co-culture of monocytes and CAFs showed induction of M2 polarization of macrophages by CAFs through M-CSF secretion. Similarly, CAF-dependent M2 macrophage polarization has been observed in a PDAC mouse model, in which CAFs induced M2 macrophage differentiation in an IL-33-dependent manner [62]. Finally, in vivo deletion of hypoxia-inducible factor 2 (HIF2) in CAFs reduced M2 macrophage differentiation and regulatory T cell recruitment. These findings evidence the ability of CAFs to suppress the immune system via regulating TAM polarization.

3.3. CAF-Neutrophil Interactions

Neutrophils are the most common systemic immune cells and play a key role in the innate immune response. In PDAC, neutrophil infiltration is increased compared to the non-cancerous pancreas and during pancreatic inflammation (pancreatitis) [63]. As for macrophages, subclasses N1 and N2 TANs have been reported. TGF-β, which also activates CAFs, is proposed to induce N2 TAN recruitment and activation. N2 TANs display immunosuppressive and pro-tumourigenic effects. As for N1 TANs, they are related to increased T cell cytotoxicity and pro-inflammatory cytokine expression [64]. In PDAC, TAN infiltration has been negatively associated with patient prognosis [63,65]. The effect of CAFs on neutrophils has been studied to a significantly lesser extent. One study proffers that CAFs can drive NETosis in neutrophils via amyloid-β expression in PDAC [66]. NETosis is the formation of neutrophil extracellular traps composed of histone-bound nuclear DNA and cytotoxic granules. Tumour-induced NETosis was linked to cancer progression, and high amyloid-β expression was associated with poor prognosis in human cancer. Inhibition of NETosis in vivo by inhibiting amyloid-β release restricted PDAC tumour growth in mice. Despite the limited number of reports thus far, CAF–neutrophil interactions seem to enhance the pro-tumourigenic effects of TANs and represent an interesting venue for further studies.

3.4. CAF-T Cells

T cell infiltration and T cell subset ratio differ across pancreatic tumours [67,68]. These differences in T cell infiltration have been correlated with PDAC patient outcomes. Specifically, high cytotoxic T cell or helper T cell infiltration is associated with improved survival, whereas high and regulatory T cell infiltration is associated with lower survival [48,63]. When spatial information was included in the evaluation, the proximity of cytotoxic T cells to cancer cells was associated with improved survival, suggesting that these cells play a crucial role in anti-tumour efficacy [48]. Nevertheless, many PDACs are characterized by the exclusion of cytotoxic T cells from the tumour, and low T cell density is believed to be one of the main barriers to the efficacy of immunotherapy.

Several reports have shown that CAFs can actively contribute to T cell exclusion. One study pointed towards a role for CXCL12 produced by FAP+ CAFs in T cell exclusion [48,49]. Chemical inhibition of CXCR4, the receptor of CXCL12, resulted in a rapid influx of T cells in the tumour beds. It remained unclear how CXCL12 contributes to T cell exclusion. In another study, it was demonstrated that CAFs that express FAK1 generated increased deposition of collagen and reduced cytotoxic T cell infiltration [69]. Hence, CAFs may hinder T cell infiltration via the release of CXCL12 and ECM remodelling through collagen deposition. In contrast, a relation between high cytotoxic T cell infiltration and a high stromal density was reported [70]. Possibly, parts of the ECM remodelling by CAFs that arise are beneficial, while other parts such as collagen are detrimental to anti-tumour immunity.

However, the general role of CAFs in mediating T cell suppression has now been challenged by several more recent studies, suggesting that T cell paucity cannot simply be explained by CAF-induced desmoplasia. A study investigating spatial relationships between cancer cells, cytotoxic T cells, CAFs and collagen showed that α-SMA+ CAFs or collagen-I were not enriched in areas with low cytotoxic T cells [48]. Furthermore, the shaping of the intra-tumoural T cell response may vary among the CAF subsets, resulting in immunosuppressive or immunostimulatory signals. myCAFs are mostly involved in inducing desmoplasia (1), iCAFs secrete cytokines (2) and apCAFs are involved in antigen presentation (3). As previously mentioned, decreased overall immune infiltration, yet increased regulatory T cell infiltration, was found when α-SMA+ myCAFs were selectively depleted in a transgenic mouse model of PDAC [30]. In another study to investigate regulatory T cells in PDAC in vivo, conditional cell depletion of regulatory T cells caused a loss of TGF-β expression and reduced α-SMA expression, the marker linked to the myCAF subset [71]. Hence, myCAFs and regulatory T cells seem to mutually influence each other. The studies mentioned above together suggest that myCAFs are less decisive in T cell exclusion and immunosuppression than previously assumed. As for iCAFs, the expression of IL-6 by the subset is associated with an increase in regulatory T cells [32]. CXCL12, the molecule earlier discussed that mediates T cell exclusion, is primarily associated with iCAFs [51]. In addition, an in vivo study on the signalling of CAFs and the effects on immune cells in KPC mice revealed that a knockout of IL-17A, a cytokine secreted by CAFs, reduced regulatory T cells and increased cytotoxic T cells [60]. The apCAF subset can influence the immune system via the presentation of antigens via MHC II, yet they do not have the co-stimulatory molecules necessary for immune cell activation. This causes T cell anergy and an increase in regulatory T cells [33,39]. Similarly, in colorectal cancer, it was shown that apCAFs reduced cytotoxic T cell activation, overall cytotoxicity and enhanced exhaustion [41]. Moreover, another study provided evidence that apCAFs can kill cytotoxic T cells directly in an antigen-dependent manner via the expression of PD-L2 and FASL [40].

In summary, CAF can influence T cells in various ways, impacting their differentiation, function or infiltration. Since T cells are the main effector cell of most immunotherapies, the effect CAFs have on T cells could impact their efficacy.

4. The Road to Immunotherapy for PDAC with CAFs as a Potential Hurdle

Immunotherapy has proven to be an effective therapeutic agent, especially in melanoma, non-small cell lung cancer and tumours with microsatellite instability [72,73]. The most abundantly studied immunotherapies are currently immune checkpoint inhibitors, which block checkpoint proteins from binding to their receptor. Other immunotherapeutic approaches investigated for PDAC include cancer vaccines, CD40 agonists, chimeric antigen receptor (CAR), T cell therapy and oncolytic viruses. The main mechanisms of these immunotherapies enhance overall immune activity or direct cytotoxic T cells to target tumour cells. As described above, interactions between CAFs and immune cells likely cause an immunosuppressive phenotype in PDAC, which could play a key part in the setbacks in immunotherapy. Below, we will discuss commonly used immunotherapeutic approaches and the current data on how CAFs might affect their efficiency.

4.1. Checkpoint Inhibitors

The most applied immunotherapy is immune checkpoint inhibition, which releases the 'brakes' put on the host T cell response against the tumour. Two of the most studied immune checkpoint molecules are cytotoxic T lymphocyte-associated protein 4 (CTLA-4) and programmed cell death protein 1 (PD-1), the latter binding to its ligand PD-L1 on cancer cells and immune cells. [74]. Checkpoint inhibitors are approved as first- or second-line therapies for an increasing number of cancer types, such as melanoma and MSI-high tumours [75]. Unfortunately, clinical trials with checkpoint inhibitors did not show clinical benefit in patients with PDAC [76–78]. Given the fact that the mutational burden is positively correlated with the response rates to immune checkpoint inhibitors, the low mutational burden of PDAC could explain the low efficacy [8]. Furthermore, the desmoplastic and immunosuppressive TME, and especially CAFs, could contribute to immunotherapy resistance [79].

As previously mentioned, CAFs can reduce cytotoxic T cell infiltration, and since checkpoint inhibitors target T cells, their presence is instrumental [33,80]. In PDAC patients with a low degree of desmoplasia, the response to checkpoint inhibitors was profoundly better [43], probably due to T cells being more able to infiltrate the TME. Likewise, the diphtheria toxin-mediated depletion of FAP+ CAFs that contributed to T cell exclusion in a KPC mouse model resulted in improved response to immune checkpoint inhibitors in otherwise non-responsive mice [49]. To understand ECM remodelling, a pan-cancer analysis has been executed that compares genomic and cellular alterations across cancer types between malignant and normal tissue [81]. The created ECM dysregulation signature was related to TGF-β signalling in CAFs and was linked to the failure of anti-PD-1 treatment. Accordingly, the data obtained in this study suggest that CAFs dysregulate ECM organisation via TGF-β dependent effects, thereby inhibiting T cell infiltration and thus the effects of checkpoint inhibitor therapy. In another study, a correlation between a subset of myCAFs and checkpoint inhibitor efficacy was reported in PDAC patients [82]. An elevated level of leucine-rich repeat containing 15 (LLRC15)+ myCAFs was associated with a poor response to anti-PD-L1 treatment. In murine breast cancer, a high abundance of CAFs was associated with insensitivity to anti-CTLA-4 and anti-PD-L1 therapy [83]. Furthermore, genetic depletion of a subset of myCAFs exhibiting the CAF receptor Endo180 (MRC2) led to increased cytotoxic T cell infiltration and improved checkpoint inhibitor sensitivity. In conclusion, CAFs conceivably seem to suppress the efficacy of checkpoint inhibitors via the inhibition of T cell infiltration.

4.2. Activating Immune Checkpoint Agonists

In addition to targeting inhibitory molecules, the activation of immunostimulatory molecules has been investigated, such as the CD40 agonists. CD40 is part of the tumour necrosis factor (TNF) receptor family and promotes T cell activation and M1 TAM polarization [84]. The first clinical trials with a CD40 agonist showed promising results in patients with non-operable PDAC [85]. Nevertheless, a phase II trial on the combination of chemotherapy, a CD40 agonist and a PD-1 inhibitor did not show increased therapeutic efficacy in the groups that received the combination with the CD40 agonist [86]. In a mouse model for colorectal cancer, the combination of a CD40 agonist with an M-CSF-1 receptor antibody increased survival compared to monotherapy [87]. This indicates that inhibiting CAF-mediated immunosuppressive M-CSF signalling might improve CD40 therapies.

Nevertheless, the presence of tumour-specific T cells is still a requisite for the efficacy of immune checkpoint inhibitors or agonists.

4.3. Cancer Vaccines

Cancer vaccines are established anti-cancer strategies when it comes to inducing novel cancer-specific T cell responses directed against tumour-specific antigens [88]. Typical vaccines contain an immunostimulatory adjuvant and antigens conveyed in the form of mRNA, DNA, peptides or whole cells. Cancer vaccine-induced immunization starts with the uptake of the antigens by dendritic cells (DC), which sets the proliferation and differentiation of T cells in motion. In pancreatic cancer, initial results with a whole-cell vaccine seemed promising. However, the addition of the vaccine to standard care chemotherapy in a phase III trial did not improve PDAC patient survival [89]. Another approach was taken with a DC vaccine loaded with whole tumour lysate. Promising results were obtained in a mouse model for PDAC, in particular when the vaccine was combined with a CD40 agonist [90]. Data from a phase I trial on this combination in PDAC patients revealed that no adverse events occurred, and T cell activity increased [91] Although research on the effects of CAFs on cancer vaccination is limited, there are indications that CAFs can inhibit DC and T cell function, thereby hampering effective immune priming. In solid tumours, the release of TGF-β and IL-6 likely suppresses the proliferation and migratory properties of DCs, resulting in interference with tumour-directed priming of cytotoxic T cells [24]. In hepatocellular carcinomas, it was shown that activation of the IL-6-mediated STAT pathway in CAFs causes them to recruit DCs and inhibit antigen presentation [92]. In addition to directly influencing DCs, CAFs also inhibit, as described before the ultimate effector cells of immunization, cytotoxic T cells [93]. In summary, both agonistic CD40 therapies and cancer vaccines are in the early stages of clinical development and show promise, in particular in combination with other immunotherapies to overcome immunosuppression.

4.4. CAR T Cells

Cytotoxic T cells of patients can be genetically engineered to express receptors that specifically recognize tumour cells, thereby creating CAR T cells. CAR T cells targeting antigen CD19 were the first approved second-line treatment for lymphoma and leukaemia patients [94]. In solid tumours, CAR T cell therapy is less efficient, which could be due to the difficulty of CAR T cells to enter desmoplastic, stroma-rich PDAC tumours. The impact of CAFs on CAR T cell therapy is still speculatory. As previously described, CAFs are linked to T cell exclusion [69]. CAR T cell therapy could be considered to target CAFs. An interesting opportunity could arise from a novel CAR T cell product directed to FAP. At present, one clinical trial has been conducted in mesothelioma patients to assess the safety and feasibility of local CAR T cell therapy targeting FAP on the mesothelioma cells [95]. No treatment-related toxicities were observed, and two out of three patients were still alive after a median follow-up of 18 months. These encouraging data may open up opportunities for using FAP-targeting CAR T cells to target FAP+ CAFs in the future.

4.5. Oncolytic Virus Therapy

Oncolytic viruses can be used to induce oncolysis of tumour cells directly or indirectly via activation of anti-tumour immunity. Adenoviruses, herpesviruses, reoviruses and parvoviruses have been tested as oncolytic viruses in clinical trials in PDAC, but until now, oncolytic viruses as monotherapy have not shown sufficient efficiency [7]. The desmoplastic nature of the TME could play a role in reduced penetration of the virus into the tumour. The ability of oncolytic viruses to infect and kill CAFs and other stromal cells besides tumour cells could be a way to decrease the pro-tumourigenic effects of CAFs [96]. Recently, tropism by the reovirus towards human and mouse pancreatic CAFs was observed, while the same work reported that the crosstalk between CAFs and tumour cells can impact the viral infection [97]. Even though oncolytic viruses are not successful as monotherapy yet, their capacity to transform an immunosuppressive TME of solid tumours into an immune active TME with enhanced T cell influx can be used to boost the efficacy of other immunotherapeutic strategies. Proof-of-concept has been demonstrated with a combination of oncolytic reovirus and CD3-bispecific T cell engagers in a mouse model of PDAC [98].

In conclusion, it has become evident that immunotherapies in PDAC have potential but do not show desired efficiency, possibly in part due to CAFs and the immunosuppressed environment they generate.

5. Targeting CAFs and Their Products to Advance Immunotherapy Responses

Given the potentially important role that CAFs play in modulating responses to immunotherapy, a solution to low efficiency could be to modulate the CAF population or the CAF-mediated downstream effects in PDAC. Given the distinct subsets and their specific role in regulating immune cell trafficking and activity, this has to be considered carefully. As indicated before, the depletion of all CAFs from PDAC tumour models in mice, contrary to the expectations, generated an immunosuppressed environment [30]. These data indicate that a more selective approach is warranted. Table 1 summarizes the effects of targeting CAFs to improve the immunological effects of immunotherapies.

Table 1. Overview of studied CAF targets in the PDAC TME with the aim to improve immunotherapy. Results from preclinical models as well as clinical trials are summarized. Effects on CAFs, immunity, tumour growth and survival are separately indicated. ↓: a decrease; ↑: an increase; -: no observed difference; n.d.: no data; CAF: cancer-associated fibroblast; PIN1: peptidyl-prolyl isomerase NIMA-interacting 1; PD-1: programmed cell death protein 1; KPC: $Kras^{LSL-G12D}$, $Trp53^{LSL-R172H}$, Pdx1-cre; hsp90: heat shock protein 90; IL-6: interleukin 6; PD-L1: programmed cell death protein ligand 1 (PD-L1); KPC-brca2: $Kras^{LSL-G12D}$, $Trp53^{LSL-R270H}$, Pdx1-cre, $Brca2^{F/F}$; CXCR2: chemokine receptor 2; FAP: fibroblast activation protein; TGF-β: transforming growth factor β; CAR: chimeric antigen receptor; NSG: NOD/scid/IL2rg−/−; ADC: antibody drug conjugate.

Immunotherapy	Target	Design	CAFs	Immunity	Tumour Growth	Survival	Reference
Preclinical studies							
Checkpoint inhibitor	PIN1 + PD-1	KPC	↓	↑	↓	↑	[99]
	hsp90 + PD-1	KPC	↓	↑	↓	n.d.	[100]
	IL-6 + PD-L1	KPC-Brca2	n.d.	↑	↓	↑	[101]
	IL-1 + PD-1	KPC	n.d.	↑	↓	↑	[102]
	CXCR2 + PD-1	KPC	n.d.	↑	↓	↑	[103,104]
	Vitamin D analogue + PD-L1	2D + 3D culture	↓	↓	-	-	[105,106]
	FAP + PD-L1	KPC	↓	↑	↓	n.d.	[49]
Vaccine	TGF-β	C57BL/6	↓	↑	↓	n.d.	[107]
	Hyaluronan + GVAX	C57BL/6	↓	↑	↓	↑	[108]
	FAP + survivin	C57BL/6	↓	↑	↓	↑	[109]
CD40 agonist	IL-6 + CD40 agonist	C57BL/6	n.d.	↑	↓	↑	[110]
	FAP + CD40 agonist	KPC-huCEA	↓	↑	↓	n.d.	[111]
CAR T cell	FAP	C57BL/6	n.d.	↑	n.d.	n.d.	[112]
	FAP	KPC	↓	↑	↓	↑	[113]
Oncolytic virus	FAP	NSG	↓	↑	↓	↑	[114]
	CAF + tumour cells	BALB/c-nu/nu	n.d.	↑	↓	↑	[115]
	TGF-β	KPC	n.d.	-	-	-	[116]
	TGF-β + CD3 + tumour cells	KPC	↓	-	-	-	[98]
ADC	FAP + tumour cells	Foxn1 nu/nu	↓	↑	↓	↑	[117]
Clinical trials							
Checkpoint inhibitor	CXCR4 + PD-1	Phase II	n.d.	↑	↓	↑	[118]
	TGF-β + PD-L1	Phase I	n.d.	n.d.	n.d.	n.d.	[119]
Oncolytic virus	Hyaluronan + tumour cells	Phase I	n.d.	↑	↓	↑	[120]

5.1. The Fight against Immunosuppression

Taking into consideration markers of the currently known CAF subsets can help identify the right targets for therapy. A decrease in iCAF differentiation or abundance and thereby iCAF-induced immunosuppression could improve immunotherapy. An inhibitor of IL-1, a promotor of iCAF differentiation, was administered together with the checkpoint inhibitor PD-1 to PDAC-bearing mice [92]. The combination therapy improved anti-PD-1 blockade sensitivity by enabling cytotoxic T cell infiltration and by attenuating tumour growth [102]. Another approach is to inhibit inflammatory cytokines secreted by iCAFs, such as IL-6. A combination of an IL-6 inhibitor and anti-PD-L1 resulted in increased cytotoxic T cell infiltration and impaired tumour progression in the PDAC KPC-Brca2 mouse model [101]. Next to directly targeting IL-6, an indirect approach has been tested by inhibition of heat shock protein 90 (hsp90) with small molecule XL-888, which decreased IL-6 secretion by CAFs [100]. Consequently, the combination of XL-888 and anti-PD-1 increased T cell infiltration and enhanced therapeutic efficacy in KPC mice [100].

IL-6 inhibition has also been applied in combination with a CD40 agonist. The treatment increased the anti-tumour effects by reduction of TGF-β, collagen, PD-L1 and PD-1 expression indicative of targeting the CAFs in the TME [110]. In addition to IL-6, iCAFs have also been targeted by inhibition of CXCL12 and its receptor C-X-C motif receptor 4 (CXCR4) [33]. In a phase II trial, the combination of a CXCR4 antagonist, anti-PD-1 and chemotherapy [118] improved cytotoxic T cell infiltration, reduced MDSCs and decreased circulating regulatory T cells [100]. These observations highlight the possibilities of inhibiting cytokines or chemokines secreted by iCAFs to reduce immunosuppression.

Recently, it was shown that iCAF-derived cytokines can activate CXCR2 on CAFs, which causes conversion into myCAFs. This phenomenon is associated with increased PDAC metastasis [121]. CXCR2 inhibition in combination with anti-PD-1 led to increased helper and cytotoxic T cell infiltration, while also reducing neutrophil and regulatory T cells [103,104]. Therefore, CAF subtype conversion might be a novel and challenging but interesting approach to improve immunotherapy.

5.2. Targeting CAFs Directly

Another approach would be to reduce the number of CAFs in the TME by therapeutically targeting these cells. The most abundantly studied molecule on CAFs that has been used as a target is FAP. In PDAC mouse models, CAR T cell therapy targeting FAP+ CAFs led to toxicity, while in another study, it was effective without inducing toxicity [112,113]. The fact that FAP+ is also expressed scarcely in other cells, such as bone marrow cells, must be taken into consideration. A second approach to target FAP used an adenovirus that expressed a bispecific T cell engager (biTe) that binds to FAP. Adenoviral therapy stimulated T cell activation and T cell infiltration and decreased tumour growth in PDAC-bearing mice [114]. A third approach to target FAP+ CAFs that has been evaluated is a vaccine. A vaccine against the tumour antigens survivin and FAP decreased the proportion of immunosuppressive cells and increased lymphocyte infiltration in a pancreatic cancer murine model [109]. An alternative way to target FAP+ CAFs is via antibody drug conjugates (ADCs) containing a cytotoxic payload. OMTX705, an ADC including a humanized anti-FAP antibody and the cytolytic compound TAM470, demonstrated potent tumour growth inhibition in PDX mouse models with various solid tumours, including PDAC [117]. Mechanistically, OMTX705 resulted in stromal cell depletion in PDAC models, as indicated by reduced α-SMA and Col11a1 staining and increased cytotoxic T cell infiltration. Lastly, a FAP-CD40 antibody that exclusively initiated anti-tumour immune activity in the presence of FAP reduced tumour growth in PDAC-bearing mice [111]. Besides FAP, a marker of interest expressed by CAFs could be peptidyl-prolyl isomerase NIMA-interacting 1 (PIN1). Upregulated PIN1 in CAFs leads to desmoplasia and thereby poor cytotoxic T cell infiltration. Inhibitors of PIN1 improved anti-PD-1 response in KPC mice and increases survival [99].

In conclusion, there are attractive targets in PDAC that can be used to relatively, specifically inhibit CAFs or as a docking site for therapeutic antibodies directed towards the TME. Further studies in PDAC patients remain warranted to elucidate potential mechanisms and efficiency.

5.3. Targeting CAFs from Multiple Angles

Besides immunosuppressive influences in the TME, CAFs are known for remodelling the ECM and inducing desmoplasia. Consequently, approaches have been evaluated to target both the immunosuppressive and desmoplastic effects of CAFs. Hyaluronan is a component of the ECM and is associated with α-SMA+ CAFs [122]. Hyaluronan inhibition could pave the way for immunotherapy by reducing desmoplasia and allowing for anti-tumour immunity. Administration of a hyaluronan inhibitor and whole cell PDAC vaccine, GVAX, in vivo inhibited immunosuppression via the reduction of CXCL12/CXR4 signalling, decreased desmoplasia and conferred a survival advantage in comparison with single-agent therapies [108]. Along similar lines, oncolytic adenovirus VCN-01, which includes hyaluronidase to improve virus intra-tumoural spread and anti-tumour immunity, was administered in combination with chemotherapy to PDAC patients in a phase I trial [120]. No dose-limiting toxicity was observed, and an overall response rate of 50% was detected. These results support the premise that reducing desmoplasia improves anti-tumour immunity.

TGF-β is involved in myCAF promotion and promoting desmoplasia while also having immunosuppressive effects. Therefore, TGF-β inhibition is an attractive approach to improve the TME of PDAC on multiple levels, including the reduction of desmoplasia and immunosuppression. A number of studies have used TGF-β inhibition in PDAC mouse models. In KPC mice, the combination of PD-L1 and a TGF-β receptor small molecule inhibitor delayed tumour growth relative to untreated mice but did not increase anti-tumour immune response [116]. In contrast, the combination therapy elicited an immune response in colorectal cancer-bearing mice in this study. As an alternative tactic, TGF-β-derived peptide vaccination was tested, which enhanced cytotoxic T cell infiltration, M1 TAM polarization and reduced myCAF abundance in PDAC-bearing mice [107]. Another study intended to enhance the efficacy of immunotherapy via the addition of a TGF-β-blocking antibody to a reovirus and via CD3-bispecific antibody combination therapy [123]. The TGF-β-blocking antibody antagonized the therapeutic efficacy of the two other treatments. The TGF-β blockade has been investigated in metastatic PDAC patients in a phase I study as well. Treatment with a TGF-β inhibitor and anti-PD-L1 did not give rise to serious adverse events, but the OS remained low, which might be related to the advanced tumour stage [119]. Accordingly, different responses are encountered after TGF-β inhibition, possibly due to the pleiotropic role of TGF-β in the TME.

Besides TGF-β, vitamin D could be targeted to reprogram the TME. Three-dimensional in vitro models showed that vitamin D decreases CAF proliferation and migration yet upregulates PD-L1 [105]. In a PDAC mouse model, activation of the vitamin D receptor decreased desmoplasia in vivo [124]. Although, vitamin D can initiate Th2 and inhibit Th1 immune responses, which further enables the immunosuppressive state of the TME [125]. Thus, vitamin D therapy could reduce desmoplasia promoted by CAFs but restrict the activity of anti-tumourigenic T cells via promoting immune suppression.

This evidence emphasizes the challenging interconnections that signalling cascades have. One pathway could both increase pro-tumourigenic immunosuppression and decrease pro-tumourigenic desmoplasia. Consequently, the network of CAFs must first be understood to know its entire effect on immunotherapy.

6. Concluding Remarks

To improve PDAC treatment options and survival, it is essential to understand the intricacies of the TME, including immune cell and CAF interactions. The current literature provides a solid foundation for the notion that CAFs are instrumental in evoking immunotherapy resistance, but the underlying mechanisms are insufficiently explored. The major question remains: are CAFs the good or the bad guys in PDAC? Pathways associated with iCAFs are predominantly, but not exclusively, deemed as pro-tumourigenic. CAFs are involved in desmoplasia, but depletion of myCAFs can lead to tumour progression. This poses a major hurdle for current and future efforts. Mapping CAF subsets and exploring their function and plasticity are key before exploring or excluding therapeutic opportunities. Thus far, the combination of CAF-targeting therapies with immunotherapies shows promising effects in murine models and some clinical trials. These studies pave the way for the future of PDAC treatment.

Author Contributions: Conceptualization, N.v.M. and L.J.A.C.H.; writing—original draft preparation, I.S.; writing—review and editing, N.v.M. and L.J.A.C.H.; visualization, I.S.; supervision, N.v.M. and L.J.A.C.H.; funding acquisition, N.v.M. and L.J.A.C.H. All authors have read and agreed to the published version of the manuscript.

Funding: Imke Stouten is supported by a research grant from Bontius Stichting (816092114).

Institutional Review Board Statement: Not applicable.

Informed Consent Statement: Not applicable.

Data Availability Statement: Not applicable.

Acknowledgments: All figures have been made with BioRender (www.biorender.com, accessed on 3 April 2023).

Conflicts of Interest: The authors declare no conflict of interest.

References

1. Kleeff, J.; Korc, M.; Apte, M.; La Vecchia, C.; Johnson, C.D.; Biankin, A.V.; Neale, R.E.; Tempero, M.; Tuveson, D.A.; Hruban, R.H.; et al. Pancreatic Cancer. *Nat. Rev. Dis. Primers* **2016**, *2*, 16022. [CrossRef] [PubMed]
2. Siegel, R.L.; Miller, K.D.; Jemal, A. Cancer Statistics, 2018. *CA Cancer J. Clin.* **2018**, *68*, 7–30. [CrossRef] [PubMed]
3. Shinde, R.S.; Bhandare, M.; Chaudhari, V.; Shrikhande, S.V. Cutting-Edge Strategies for Borderline Resectable Pancreatic Cancer. *Ann. Gastroenterol. Surg.* **2019**, *3*, 368–372. [CrossRef] [PubMed]
4. Principe, D.R.; Underwood, P.W.; Korc, M.; Trevino, J.G.; Munshi, H.G.; Rana, A. The Current Treatment Paradigm for Pancreatic Ductal Adenocarcinoma and Barriers to Therapeutic Efficacy. *Front. Oncol.* **2021**, *11*, 688377. [CrossRef] [PubMed]
5. Zeng, S.; Pöttler, M.; Lan, B.; Grützmann, R.; Pilarsky, C.; Yang, H. Chemoresistance in Pancreatic Cancer. *Int. J. Mol. Sci.* **2019**, *20*, 4504. [CrossRef] [PubMed]
6. Yoon, J.H.; Jung, Y.J.; Moon, S.H. Immunotherapy for Pancreatic Cancer. *World J. Clin. Cases* **2021**, *9*, 2969–2982. [CrossRef] [PubMed]
7. Tassone, E.; Muscolini, M.; van Montfoort, N.; Hiscott, J. Oncolytic Virotherapy for Pancreatic Ductal Adenocarcinoma: A Glimmer of Hope after Years of Disappointment? *Cytokine Growth Factor Rev.* **2020**, *56*, 141–148. [CrossRef]
8. Yarchoan, M.; Hopkins, A.; Jaffee, E.M. Tumor Mutational Burden and Response Rate to PD-1 Inhibition. *N. Engl. J. Med.* **2017**, *377*, 2500. [CrossRef]
9. Apte, M.V.; Park, S.; Phillips, P.A.; Santucci, N.; Goldstein, D.; Kumar, R.K.; Ramm, G.A.; Buchler, M.; Friess, H.; McCarroll, J.A.; et al. Desmoplastic Reaction in Pancreatic Cancer: Role of Pancreatic Stellate Cells. *Pancreas* **2004**, *29*, 179–187. [CrossRef]
10. Herting, C.J.; Karpovsky, I.; Lesinski, G.B. The Tumor Microenvironment in Pancreatic Ductal Adenocarcinoma: Current Perspectives and Future Directions. *Cancer Metastasis Rev.* **2021**, *40*, 675–689. [CrossRef]
11. Lafaro, K.J.; Melstrom, L.G. The Paradoxical Web of Pancreatic Cancer Tumor Microenvironment. *Am. J. Pathol.* **2019**, *189*, 44–57. [CrossRef] [PubMed]
12. Maia, A.; Schöllhorn, A.; Schuhmacher, J.; Gouttefangeas, C. CAF-Immune Cell Crosstalk and Its Impact in Immunotherapy. *Semin. Immunopathol.* **2022**, *45*, 203–214. [CrossRef] [PubMed]
13. Desbois, M.; Wang, Y. Cancer-Associated Fibroblasts: Key Players in Shaping the Tumor Immune Microenvironment. *Immunol. Rev.* **2021**, *302*, 241–258. [CrossRef] [PubMed]
14. Sahai, E.; Astsaturov, I.; Cukierman, E.; DeNardo, D.G.; Egeblad, M.; Evans, R.M.; Fearon, D.; Greten, F.R.; Hingorani, S.R.; Hunter, T.; et al. A Framework for Advancing Our Understanding of Cancer-Associated Fibroblasts. *Nat. Rev. Cancer* **2020**, *20*, 174. [CrossRef]
15. Arina, A.; Idel, C.; Hyjek, E.M.; Alegre, M.L.; Wang, Y.; Bindokas, V.P.; Weichselbaum, R.R.; Schreiber, H. Tumor-Associated Fibroblasts Predominantly Come from Local and Not Circulating Precursors. *Proc. Natl. Acad. Sci. USA* **2016**, *113*, 7551–7556. [CrossRef]

16. Vaish, U.; Jain, T.; Are, A.C.; Dudeja, V. Cancer-Associated Fibroblasts in Pancreatic Ductal Adenocarcinoma: An Update on Heterogeneity and Therapeutic Targeting. *Int. J. Mol. Sci.* **2021**, *22*, 13408. [CrossRef]

17. Gascard, P.; Tlsty, T.D. Carcinoma-Associated Fibroblasts: Orchestrating the Composition of Malignancy. *Genes Dev.* **2016**, *30*, 1002–1019. [CrossRef]

18. Ping, Q.; Yan, R.; Cheng, X.; Wang, W.; Zhong, Y.; Hou, Z.; Shi, Y.; Wang, C.; Li, R. Cancer-Associated Fibroblasts: Overview, Progress, Challenges, and Directions. *Cancer Gene Ther.* **2021**, *28*, 984–999. [CrossRef]

19. Monteran, L.; Erez, N. The Dark Side of Fibroblasts: Cancer-Associated Fibroblasts as Mediators of Immunosuppression in the Tumor Microenvironment. *Front. Immunol.* **2019**, *10*, 1835. [CrossRef]

20. von Hoerschelmann, E.; Neumann, C.C.M.; Felsenstein, M.; Gogolok, J.; Reutzel-Selke, A.; Sauer, I.M.; Pratschke, J.; Bahra, M.; Schmuck, R.B. Cancer Associated Fibroblasts Derived from Pancreatic Adenocarcinoma and Their Role in Cell Migration. *Anticancer Res.* **2021**, *41*, 4229–4238. [CrossRef]

21. Ligorio, M.; Sil, S.; Malagon-Lopez, J.; Nieman, L.T.; Misale, S.; di Pilato, M.; Ebright, R.Y.; Karabacak, M.N.; Kulkarni, A.S.; Liu, A.; et al. Stromal Microenvironment Shapes the Intratumoral Architecture of Pancreatic Cancer. *Cell* **2019**, *178*, 160. [CrossRef] [PubMed]

22. Nielsen, M.F.B.; Mortensen, M.B.; Detlefsen, S. Key Players in Pancreatic Cancer-Stroma Interaction: Cancer-Associated Fibroblasts, Endothelial and Inflammatory Cells. *World J. Gastroenterol.* **2016**, *22*, 2678–2700. [CrossRef] [PubMed]

23. Masamune, A.; Kikuta, K.; Watanabe, T.; Satoh, K.; Hirota, M.; Shimosegawa, T. Hypoxia Stimulates Pancreatic Stellate Cells to Induce Fibrosis and Angiogenesis in Pancreatic Cancer. *Am. J. Physiol. Gastrointest. Liver Physiol.* **2008**, *295*, 709–717. [CrossRef] [PubMed]

24. Harryvan, T.J.; Verdegaal, E.M.E.; Hardwick, J.C.H.; Hawinkels, L.J.A.C.; van der Burg, S.H. Targeting of the Cancer-Associated Fibroblast-T-Cell Axis in Solid Malignancies. *J. Clin. Med.* **2019**, *8*, 1989. [CrossRef]

25. Lee, H.O.; Mullins, S.R.; Franco-Barraza, J.; Valianou, M.; Cukierman, E.; Cheng, J.D. FAP-Overexpressing Fibroblasts Produce an Extracellular Matrix That Enhances Invasive Velocity and Directionality of Pancreatic Cancer Cells. *BMC Cancer* **2011**, *11*, 245. [CrossRef] [PubMed]

26. Vonlaufen, A.; Joshi, S.; Qu, C.; Phillips, P.A.; Xu, Z.; Parker, N.R.; Toi, C.S.; Pirola, R.C.; Wilson, J.S.; Goldstein, D.; et al. Pancreatic Stellate Cells: Partners in Crime with Pancreatic Cancer Cells. *Cancer Res.* **2008**, *68*, 2085–2093. [CrossRef] [PubMed]

27. Kraman, M.; Bambrough, P.J.; Arnold, J.N.; Roberts, E.W.; Magiera, L.; Jones, J.O.; Gopinathan, A.; Tuveson, D.A.; Fearon, D.T. Suppression of Antitumor Immunity by Stromal Cells Expressing Fibroblast Activation Protein-Alpha. *Science* **2010**, *330*, 827–830. [CrossRef]

28. Erdogan, B.; Ao, M.; White, L.M.; Means, A.L.; Brewer, B.M.; Yang, L.; Washington, M.K.; Shi, C.; Franco, O.E.; Weaver, A.M.; et al. Cancer-Associated Fibroblasts Promote Directional Cancer Cell Migration by Aligning Fibronectin. *J. Cell Biol.* **2017**, *216*, 3799–3816. [CrossRef]

29. Zhang, J.; Chen, M.; Fang, C.; Luo, P. A Cancer-Associated Fibroblast Gene Signature Predicts Prognosis and Therapy Response in Patients with Pancreatic Cancer. *Front. Oncol.* **2022**, *12*, 1052132. [CrossRef]

30. Özdemir, B.C.; Pentcheva-Hoang, T.; Carstens, J.L.; Zheng, X.; Wu, C.-C.; Simpson, T.R.; Laklai, H.; Sugimoto, H.; Kahlert, C.; Novitskiy, S.V.; et al. Depletion of Carcinoma-Associated Fibroblasts and Fibrosis Induces Immunosuppression and Accelerates Pancreas Cancer with Reduced Survival. *Cancer Cell* **2014**, *25*, 719–734. [CrossRef]

31. Rhim, A.D.; Oberstein, P.E.; Thomas, D.H.; Mirek, E.T.; Palermo, C.F.; Sastra, S.A.; Dekleva, E.N.; Saunders, T.; Becerra, C.P.; Tattersall, I.W.; et al. Stromal Elements Act to Restrain, Rather than Support, Pancreatic Ductal Adenocarcinoma. *Cancer Cell* **2014**, *25*, 735–747. [CrossRef] [PubMed]

32. Öhlund, D.; Handly-Santana, A.; Biffi, G.; Elyada, E.; Almeida, A.S.; Ponz-Sarvise, M.; Corbo, V.; Oni, T.E.; Hearn, S.A.; Lee, E.J.; et al. Distinct Populations of Inflammatory Fibroblasts and Myofibroblasts in Pancreatic Cancer. *J. Exp. Med.* **2017**, *214*, 579–596. [CrossRef] [PubMed]

33. Elyada, E.; Bolisetty, M.; Laise, P.; Flynn, W.F.; Courtois, E.T.; Burkhart, R.A.; Teinor, J.A.; Belleau, P.; Biffi, G.; Lucito, M.S.; et al. Cross-Species Single-Cell Analysis of Pancreatic Ductal Adenocarcinoma Reveals Antigen-Presenting Cancer-Associated Fibroblasts. *Cancer Discov.* **2019**, *9*, 1102–1123. [CrossRef]

34. Hingorani, S.R.; Wang, L.; Multani, A.S.; Combs, C.; Deramaudt, T.B.; Hruban, R.H.; Rustgi, A.K.; Chang, S.; Tuveson, D.A. Trp53R172H and KrasG12D Cooperate to Promote Chromosomal Instability and Widely Metastatic Pancreatic Ductal Adenocarcinoma in Mice. *Cancer Cell* **2005**, *7*, 469–483. [CrossRef] [PubMed]

35. Biffi, G.; Oni, T.E.; Spielman, B.; Hao, Y.; Elyada, E.; Park, Y.; Preall, J.; Tuveson, D.A. IL1-Induced JAK/STAT Signaling Is Antagonized by TGFβ to Shape CAF Heterogeneity in Pancreatic Ductal Adenocarcinoma. *Cancer Discov.* **2019**, *9*, 282–301. [CrossRef]

36. Galbo, P.M.; Zang, X.; Zheng, D. Molecular Features of Cancer-Associated Fibroblast Subtypes and Their Implication on Cancer Pathogenesis, Prognosis, and Immunotherapy Resistance. *Clin. Cancer Res.* **2021**, *27*, 2636–2647. [CrossRef]

37. Costa, A.; Kieffer, Y.; Scholer-Dahirel, A.; Pelon, F.; Bourachot, B.; Cardon, M.; Sirven, P.; Magagna, I.; Fuhrmann, L.; Bernard, C.; et al. Fibroblast Heterogeneity and Immunosuppressive Environment in Human Breast Cancer. *Cancer Cell* **2018**, *33*, 463–479.e10. [CrossRef]

38. Huang, H.; Wang, Z.; Zhang, Y.; Pradhan, R.N.; Ganguly, D.; Chandra, R.; Murimwa, G.; Wright, S.; Gu, X.; Maddipati, R.; et al. Mesothelial Cell-Derived Antigen-Presenting Cancer-Associated Fibroblasts Induce Expansion of Regulatory T Cells in Pancreatic Cancer. *Cancer Cell* **2022**, *40*, 656–673. [CrossRef]
39. Lakins, M.A.; Ghorani, E.; Munir, H.; Martins, C.P.; Shields, J.D. Cancer-Associated Fibroblasts Induce Antigen-Specific Deletion of CD8 + T Cells to Protect Tumour Cells. *Nat. Commun.* **2018**, *9*, 948. [CrossRef]
40. Harryvan, T.J.; Visser, M.; de Bruin, L.; Plug, L.; Griffioen, L.; Mulder, A.; van Veelen, P.A.; van der Heden Van Noort, G.J.; Jongsma, M.L.M.; Meeuwsen, M.H.; et al. Original Research: Enhanced Antigen Cross-Presentation in Human Colorectal Cancer-Associated Fibroblasts through Upregulation of the Lysosomal Protease Cathepsin S. *J. Immunother. Cancer* **2022**, *10*, 3591. [CrossRef]
41. Hutton, C.; Heider, F.; Blanco-Gomez, A.; Banyard, A.; Kononov, A.; Zhang, X.; Karim, S.; Paulus-Hock, V.; Watt, D.; Steele, N.; et al. Single-Cell Analysis Defines a Pancreatic Fibroblast Lineage That Supports Anti-Tumor Immunity. *Cancer Cell* **2021**, *39*, 1227–1244. [CrossRef] [PubMed]
42. Schoonderwoerd, M.J.A.; Hakuno, S.K.; Sassen, M.; Kuhlemaijer, E.B.; Paauwe, M.; Slingerland, M.; Fransen, M.F.; Hawinkels, L.J.A.C. Targeting Endoglin Expressing Cells in the Tumor Microenvironment Does Not Inhibit Tumor Growth in a Pancreatic Cancer Mouse Model. *Onco Targets Ther.* **2021**, *14*, 5205–5220. [CrossRef]
43. Wang, Y.; Liang, Y.; Xu, H.; Zhang, X.; Mao, T.; Cui, J.; Yao, J.; Wang, Y.; Jiao, F.; Xiao, X.; et al. Single-Cell Analysis of Pancreatic Ductal Adenocarcinoma Identifies a Novel Fibroblast Subtype Associated with Poor Prognosis but Better Immunotherapy Response. *Cell Discov.* **2021**, *7*, 36. [CrossRef] [PubMed]
44. Zhang, Y.; Wei, C.; Guo, C.C.; Bi, R.X.; Xie, J.; Guan, D.H.; Yang, C.H.; Jiang, Y.H. Prognostic Value of MicroRNAs in Hepatocellular Carcinoma: A Meta-Analysis. *Oncotarget* **2017**, *8*, 107237. [CrossRef] [PubMed]
45. Pausch, T.M.; Aue, E.; Wirsik, N.M.; Freire Valls, A.; Shen, Y.; Radhakrishnan, P.; Hackert, T.; Schneider, M.; Schmidt, T. Metastasis-Associated Fibroblasts Promote Angiogenesis in Metastasized Pancreatic Cancer via the CXCL8 and the CCL2 Axes. *Sci. Rep.* **2020**, *10*, 5420. [CrossRef]
46. Kemp, S.B.; di Magliano, M.P.; Crawford, H.C. Myeloid Cell Mediated Immune Suppression in Pancreatic Cancer. *Cell. Mol. Gastroenterol. Hepatol.* **2021**, *12*, 1531–1542. [CrossRef]
47. Li, K.Y.; Yuan, J.L.; Trafton, D.; Wang, J.X.; Niu, N.; Yuan, C.H.; Liu, X.B.; Zheng, L. Pancreatic Ductal Adenocarcinoma Immune Microenvironment and Immunotherapy Prospects. *Chronic Dis. Transl. Med.* **2020**, *6*, 6. [CrossRef] [PubMed]
48. Carstens, J.L.; De Sampaio, P.C.; Yang, D.; Barua, S.; Wang, H.; Rao, A.; Allison, J.P.; Le Bleu, V.S.; Kalluri, R. Spatial Computation of Intratumoral T Cells Correlates with Survival of Patients with Pancreatic Cancer. *Nat. Commun.* **2017**, *8*, 15095. [CrossRef]
49. Feig, C.; Jones, J.O.; Kraman, M.; Wells, R.J.B.; Deonarine, A.; Chan, D.S.; Connell, C.M.; Roberts, E.W.; Zhao, Q.; Caballero, O.L.; et al. Targeting CXCL12 from FAP-Expressing Carcinoma-Associated Fibroblasts Synergizes with Anti-PD-L1 Immunotherapy in Pancreatic Cancer. *Proc. Natl. Acad. Sci. USA* **2013**, *110*, 20212–20217. [CrossRef]
50. Markowitz, J.; Brooks, T.R.; Duggan, M.C.; Paul, B.K.; Pan, X.; Wei, L.; Abrams, Z.; Luedke, E.; Lesinski, G.B.; Mundy-Bosse, B.; et al. Patients with Pancreatic Adenocarcinoma Exhibit Elevated Levels of Myeloid-Derived Suppressor Cells upon Progression of Disease. *Cancer Immunol. Immunother.* **2015**, *64*, 149–159. [CrossRef]
51. Malik, S.; Westcott, J.M.; Brekken, R.A.; Burrows, F.J. CXCL12 in Pancreatic Cancer: Its Function and Potential as a Therapeutic Drug Target. *Cancers* **2021**, *14*, 86. [CrossRef] [PubMed]
52. Mace, T.A.; Ameen, Z.; Collins, A.; Wojcik, S.; Mair, M.; Young, G.S.; Fuchs, J.R.; Eubank, T.D.; Frankel, W.L.; Bekaii-Saab, T.; et al. Pancreatic Cancer-Associated Stellate Cells Promote Differentiation of Myeloid-Derived Suppressor Cells in a STAT3-Dependent Manner. *Cancer Res.* **2013**, *73*, 3007–3018. [CrossRef] [PubMed]
53. Gabrilovich, D.I.; Nagaraj, S. Myeloid-Derived Suppressor Cells as Regulators of the Immune System. *Nat. Rev. Immunol.* **2009**, *9*, 162–174. [CrossRef] [PubMed]
54. Bronte, V.; Zanovello, P. Regulation of Immune Responses by L-Arginine Metabolism. *Nat. Rev. Immunol.* **2005**, *5*, 641–654. [CrossRef]
55. Trovato, R.; Fiore, A.; Sartori, S.; Canè, S.; Giugno, R.; Cascione, L.; Paiella, S.; Salvia, R.; de Sanctis, F.; Poffe, O.; et al. Immunosuppression by Monocytic Myeloid-Derived Suppressor Cells in Patients with Pancreatic Ductal Carcinoma Is Orchestrated by STAT3. *J. Immunother. Cancer* **2019**, *7*, 255. [CrossRef]
56. Zhu, Y.; Herndon, J.M.; Sojka, D.K.; Kim, K.W.; Knolhoff, B.L.; Zuo, C.; Cullinan, D.R.; Luo, J.; Bearden, A.R.; Lavine, K.J.; et al. Tissue-Resident Macrophages in Pancreatic Ductal Adenocarcinoma Originate from Embryonic Hematopoiesis and Promote Tumor Progression. *Immunity* **2017**, *47*, 323–338.e6. [CrossRef]
57. Bolli, E.; Movahedi, K.; Laoui, D.; Ginderachter, J.A.V. Novel Insights in the Regulation and Function of Macrophages in the Tumor Microenvironment. *Curr. Opin. Oncol.* **2017**, *29*, 55–61. [CrossRef]
58. Hu, H.; Hang, J.J.; Han, T.; Zhuo, M.; Jiao, F.; Wang, L.W. The M2 Phenotype of Tumor-Associated Macrophages in the Stroma Confers a Poor Prognosis in Pancreatic Cancer. *Tumour Biol.* **2016**, *37*, 8657–8664. [CrossRef]
59. Gordon, S.; Taylor, P.R. Monocyte and Macrophage Heterogeneity. *Nat. Rev. Immunol.* **2005**, *5*, 953–964. [CrossRef]

60. Mucciolo, G.; Curcio, C.; Roux, C.; Li, W.Y.; Capello, M.; Curto, R.; Chiarle, R.; Giordano, D.; Satolli, M.A.; Lawlor, R.; et al. IL17A Critically Shapes the Transcriptional Program of Fibroblasts in Pancreatic Cancer and Switches on Their Protumorigenic Functions. *Proc. Natl. Acad. Sci. USA* **2021**, *118*, e2020395118. [CrossRef]
61. Kuen, J.; Darowski, D.; Kluge, T.; Majety, M. Pancreatic Cancer Cell/Fibroblast Co-Culture Induces M2 like Macrophages That Influence Therapeutic Response in a 3D Model. *PLoS ONE* **2017**, *12*, e0182039. [CrossRef] [PubMed]
62. Andersson, P.; Yang, Y.; Hosaka, K.; Zhang, Y.; Fischer, C.; Braun, H.; Liu, S.; Yu, G.; Liu, S.; Beyaert, R.; et al. Molecular Mechanisms of IL-33-Mediated Stromal Interactions in Cancer Metastasis. *JCI Insight.* **2018**, *3*, e122375. [CrossRef] [PubMed]
63. Ino, Y.; Yamazaki-Itoh, R.; Shimada, K.; Iwasaki, M.; Kosuge, T.; Kanai, Y.; Hiraoka, N. Immune Cell Infiltration as an Indicator of the Immune Microenvironment of Pancreatic Cancer. *Br. J. Cancer* **2013**, *108*, 914–923. [CrossRef] [PubMed]
64. Fridlender, Z.G.; Sun, J.; Kim, S.; Kapoor, V.; Cheng, G.; Ling, L.; Worthen, G.S.; Albelda, S.M. Polarization of Tumor-Associated Neutrophil Phenotype by TGF-Beta: "N1" versus "N2" TAN. *Cancer Cell* **2009**, *16*, 183–194. [CrossRef] [PubMed]
65. Wang, W.Q.; Liu, L.; Xu, H.X.; Wu, C.T.; Xiang, J.F.; Xu, J.; Liu, C.; Long, J.; Ni, Q.X.; Yu, X.J. Infiltrating Immune Cells and Gene Mutations in Pancreatic Ductal Adenocarcinoma. *Br. J. Surg.* **2016**, *103*, 1189–1199. [CrossRef]
66. Munir, H.; Jones, J.O.; Janowitz, T.; Hoffmann, M.; Euler, M.; Martins, C.P.; Welsh, S.J.; Shields, J.D. Stromal-Driven and Amyloid β-Dependent Induction of Neutrophil Extracellular Traps Modulates Tumor Growth. *Nat. Commun.* **2021**, *12*, 683. [CrossRef]
67. Stromnes, I.M.; Hulbert, A.; Pierce, R.H.; Greenberg, P.D.; Hingorani, S.R. T-Cell Localization, Activation, and Clonal Expansion in Human Pancreatic Ductal Adenocarcinoma. *Cancer Immunol. Res.* **2017**, *5*, 978–991. [CrossRef]
68. Li, J.; Byrne, K.T.; Yan, F.; Yamazoe, T.; Chen, Z.; Baslan, T.; Richman, L.P.; Lin, J.H.; Sun, Y.H.; Rech, A.J.; et al. Tumor Cell-Intrinsic Factors Underlie Heterogeneity of Immune Cell Infiltration and Response to Immunotherapy. *Immunity* **2018**, *49*, 178–193. [CrossRef]
69. Hosein, A.N.; Brekken, R.A.; Maitra, A. Pancreatic Cancer Stroma: An Update on Therapeutic Targeting Strategies. *Nat. Rev. Gastroenterol. Hepatol.* **2020**, *17*, 487. [CrossRef]
70. Diana, A.; Wang, L.M.; D'Costa, Z.; Allen, P.; Azad, A.; Silva, M.A.; Soonawalla, Z.; Liu, S.; McKenna, W.G.; Muschel, R.J.; et al. Prognostic Value, Localization and Correlation of PD-1/PD-L1, CD8 and FOXP3 with the Desmoplastic Stroma in Pancreatic Ductal Adenocarcinoma. *Oncotarget* **2016**, *7*, 40992–41004. [CrossRef]
71. Zhang, Y.; Lazarus, J.; Steele, N.G.; Yan, W.; Lee, H.J.; Nwosu, Z.C.; Halbrook, C.J.; Menjivar, R.E.; Kemp, S.B.; Sirihorachai, V.R.; et al. Regulatory T-Cell Depletion Alters the Tumor Microenvironment and Accelerates Pancreatic Carcinogenesis. *Cancer Discov.* **2020**, *10*, 422–439. [CrossRef] [PubMed]
72. Robinson, C.; Xu, M.M.; Nair, S.K.; Beasley, G.M.; Rhodin, K.E. Oncolytic Viruses in Melanoma. *Front. Biosci.* **2022**, *27*, 63. [CrossRef] [PubMed]
73. Marin-Acevedo, J.A.; Kimbrough, E.M.O.; Lou, Y. Next Generation of Immune Checkpoint Inhibitors and Beyond. *J. Hematol. Oncol.* **2021**, *14*, 45. [CrossRef] [PubMed]
74. Kabacaoglu, D.; Ciecielski, K.J.; Ruess, D.A.; Algül, H. Immune Checkpoint Inhibition for Pancreatic Ductal Adenocarcinoma: Current Limitations and Future Options. *Front. Immunol.* **2018**, *9*, 1878. [CrossRef] [PubMed]
75. Gong, J.; Chehrazi-Raffle, A.; Reddi, S.; Salgia, R. Development of PD-1 and PD-L1 Inhibitors as a Form of Cancer Immunotherapy: A Comprehensive Review of Registration Trials and Future Considerations. *J. Immunother. Cancer* **2018**, *6*, 8. [CrossRef] [PubMed]
76. Herbst, R.S.; Soria, J.C.; Kowanetz, M.; Fine, G.D.; Hamid, O.; Gordon, M.S.; Sosman, J.A.; McDermott, D.F.; Powderly, J.D.; Gettinger, S.N.; et al. Predictive Correlates of Response to the Anti-PD-L1 Antibody MPDL3280A in Cancer Patients. *Nature* **2014**, *515*, 563. [CrossRef] [PubMed]
77. Brahmer, J.R.; Tykodi, S.S.; Chow, L.Q.M.; Hwu, W.-J.; Topalian, S.L.; Hwu, P.; Drake, C.G.; Camacho, L.H.; Kauh, J.; Odunsi, K.; et al. Safety and Activity of Anti-PD-L1 Antibody in Patients with Advanced Cancer. *N. Engl. J. Med.* **2012**, *366*, 2455–2465. [CrossRef] [PubMed]
78. Patnaik, A.; Kang, S.P.; Rasco, D.; Papadopoulos, K.P.; Elassaiss-Schaap, J.; Beeram, M.; Drengler, R.; Chen, C.; Smith, L.; Espino, G.; et al. Phase I Study of Pembrolizumab (MK-3475; Anti-PD-1 Monoclonal Antibody) in Patients with Advanced Solid Tumors. *Clin. Cancer Res.* **2015**, *21*, 4286–4293. [CrossRef]
79. Schmiechen, Z.C.; Stromnes, I.M. Mechanisms Governing Immunotherapy Resistance in Pancreatic Ductal Adenocarcinoma. *Front. Immunol.* **2021**, *11*, 613815. [CrossRef]
80. Ene-Obong, A.; Clear, A.J.; Watt, J.; Wang, J.; Fatah, R.; Riches, J.C.; Marshall, J.F.; Chin-Aleong, J.; Chelala, C.; Gribben, J.G.; et al. Activated Pancreatic Stellate Cells Sequester CD8+ T-Cells to Reduce Their Infiltration of the Juxtatumoral Compartment of Pancreatic Ductal Adenocarcinoma. *Gastroenterology* **2013**, *145*, 1121. [CrossRef]
81. Chakravarthy, A.; Khan, L.; Bensler, N.P.; Bose, P.; de Carvalho, D.D. TGF-β-Associated Extracellular Matrix Genes Link Cancer-Associated Fibroblasts to Immune Evasion and Immunotherapy Failure. *Nat. Commun.* **2018**, *9*, 4692. [CrossRef]
82. Dominguez, C.X.; Müller, S.; Keerthivasan, S.; Koeppen, H.; Hung, J.; Gierke, S.; Breart, B.; Foreman, O.; Bainbridge, T.W.; Castiglioni, A.; et al. Single-Cell RNA Sequencing Reveals Stromal Evolution into LRRC15+ Myofibroblasts as a Determinant of Patient Response to Cancer Immunotherapy. *Cancer Discov.* **2020**, *10*, 232–253. [CrossRef] [PubMed]
83. Jenkins, L.; Jungwirth, U.; Avgustinova, A.; Iravani, M.; Mills, A.; Haider, S.; Harper, J.; Isacke, C.M. Cancer-Associated Fibroblasts Suppress CD8+ T-Cell Infiltration and Confer Resistance to Immune-Checkpoint Blockade. *Cancer Res.* **2022**, *82*, 2904. [CrossRef] [PubMed]

84. Vonderheide, R.H. CD40 Agonist Antibodies in Cancer Immunotherapy. *Annu. Rev. Med.* **2020**, *71*, 47–58. [CrossRef] [PubMed]

85. Beatty, G.L.; Chiorean, E.G.; Fishman, M.P.; Saboury, B.; Teitelbaum, U.R.; Sun, W.; Huhn, R.D.; Song, W.; Li, D.; Sharp, L.L.; et al. CD40 Agonists Alter Tumor Stroma and Show Efficacy against Pancreatic Carcinoma in Mice and Humans. *Science* **2011**, *331*, 1612–1616. [CrossRef] [PubMed]

86. Padrón, L.J.; Maurer, D.M.; O'Hara, M.H.; O'Reilly, E.M.; Wolff, R.A.; Wainberg, Z.A.; Ko, A.H.; Fisher, G.; Rahma, O.; Lyman, J.P.; et al. Sotigalimab and/or Nivolumab with Chemotherapy in First-Line Metastatic Pancreatic Cancer: Clinical and Immunologic Analyses from the Randomized Phase 2 PRINCE Trial. *Nat. Med.* **2022**, *28*, 1167–1177. [CrossRef] [PubMed]

87. Wiehagen, K.R.; Girgis, N.M.; Yamada, D.H.; Smith, A.A.; Chan, S.R.; Grewal, I.S.; Quigley, M.; Verona, R.I. Combination of CD40 Agonism and CSF-1R Blockade Reconditions Tumor-Associated Macrophages and Drives Potent Antitumor Immunity. *Cancer Immunol. Res.* **2017**, *5*, 1109–1121. [CrossRef]

88. Van Der Burg, S.H.; Arens, R.; Ossendorp, F.; Van Hall, T.; Melief, C.J.M. Vaccines for Established Cancer: Overcoming the Challenges Posed by Immune Evasion. *Nat. Rev. Cancer* **2016**, *16*, 219–233. [CrossRef]

89. Hewitt, D.B.; Nissen, N.; Hatoum, H.; Musher, B.; Seng, J.; Coveler, A.L.; Al-Rajabi, R.; Yeo, C.J.; Leiby, B.; Banks, J.; et al. A Phase 3 Randomized Clinical Trial of Chemotherapy with or without Algenpantucel-L (HyperAcute-Pancreas) Immunotherapy in Subjects with Borderline Resectable or Locally Advanced Unresectable Pancreatic Cancer. *Ann. Surg.* **2022**, *275*, 45–53. [CrossRef]

90. Lau, S.P.; van Montfoort, N.; Kinderman, P.; Lukkes, M.; Klaase, L.; van Nimwegen, M.; van Gulijk, M.; Dumas, J.; Mustafa, D.A.M.; Lievense, S.L.A.; et al. Dendritic Cell Vaccination and CD40-Agonist Combination Therapy Licenses T Cell-Dependent Antitumor Immunity in a Pancreatic Carcinoma Murine Model. *J. Immunother. Cancer* **2020**, *8*, e000772. [CrossRef]

91. Lau, S.P.; Klaase, L.; Vink, M.; Dumas, J.; Bezemer, K.; van Krimpen, A.; van der Breggen, R.; Wismans, L.V.; Doukas, M.; de Koning, W.; et al. Autologous Dendritic Cells Pulsed with Allogeneic Tumour Cell Lysate Induce Tumour-Reactive T-Cell Responses in Patients with Pancreatic Cancer: A Phase I Study. *Eur. J. Cancer* **2022**, *169*, 20–31. [CrossRef] [PubMed]

92. Cheng, J.T.; Deng, Y.N.; Yi, H.M.; Wang, G.Y.; Fu, B.S.; Chen, W.J.; Liu, W.; Tai, Y.; Peng, Y.W.; Zhang, Q. Hepatic Carcinoma-Associated Fibroblasts Induce IDO-Producing Regulatory Dendritic Cells through IL-6-Mediated STAT3 Activation. *Oncogenesis* **2016**, *5*, e198. [CrossRef] [PubMed]

93. Liu, J.; Fu, M.; Wang, M.; Wan, D.; Wei, Y.; Wei, X. Cancer Vaccines as Promising Immuno-Therapeutics: Platforms and Current Progress. *J. Hematol. Oncol.* **2022**, *15*, 28. [CrossRef] [PubMed]

94. Arabi, F.; Torabi-Rahvar, M.; Shariati, A.; Ahmadbeigi, N.; Naderi, M. Antigenic Targets of CAR T Cell Therapy. A Retrospective View on Clinical Trials. *Exp. Cell Res.* **2018**, *369*, 1–10. [CrossRef]

95. Curioni, A.; Britschgi, C.; Hiltbrunner, S.; Bankel, L.; Gulati, P.; Weder, W.; Opitz, I.; Lauk, O.; Caviezel, C.; Knuth, A.; et al. A Phase I Clinical Trial of Malignant Pleural Mesothelioma Treated with Locally Delivered Autologous Anti-FAP-Targeted CAR T-Cells. *Ann. Oncol.* **2019**, *30*, v501. [CrossRef]

96. Everts, A.; Bergeman, M.; McFadden, G.; Kemp, V. Simultaneous Tumor and Stroma Targeting by Oncolytic Viruses. *Biomedicines* **2020**, *8*, 474. [CrossRef]

97. Harryvan, T.J.; Golo, M.; Dam, N.; Schoonderwoerd, M.J.A.; Farshadi, E.A.; Hornsveld, M.; Hoeben, R.C.; Hawinkels, L.J.A.C.; Kemp, V. Gastrointestinal Cancer-Associated Fibroblasts Expressing Junctional Adhesion Molecule-A Are Amenable to Infection by Oncolytic Reovirus. *Cancer Gene Ther.* **2022**, *29*, 1918–1929. [CrossRef]

98. Groeneveldt, C.; Kinderman, P.; Van Den Wollenberg, D.J.M.; Van Den Oever, R.L.; Middelburg, J.; Mustafa, D.A.M.; Hoeben, R.C.; Van Der Burg, S.H.; Van Hall, T.; Van Montfoort, N. Original Research: Preconditioning of the Tumor Microenvironment with Oncolytic Reovirus Converts CD3-Bispecific Antibody Treatment into Effective Immunotherapy. *J. Immunother. Cancer* **2020**, *8*, 1191. [CrossRef]

99. Koikawa, K.; Kibe, S.; Suizu, F.; Sekino, N.; Kim, N.; Manz, T.D.; Pinch, B.J.; Akshinthala, D.; Verma, A.; Gaglia, G.; et al. Targeting Pin1 Renders Pancreatic Cancer Eradicable by Synergizing with Immunochemotherapy. *Cell* **2021**, *184*, 4753–4771.e27. [CrossRef]

100. Zhang, Y.; Ware, M.B.; Zaidi, M.Y.; Ruggieri, A.N.; Olson, B.M.; Komar, H.; Farren, M.R.; Nagaraju, G.P.; Zhang, C.; Chen, Z.; et al. Heat Shock Protein-90 Inhibition Alters Activation of Pancreatic Stellate Cells and Enhances the Efficacy of PD-1 Blockade in Pancreatic Cancer. *Mol. Cancer Ther.* **2021**, *20*, 150–160. [CrossRef]

101. MacE, T.A.; Shakya, R.; Pitarresi, J.R.; Swanson, B.; McQuinn, C.W.; Loftus, S.; Nordquist, E.; Cruz-Monserrate, Z.; Yu, L.; Young, G.; et al. IL-6 and PD-L1 Antibody Blockade Combination Therapy Reduces Tumour Progression in Murine Models of Pancreatic Cancer. *Gut* **2018**, *67*, 320–332. [CrossRef] [PubMed]

102. Das, S.; Shapiro, B.; Vucic, E.A.; Vogt, S.; Bar-Sagi, D. Tumor Cell-Derived IL1β Promotes Desmoplasia and Immune Suppression in Pancreatic Cancer. *Cancer Res.* **2020**, *80*, 1088–1101. [CrossRef] [PubMed]

103. Steele, C.W.; Karim, S.A.; Leach, J.D.G.; Bailey, P.; Upstill-Goddard, R.; Rishi, L.; Foth, M.; Bryson, S.; McDaid, K.; Wilson, Z.; et al. CXCR2 Inhibition Profoundly Suppresses Metastases and Augments Immunotherapy in Pancreatic Ductal Adenocarcinoma. *Cancer Cell* **2016**, *29*, 832. [CrossRef] [PubMed]

104. Chao, T.; Furth, E.E.; Vonderheide, R.H. CXCR2-Dependent Accumulation of Tumor-Associated Neutrophils Regulates T-Cell Immunity in Pancreatic Ductal Adenocarcinoma. *Cancer Immunol. Res.* **2016**, *4*, 968–982. [CrossRef] [PubMed]

105. Gorchs, L.; Ahmed, S.; Mayer, C.; Knauf, A.; Moro, C.F.; Svensson, M.; Heuchel, R.; Rangelova, E.; Bergman, P.; Kaipe, H. The Vitamin D Analogue Calcipotriol Promotes an Anti-Tumorigenic Phenotype of Human Pancreatic CAFs but Reduces T Cell Mediated Immunity. *Sci. Rep.* **2020**, *10*, 1–15. [CrossRef] [PubMed]

106. Dimitrov, V.; Bouttier, M.; Boukhaled, G.; Salehi-Tabar, R.; Avramescu, R.G.; Memari, B.; Hasaj, B.; Lukacs, G.L.; Krawczyk, C.M.; White, J.H. Hormonal Vitamin D Up-Regulates Tissue-Specific PD-L1 and PD-L2 Surface Glycoprotein Expression in Humans but Not Mice. *J. Biol. Chem.* **2017**, *292*, 20657–20668. [CrossRef]

107. Perez-Penco, M.; Weis-Banke, S.E.; Schina, A.; Siersbæk, M.; Hübbe, M.L.; Jørgensen, M.A.; Lecoq, I.; de la Torre, L.L.; Bendtsen, S.K.; Martinenaite, E.; et al. TGFβ-Derived Immune Modulatory Vaccine: Targeting the Immunosuppressive and Fibrotic Tumor Microenvironment in a Murine Model of Pancreatic Cancer. *J. Immunother. Cancer* **2022**, *10*, e005491. [CrossRef]

108. Blair, A.B.; Kim, V.M.; Muth, S.T.; Saung, M.T.; Lokker, N.; Blouw, B.; Armstrong, T.D.; Jaffee, E.M.; Tsujikawa, T.; Coussens, L.M.; et al. Dissecting the Stromal Signalingandregulation of Myeloid Cells and Memory Effector T Cells in Pancreatic Cancer. *Clin. Cancer Res.* **2019**, *25*, 5351–5363. [CrossRef] [PubMed]

109. Geng, F.; Dong, L.; Bao, X.; Guo, Q.; Guo, J.; Zhou, Y.; Yu, B.; Wu, H.; Wu, J.; Zhang, H.; et al. CAFs/Tumor Cells Co-Targeting DNA Vaccine in Combination with Low-Dose Gemcitabine for the Treatment of Panc02 Murine Pancreatic Cancer. *Mol. Ther. Oncolytics* **2022**, *26*, 304–313. [CrossRef]

110. Eriksson, E.; Milenova, I.; Wenthe, J.; Moreno, R.; Alemany, R.; Loskog, A. IL-6 Signaling Blockade during CD40-Mediated Immune Activation Favors Antitumor Factors by Reducing TGF-β, Collagen Type I, and PD-L1/PD-1. *J. Immunol.* **2019**, *202*, 787–798. [CrossRef]

111. Sum, E.; Rapp, M.; Frobel, P.; le Clech, M.; Durr, H.; Giusti, A.M.; Perro, M.; Speziale, D.; Kunz, L.; Menietti, E.; et al. Fibroblast Activation Protein α-Targeted CD40 Agonism Abrogates Systemic Toxicity and Enables Administration of High Doses to Induce Effective Antitumor Immunity. *Clin. Cancer Res.* **2021**, *27*, 4036–4053. [CrossRef] [PubMed]

112. Tran, E.; Chinnasamy, D.; Yu, Z.; Morgan, R.A.; Lee, C.C.R.; Restifo, N.P.; Rosenberg, S.A. Immune Targeting of Fibroblast Activation Protein Triggers Recognition of Multipotent Bone Marrow Stromal Cells and Cachexia. *J. Exp. Med.* **2013**, *210*, 1065–1068. [CrossRef] [PubMed]

113. Lo, A.; Wang, L.C.S.; Scholler, J.; Monslow, J.; Avery, D.; Newick, K.; O'Brien, S.; Evans, R.A.; Bajor, D.J.; Clendenin, C.; et al. Tumor-Promoting Desmoplasia Is Disrupted by Depleting FAP-Expressing Stromal Cells. *Cancer Res.* **2015**, *75*, 2800–2810. [CrossRef] [PubMed]

114. de Sostoa, J.; Fajardo, C.A.; Moreno, R.; Ramos, M.D.; Farrera-Sal, M.; Alemany, R. Targeting the Tumor Stroma with an Oncolytic Adenovirus Secreting a Fibroblast Activation Protein-Targeted Bispecific T-Cell Engager. *J. Immunother. Cancer* **2019**, *7*, 19. [CrossRef]

115. Puig-Saus, C.; Laborda, E.; Rodríguez-García, A.; Cascalló, M.; Moreno, R.; Alemany, R. The Combination of I-Leader Truncation and Gemcitabine Improves Oncolytic Adenovirus Efficacy in an Immunocompetent Model. *Cancer Gene Ther.* **2014**, *21*, 68–73. [CrossRef]

116. Sow, H.S.; Ren, J.; Camps, M.; Ossendorp, F.; Dijke, P.T. Combined Inhibition of TGF-β Signaling and the PD-L1 Immune Checkpoint Is Differentially Effective in Tumor Models. *Cells* **2019**, *8*, 320. [CrossRef]

117. Fabre, M.; Ferrer, C.; Domínguez-Hormaetxe, S.; Bockorny, B.; Murias, L.; Seifert, O.; Eisler, S.A.; Kontermann, R.E.; Pfizenmaier, K.; Lee, S.Y.; et al. OMTX705, a Novel FAP-Targeting ADC Demonstrates Activity in Chemotherapy and Pembrolizumab-Resistant Solid Tumor Models. *Clin. Cancer Res.* **2020**, *26*, 3420–3430. [CrossRef]

118. Bockorny, B.; Semenisty, V.; Macarulla, T.; Borazanci, E.; Wolpin, B.M.; Stemmer, S.M.; Golan, T.; Geva, R.; Borad, M.J.; Pedersen, K.S.; et al. BL-8040, a CXCR4 Antagonist, in Combination with Pembrolizumab and Chemotherapy for Pancreatic Cancer: The COMBAT Trial. *Nat. Med.* **2020**, *26*, 878–885. [CrossRef]

119. Melisi, D.; Oh, D.Y.; Hollebecque, A.; Calvo, E.; Varghese, A.; Borazanci, E.; Macarulla, T.; Merz, V.; Zecchetto, C.; Zhao, Y.; et al. Safety and Activity of the TGFβ Receptor I Kinase Inhibitor Galunisertib plus the Anti-PD-L1 Antibody Durvalumab in Metastatic Pancreatic Cancer. *J. Immunother. Cancer* **2021**, *9*, e002068. [CrossRef]

120. Garcia-Carbonero, R.; Bazan-Peregrino, M.; Gil-Martín, M.; Álvarez, R.; Macarulla, T.; Riesco-Martinez, M.C.; Verdaguer, H.; Guillén-Ponce, C.; Farrera-Sal, M.; Moreno, R.; et al. Phase I, Multicenter, Open-Label Study of Intravenous VCN-01 Oncolytic Adenovirus with or without Nab-Paclitaxel plus Gemcitabine in Patients with Advanced Solid Tumors. *J. Immunother. Cancer* **2022**, *10*, e003255. [CrossRef]

121. Sun, X.; He, X.; Zhang, Y.; Hosaka, K.; Andersson, P.; Wu, J.; Wu, J.; Jing, X.; Du, Q.; Hui, X.; et al. Inflammatory Cell-Derived CXCL3 Promotes Pancreatic Cancer Metastasis through a Novel Myofibroblast-Hijacked Cancer Escape Mechanism. *Gut* **2022**, *71*, 129–147. [CrossRef] [PubMed]

122. Li, X.; Shepard, H.M.; Cowell, J.A.; Zhao, C.; Osgood, R.J.; Rosengren, S.; Blouw, B.; Garrovillo, S.A.; Pagel, M.D.; Whatcott, C.J.; et al. Parallel Accumulation of Tumor Hyaluronan, Collagen, and Other Drivers of Tumor Progression. *Clin. Cancer Res.* **2018**, *24*, 4798–4807. [CrossRef] [PubMed]

123. Groeneveldt, C.; van Ginkel, J.Q.; Kinderman, P.; Sluijter, M.; Griffioen, L.; Labrie, C.; van den Wollenberg, D.J.M.; Hoeben, R.C.; van der Burg, S.H.; ten Dijke, P.; et al. Intertumoral Differences Dictate the Outcome of TGF-β Blockade on the Efficacy of Viro-Immunotherapy. *Cancer Res. Commun.* **2023**, *3*, 325–337. [CrossRef]

124. Sherman, M.H.; Yu, R.T.; Engle, D.D.; Ding, N.; Atkins, A.R.; Tiriac, H.; Collisson, E.A.; Connor, F.; van Dyke, T.; Kozlov, S.; et al. Vitamin D Receptor-Mediated Stromal Reprogramming Suppresses Pancreatitis and Enhances Pancreatic Cancer Therapy. *Cell* **2014**, *159*, 80–93. [CrossRef] [PubMed]
125. Mora, J.R.; Iwata, M.; von Andrian, U.H. Vitamin Effects on the Immune System: Vitamins A and D Take Centre Stage. *Nat. Rev. Immunol.* **2008**, *8*, 685–698. [CrossRef]

Disclaimer/Publisher's Note: The statements, opinions and data contained in all publications are solely those of the individual author(s) and contributor(s) and not of MDPI and/or the editor(s). MDPI and/or the editor(s) disclaim responsibility for any injury to people or property resulting from any ideas, methods, instructions or products referred to in the content.

International Journal of
Molecular Sciences

Opinion

Inhibition of the RAF/MEK/ERK Signaling Cascade in Pancreatic Cancer: Recent Advances and Future Perspectives

Christos Adamopoulos [1,2], Donatella Delle Cave [3,*] and Athanasios G. Papavassiliou [1,*]

[1] Department of Biological Chemistry, Medical School, National and Kapodistrian University of Athens, 11527 Athens, Greece; cadamop@med.uoa.gr
[2] Department of Oncological Sciences, Icahn School of Medicine at Mount Sinai, New York, NY 10029, USA
[3] Institute of Genetics and Biophysics 'Adriano Buzzati-Traverso', CNR, 80131 Naples, Italy
* Correspondence: donatella.dellecave@igb.cnr.it (D.D.C.); papavas@med.uoa.gr (A.G.P.); Tel.: +30-210-746-2508 (A.G.P.)

Abstract: Pancreatic cancer represents a formidable challenge in oncology, primarily due to its aggressive nature and limited therapeutic options. The prognosis of patients with pancreatic ductal adenocarcinoma (PDAC), the main form of pancreatic cancer, remains disappointingly poor with a 5-year overall survival of only 5%. Almost 95% of PDAC patients harbor Kirsten rat sarcoma virus (KRAS) oncogenic mutations. KRAS activates downstream intracellular pathways, most notably the rapidly accelerated fibrosarcoma (RAF)/mitogen-activated protein kinase kinase (MEK)/extracellular signal-regulated kinase (ERK) signaling axis. Dysregulation of the RAF/MEK/ERK pathway is a crucial feature of pancreatic cancer and therefore its main components, RAF, MEK and ERK kinases, have been targeted pharmacologically, largely by small-molecule inhibitors. The recent advances in the development of inhibitors not only directly targeting the RAF/MEK/ERK pathway but also indirectly through inhibition of its regulators, such as Src homology-containing protein tyrosine phosphatase 2 (SHP2) and Son of sevenless homolog 1 (SOS1), provide new therapeutic opportunities. Moreover, the discovery of allele-specific small-molecule inhibitors against mutant KRAS variants has brought excitement for successful innovations in the battle against pancreatic cancer. Herein, we review the recent advances in targeted therapy and combinatorial strategies with focus on the current preclinical and clinical approaches, providing critical insight, underscoring the potential of these efforts and supporting their promise to improve the lives of patients with PDAC.

Keywords: pancreatic cancer; RAF/MEK/ERK pathway; small-molecule inhibitors; KRAS; targeted therapy

Citation: Adamopoulos, C.; Cave, D.D.; Papavassiliou, A.G. Inhibition of the RAF/MEK/ERK Signaling Cascade in Pancreatic Cancer: Recent Advances and Future Perspectives. *Int. J. Mol. Sci.* **2024**, *25*, 1631. https://doi.org/10.3390/ijms25031631

Academic Editor: Evgeny Imyanitov

Received: 20 December 2023
Revised: 22 January 2024
Accepted: 26 January 2024
Published: 28 January 2024

Copyright: © 2024 by the authors. Licensee MDPI, Basel, Switzerland. This article is an open access article distributed under the terms and conditions of the Creative Commons Attribution (CC BY) license (https://creativecommons.org/licenses/by/4.0/).

1. Introduction

Pancreatic cancer is one of the deadliest tumors and is expected to become the second leading cause of cancer-related mortality in the US. The 5-year overall survival (OS) of patients with pancreatic ductal adenocarcinoma (PDAC), the most common form of pancreatic cancer, has only minimally improved to only 11%, presenting a modest improvement compared to other malignancies [1,2]. The role of the rapidly accelerated fibrosarcoma (RAF)/mitogen-activated protein kinase kinase (MEK)/extracellular signal-regulated kinase (ERK) pathway as the main RAS effector pathway in initiation and progression of pancreatic cancer is well established [3]. Targeted therapy using small-molecule inhibitors against components of the RAF/MEK/ERK pathway has shown significant potential for PDAC. The Kirsten rat sarcoma virus (KRAS) mutation is a hallmark of PDAC, and only recently has there been progress in drug development, with compounds that directly target the once considered "undruggable" RAS. These compounds include KRAS-mutant-specific inhibitors, that are foreseen to change the landscape in PDAC management [4] (Figure 1).

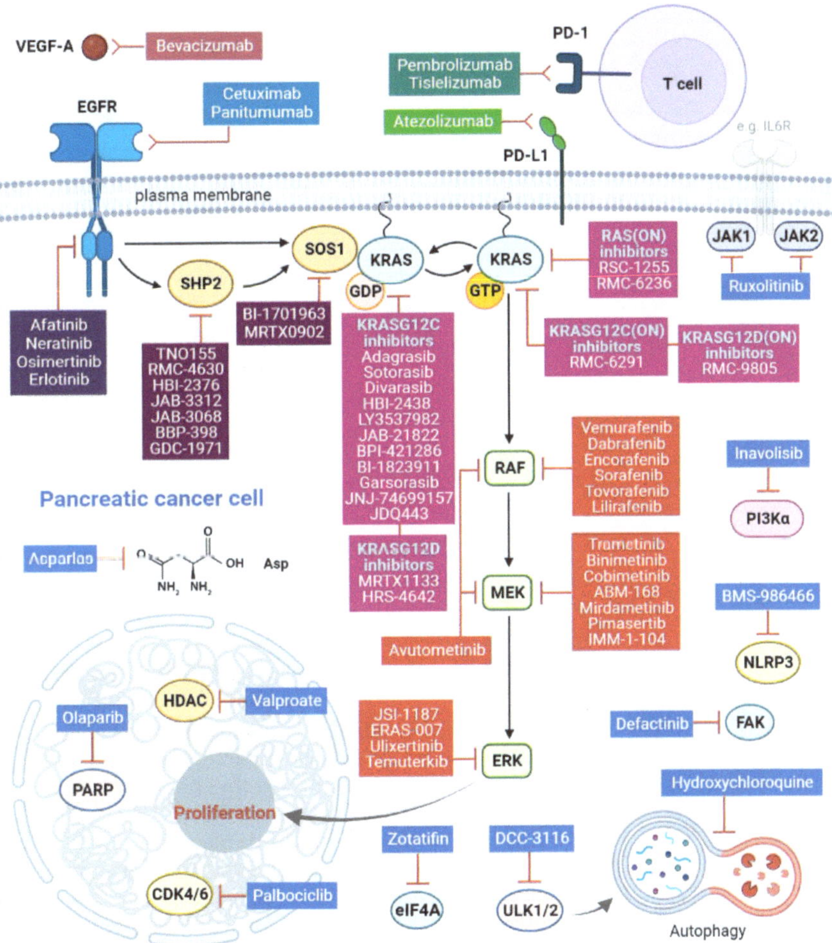

Figure 1. RAF/MEK/ERK pathway inhibitors and combination therapies under clinical evaluation for pancreatic cancer. EGFR, epidermal growth factor receptor; SHP2, Src homology 2 domain-containing phosphatase 2; SOS1, Son of sevenless homolog 1; KRAS, Kirsten rat sarcoma viral oncogene homolog; GDP, guanosine diphosphate; GTP, guanosine triphosphate; RAF, rapidly accelerated fibrosarcoma; MEK, mitogen-activated protein kinase kinase; ERK, extracellular signal-regulated kinase; CDK4/6, cyclin-dependent kinase 4/6; HDAC, histone deacetylase; PARP, poly-adenosine diphosphate (ADP) ribose polymerase; eIF4A, eukaryotic translation initiation factor 4A; ULK1/2, unc-51-like autophagy-activating kinases 1 and 2; NLRP3, Nod-like receptor protein 3; FAK, focal adhesion kinase; JAK1/2, Janus kinase 1/2; PD-1, programmed cell death protein 1; PD-L1, programmed death ligand 1; VEGF-A, vascular endothelial growth factor A; PI3Kα, phosphoinositide 3-kinase α; IL6R, interleukin 6 receptor; Asp, asparagine. This figure was created using the tools provided by BioRender.com (accessed on 21 January 2024).

Here, we discuss rational treatment approaches with the currently available therapeutic options for PDAC patients, including novel targeting strategies using current and new compounds. We focus on the combinatorial strategies and the current clinical attempts for evaluation of the RAF/MEK/ERK pathway inhibitors that are currently in clinical development. This is significant because there is an urgent need to establish new frameworks and improve future treatments. Our aim is to contribute to understanding the complexity of the RAF/MEK/ERK pathway inhibition, which holds substantial promise for developing effective treatment modalities against this aggressive malignancy.

2. RAF/MEK/ERK Signaling Pathway in Pancreatic Cancer

The RAF/MEK/ERK pathway which controls cell growth, differentiation and survival is often upregulated in pancreatic cancer. The orchestrator of this upregulation is the small GTPase KRAS, which is mutated in 95% of patients with pancreatic cancer [2]. The most common KRAS mutations in PDAC are substitutions in position G12, with KRASG12D (41%), KRASG12V (34%) and KRASG12R (16%) being the most frequent and G12C (1–2%) the least [5]. KRAS in its active GTP-bound form promotes RAF kinase activation through dimerization and phosphorylation, resulting in phosphorylation of its substrate MEK kinase. MEK phosphorylates and activates the terminal kinase ERK. Activated ERK regulates growth-promoting transcription [2]. The RAF/MEK/ERK pathway is the key effector pathway for initiation and progression of KRAS-driven PDAC [3]. Therefore, apart from targeted efforts against the key members of the RAF/MEK/ERK pathway, several drugs, targeting different components of this pathway, including the upstream epidermal growth factor receptor (EGFR) family members and the RAF/MEK/ERK pathway regulators Src homology-containing protein tyrosine phosphatase 2 (SHP2) and Son of sevenless homolog 1 (SOS1), have been explored extensively for therapeutic intervention in PDAC [2–5] (Figure 1).

3. Targeting Strategies

3.1. EGFR Family Inhibition

Initial efforts were directed against the upstream frequently dysregulated EGFR/human epidermal growth factor receptor 2 (HER2 or ERBB2) signaling. Erlotinib, an EGFR inhibitor, combined with gemcitabine, a first-line chemotherapy, in patients with advanced PDAC, showed modest survival benefits [6]. When erlotinib was combined with gemcitabine together with nab-paclitaxel, a tubulin-polymerization stabilizer, it exhibited some clinical activity despite the observed toxicities [7]. However, combining cetuximab, a monoclonal antibody against EGFR, with gemcitabine did not show improved outcome [8]. Similarly, the combination of lapatinib, a dual tyrosine kinase inhibitor against both HER2 and EGFR, with gemcitabine or capecitabine did not demonstrate any efficacy [9,10]. To improve these results, a second-generation ERBB family inhibitor, afatinib, was used in combination with gemcitabine. Afatinib binds covalently to cysteine 797 of the EGFR and the corresponding cysteines 805 and 803 in HER2 and human epidermal growth factor receptor 4 (ErB4/HER4), respectively, inhibiting downstream signaling from all homo- and heterodimers formed by ERBB family members. However, again, this combination did not show any efficacy [11]. Subsequent clinical studies, based on preclinical synergistic evidence, assessed the EGFR inhibition in combination with components of the RAF/MEK/ERK pathway such as BRAF and MEK. These studies combined erlotinib with either sorafenib [12], a multikinase RAF inhibitor, or selumetinib [13], a MEK inhibitor, but showed modest activity. More recently, the addition of panitumumab, an EGFR monoclonal antibody, to erlotinib and gemcitabine demonstrated a small but significantly prolonged overall survival, despite the observed toxicities [14]. Overall, these studies did not show sufficient evidence of effectiveness. This agrees with the conclusions from a retrospective analysis showing that EGFR and KRAS alterations were not predictive for patient benefit from anti-EGFR therapy [15]. However, in a preclinical study, the highly selective irreversible EGFR/HER2 inhibitor neratinib suppressed KRAS mutant levels in PDAC cells [16]. The efficacy of neratinib in combination with valproate, a histone deacetylase (HDAC) inhibitor, is being evaluated in a clinical trial in patients with advanced RAS-mutated solid tumors (Table 1).

Table 1. RAF/MEK/ERK pathway inhibitors currently in clinical evaluation for pancreatic cancer. EGFR, epidermal growth factor receptor; ERBB2/HER2, receptor tyrosine-protein kinase erbB 2; RAF, rapidly accelerated fibrosarcoma; SHP2, Src homology 2 domain-containing phosphatase 2; SOS1, Son of sevenless homolog 1; KRAS, Kirsten rat sarcoma viral oncogene homolog; RAS, rat sarcoma viral oncogene homolog; RAF, rapidly accelerated fibrosarcoma; MEK, mitogen-activated protein kinase kinase; ERK, extracellular signal-regulated kinase; CDK4/6, cyclin-dependent kinase 4/6; HDAC, histone deacetylase; PARP, poly-adenosine diphosphate (ADP) ribose polymerase; eIF4A, eukaryotic translation initiation factor 4A; ULK1/2, unc-51-like autophagy-activating kinases 1 and 2; NLRP3, Nod-like receptor protein 3; PD-1, programmed cell death protein 1; PD-L1, programmed death ligand 1; VEGF-A, vascular endothelial growth factor A; PI3Kα, phosphoinositide 3-kinase α; FAK, focal adhesion kinase; JAK1/2, Janus kinase 1/2.

Drug(s)	Target(s)	Second Drug(s)	Second Target(s)	Phase	Clinical Study Code
Neratinib	EGFR, ERBB2/HER2	Divalproex sodium (Valproate)	HDAC	I/II	NCT03919292
Vemurafenib	BRAFV600E/K	Sorafenib	RAF	II	NCT05068752
Lilirafenib	BRAF	Mirdametinib	MEK	I	NCT03905148
Tovorafenib	RAF	Pimasertib	MEK	I/II	NCT04985604
Avutometinib	MEK, RAF	Defactinib	FAK	I/II	NCT05669482
ABM-168	MEK			I	NCT05831995
Binimetinib	MEK	Hydroxychloroquine	Autophagy	I	NCT04132505
Binimetinib	MEK	Encorafenib	RAFV600E/K	II	NCT04390243
Binimetinib	MEK	Palbociclib	CDK4/6	II	NCT05554367
Trametinib	MEK	Hydroxychloroquine	Autophagy	I	NCT03825289
Trametinib	MEK	Ruxolitinib	JAK1/JAK2	I	NCT04303403
Cobimetinib	MEK	Calaspargase pegol-mnkl (Asparlas)	Asparagine	I	NCT05034627
IMM-1-104	MEK			I/II	NCT05585320
Temuterkib	ERK	RMC-4630	SHP2	I	NCT04916236
Temuterkib	ERK	Hydroxychloroquine sulfate	Autophagy	II	NCT04386057
Ulixertinib	ERK	Palbociclib	CDK4/6	I	NCT03454035
ERAS-007	ERK	Encorafenib Palbociclib Cetuximab *	BRAFV600E/K CDK4/6 EGFR	I/II	NCT05039177
BI-1701963	SOS1	Adagrasib	KRASG12C	I	NCT04975256
BI-1701963	SOS1	Trametinib	MEK	I	NCT04111458
HBI-2376	SHP2			I	NCT05163028
JAB-3068	SHP2			I/II	NCT03565003
JAB-3312	SHP2			I	NCT04045496
JAB-3312	SHP2	Binimetinib Pembrolizumab * Sotorasib Osimertinib	MEK PD-1 KRASG12C EGFR	I/II	NCT04720976
BBP-398	SHP2	Sotorasib	KRASG12C	I	NCT05480865
RMC-6291	KRASG12C			I	NCT05462717

Table 1. *Cont.*

Drug(s)	Target(s)	Second Drug(s)	Second Target(s)	Phase	Clinical Study Code
RMC-6291	KRASG12C	RMC-6236	RAS (pan-mutant and wild-type)	I	NCT06128551
HBI-2438	KRASG12C			I	NCT05485974
LY3537982	KRASG12C			I	NCT04956640
JAB-21822	KRASG12C			II	NCT06008288
JAB-21822	KRASG12C	Cetuximab *	EGFR	I/II	NCT05002270
JAB-21822	KRASG12C	JAB-3312	SHP2	I/II	NCT05288205
Adagrasib	KRASG12C			I	NCT05634525
Adagrasib	KRASG12C	TNO155	SHP2	I/II	NCT04330664
Adagrasib	KRASG12C	Afatinib Cetuximab * Pembrolizumab *	EGFR/HER2 EGFR PD-1	I	NCT03785249
Adagrasib	KRASG12C	Olaparib	PARP	I	NCT06130254
Adagrasib	KRASG12C	BMS-986466 [†] −/+ cetuximab *	NLRP3 EGFR	I/II	NCT06024174
Adagrasib	KRASG12C	MRTX0902	SOS1	I/II	NCT05578092
BPI-421286	KRASG12C			I	NCT05315180
BI-1823911	KRASG12C	BI-1701963	SOS1	I	NCT04973163
Divarasib	KRASG12C	Atezolizumab * Cetuximab * Bevacizumab * Erlotinib GDC-1971 Inavolisib	PD-L1 EGFR VEGFA EGFR SHP2 PI3Kα	I	NCT04449874
Garsorasib	KRASG12C			I	NCT04585035
JNJ-74699157	KRASG12C			I	NCT04006301
JDQ443	KRASG12C	TNO155 Tislelizumab *	SHP2 PD-1	I/II	NCT04699188
MK-1084	KRASG12C	Pembrolizumab *	PD-1	I	NCT05067283
Sotorasib	KRASG12C			I/II	NCT03600883
Sotorasib [§]	KRASG12C			II	NCT04185883
Sotorasib	KRASG12C			I	NCT04380753
Sotorasib	KRASG12C	Panitumumab *	EGFR	II	NCT05638295
Sotorasib	KRASG12C	Panitumumab*	EGFR	II	NCT05993455
Sotorasib	KRASG12C	DCC-3116 [†]	ULK1/2	I/II	NCT04892017
Sotorasib	KRASG12C	Zotatifin [†]	eIF4A	I/II	NCT04092673
MRTX1133	KRASG12D			I/II	NCT05737706
RMC-9805	KRASG12D			I	NCT06040541
HRS-4642	KRASG12D			I	NCT05533463

Table 1. *Cont.*

Drug(s)	Target(s)	Second Drug(s)	Second Target(s)	Phase	Clinical Study Code
RMC-6236	RAS (pan-mutant and wild-type)			I	NCT05379985
RSC-1255	RAS (pan-mutant and wild-type)			I	NCT04678648

* Monoclonal antibody; § as monotherapy or in combination with various anti-cancer agents; † main test drug of the trial.

3.2. RAF/MEK/ERK Pathway Component Inhibition

3.2.1. RAF Inhibition

Early attempts to target BRAF in unselected patients with advanced PDAC, using the BRAF inhibitor sorafenib in combination with gemcitabine did not show any benefit [17]. This is possibly explained by the fact that sorafenib is a multikinase inhibitor and its clinical activity is generally attributed to off-target inhibition. The use of the current clinically available BRAF inhibitors (vemurafenib, dabrafenib and encorafenib) is FDA approved for BRAFV600E-mutant metastatic melanoma but not for RAS-mutant tumors. BRAFV600E oncoprotein signals as a monomer and current BRAF inhibitors target and inhibit BRAF monomers. However, this selectivity limits their effectiveness in RAS-driven tumors, where RAFs (BRAF and CRAF) signal as dimers [18]. Additionally, in RAS-mutant tumors these RAF inhibitors promote paradoxical activation of the mitogen-activated protein kinase (MAPK) signaling by inducing wild-type RAF dimerization [19]. Next-generation RAF inhibitors that inhibit both dimers and monomers are currently in clinical development. These RAF inhibitors induce minimal paradoxical activation and show preclinical activity in RAS-mutant tumors [18,20–22]. A clinical trial testing the efficacy of the next-generation RAF inhibitor lilirafenib, including KRAS-mutant PDAC patients, reported stable disease as best response [23]. Additionally, a second current clinical trial is assessing the combination of lilirafenib with the MEK inhibitor mirdametinib in patients with advanced or refractory solid tumors (Table 1). Another clinical trial is evaluating combined vemurafenib and sorafenib treatment in individuals with KRAS-mutant PDAC who have progressed on standard chemotherapy (Table 1).

Activating BRAF alterations make up approximately 30% of KRAS wild-type PDAC and 2% of all PDAC cases [24]. These most commonly include substitutions in position V600, most commonly BRAFV600E. BRAF mutations are mutually exclusive with KRAS mutations and are typically associated with poor prognosis [25]. Multiple preclinical BRAF-mutated models suggest that these alterations can be targeted with combination of BRAF and MEK inhibitors [20,26,27]. Furthermore, BRAFV600E expression in a genetically engineered mouse model of PDAC was sufficient to induce the formation of pancreatic intraepithelial neoplasia lesions, revealing the central role of the RAF/MEK/ERK pathway in PDAC tumorigenesis [3]. Additionally, in a patient-derived orthotopic mouse model of PDAC, treatment with the MEK inhibitors trametinib or cobimetinib resulted in tumor suppression [28]. Molecular targeting of BRAFV600E in KRAS wild-type PDAC, using BRAF and MEK inhibitor, has been reported in several case reports in which patients progressed after first lines of chemotherapy. A case report with a patient with BRAF-mutant advanced PDAC reported objective tumor response to combined vemurafenib plus trametinib treatment [29]. Li et al. reported a case of a patient with metastatic BRAFV600E-mutant PDAC who achieved almost a complete response to dabrafenib plus trametinib treatment. Notably, the patient was rechallenged successfully with the regimen after relapse [30]. Two BRAF-mutant PDAC patients showed a significant reduction in carbohydrate antigen 19-9 levels, a PDAC-associated tumor antigen, following co-treatment with dabrafenib plus trametinib [31]. Wang et al. reported a partial response in the case of advanced metastatic PDAC, after vemurafenib plus trametinib administration [32]. In a recent case report, two patients with BRAFV600E-mutant PDAC exhibited a favorable response to dabrafenib and trametinib co-treatment [33]. Ardalan et al. reported that the addition of the MEK inhibitor cobimetinib to gemcitabine and nab-paclitaxel in BRAF-mutant patients was followed by a complete response to therapy for 16 months [34]. Furthermore, a clinical trial is underway evaluating the combination of encorafenib with the MEK inhibitor binimetinib in BRAFV600E-mutant PDAC patients (Table 1).

3.2.2. MEK Inhibition

Despite the promising preclinical evidence suggesting potent MAPK pathway inhibition, using MEK inhibitors in PDAC, early clinical trials showed limited efficacy. MEK inhibitors, trametinib or pimasertib, in combination with gemcitabine did not show any benefit when compared with gemcitabine alone [35,36]. Of note, the combination of the MEK inhibitor refametinib with gemcitabine was well tolerated and resulted in an objective response rate of 23%, with improved outcomes for KRAS wild-type patients [37]. In contrast, the assessment of selumetinib versus chemotherapy with capecitabine or the dual MEK and protein kinase B (AKT) kinase inhibition with selumetinib and the AKT inhibitor MK-2206 versus oxaliplatin-5-flourouracil-based chemotherapy in patients with advanced PDAC did not show any efficacy [38,39]. Another study on the combination of trametinib with the mammalian target of rapamycin (mTOR) inhibitor everolimus showed modest clinical efficacy, although it was unable to define optimal doses for the two compounds [40]. Moreover, when trametinib was combined with the CDK4/6 inhibitor ribociclib there was no benefit, and the study was terminated [41]. Likewise, in a combination of binimetinib with either the poly-adenosine diphosphate (ADP) ribose polymerase (PARP) inhibitor talazoparib or the programmed death 1 (PD-1) ligand 1 (PD-L1) inhibitor avelumab in patients with metastatic PDAC, no objective responses were observed [42]. Interestingly, preclinical evidence suggests that pancreatic tumors with KRASG12R, the third most common KRAS mutation in PDAC (16%), are more sensitive to MEK or ERK inhibition [43]. This is supported by the documented clinical benefit for patients with KRASG12R-mutant PDAC treated with MEK inhibitors [44,45]. More recently, a phase I clinical trial evaluated ABM-168, a novel small-molecule, allosteric, highly selective MEK inhibitor in adults with advanced solid tumors, including pancreatic carcinoma, who had confirmed RAS, RAF or neurofibromatosis type 1 (NF-1) mutations (Table 1). Another ongoing clinical study is testing the MEK inhibitor cobimetinib in combination with the enzyme calaspargase pegol-mnkl (asparlas) that blocks the biosynthesis of the non-essential amino acid asparagine, leading to starvation of cancer cells (Table 1) [46].

3.2.3. ERK Inhibition

The clinical development of ERK inhibitors raised the hope that direct ERK inhibition could block the MAPK pathway oncogenic transcriptional output. However, early clinical trials using ERK inhibitors against RAS-mutant tumors, including PDAC, were unsuccessful [47,48]. In a recent study, ERK inhibition induced autophagy in KRAS-mutant PDAC and the dual ERK and autophagy inhibition, using SCH772984 and hydrochloroquine, respectively, resulted in enhanced anti-tumor activity in PDAC preclinical models [43,49]. Several clinical studies are testing the synergistic effect of combining hydroxychloroquine with binimetinb, trametinib or the ERK inhibitor temuterkib (Table 1). Cyclin-dependent kinase 4/6 (CDK4/6) is a downstream target of activated/phosphorylated ERK and there is evidence for anti-tumor activity of the dual ERK and CDK4/6 inhibition, using ulixertinib and palbociclib, respectively (Table 1) [50,51] (Figure 1).

3.3. RAF/MEK/ERK Pathway Regulator Inhibition

3.3.1. SHP2 Inhibition

The discovery of SHP2 inhibitors revealed the dependency of KRAS-mutant tumors in SHP2 [52]. In addition, several studies demonstrated that SHP2 inhibition prevents the receptor tyrosine kinase (RTK)-mediated development of adaptive resistance caused by MEK or BRAF inhibitors [53,54]. Accordingly, co-targeting SHP2 and MEK or ERK using small-molecule inhibitors has been investigated preclinically, showing promising results in various KRAS-mutant tumors, including PDAC [52–56]. In another strategy, SHP2 inhibitors are combined with the recently developed allele-specific KRASG12C inhibitors to overcome the development of adaptive resistance mediated by wild-type RAS [57]. This approach has been evaluated by two clinical trials testing the combination of a SHP2 inhibitor with a KRASG12C inhibitor, TNO155 with MRTX849 and JAB-3312 with JAB-21822, respectively, in KRASG12C-mutant patients with advanced solid tumors (Table 1). Another aspect of SHP2 inhibition is its reported immunomodulatory function [58]. Based on this evidence, the combination of SHP2 and KRASG12C inhibitors can promote anti-tumor immunity by disrupting MAPK-activating signals from the tumor microenvironment to cancer cells [59].

3.3.2. SOS1 Inhibition

In preclinical PDAC models, it has been shown that SOS1 is essential for the survival of RAS-mutated cancer cells [60]. Small-molecule SOS1 inhibitors that disrupt the SOS1–RAS interaction have been under development for the treatment of KRAS-mutated cancers. Recently, a selective SOS1 inhibitor, BI-3406, has been reported to reduce GTP-bound RAS levels and tumor growth across KRAS-driven cancer models [61,62]. Moreover, the combined treatment of BI-3406 with trametinib resulted in sustained RAF/MEK/ERK pathway inhibition and suppression of tumors in KRAS-mutated xenograft models, overcoming pathway feedback reactivation [62]. The corresponding clinical compound BI-1701963 was tested in a clinical trial for KRAS-mutated solid tumors, including PDAC, with preliminary data demonstrating good tolerability and modest activity [63]. The current second phase of the study is evaluating the effectiveness of the combination of BI-1701963 with trametinib (Table 1).

3.4. KRAS Inhibition

Sotorasib is a first-in-class small-molecule inhibitor developed to selectively target KRASG12C, providing evidence for in vivo activity [64]. Adagrasib, another KRASG12C inhibitor, has shown clinical activity in KRASG12C-mutated tumors, including PDAC [65]. Both inhibitors, FDA-approved for *KRAS*G12C-mutant non-small cell lung cancer (NSCLC), trap *KRAS*G12C in its inactive GDP-bound state and are now listed in the national comprehensive cancer network (NCCN) clinical practice guidelines as for additional *KRASG12C*-mutant histologies, including pancreatic and colorectal cancers [66]. Another more potent GDP-bound *KRASG12C* inhibitor, divarasib, in combination with various anti-cancer therapies (Table 1), has shown promising clinical benefit in a small cohort of patients with pancreatic adenocarcinoma harboring the KRASG12C mutation [67]. However, the low prevalence of KRASG12C mutation in PDAC (1–2%) limits the applicability of this approach. Luckily, MRTX1133, a "game-changer" compound, has been developed selectively targeting KRASG12D [68]. MRTX1133 is currently under clinical evaluation, while other novel compounds targeting KRASG12D as well (HRS-4642, RMC-9805) are being assessed in phase I clinical trials (Table 1). A novel non-covalent pan-KRAS inhibitor prevents the activation of wild-type KRAS and a range of KRAS mutants, excluding G12R and Q61L/K/R while sparing NRAS and HRAS isoforms (Kim). This pan-KRAS inhibitor showed preclinical anti-tumor activity in various models, indicating broad therapeutic implications in patients with KRAS-driven cancers, including pancreatic cancer [69]. ADT-007, another pan-KRAS inhibitor, that inhibits GTP binding to both mutated and wild-type KRAS, blocks oncogenic KRAS signaling and modulates T cell activation in preclinical PDAC in vitro and in vivo models [70]. Recently, tricomplex inhibitors that target the active GTP-bound state RAS(ON) for both mutant and wild-type RAS have shown promising results for KRASG12V-mutant cancers [71]. The first in class of these inhibitors, RMC-6232, forms a tricomplex with RAS(ON) and an abundant intracellular chaperon protein cyclophilin A, sterically inhibiting RAS binding to its effectors [72,73]. RMC-6232 is being assessed in a phase I clinical trial for KRASG12-mutant tumors (Table 1) and appears effective against KRAS position 12 (G12X) mutants, including G12D, G12V and G12R, inducing durable suppression of the RAS pathway activation in preclinical cellular in vitro and in vivo xenograft PDAC models [66,73].

3.5. Toxicity Challenges

As researchers explore the dynamic space of targeted therapy combinations in pancreatic cancer, the optimism of the recent advancements is tempered by the potential for overlapping toxicities in regimens incorporating two inhibitors targeting within the RAF/MEK/ERK pathway or combined with other targets, like in the case of dual inhibition with afatinib and trametinib [74]. Targeting the upstream regulators of the RAF/MEK/ERK pathway, SHP2 and SOS1, holds promise but at the same time raises concerns about unanticipated on-target toxicities, as evidenced by clear dose-associated cytopenias [75,76]. The clinical trial combining the KRASG12C inhibitor, sotorasib, with the anti-PD1 and anti-PD-L1 monoclonal antibodies pembrolizumab and atezolizumab, respectively, revealed increased liver toxicities [77]. The mechanisms driving these toxicities remain elusive, prompting hypotheses ranging from enhanced immune-mediated effects triggered by targeted therapies to potential off-target covalent protein–drug conjugates causing liver damage, exacerbated by systemic immune activation. Interestingly, Genentech's GDC-6036, a KRASG12C inhibitor administered at lower doses, has shown reduced liver toxicities in phase I testing, suggesting that dosage adjustments may play a crucial role in mitigating adverse effects [78]. Preclinical studies of the RMC-6236 tricomplex have shown success in inhibiting active RAS(ON), including cases of acquired resistance by KRASG12C inhibitors, but the ubiquitous nature of cyclophilin A introduces uncertainties about the therapeutic window and potential off-target activity [76,79]. Amidst the hope for breakthroughs in treating pancreatic cancer, the toxicity challenges underscore the critical need for meticulous exploration of treatment schedules, adjustments and a deep understanding of the intricate interplay within the complex signaling pathways [76].

4. Discussion—Future Perspectives

The presence of KRAS mutations in pancreatic cancer has significant prognostic implications, influencing both overall survival (OS) and treatment response. Patients with KRAS-mutated PDAC generally exhibit a poorer prognosis [80]. Furthermore, recent findings indicate distinct survival outcomes related to specific KRAS mutations. For instance, patients with KRASG12D-mutated PDAC demonstrated a significantly shorter median overall survival compared to those with KRASG12R mutations, indicating a prognostic value for KRASG12D mutation [81,82]. Notably, the type of KRAS mutation may also impact the response to first-line chemotherapy. Thus, FOLFIRINOX showed improved survival in patients with KRASG12D and KRASG12V mutations, while the patients with KRASG12C-mutated tumors exhibited longer overall survival when treated with gemcitabine plus nab-paclitaxel [83]. Additionally, the variant allele frequency (VAF) and allelic imbalance of KRAS further contribute to prognosis. Higher KRAS VAF is associated with shorter survival, and allelic imbalance, leading to increased mutant KRAS dosage, correlates with a more aggressive clinical behavior [84,85]. Beyond KRAS, BRAF mutational status seems to have prognostic value. In a case report, two PDAC patients who had not responded to initial systemic chemotherapy, after identification of BRAFV600E mutation through next generation sequencing, were treated with combined dabrafenib and trametinib and sustained a favorable response [33]. These findings underscore the importance of molecular profiling, specifically KRAS mutation characterization, in guiding prognosis and tailoring therapeutic strategies for pancreatic cancer patients.

The omnipresent *KRAS* mutation in PDAC and the progress in drug development of small-molecule inhibitors led the early therapeutic efforts targeting the main components downstream of the RAF/MEK/ERK pathway. Although preclinical studies demonstrated promising findings, the clinical attempts were unsuccessful due to low efficacy and dose-limiting toxicities [86]. However, when a precision medicine approach was followed the paradigm was shifted. Case reports indicate benefits when the BRAF mutational status was confirmed, in a wild-type KRAS context, before therapeutic intervention [29–34]. Pharmacological targeting of the canonical components of RAF/MEK/ERK signaling in RAS-dependent tumors is often limited by the development of adaptive resistance, which is usually mediated by feedback activation of RTK signaling, resulting in reactivation of the RAF/MEK/ERK pathway activity [53]. Thus, the strategy of targeting additional effectors downstream of RTKs and upstream of RAS, such as SHP2 and SOS1, is attractive. Therapies that directly target the mutated components of the pathway such as KRAS or BRAF could be combined with inhibitors against upstream regulators such as SHP2 and SOS1 and downstream pathway components such as MEK or ERK for a sustained inhibition (Table 1). The concept of dual pathway inhibition has been successfully tested in the case of *BRAFV600E*-mutated melanoma, where vertical double BRAF/MEK inhibition has gained FDA approval. Furthermore, in the context of *BRAFV600E*-mutant tumors there is evidence for effectiveness of a triple inhibitory strategy within the RAF/MEK/ERK pathway [87]. A recent clinical trial, based on promising preclinical data, tested the combination of avutometinib, a first-in-class RAF/MEK clamp and a compound designed to inhibit MEK and block RAF-mediated phosphorylation of MEK in combination with the focal adhesion kinase (FAK) inhibitor defactinib (Table 1, Figure 1). The recent approval of the *KRASG12C* inhibitor sotorasib for *KRASG12C*-mutated non-small cell lung cancer (NSCLC) allowed the enrolment of low-frequency *KRASG12C*-PDAC cases in clinical trials for evaluation (Table 1). Currently, direct inhibition of mutant RAS through allele-specific inhibitors provides a therapeutic opportunity. Interestingly, inhibition of KRASG12D, using MRTX1133, in immunocompetent PDAC models resulted in tumor suppression by increasing tumor-associated macrophages (TAMs) and tumor-infiltrating cytotoxic T cells [88]. This argues that KRASG12D inhibition has a potential immunomodulatory function, which may be beneficial especially for patients with pancreatic cancer, an immunologically "cold" malignancy.

5. Conclusions

Despite the progress in drug discovery, there is an additional need to develop novel, more potent and broader RAF/MEK/ERK pathway inhibitors, including KRAS-mutant inhibitors, for improved tailored targeted therapy [89]. Developing effective and mechanism-based combination therapy regimens is essential to maximizing the efficacy of RAF/MEK/ERK pathway inhibition, which holds great promise for pancreatic cancer control.

Author Contributions: Conceptualization, C.A., D.D.C. and A.G.P.; writing—original draft preparation, C.A.; literature search, C.A.; supervision, D.D.C. and A.G.P.; writing—review and editing, A.G.P. All authors have read and agreed to the published version of the manuscript.

Funding: This research received no external funding.

Institutional Review Board Statement: Not applicable.

Informed Consent Statement: Not applicable.

Data Availability Statement: Data are contained within the article.

Acknowledgments: D.D.C. was supported by Fondazione Umberto Veronesi (FUV) and Fondazione Italiana per la ricerca sulle Malattie del Pancreas (FIMP).

Conflicts of Interest: The authors declare no conflicts of interest.

References

1. Nevala-Plagemann, C.; Hidalgo, M.; Garrido-Laguna, I. From state-of-the-art treatments to novel therapies for advanced-stage pancreatic cancer. *Nat. Rev. Clin. Oncol.* **2020**, *17*, 108–123. [CrossRef]
2. Mollinedo, F.; Gajate, C. Novel therapeutic approaches for pancreatic cancer by combined targeting of RAF→MEK→ERK signaling and autophagy survival response. *Ann. Transl. Med.* **2019**, *7*, S153. [CrossRef] [PubMed]
3. Collisson, E.A.; Trejo, C.L.; Silva, J.M.; Gu, S.; Korkola, J.E.; Heiser, L.M.; Charles, R.P.; Rabinovich, B.A.; Hann, B.; Dankort, D.; et al. A central role for RAF→MEK→ERK signaling in the genesis of pancreatic ductal adenocarcinoma. *Cancer Discov.* **2012**, *2*, 685–693. [CrossRef] [PubMed]
4. Rudloff, U. Emerging kinase inhibitors for the treatment of pancreatic ductal adenocarcinoma. *Expert Opin. Emerg. Drugs* **2022**, *27*, 345–368. [CrossRef]
5. Waters, A.M.; Der, C.J. KRAS: The Critical Driver and Therapeutic Target for Pancreatic Cancer. *Cold Spring Harb. Perspect. Med.* **2018**, *8*, a031435. [CrossRef] [PubMed]
6. Moore, M.J.; Goldstein, D.; Hamm, J.; Figer, A.; Hecht, J.R.; Gallinger, S.; Au, H.J.; Murawa, P.; Walde, D.; Wolff, R.A.; et al. Erlotinib plus gemcitabine compared with gemcitabine alone in patients with advanced pancreatic cancer: A phase III trial of the National Cancer Institute of Canada Clinical Trials Group. *J. Clin. Oncol.* **2007**, *41*, 1960–1966. [CrossRef] [PubMed]
7. Cohen, S.J.; O'Neil, B.H.; Berlin, J.; Ames, P.; McKinley, M.; Horan, J.; Catalano, P.M.; Davies, A.; Weekes, C.D.; Leichman, L. A phase 1b study of erlotinib in combination with gemcitabine and nab-paclitaxel in patients with previously untreated advanced pancreatic cancer: An Academic Oncology GI Cancer Consortium study. *Cancer Chemother. Pharmacol.* **2016**, *77*, 693–701. [CrossRef] [PubMed]
8. Philip, P.A.; Benedetti, J.; Corless, C.L.; Wong, R.; O'Reilly, E.M.; Flynn, P.J.; Rowland, K.M.; Atkins, J.N.; Mirtsching, B.C.; Rivkin, S.E.; et al. Phase III study comparing gemcitabine plus cetuximab versus gemcitabine in patients with advanced pancreatic adenocarcinoma: Southwest Oncology Group-directed intergroup trial S0205. *J. Clin. Oncol.* **2010**, *28*, 3605. [CrossRef] [PubMed]
9. Safran, H.; Miner, T.; Bahary, N.; Whiting, S.; Lopez, C.D.; Sun, W.; Charpentier, K.; Shipley, J.; Anderson, E.; McNulty, B.; et al. Lapatinib and gemcitabine for metastatic pancreatic cancer. A phase II study. *Am. J. Clin. Oncol.* **2011**, *34*, 50–52. [CrossRef]
10. Wu, Z.; Gabrielson, A.; Hwang, J.J.; Pishvaian, M.J.; Weiner, L.M.; Zhuang, T.; Ley, L.; Marshall, J.L.; He, A.R. Phase II study of lapatinib and capecitabine in second-line treatment for metastatic pancreatic cancer. *Cancer Chemother. Pharmacol.* **2015**, *76*, 1309–1314. [CrossRef]
11. Haas, M.; Waldschmidt, D.T.; Stahl, M.; Reinacher-Schick, A.; Freiberg-Richter, J.; Fischer von Weikersthal, L.; Kaiser, F.; Kanzler, S.; Frickhofen, N.; Seufferlein, T.; et al. Afatinib plus gemcitabine versus gemcitabine alone as first-line treatment of metastatic pancreatic cancer: The randomised, open-label phase II ACCEPT study of the Arbeitsgemeinschaft Internistische Onkologie with an integrated analysis of the 'burden of therapy' method. *Eur. J. Cancer* **2021**, *146*, 95–106.
12. Cardin, D.B.; Goff, L.; Li, C.I.; Shyr, Y.; Winkler, C.; DeVore, R.; Schlabach, L.; Holloway, M.; McClanahan, P.; Meyer, K.; et al. Phase II trial of sorafenib and erlotinib in advanced pancreatic cancer. *Cancer Med.* **2014**, *3*, 572–579. [CrossRef]
13. Ko, A.H.; Bekaii-Saab, T.; Van Ziffle, J.; Mirzoeva, O.M.; Joseph, N.M.; Talasaz, A.; Kuhn, P.; Tempero, M.A.; Collisson, E.A.; Kelley, R.K.; et al. A Multicenter, Open-Label Phase II Clinical Trial of Combined MEK plus EGFR Inhibition for Chemotherapy-Refractory Advanced Pancreatic Adenocarcinoma. *Clin. Cancer Res.* **2016**, *22*, 61–68. [CrossRef] [PubMed]
14. Halfdanarson, T.R.; Foster, N.R.; Kim, G.P.; Meyers, J.P.; Smyrk, T.C.; McCullough, A.E.; Ames, M.M.; Jaffe, J.P.; Alberts, S.R. A Phase II Randomized Trial of Panitumumab, Erlotinib, and Gemcitabine versus Erlotinib and Gemcitabine in Patients with Untreated, Metastatic Pancreatic Adenocarcinoma: North Central Cancer Treatment Group Trial N064B. *Oncologist* **2019**, *24*, 589–e160. [CrossRef] [PubMed]
15. Vickers, M.M.; Powell, E.D.; Asmis, T.R.; Jonker, D.J.; Hilton, J.F.; O'Callaghan, C.J.; Tu, D.; Parulekar, W.; Moore, M.J. Comorbidity, age and overall survival in patients with advanced pancreatic cancer—Results from NCIC CTG PA.3: A phase III trial of gemcitabine plus erlotinib or placebo. *Eur. J. Cancer* **2012**, *48*, 1434–1442. [CrossRef] [PubMed]
16. Dent, P.; Booth, L.; Roberts, J.L.; Liu, J.; Poklepovic, A.; Lalani, A.S.; Tuveson, D.; Martinez, J.; Hancock, J.F. Neratinib inhibits Hippo/YAP signaling, reduces mutant K-RAS expression, and kills pancreatic and blood cancer cells. *Oncogene* **2019**, *38*, 5890–5904. [CrossRef] [PubMed]
17. Gonçalves, A.; Gilabert, M.; François, E.; Dahan, L.; Perrier, H.; Lamy, R.; Re, D.; Largillier, R.; Gasmi, M.; Tchiknavorian, X.; et al. BAYPAN study: A double-blind phase III randomized trial comparing gemcitabine plus sorafenib and gemcitabine plus placebo in patients with advanced pancreatic cancer. *Ann. Oncol.* **2012**, *23*, 2799–2805. [CrossRef]
18. Moore, A.R.; Rosenberg, S.C.; McCormick, F.; Malek, S. RAS-targeted therapies: Is the undruggable drugged? *Nat. Rev. Drug Discov.* **2020**, *19*, 533–552. [CrossRef] [PubMed]
19. Poulikakos, P.I.; Zhang, C.; Bollag, G.; Shokat, K.M.; Rosen, N. RAF inhibitors transactivate RAF dimers and ERK signalling in cells with wild-type BRAF. *Nature* **2010**, *464*, 427–430. [CrossRef] [PubMed]
20. Karoulia, Z.; Wu, Y.; Ahmed, T.A.; Xin, Q.; Bollard, J.; Krepler, C.; Wu, X.; Zhang, C.; Bollag, G.; Herlyn, M.; et al. An Integrated Model of RAF Inhibitor Action Predicts Inhibitor Activity against Oncogenic BRAF Signaling. *Cancer Cell* **2016**, *30*, 485–498. [CrossRef] [PubMed]
21. Peng, S.B.; Henry, J.R.; Kaufman, M.D.; Lu, W.P.; Smith, B.D.; Vogeti, S.; Rutkoski, T.J.; Wise, S.; Chun, L.; Zhang, Y.; et al. Inhibition of RAF Isoforms and Active Dimers by LY3009120 Leads to Anti-tumor Activities in RAS or BRAF Mutant Cancers. *Cancer Cell* **2015**, *28*, 384–398. [CrossRef] [PubMed]

22. Vakana, E.; Pratt, S.; Blosser, W.; Dowless, M.; Simpson, N.; Yuan, X.J.; Jaken, S.; Manro, J.; Stephens, J.; Zhang, Y.; et al. LY3009120, a panRAF inhibitor, has significant anti-tumor activity in BRAF and KRAS mutant preclinical models of colorectal cancer. *Oncotarget* **2017**, *8*, 9251–9266. [CrossRef] [PubMed]

23. Desai, J.; Gan, H.; Barrow, C.; Jameson, M.; Atkinson, V.; Haydon, A.; Millward, M.; Begbie, S.; Brown, M.; Markman, B.; et al. Phase I, Open-Label, Dose-Escalation/Dose-Expansion Study of Lifirafenib (BGB-283), an RAF Family Kinase Inhibitor, in Patients with Solid Tumors. *J. Clin. Oncol.* **2020**, *38*, 2140–2150. [CrossRef]

24. Singhi, A.D.; George, B.; Greenbowe, J.R.; Chung, J.; Suh, J.; Maitra, A.; Klempner, S.J.; Hendifar, A.; Milind, J.M.; Golan, T.; et al. Real-Time Targeted Genome Profile Analysis of Pancreatic Ductal Adenocarcinomas Identifies Genetic Alterations That Might Be Targeted with Existing Drugs or Used as Biomarkers. *Gastroenterology* **2019**, *156*, 2242–2253.e4. [CrossRef] [PubMed]

25. Hyman, D.M.; Puzanov, I.; Subbiah, V.; Faris, J.E.; Chau, I.; Blay, J.Y.; Wolf, J.; Raje, N.S.; Diamond, E.L.; Hollebecque, A.; et al. Vemurafenib in Multiple Nonmelanoma Cancers with BRAF V600 Mutations. *N. Engl. J. Med.* **2015**, *373*, 726–736. [CrossRef] [PubMed]

26. Aguirre, A.J.; Nowak, J.A.; Camarda, N.D.; Moffitt, R.A.; Ghazani, A.A.; Hazar-Rethinam, M.; Raghavan, S.; Kim, J.; Brais, L.K.; Ragon, D.; et al. Real-time Genomic Characterization of Advanced Pancreatic Cancer to Enable Precision Medicine. *Cancer Discov.* **2018**, *8*, 1096–1111. [CrossRef] [PubMed]

27. Chen, S.H.; Zhang, Y.; Van Horn, R.D.; Yin, T.; Buchanan, S.; Yadav, V.; Mochalkin, I.; Wong, S.S.; Yue, Y.G.; Huber, L.; et al. Oncogenic BRAF Deletions That Function as Homodimers and Are Sensitive to Inhibition by RAF Dimer Inhibitor LY3009120. *Cancer Discov.* **2016**, *6*, 300–315. [CrossRef] [PubMed]

28. Kawaguchi, K.; Igarashi, K.; Murakami, T.; Kiyuna, T.; Lwin, T.M.; Hwang, H.K.; Delong, J.C.; Clary, B.M.; Bouvet, M.; Unno, M.; et al. MEK inhibitors cobimetinib and trametinib, regressed a gemcitabine-resistant pancreatic-cancer patient-derived orthotopic xenograft (PDOX). *Oncotarget* **2017**, *8*, 47490–47496. [CrossRef]

29. Seghers, A.K.; Cuyle, P.J.; Van Cutsem, E. Molecular Targeting of a BRAF Mutation in Pancreatic Ductal Adenocarcinoma: Case Report and Literature Review. *Target. Oncol.* **2020**, *15*, 407–410. [CrossRef]

30. Li, H.S.; Yang, K.; Wang, Y. Remarkable response of BRAF (V600E)-mutated metastatic pancreatic cancer to BRAF/MEK inhibition: A case report. *Gastroenterol. Rep.* **2021**, *10*, goab031. [CrossRef]

31. Grinshpun, A.; Zarbiv, Y.; Roszik, J.; Subbiah, V.; Hubert, A. Beyond KRAS: Practical molecular targets in pancreatic adenocarcinoma. *Case Rep. Oncol.* **2019**, *12*, 7–13. [CrossRef] [PubMed]

32. Wang, Z.; He, D.; Chen, C.; Liu, X.; Ke, N. Vemurafenib Combined with Trametinib Significantly Benefits the Survival of a Patient with Stage IV Pancreatic Ductal Adenocarcinoma with BRAF V600E Mutation: A Case Report. *Front. Oncol.* **2022**, *11*, 801320. [CrossRef] [PubMed]

33. Shah, S.; Rana, T.; Kancharla, P.; Monga, D. Targeted Therapy for BRAF V600E Positive Pancreatic Adenocarcinoma: Two Case Reports. *Cancer Genom. Proteom.* **2023**, *20*, 398–403. [CrossRef] [PubMed]

34. Ardalan, B.; Azqueta, J.I.; England, J.; Eatz, T.A. Potential benefit of treatment with MEK inhibitors and chemotherapy in BRAF-mutated KRAS wild-type pancreatic ductal adenocarcinoma patients: A case report. *Cold Spring Harb. Mol. Case Stud.* **2021**, *7*, a006108. [CrossRef] [PubMed]

35. Infante, J.R.; Somer, B.G.; Park, J.O.; Li, C.P.; Scheulen, M.E.; Kasubhai, S.M.; Oh, D.Y.; Liu, Y.; Redhu, S.; Steplewski, K.; et al. A randomised, double-blind, placebo-controlled trial of trametinib, an oral MEK inhibitor, in combination with gemcitabine for patients with untreated metastatic adenocarcinoma of the pancreas. *Eur. J. Cancer* **2014**, *50*, 2072–2081. [CrossRef] [PubMed]

36. Van Cutsem, E.; Hidalgo, M.; Canon, J.L.; Macarulla, T.; Bazin, I.; Poddubskaya, E.; Manojlovic, N.; Radenkovic, D.; Verslype, C.; Raymond, E.; et al. Phase I/II trial of pimasertib plus gemcitabine in patients with metastatic pancreatic cancer. *Int. J. Cancer* **2018**, *143*, 2053–2064. [CrossRef] [PubMed]

37. Van Laethem, J.L.; Riess, H.; Jassem, J.; Haas, M.; Martens, U.M.; Weekes, C.; Peeters, M.; Ross, P.; Bridgewater, J.; Melichar, B.; et al. Phase I/II Study of Refametinib (BAY 86-9766) in Combination with Gemcitabine in Advanced Pancreatic cancer. *Target. Oncol.* **2017**, *12*, 97–109. [CrossRef] [PubMed]

38. Bodoky, G.; Timcheva, C.; Spigel, D.R.; La Stella, P.J.; Ciuleanu, T.E.; Pover, G.; Tebbutt, N.C. A phase II open-label randomized study to assess the efficacy and safety of selumetinib (AZD6244 [ARRY-142886]) versus capecitabine in patients with advanced or metastatic pancreatic cancer who have failed first-line gemcitabine therapy. *Investig. New Drugs* **2012**, *30*, 1216–1223. [CrossRef] [PubMed]

39. Chung, V.; McDonough, S.; Philip, P.A.; Cardin, D.; Wang-Gillam, A.; Hui, L.; Tejani, M.A.; Seery, T.E.; Dy, I.A.; Al Baghdadi, T.; et al. Effect of Selumetinib and MK-2206 vs Oxaliplatin and Fluorouracil in Patients with Metastatic Pancreatic Cancer after Prior Therapy: SWOG S1115 Study Randomized Clinical Trial. *JAMA Oncol.* **2017**, *3*, 516–522. [CrossRef] [PubMed]

40. Tolcher, A.W.; Bendell, J.C.; Papadopoulos, K.P.; Burris, H.A., 3rd; Patnaik, A.; Jones, S.F.; Rasco, D.; Cox, D.S.; Durante, M.; Bellew, K.M.; et al. A phase IB trial of the oral MEK inhibitor trametinib (GSK1120212) in combination with everolimus in patients with advanced solid tumors. *Ann. Oncol.* **2015**, *26*, 58–64. [CrossRef] [PubMed]

41. LoRusso, P.; Fakih, M.; De Vos, F.Y.F.L.; Beck, J.T.; Merchan, J.; Shapiro, G.; Lin, C.-C.; Spratlin, J.; Cascella, T.; Sandalic, L.; et al. Phase Ib study of ribociclib (R) + trametinib (T) in patients (pts) with metastatic/advanced solid tumours. *Ann. Oncol.* **2020**, *31*, S484. [CrossRef]

42. Rodon Ahnert, J.; Tan, D.S.; Garrido-Laguna, I.; Harb, W.; Bessudo, A.; Beck, J.T.; Rottey, S.; Bahary, N.; Kotecki, N.; Zhu, Z.; et al. Avelumab or talazoparib in combination with binimetinib in metastatic pancreatic ductal adenocarcinoma: Dose-finding results from phase Ib of the JAVELIN PARP MEKi trial. *ESMO Open* **2023**, *8*, 101584. [CrossRef] [PubMed]

43. Hobbs, G.A.; Baker, N.M.; Miermont, A.M.; Thurman, R.D.; Pierobon, M.; Tran, T.H.; Anderson, A.O.; Waters, A.M.; Diehl, J.N.; Papke, B.; et al. Atypical KRASG12R Mutant Is Impaired in PI3K Signaling and Macropinocytosis in Pancreatic Cancer. *Cancer Discov.* **2020**, *10*, 104–123. [CrossRef] [PubMed]

44. Kenney, C.; Kunst, T.; Webb, S.; Christina, D., Jr.; Arrowood, C.; Steinberg, S.M.; Mettu, N.B.; Kim, E.J.; Rudloff, U. Phase II study of selumetinib, an orally active inhibitor of MEK1 and MEK2 kinases, in KRASG12R-mutant pancreatic ductal adenocarcinoma. *Investig. New Drugs* **2021**, *39*, 821–828. [CrossRef] [PubMed]

45. Ardalan, B.; Azqueta, J.; Sleeman, D. Cobimetinib Plus Gemcitabine: An Active Combination in KRAS G12R-Mutated Pancreatic Ductal Adenocarcinoma Patients in Previously Treated and Failed Multiple Chemotherapies. *J. Pancreat. Cancer* **2021**, *7*, 65–70. [CrossRef]

46. Lopez, C.D.; Kardosh, A.; Chen, E.Y.; Pegna, G.J.; Goodyear, S.; Taber, E.; Rajagopalan, B.; Edmerson, E.; Vo, J.; Jackson, A.; et al. Casper: A phase I, open-label, dose finding study of calaspargase pegol-mnkl (cala) in combination with cobimetinib (cobi) in locally advanced or metastatic pancreatic ductal adenocarcinoma (PDAC). *J. Clin. Oncol.* **2023**, *41*, TPS772. [CrossRef]

47. Grierson, P.M.; Tan, B.; Pedersen, K.S.; Park, H.; Suresh, R.; Amin, M.A.; Trikalinos, N.A.; Knoerzer, D.; Kreider, B.; Reddy, A.; et al. Phase Ib Study of Ulixertinib Plus Gemcitabine and Nab-Paclitaxel in Patients with Metastatic Pancreatic Adenocarcinoma. *Oncologist* **2023**, *28*, e115–e123. [CrossRef]

48. Wang, J.; Johnson, M.; Barve, M.; Pelster, M.; Chen, X.; Li, Z.; Gordon, J.; Reiss, M.; Pai, S.; Falchook, G.; et al. Preliminary results from HERKULES-1: A phase 1b/2, open-label, multicenter study of ERAS-007, an oral ERK1/2 inhibitor, in patients with advanced or metastatic solid tumors. *Eur. J. Cancer* **2022**, *174*, S80–S81. [CrossRef]

49. Bryant, K.L.; Stalnecker, C.A.; Zeitouni, D.; Klomp, J.E.; Peng, S.; Tikunov, A.P.; Gunda, V.; Pierobon, M.; Waters, A.M.; George, S.D.; et al. Combination of ERK and autophagy inhibition as a treatment approach for pancreatic cancer. *Nat. Med.* **2019**, *25*, 628–640. [CrossRef] [PubMed]

50. Franco, J.; Witkiewicz, A.K.; Knudsen, E.S. CDK4/6 inhibitors have potent activity in combination with pathway selective therapeutic agents in models of pancreatic cancer. *Oncotarget* **2014**, *5*, 6512–6525. [CrossRef] [PubMed]

51. Franco, J.; Balaji, U.; Freinkman, E.; Witkiewicz, A.K.; Knudsen, E.S. Metabolic Reprogramming of Pancreatic Cancer Mediated by CDK4/6 Inhibition Elicits Unique Vulnerabilities. *Cell Rep.* **2016**, *14*, 979–990. [CrossRef]

52. Ruess, D.A.; Heynen, G.J.; Ciecielski, K.J.; Ai, J.; Berninger, A.; Kabacaoglu, D.; Görgülü, K.; Dantes, Z.; Wörmann, S.M.; Diakopoulos, K.N.; et al. Mutant KRAS-driven cancers depend on PTPN11/SHP2 phosphatase. *Nat. Med.* **2018**, *24*, 954–960. [CrossRef]

53. Ahmed, T.A.; Adamopoulos, C.; Karoulia, Z.; Wu, X.; Sachidanandam, R.; Aaronson, S.A.; Poulikakos, P.I. SHP2 Drives Adaptive Resistance to ERK Signaling Inhibition in Molecularly Defined Subsets of ERK-Dependent Tumors. *Cell Rep.* **2019**, *26*, 65–78.e5. [CrossRef] [PubMed]

54. Fedele, C.; Ran, H.; Diskin, B.; Wei, W.; Jen, J.; Geer, M.J.; Araki, K.; Ozerdem, U.; Simeone, D.M.; Miller, G.; et al. SHP2 Inhibition Prevents Adaptive Resistance to MEK Inhibitors in Multiple Cancer Models. *Cancer Discov.* **2018**, *8*, 1237–1249. [CrossRef]

55. Mainardi, S.; Mulero-Sánchez, A.; Prahallad, A.; Germano, G.; Bosma, A.; Krimpenfort, P.; Lieftink, C.; Steinberg, J.D.; de Wit, N.; Gonçalves-Ribeiro, S.; et al. SHP2 is required for growth of KRAS-mutant non-small-cell lung cancer in vivo. *Nat. Med.* **2018**, *24*, 961–967. [CrossRef] [PubMed]

56. Frank, K.J.; Mulero-Sánchez, A.; Berninger, A.; Ruiz-Cañas, L.; Bosma, A.; Görgülü, K.; Wu, N.; Diakopoulos, K.N.; Kaya-Aksoy, E.; Ruess, D.A.; et al. Extensive preclinical validation of combined RMC-4550 and LY3214996 supports clinical investigation for KRAS mutant pancreatic cancer. *Cell Rep. Med.* **2022**, *3*, 100815. [CrossRef] [PubMed]

57. Ryan, M.B.; Fece de la Cruz, F.; Phat, S.; Myers, D.T.; Wong, E.; Shahzade, H.A.; Hong, C.B.; Corcoran, R.B. Vertical Pathway Inhibition Overcomes Adaptive Feedback Resistance to KRASG12C Inhibition. *Clin. Cancer Res.* **2020**, *26*, 1633–1643. [CrossRef] [PubMed]

58. Fedele, C.; Li, S.; Teng, K.W.; Foster, C.J.R.; Peng, D.; Ran, H.; Mita, P.; Geer, M.J.; Hattori, T.; Koide, A.; et al. SHP2 inhibition diminishes KRASG12C cycling and promotes tumor microenvironment remodeling. *J. Exp. Med.* **2021**, *218*, e20201414. [CrossRef] [PubMed]

59. Canon, J.; Rex, K.; Saiki, A.Y.; Mohr, C.; Cooke, K.; Bagal, D.; Gaida, K.; Holt, T.; Knutson, C.G.; Koppada, N.; et al. The clinical KRAS(G12C) inhibitor AMG 510 drives anti-tumour immunity. *Nature* **2019**, *575*, 217–223. [CrossRef]

60. Jeng, H.H.; Taylor, L.J.; Bar-Sagi, D. Sos-mediated cross-activation of wild-type Ras by oncogenic Ras is essential for tumorigenesis. *Nat. Commun.* **2012**, *3*, 1168. [CrossRef]

61. Ma, Y.; Schulz, B.; Trakooljul, N.; Al Ammar, M.; Sekora, A.; Sender, S.; Hadlich, F.; Zechner, D.; Weiss, F.U.; Lerch, M.M.; et al. Inhibition of KRAS, MEK and PI3K Demonstrate Synergistic Anti-Tumor Effects in Pancreatic Ductal Adenocarcinoma Cell Lines. *Cancers* **2022**, *14*, 4467. [CrossRef] [PubMed]

62. Hofmann, M.H.; Gmachl, M.; Ramharter, J.; Savarese, F.; Gerlach, D.; Marszalek, J.R.; Sanderson, M.P.; Kessler, D.; Trapani, F.; Arnhof, H.; et al. BI-3406, a Potent and Selective SOS1-KRAS Interaction Inhibitor, Is Effective in KRAS-Driven Cancers through Combined MEK Inhibition. *Cancer Discov.* **2021**, *11*, 142–157. [CrossRef] [PubMed]

63. Johnson, M.; Gort, E.; Pant, S.; Lolkema, M.; Sebastian, M.; Scheffler, M.; Hwang, J.; Dünzinger, U.; Riemann, K.; Kitzing, T.; et al. 524P A phase I, open-label, dose-escalation trial of BI 1701963 in patients (pts) with KRAS mutated solid tumours: A snapshot analysis. *Ann. Oncol.* **2021**, *32*, S591–S592. [CrossRef]

64. Janes, M.R.; Zhang, J.; Li, L.S.; Hansen, R.; Peters, U.; Guo, X.; Chen, Y.; Babbar, A.; Firdaus, S.J.; Darjania, L.; et al. Targeting KRAS Mutant Cancers with a Covalent G12C-Specific Inhibitor. *Cell* **2018**, *172*, 578–589.e17. [CrossRef] [PubMed]

65. Jänne, P.A.; Riely, G.J.; Gadgeel, S.M.; Heist, R.S.; Ou, S.I.; Pacheco, J.M.; Johnson, M.L.; Sabari, J.K.; Leventakos, K.; Yau, E.; et al. Adagrasib in Non-Small-Cell Lung Cancer Harboring a KRASG12C Mutation. *N. Engl. J. Med.* **2022**, *387*, 120–131. [CrossRef] [PubMed]

66. Murciano-Goroff, Y.R.; Suehnholz, S.P.; Drilon, A.; Chakravarty, D. Precision Oncology: 2023 in Review. *Cancer Discov.* **2023**, *13*, 2525–2531. [CrossRef]

67. Sacher, A.; LoRusso, P.; Patel, M.R.; Miller, W.H., Jr.; Garralda, E.; Forster, M.D.; Santoro, A.; Falcon, A.; Kim, T.W.; Paz-Ares, L.; et al. Single-Agent Divarasib (GDC-6036) in Solid Tumors with a KRAS G12C Mutation. *N. Engl. J. Med.* **2023**, *389*, 710–721. [CrossRef]

68. Wang, X.; Allen, S.; Blake, J.F.; Bowcut, V.; Briere, D.M.; Calinisan, A.; Dahlke, J.R.; Fell, J.B.; Fischer, J.P.; Gunn, R.J.; et al. Identification of MRTX1133, a Noncovalent, Potent, and Selective KRASG12D Inhibitor. *J. Med. Chem.* **2022**, *65*, 3123–3133. [CrossRef] [PubMed]

69. Kim, D.; Herdeis, L.; Rudolph, D.; Zhao, Y.; Böttcher, J.; Vides, A.; Ayala-Santos, C.I.; Pourfarjam, Y.; Cuevas-Navarro, A.; Xue, J.Y.; et al. Pan-KRAS inhibitor disables oncogenic signalling and tumour growth. *Nature* **2023**, *619*, 160–166. [CrossRef]

70. Foote, J.B.; Mattox, T.E.; Keeton, A.B.; Purnachandra, G.N.; Maxuitenko, Y.; Chen, X.; Valiyaveettil, J.; Buchsbaum, D.J.; Piazza, G.A.; El-Rayes, B.F. Abstract 4140: Oncogenic KRAS inhibition with ADT-007 primes T cell responses in pancreatic ductal adenocarcinoma. *Cancer Res.* **2023**, *83* (Suppl. 7), 4140. [CrossRef]

71. Koltun, E.; Cregg, J.; Rice, M.A.; Whalen, D.M.; Freilich, R.; Jiang, J.; Hansen, R.; Bermingham, A.; Knox, D.; Dinglasan, J.; et al. Abstract 1260: First-in-class, orally bioavailable KRASG12V(ON) tri-complex inhibitors, as single agents and in combinations, drive profound anti-tumor activity in preclinical models of KRASG12V mutant cancers. *Cancer Res.* **2021**, *13*, 1260. [CrossRef]

72. de Jesus, V.H.F.; Mathias-Machado, M.C.; de Farias, J.P.F.; Aruquipa, M.P.S.; Jácome, A.A.; Peixoto, R.D. Targeting KRAS in Pancreatic Ductal Adenocarcinoma: The Long Road to Cure. *Cancers* **2023**, *15*, 5015. [CrossRef] [PubMed]

73. Koltun, E.S.; Rice, M.A.; Gustafson, W.C.; Wilds, D.; Jiang, J.; Lee, B.J.; Wang, Z.; Chang, S.; Flagella, M.; Mu, Y.; et al. Direct targeting of KRASG12X mutant cancers with RMC-6236, a first-in-class, RAS-selective, orally bioavailable, tri-complex RASMULTI(ON) inhibitor. *Cancer Res.* **2022**, *82* (Suppl. S12), 3597. [CrossRef]

74. Park, S.R.; Davis, M.; Doroshow, J.H.; Kummar, S. Safety and feasibility of targeted agent combinations in solid tumours. *Nat. Rev. Clin. Oncol.* **2013**, *10*, 154–168. [CrossRef] [PubMed]

75. Liu, C.; Lu, H.; Wang, H.; Loo, A.; Zhang, X.; Yang, G.; Kowal, C.; Delach, S.; Wang, Y.; Goldoni, S.; et al. Combinations with Allosteric SHP2 Inhibitor TNO155 to Block Receptor Tyrosine Kinase Signaling. *Clin. Cancer Res.* **2021**, *27*, 342–354. [CrossRef]

76. Akhave, N.S.; Biter, A.B.; Hong, D.S. The Next Generation of KRAS Targeting: Reasons for Excitement and Concern. *Mol. Cancer Ther.* **2022**, *21*, 1645–1651. [CrossRef] [PubMed]

77. Li, B.T.; Falchook, G.S.; Durm, G.A.; Burns, T.F.; Skoulidis, F.; Ramalingam, S.S.; Spira, A.; Bestvina, C.M.; Goldberg, S.B.; Veluswamy, R.; et al. CodeBreaK 100/101: First report of safety/efficacy of sotorasib in combination with pembrolizumab or atezolizumab in advanced KRAS p.G12C NSCLC. *J. Thorac. Oncol.* **2022**, *17*, S10–S11. [CrossRef]

78. Sacher, A.; Patel, M.R.; Miller, W.H., Jr.; Desai, J.; Garralda, E.; Bowyer, S.; Kim, T.W.; De Miguel, M.; Falcon, A.; Krebs, M.G.; et al. OA03.04 phase I A study to evaluate GDC-6036 monotherapy in patients with Non-small Cell Lung Cancer (NSCLC) with KRAS G12C mutation. *J. Thorac. Oncol.* **2022**, *17*, S8–S9. [CrossRef]

79. Tanaka, N.; Lin, J.J.; Li, C.; Ryan, M.B.; Zhang, J.; Kiedrowski, L.A.; Michel, A.G.; Syed, M.U.; Fella, K.A.; Sakhi, M.; et al. Clinical Acquired Resistance to KRASG12C Inhibition through a Novel KRAS Switch-II Pocket Mutation and Polyclonal Alterations Converging on RAS-MAPK Reactivation. *Cancer Discov.* **2021**, *11*, 1913–1922. [CrossRef] [PubMed]

80. Zhang, J.; Darman, L.; Hassan, M.S.; Von Holzen, U.; Awasthi, N. Targeting KRAS for the potential treatment of pancreatic ductal adenocarcinoma: Recent advancements provide hope (Review). *Oncol. Rep.* **2023**, *50*, 206. [CrossRef] [PubMed]

81. Philip, P.A.; Azar, I.; Xiu, J.; Hall, M.J.; Hendifar, A.E.; Lou, E.; Hwang, J.J.; Gong, J.; Feldman, R.; Ellis, M.; et al. Molecular Characterization of KRAS Wild-type Tumors in Patients with Pancreatic Adenocarcinoma. *Clin. Cancer Res.* **2022**, *28*, 2704–2714. [CrossRef] [PubMed]

82. Ardalan, B.; Ciner, A.; Baca, Y.; Darabi, S.; Kasi, A.; Lou, E.; Azqueta, J.I.; Xiu, J.; Nabhan, C.; Shields, A.F.; et al. Not all treated KRAS-mutant pancreatic adenocarcinomas are equal: KRAS G12D and survival outcome. *J. Clin. Oncol.* **2023**, *41*, 4020. [CrossRef]

83. Ardalan, B.; Ciner, A.; Baca, Y.; Darabi, S.; Kasi, A.; Lou, E.; Azqueta, J.I.; Xiu, J.; Nabhan, C.; Shields, A.F.; et al. Prognostic indicators of KRAS G12X mutations in pancreatic cancer. *J. Clin. Oncol.* **2023**, *41*, 735. [CrossRef]

84. Ciner, A.; Ardalan, B.; Baca, Y.; Darabi, S.; Kasi, A.; Lou, E.; Azqueta, J.I.; Xiu, J.; Nabhan, C.; Shields, A.F.; et al. KRAS G12C-mutated pancreatic cancer: Clinical outcomes based on chemotherapeutic regimen. *J. Clin. Oncol.* **2023**, *41*, 4150. [CrossRef]

85. Suzuki, T.; Masugi, Y.; Inoue, Y.; Hamada, T.; Tanaka, M.; Takamatsu, M.; Arita, J.; Kato, T.; Kawaguchi, Y.; Kunita, A.; et al. KRAS variant allele frequency, but not mutation positivity, associates with survival of patients with pancreatic cancer. *Cancer Sci.* **2022**, *113*, 3097–3109. [CrossRef] [PubMed]

86. Mueller, S.; Engleitner, T.; Maresch, R.; Zukowska, M.; Lange, S.; Kaltenbacher, T.; Konukiewitz, B.; Öllinger, R.; Zwiebel, M.; Strong, A.; et al. Evolutionary routes and KRAS dosage define pancreatic cancer phenotypes. *Nature* **2018**, *554*, 62–68. [CrossRef]
87. Adamopoulos, C.; Ahmed, T.A.; Tucker, M.R.; Ung, P.M.U.; Xiao, M.; Karoulia, Z.; Amabile, A.; Wu, X.; Aaronson, S.A.; Ang, C.; et al. Exploiting Allosteric Properties of RAF and MEK Inhibitors to Target Therapy-Resistant Tumors Driven by Oncogenic BRAF Signaling. *Cancer Discov.* **2021**, *11*, 1716–1735. [CrossRef] [PubMed]
88. Kemp, S.B.; Cheng, N.; Markosyan, N.; Sor, R.; Kim, I.K.; Hallin, J.; Shoush, J.; Quinones, L.; Brown, N.V.; Bassett, J.B.; et al. Efficacy of a Small-Molecule Inhibitor of KrasG12D in Immunocompetent Models of Pancreatic Cancer. *Cancer Discov.* **2023**, *13*, 298–311. [CrossRef] [PubMed]
89. Mullard, A. The KRAS crowd targets its next cancer mutations. *Nat. Rev. Drug Discov.* **2023**, *22*, 167–171. [CrossRef] [PubMed]

Disclaimer/Publisher's Note: The statements, opinions and data contained in all publications are solely those of the individual author(s) and contributor(s) and not of MDPI and/or the editor(s). MDPI and/or the editor(s) disclaim responsibility for any injury to people or property resulting from any ideas, methods, instructions or products referred to in the content.

MDPI AG
Grosspeteranlage 5
4052 Basel
Switzerland
Tel.: +41 61 683 77 34

International Journal of Molecular Sciences Editorial Office
E-mail: ijms@mdpi.com
www.mdpi.com/journal/ijms

Disclaimer/Publisher's Note: The statements, opinions and data contained in all publications are solely those of the individual author(s) and contributor(s) and not of MDPI and/or the editor(s). MDPI and/or the editor(s) disclaim responsibility for any injury to people or property resulting from any ideas, methods, instructions or products referred to in the content.

www.ingramcontent.com/pod-product-compliance
Lightning Source LLC
LaVergne TN
LVHW071933160726
843515LV00011B/2599

9 783725 822898